The U.S.-CHINA Military Scorecard

Forces, Geography, and the Evolving Balance of Power 1996–2017

Eric Heginbotham

Michael Nixon, Forrest E. Morgan, Jacob L. Heim, Jeff Hagen, Sheng Li, Jeffrey Engstrom, Martin C. Libicki, Paul DeLuca, David A. Shlapak, David R. Frelinger, Burgess Laird, Kyle Brady, Lyle J. Morris

Prepared for the United States Air Force

For more information on this publication, visit www.rand.org/t/rr392

Library of Congress Cataloging-in-Publication Data

Heginbotham, Eric, author.
The U.S.-China military scorecard : forces, geography, and the evolving balance of power, 1996-2017 / Eric Heginbotham ... [and thirteen others].
pages cm
Includes bibliographical references.
ISBN 978-0-8330-8219-0 (pbk. : alk. paper)
1. China—Strategic aspects. 2. China—Military policy. 3. United States—Military policy. 4. Spratly Islands—Strategic aspects. 5. Taiwan—Strategic aspects. 6. National security—China. 7. National security—United States. 8. National security—Pacific Area. 9. China—Armed Forces. 10. United States—Armed Forces. I. Title.

UA835.H427 2015
355'.033551—dc23

2015031303

Published by the RAND Corporation, Santa Monica, Calif.

Cover design by Eileen Delson La Russo

Cover map photo by FrankRamspott/iStock; U.S. flag by tacktack/iStock; Chinese flag by chokkicx/iStock

Preface

Over the past two decades, the Chinese People's Liberation Army (PLA) has transformed itself from a large but antiquated force into a capable, modern military. In many areas, its technology and skill levels lag behind those of the United States, but it has narrowed the gap. Moreover, it enjoys the advantage of proximity in most plausible Asian conflict scenarios and has developed capabilities that capitalize on that advantage. How would Chinese and U.S. forces perform in operations against one another in such a conflict? What is the balance of power? What are the prospects for deterrence, and what can be done to strengthen them?

This volume examines relative U.S. and Chinese military capabilities in ten operational areas, covering the air and missile, maritime, space and counterspace, cyber, and nuclear domains. It looks at trends across time, from 1996 to the present, as well as potential developments through 2017. And it examines the impact of distance and geography on military power by assessing capabilities in the context of two scenarios at different distances from China: one centered on Taiwan and the other on the Spratly Islands. This research should be of interest to defense analysts, Asian foreign policy and security specialists, policymakers, military officers, and anyone interested in Chinese military modernization and the balance of power in Asia.

Contribution of Open-Source Analysis

Although much has been written on Chinese military issues in recent years, most of this work has revolved around equipment inventories and discussions of individual weapon systems. Conducted by the RAND Corporation at the behest of the U.S. Air Force, this research is intended to advance the public discussion of China-related defense issues by introducing dynamic analyses that account for the operational context of military conflict in East Asia.

Information and data are drawn from well-established public sources. For example, *The Military Balance* by the International Institute for Strategic Studies provided most equipment numbers, while Jane's databases were consulted for equipment capabilities. In cases where unclassified information on system capabilities was not available, we considered a range of possible parameters. All of the dynamic modeling methodology (which involved a mix of statistical analysis, Monte Carlo simulation, and modified Lanchester equations) is publicly available and widely used by specialists at U.S. and foreign civilian and military universities.

While the data and methodology are publicly available and have been combined elsewhere to address various military problems, this report offers the broadest and most systematically developed set of dynamic analyses available in the unclassified literature on contemporary East Asian military issues. It is our hope that it will prompt others to conduct similar work and further probe and extend the research and its conclusions. Although the work was sponsored by the U.S. Air Force, the analysis was conducted independently and does not necessarily conform to the views or analyses of that service, the U.S. military, or the U.S. government.

The research reported here was commissioned by the U.S. Air Force and conducted within the Strategy and Doctrine Program of RAND Project AIR FORCE.

RAND Project AIR FORCE

RAND Project AIR FORCE (PAF), a division of the RAND Corporation, is the U.S. Air Force's federally funded research and development center for studies and analyses. PAF provides the Air Force with independent analyses of policy alternatives affecting the development, employment, combat readiness, and support of current and future air, space, and cyber forces. Research is conducted in four programs: Force Modernization and Employment; Manpower, Personnel, and Training; Resource Management; and Strategy and Doctrine.

Additional information about PAF is available on our website:
http://www.rand.org/paf/

Contents

Figures

Tables

Summary

Introduction

Over the past two decades, China's People's Liberation Army (PLA) has transformed itself from a large but antiquated force into a capable, modern military. In most areas, its technology and skill levels lag behind those of the United States, but it has narrowed the gap. Moreover, it enjoys the advantage of proximity in most plausible scenarios and has developed capabilities that capitalize on that advantage. In the years ahead, sound understanding of regional military issues will become increasingly important in establishing appropriate U.S. political and military policies in Asia. Yet there are few sources available in the public realm that move beyond equipment counts and address these issues in ways that account for interactive operational dynamics and, especially, the effects of geography.[1]

To advance the public debate, this report examines U.S. and Chinese military capabilities in ten operational areas, producing a "scorecard" for each. None of the scorecards is exhaustive in scope or detail. Rather, each develops indicative metrics that we apply consistently to the assessment of relative capabilities in four snapshot years: 1996, 2003, 2010, and 2017 (projected). By employing a consistent methodology, the scorecards provide a portrait of trends over time. To provide insight into the impact of geography and distance, each of the scorecards evaluates capabilities in the context of two scenarios: a Taiwan invasion and a Spratly Islands campaign. These scenarios center on locations that lie roughly 160 km and 940 km, respectively, from the Chinese coast.

To be clear, the goal is to avoid war. The authors do not hope for or anticipate armed conflict with China. The scenarios and the operational activities depicted in them are not meant to signify anything about either the likelihood of a future conflict or the course of events should one occur. Nor do they represent U.S. national or military policy about whether or how such a war would be fought. The scenarios are,

[1] All of the information on equipment numbers and capabilities, as well as the analytical methodology applied, was taken from publicly available sources. To maintain a scrupulously open-source approach, none of the information in those sources was vetted or checked against non-publicly available government sources.

rather, a means to evaluate relative capabilities, providing notional distances, geography, and other situation-specific factors necessary to make such an assessment.

The remainder of this summary follows the chapter structure of the larger report. It first addresses broad trends in Chinese and U.S. force development. It then briefly reviews the results of each of the ten scorecards. Finally, it introduces the overall conclusions of the study and offers an assessment of the implications for the United States and Asia.

Chinese and U.S. Military Development

The Chinese and U.S. militaries have followed very different developmental paths. When Mao Zedong died in 1976, the PLA was a bloated, ineffective force designed primarily for "people's war." During the 1980s, military reforms trimmed the size of the force and refocused its energies on preparing for conventional conflict. Modernization accelerated after the Taiwan Strait crisis of 1996, and Chinese defense budgets began to grow rapidly. Annual real (inflation-adjusted) growth in China's defense spending averaged 11 percent per year between 1996 and 2015. Modernization has largely optimized capabilities for conflict across the Taiwan Strait and has focused on developing air and naval forces, conventionally armed ballistic missiles, and counter-space and cyber capabilities.

With global military responsibilities, the United States is less able to optimize its forces solely for Asian scenarios. Indeed, after the Cold War, it developed systems and capabilities that were intended more for low-intensity conflict. U.S. military budgets increased rapidly after the attacks of September 2001 before declining between 2010 and 2015. The U.S. 2015 defense budget, including supplemental funds for ongoing military operations, was $560 billion in 2015, roughly 57 percent larger (adjusted for inflation) than in 1996.[2] To the extent budgets have increased, much of the additional money has gone to fund combat operations in the Middle East, or to develop capabilities most relevant to those operations.

In recent years, the differences described above have diminished somewhat. In December 2004, then-premier of China Hu Jintao outlined "new historical missions" for the PLA, which opened the door to a wider range of operations.[3] China's 2015 defense white paper (titled, for the first time, *China's Military Strategy*) stipulates, "In response to the new requirements coming from the country's growing strategic interests, the armed forces will actively participate in both regional and international secu-

[2] For historical figures, see Office of the Under Secretary of Defense, Chief Financial Officer, *United States Department of Defense Fiscal Year 2015 Budget Request*, March 2014.

[3] James Mulvenon, "Chairman Hu and the PLA's 'New Historic Missions,'" *China Leadership Monitor*, No. 27, January 2009.

rity cooperation and effectively secure China's overseas interests."[4] Preparing for those tasks, including some far from China, will add to the country's military toolkit but will also dissipate its focus. The U.S. military, for its part, has deployed additional assets to the Pacific. This shift, under way since the mid-2000s, was reaffirmed in a January 2012 document, *Sustaining U.S. Global Leadership: Priorities for 21st Century Defense*, which stated that the U.S. military "will of necessity rebalance toward the Asia-Pacific region."[5]

Paired system-for-system or at the level of the individual service member, the United States still maintains a substantial military advantage. However, China would enjoy enormous situational and geographic advantages in any likely East Asian scenario that would largely offset these strengths. As China prepares for conflicts close to its periphery, the mainland provides large and relatively secure staging areas for operations. This enables the PLA to focus largely on "tooth" (combat forces) as opposed to "tail" (support assets). This report assesses how the United States and China would fare against each other in four snapshot years and at two different distances from China.

Scorecards

The core of the report is a set of ten scorecards, each addressing relative U.S. and Chinese capabilities in a specific operational area. The scorecards are categorized loosely into the air, maritime, space, cyber, and nuclear domains. However, many include elements that cross domains (for example, by exploring the impact of U.S. air attacks against Chinese amphibious forces). The scorecards are as follows:

Air and Missile Scorecards

- *Scorecard 1: Chinese Capability to Attack Air Bases* evaluates the PLA's capability to deny U.S. forces the use of forward air bases.
- *Scorecard 2: Air Campaigns Over Taiwan and the Spratly Islands* evaluates the relative capability of U.S. and PLA air forces to gain air superiority.
- *Scorecard 3: U.S. Penetration of Chinese Airspace* evaluates the U.S. capability to penetrate Chinese air defenses.
- *Scorecard 4: U.S. Capability to Attack Chinese Air Bases* evaluates the U.S. capability to attack Chinese air bases and degrade air base operations.

---

[4] State Council Information Office of the People's Republic of China, *China's Military Strategy*, Beijing, May 2015.

[5] U.S. Department of Defense, *Sustaining U.S. Global Leadership: Priorities for 21st Century Defense*, Washington, D.C., January 2012.

Maritime Scorecards

- *Scorecard 5: Chinese Anti-Surface Warfare* evaluates the PLA's capability to destroy or damage U.S. aircraft carriers or other warships.
- *Scorecard 6: U.S. Anti-Surface Warfare Capabilities Versus Chinese Naval Ships* evaluates the U.S. capability to destroy Chinese amphibious ships and escorts.

Space, Cyber, and Nuclear Scorecards

- *Scorecard 7: U.S. Counterspace Capabilities Versus Chinese Space Systems* evaluates the U.S. capability to deny or inhibit China's use of satellites.
- *Scorecard 8: Chinese Counterspace Capabilities Versus U.S. Space Systems* evaluates China's capability to deny or inhibit U.S. use of satellites.
- *Scorecard 9: U.S. and Chinese Cyberwarfare Capabilities* evaluates the relative capability of U.S. and PLA forces to gain a military advantage from cyber operations.
- *Scorecard 10: U.S. and Chinese Strategic Nuclear Stability* evaluates the capability of both sides to survive and retaliate against a nuclear attack.

The list of scorecards was not taken from an officially sanctioned typology of operations; however, each is intended to reflect a type or category of conflict that will be recognizable to professional military officers. They reflect both sides' discussions of key operational concepts and tasks, and they take into account Chinese and U.S. writing on the subject. The list does not include operational types that would largely be executed by allied or partner states (such as ground operations). They do not include or encompass all types of operations in which U.S. and Chinese forces might engage against one another. And there is no effort to build a unified campaign model that links the different scorecards together, such that the results from one feed into the others. Nevertheless, the scorecards do cover most of the important types of operations that would be central to combat between U.S. and Chinese forces in an East Asian conflict, and we discuss the interrelationships between them, both within each chapter and in the conclusions.

Most of the scorecards employ both qualitative and quantitative analysis and involve modeling of combat dynamics in the area addressed. (The exceptions are the two counterspace scorecards and the cyberwarfare scorecard, which, because of source limitations, rely on qualitative analysis.) Given the inherent uncertainty of warfare, as well as the limitations of the data, results should not be taken as precisely predictive of actual outcomes in a hypothetical war between the countries. Nevertheless, the analysis, which accounts for the spatial (or geographic), temporal, and material aspects of conflict—along with the dynamic interactions between these factors—is intended to capture the general magnitude of the challenges facing U.S. commanders in each area. And given that we employ the same methodology and metrics for each snapshot year, varying only the inventories held by each side at that time according to the best available information, the approach is particularly well suited to capturing the direction and speed of change in the balance of power. Summaries of the major findings of each scorecard follow.

Scorecard 1: Chinese Capability to Attack Air Bases

Given the importance of airpower in America's recent wars, it is not surprising that China has sought ways of neutralizing U.S. capabilities in this area. Of greatest significance, the PLA has developed ballistic and cruise missiles that threaten forward U.S. air bases. From a handful of conventionally armed ballistic missiles in 1996, China's inventory now numbers roughly 1,400 ballistic missiles and hundreds of cruise missiles. Although most are short-range systems, they include a growing number of intermediate-range ballistic missiles that can reach U.S. bases in Japan. Importantly, accuracy has also improved. Circular error probabilities have decreased from hundreds of meters in the 1990s to as few as five or ten meters today. Weapon ranges have increased from short (less than 1,000 km) to medium (1,000–3,000 km).

RAND models of attacks by these ballistic missiles on Kadena Air Base, the closest U.S. air base to the Taiwan Strait, suggest that even a relatively small number of accurate missiles could shut the base to flight operations for critical days at the outset of hostilities, and focused, committed attacks might close a single base for weeks. U.S. countermeasures—such as improved defenses, hardened shelters for aircraft, faster runway repair methods, or the dispersion of aircraft—can potentially mitigate the threat. But barring a major U.S. defensive technological breakthrough, the growing number and variety of Chinese missiles will almost certainly challenge the U.S. ability to operate from forward bases. As a larger proportion of U.S. aircraft are forced to fly from bases that are either susceptible to attack or farther from the scene of conflict, basing issues will greatly complicate U.S. efforts to gain air superiority over the battlefield.

Scorecard 2: Air Campaigns Over Taiwan and the Spratly Islands

In virtually any East Asian scenario, U.S. Air Force and U.S. Navy aircraft would play a critical role in blunting Chinese attacks. Since 1996, the United States has improved existing aircraft and introduced so-called fifth-generation aircraft, including the F-22 and F-35. China, meanwhile, has replaced many of its obsolete second-generation aircraft, which made up an overwhelming proportion of its force in 1996, with modern fourth-generation designs. These fourth-generation aircraft now constitute roughly half of the PLA Air Force's fighter inventory. The net effect of these changes has been to narrow, but not close, the qualitative gap between the U.S. and Chinese air forces.

To evaluate the impact of this change on the two scenarios considered, we employed tactical and operational air combat models, using the appropriate basing, flight distances, and force structure data. The models evaluate the number of fighter aircraft that the United States would need to maintain in the Western Pacific to defeat a Chinese air campaign. The results suggest that U.S. requirements have increased by several hundred percent since 1996. In the 2017 Taiwan case, U.S. commanders would probably be unable to find the basing required for U.S. forces to prevail in a seven-day campaign. They could relax their time requirement and prevail in a more extended campaign, but this would entail leaving ground and naval forces vulnerable to Chinese air operations for a correspondingly longer period. The Spratly Islands scenario would be easier, requiring roughly half the forces of the Taiwan scenario.

Scorecard 3: U.S. Penetration of Chinese Airspace

U.S. commanders are equally concerned by the development of Chinese air defenses, which would make it more difficult to operate in or near Chinese airspace in the event of a conflict. In 1996, the vast majority of China's 500+ long-range surface-to-air missile (SAM) systems were Chinese duplicates of the obsolete Russian SA-2 missile (with a range of roughly 35 km). By 2010, China had deployed roughly 200 launchers for "double-digit SAMs." The newer missiles have more sophisticated seekers and ranges of up to 200 km. Combined with more capable fighter aircraft and the addition of new airborne warning and control system–equipped aircraft, the Chinese integrated air defense system (IADS) has become a formidable obstacle. At the same time, however, U.S. air forces have made improvements to their penetration capabilities, with the addition of stealth aircraft and new SEAD (suppression of enemy air defenses) aircraft.

We used a target coverage model to evaluate the ability of U.S. strike aircraft to penetrate Chinese defenses in the Taiwan and Spratly Islands scenarios. The results show net gains for China, with its improved IADS reducing the ability of even the improved U.S. forces to penetrate Chinese airspace at moderate risk. Our airspace penetration model shows that although standoff attack capabilities, stealth, and SEAD mitigate the impact of Chinese defenses, the ability to penetrate and strike targets opposite Taiwan with minimal risk to the U.S. aircraft involved declines significantly between 1996 and 2017. However, the U.S. ability to penetrate targets in the Spratly scenario remains far more robust. This is because the same number of critical but scarce U.S. assets (such as standoff weapons and stealth aircraft) can be allocated to attack a much smaller target set and because the relevant target set is, on balance, closer to the coast.

Scorecard 4: U.S. Capability to Attack Chinese Air Bases

While penetrating Chinese airspace has become more hazardous, especially in the high-threat environment opposite Taiwan, the development of new generations of precision weapons since 1996 gives the United States new options and greater punch. Virtually all of the iron bombs used by U.S. forces today are equipped with guidance packages, such as the Joint Direct Attack Munition, which have turned them into all-weather, precision weapons. At longer ranges, U.S. forces can utilize an array of standoff weapons, which are capable of hitting their targets from hundreds of kilometers away and can be deployed from a growing variety of platforms.[6]

This larger and more varied inventory of precision and standoff weapons enables U.S. air forces to attack more targets and cause more damage with each attack. To assess the net impact of improvements to both U.S. offensive and Chinese defensive capabilities, we modeled attacks on the 40 Chinese air bases within unrefueled fighter range of Taiwan, and, separately, on the smaller number from which Chinese aircraft could range the Spratly Islands. Runway attack models suggest that, in 1996, U.S. air attacks could close Chinese runways for an average of eight hours. This figure had increased to between two and three days by 2010, and it remained roughly similar through 2017. In all four snapshot years, U.S. air forces could effectively close all of China's air bases opposite the Spratly Islands for the first week of operations. While ground attack represents a rare bright spot for relative U.S. performance, it is important to note that the inventory of standoff weapons is finite, and performance in a longer conflict would depend on a wider range of factors.

Scorecard 5: Chinese Anti-Surface Warfare Capabilities

The PLA has placed as much emphasis on putting U.S. aircraft carrier strike groups (CSGs) at risk as it has into efforts to neutralize U.S. ground-based airpower. China has developed a credible and increasingly robust over-the-horizon (OTH) intelligence, surveillance, and reconnaissance (ISR) capability. It launched its first operational military imaging satellites in 2000 and deployed its first OTH skywave radar system in 2007. The skywave system can detect targets and provide a general, though not precise, location out to 2,000 km beyond China's coastline. The development of China's space and electronics sectors has enabled it to increase the pace of satellite launches and deploy a wider range of sophisticated ISR satellites.

China's development of anti-ship ballistic missiles—the first of their kind anywhere in the world—presents a new threat dimension for U.S. naval commanders. That said, the kill chain for these missiles will pose great difficulties for the PLA, and the United States will make every effort to develop countermeasures. Anti-ship ballistic missiles therefore may not pose the kind of one-shot, one-kill threat sometimes supposed in the popular media. At the same time, however, the ongoing modernization of Chinese air and, especially, submarine capabilities represents a more certain and challenging threat to CSGs. Between 1996 and 2015, the number of modern diesel submarines in China's inventory rose from two to 37, and all but four of theses boats are armed with cruise missiles (as well as torpedoes). RAND modeling suggests that

---

[6] These weapons include the Tomahawk Land Attack Missile (TLAM), the conventional air-launched cruise missile (CALCM), the Standoff Land Attack Missile (SLAM), the SLAM–Extended Range (SLAM-ER), the Joint Air-to-Surface Standoff Missile (JASSM), the JASSM–Extended Range (JASSM-ER), and the small-diameter bomb (SDB). Unlike the others, the SDB is a free-fall weapon, but with deployed "wings," it has a glide range of 60 km, or roughly twice that of other iron bombs.

the effectiveness of the Chinese submarine fleet (as measured by the number of attack opportunities it might achieve against carriers) rose by roughly an order of magnitude between 1996 and 2010, and that it will continue to improve through 2017. Chinese submarines would present a credible threat to U.S. surface ships in a conflict over Taiwan or the South China Sea.

Scorecard 6: U.S. Anti-Surface Warfare Capabilities Versus Chinese Naval Ships

We also assessed Chinese amphibious capabilities and the ability of U.S. submarine, air, and surface forces to sink Chinese amphibious ships. We found that the U.S. ability to destroy Chinese amphibious forces has declined since 1996 but nevertheless remains formidable. China's total amphibious ship capacity is on track to double between 1996 and 2017. China has also deployed larger numbers of more sophisticated anti-submarine warfare helicopters and ships. Largely as a function of the greater number of target ships, RAND modeling suggests that the expected damage that U.S. submarines might inflict has declined since 1996. Even by 2017, however, U.S. submarines alone would be able to destroy almost 40 percent of Chinese amphibious shipping during a seven-day campaign, losses that would likely wreak havoc on the organizational integrity of a landing force.

U.S. aircraft and surface ships armed with cruise missiles would likely also participate in anti-surface warfare. The development and deployment of new classes of U.S. anti-ship cruise missiles remained a relatively low priority for some years after the end of the Cold War, and U.S. advances in this area did not keep pace with those elsewhere in the world. Over the past several years, however, the U.S. military has refocused on developing missiles better suited to the high-threat environment. Although U.S. capability against Chinese amphibious forces has declined somewhat, a combination of submarine, air, and surface attacks would nevertheless pose a serious threat to Chinese amphibious forces and their ability to conduct or sustain an amphibious invasion.

Scorecard 7: U.S. Counterspace Capabilities Versus Chinese Space Systems

The United States, with 526 operational satellites, has a far more extensive orbital infrastructure than does China, with 132 satellites (as of January 2015). However, China has been accelerating its space efforts. Its average rate of satellite launches in 2009–2014 was more than double that of 2003–2008, and more than triple that in 1997–2002. The United States has historically been hesitant to deploy operational counterspace capabilities, in part because it fears legitimating such deployments by others and because of its own dependence on space support for other types of military operations. In 2002, however, Washington changed course and approved funding for selective counterspace capabilities. In 2004, the Counter Communications System, designed to jam enemy communication satellites, reached initial operational capability.

The U.S. military could also utilize experimental or dual-use systems. Laser ranging stations could provide accurate position data to other counterspace systems. More powerful lasers, such as the High-Energy Laser system, could potentially be used to dazzle Chinese satellites' optical sensors. Finally, the U.S. military could potentially use improved ballistic missile interceptors as kinetic weapons, though practical and political considerations would weigh strongly against such destructive attacks. Overall, although the United States leads in the use of space to support terrestrial operations, its counterspace capabilities remain relatively underdeveloped.

Scorecard 8: Chinese Counterspace Capabilities Versus U.S. Space Systems

China has pursued an extensive range of counterspace capabilities. It demonstrated a kinetic anti-satellite capability in 2007 with a missile test against a nonoperational Chinese weather satellite at an altitude of 850 km. At that altitude, many U.S. satellites in low earth orbit (LEO) would be vulnerable. China has also announced three tests of ballistic missile defense interceptors, the latest in July 2014. These tests apparently took place at similar altitudes to the anti-satellite test and almost certainly employed technologies that could also be employed in anti-satellite weapons or roles. Ultimately, political considerations, the fear of escalation, and the vulnerability of Chinese systems to debris may deter the PLA from employing kinetic attacks. Arguably more worrisome are the PLA's Russian-made jamming systems and high-powered dual-use radio transmitters, which might be used against U.S. communication and ISR satellites. Like the United States, China operates laser-ranging stations, which might be able to dazzle U.S. satellites or track their orbits to facilitate other forms of attack.

In addition to Chinese offensive capabilities, the degree of threat posed to specific U.S. satellite constellations depends on the altitude, number, and orbit of satellites in those constellations and the ability of U.S. systems to maintain functionality in the face of attack. We evaluated threats posed to seven distinct U.S. space-based functions. The degree of threat to most of them is increasing. Threats to communication satellites (which are subject to jamming) and imaging systems (which are small in number, with four in LEO) are particularly severe. In two cases, the U.S. Global Positioning System and missile warning systems, upgrades or improvements to satellite function and numbers may mitigate risk substantially.

Scorecard 9: U.S. and Chinese Cyberwarfare Capabilities

China's cyber activities have become a major source of concern in the United States and allied countries. There is strong evidence that many of the hostile cyber espionage activities emanating from China are tied to the PLA. The PLA has maintained organized cyber units since at least the late-1990s, while the U.S. Cyber Command was only formed in 2009. Nevertheless, under wartime conditions, the United States might not fare as poorly in the cyber domain as many assume. Cyber Command works closely with the National Security Agency and can draw heavily on the latter's sophisticated toolkit.

Moreover, in evaluating the likely relative impact of cyber attacks, the target user's skills, network management, and general resiliency are at least as important as the attacker's capabilities. In all of these areas, the United States enjoys substantial advantages, though Chinese performance is improving. Chinese cyber security is suspect, and its civilian computers suffer from the world's highest rate of infection by malware. Both sides might nevertheless face significant surprises in the cyber domain during a conflict, and U.S. logistical efforts are particularly vulnerable, since they rely on unclassified networks that are connected to the Internet.

Scorecard 10: U.S. and Chinese Strategic Nuclear Stability

The nuclear scorecard evaluates crisis stability in the bilateral nuclear relationship rather than the advantage enjoyed by one side or the other. Specifically, the scorecard examines the survivability of both sides' second-strike capabilities in the face of a first strike by the other. When both sides maintain a survivable second-strike capability, the incentives for both the stronger and weaker parties to strike first diminish and stability is, in that sense, enhanced. The scorecard analysis considers the number, range, and accuracy of both sides' offensive weapons, as well as the number, mobility, and "hardness" of nuclear targets.

China has modernized its nuclear forces steadily since 1996, increasing their quantity as well as improving quality. It has improved survivability through the introduction of the road-mobile DF-31 (CSS-9) and DF-31A intercontinental ballistic missiles (ICBMs) and the Type 094 *Jin*-class ballistic missile submarine (SSBNs), capable of carrying 12 modern JL-2 submarine-launched ballistic missiles (SLBMs) with a range of approximately 7,400 km. In April 2015, the U.S. Department of Defense said that China has added multiple independently targetable reentry vehicles to some of its DF-5 missiles, and China is currently developing next-generation road-mobile ICBMs, SSBNs, and SLBMs. The United States has committed major funding to modernize its nuclear arsenal but, in keeping with both Strategic Arms Reduction Treaty (START) and New START commitments and in contrast to China, is reducing the number of operationally deployed warheads and strategic delivery systems (Heavy Bombers, ICBMs, SSBNs).

Despite additions to the Chinese nuclear force and U.S. reductions, even by 2017, the United States will still enjoy a numerical advantage in warheads of at least 13 to one. A Chinese first strike could not plausibly deny the United States a retaliatory capability in any of the snapshot years considered. For its part, Chinese survivability has improved significantly. Nuclear exchange modeling suggests that, as late as 2003, only a handful of Chinese systems might have survived a U.S. first strike—and even this outcome would have depended largely on China deploying its single, unreliable *Xia*-class SSBN prior to an attack. In the 2010 and 2017 cases, more Chinese warheads survive, and no foreign leader could contemplate a disarming first strike against China with any degree of confidence.

Coding the Results

Figure S.1 shows how we coded each scorecard. For the first nine scorecards, we employed a five-color stoplight approach to depict major or minor U.S. advantage (in dark and light green), major or minor Chinese advantage (in red and orange), and approximate parity (in yellow). *Advantage*, in this context, means that one side is able to achieve its primary objectives in an operationally relevant period while the other

Figure S.1
Summary Coding of Scorecard Results

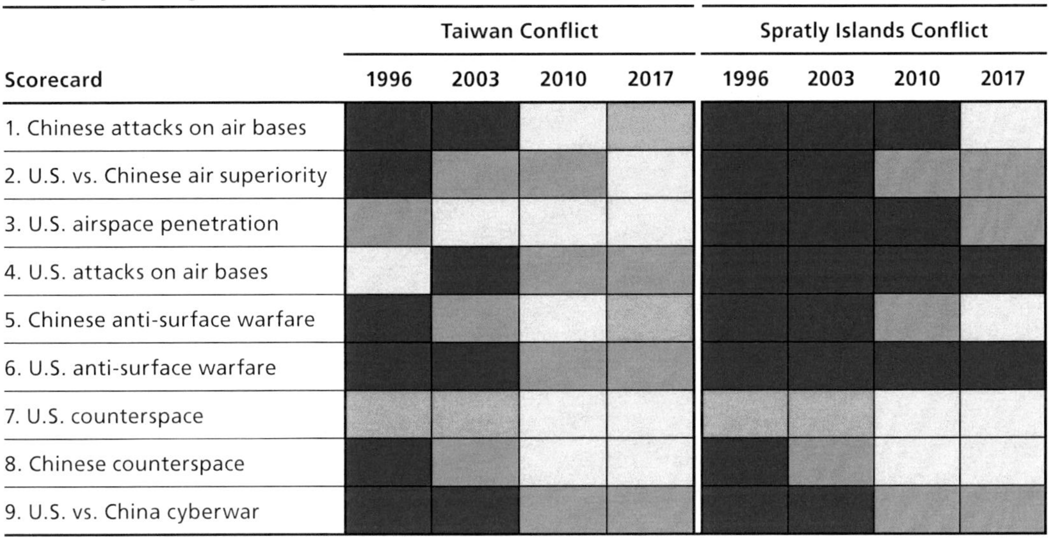

	Taiwan Conflict				Spratly Islands Conflict			
Scorecard	**1996**	**2003**	**2010**	**2017**	**1996**	**2003**	**2010**	**2017**
1. Chinese attacks on air bases								
2. U.S. vs. Chinese air superiority								
3. U.S. airspace penetration								
4. U.S. attacks on air bases								
5. Chinese anti-surface warfare								
6. U.S. anti-surface warfare								
7. U.S. counterspace								
8. Chinese counterspace								
9. U.S. vs. China cyberwar								

	Country	1996, 2003, and 2010	2017
10. Nuclear stability (confidence in secure second-strike capability)	China	Low confidence	Medium confidence
	U.S.	High confidence	

NOTES: To prevail in either Taiwan or the Spratly Islands, China's offensive goals would require it to hold advantages in nearly all operational categories simultaneously. U.S. defensive goals could be achieved by holding the advantage in only a few areas. Nevertheless, China's improved performance could raise costs, lengthen the conflict, and increase risks to the United States.

Key for Scorecards 1–9

U.S. Capabilities		Chinese Capabilities
Major advantage		Major disadvantage
Advantage		Disadvantage
Approximate parity		Approximate parity
Disadvantage		Advantage
Major disadvantage		Major advantage

side would have trouble in doing so.[7] For the nuclear scorecard, results are expressed as the degree of confidence that each side could reasonably expect to have in the survivability of its second-strike strategic nuclear capability. The summary coding suggests a number of trends and conclusions, outlined in the next section. But the coding should be considered only in the context of the fuller analysis presented in this report, as well as with the caveats discussed in Chapters One and Thirteen.

Conclusions: Receding Frontier of U.S. Dominance

Looking across the results of all ten scorecards, four broad trends emerge:

- Since 1996, the PLA has made tremendous strides, and, despite improvements to the U.S. military, the net change in capabilities is moving in favor of China. Some aspects of Chinese military modernization, such as improvements to PLA ballistic missiles, fighter aircraft, and attack submarines, have come extraordinarily quickly by any reasonable historical standard.
- The trends vary by mission area, and relative Chinese gains have not been uniform across all areas. In some areas, U.S. improvements have given the United States new options, or at least mitigated the speed at which Chinese military modernization has shifted the relative balance.
- Distances, even relatively short distances, have a major impact on the two sides' ability to achieve critical objectives. Chinese power projection capabilities are improving, but present limitations mean that the PLA's ability to influence events and win battles diminishes rapidly beyond the unrefueled range of jet fighters and diesel submarines. This is likely to change in the years beyond those considered in this report, though operating at greater distances from China will always work, on balance, against China.
- The PLA is not close to catching up to the U.S. military in terms of aggregate capabilities, but it does not need to catch up to the United States to dominate its immediate periphery. The advantages conferred by proximity severely complicate U.S. military tasks while providing major advantages to the PLA. This is the central finding of this study and highlights the value of campaign analysis, rather than more abstract assessments of capabilities.

[7] The duration of conflict considered is key to the coding of results. Our coding considers advantage primarily in the first three weeks of a conflict. The first several weeks could see immense destruction and set the conditions under which a longer war might or might not continue. Expectation about how a conflict might play out during the first few weeks will shape leadership behavior on both sides during crises. Nevertheless, it should be borne in mind that conflicts can evolve in various ways, and the results could be quite different in a more protracted fight.

Over the next five to 15 years, if U.S. and PLA forces remain on roughly current trajectories, Asia will witness a progressively receding frontier of U.S. dominance. The United States would probably still prevail in a protracted war centered in virtually any area, and Beijing should not infer from the above generalization that it stands to gain from conflict. U.S. and Chinese forces would likely face losses on a scale that neither has suffered in recent decades. But PLA forces will become more capable of establishing temporary local air and naval superiority at the outset of a conflict. In certain regional contingencies, this temporal or local superiority might enable the PLA to achieve limited objectives without "defeating" U.S. forces. Perhaps even more worrisome from a military-political perspective, the ability to contest dominance might lead Chinese leaders to believe that they could deter U.S. intervention in a conflict between it and one or more of its neighbors. This, in turn, would undermine U.S. deterrence and could, in a crisis, tip the balance of debate in Beijing as to the advisability of using force.

Although the United States will probably not have the resources to prevent all further erosion of the balance of military power over the next decade, it can adjust its force structure, operating concepts, and diplomacy in ways that will slow the process and limit the impact on deterrence and other U.S. strategic interests. In the longer term, technological and, especially, economic variables will determine whether and when the larger trend can be reversed or the balance stabilized, while political events will determine how important the military equation is in defining the relationship between Washington and Beijing.

Recommendations

Based on the findings outlined here, we make five broad recommendations designed to buttress deterrence, reduce U.S. losses at the start of a conflict, and ensure victory should war occur. All represent areas in which we believe additional thought and analysis are in order, rather than fully developed policy proposals:

- Western governments and commentators should work to shape Chinese perceptions. Trends in the balance of power are moving against the United States, but analysts should make it clear that war would carry immense risks for Beijing.
- Procurement priorities should be adjusted, with more emphasis on base redundancy and survivability; standoff systems optimized for high-intensity conflict; stealthy, survivable fighters and bombers; submarine and anti-submarine warfare; and robust space and counterspace capabilities. To pay for these priorities, more rapid cuts to legacy fighter forces should be considered, as should decreasing the emphasis on large aircraft carriers.

- The process by which the United States plans for Pacific military operations should be made as dynamic and open as possible. An active denial strategy that capitalizes on Asia's strategic depth and enable U.S. forces to absorb initial blows and fight their way back toward final objectives should be considered. Defending static positions near China may simply become unaffordable.
- U.S. political-military efforts with the Pacific island states and Southeast Asian nations should be intensified, with the goal of expanding potential access in wartime. While the greatest immediate prospects are in deepening defense relations with the Philippines and Vietnam, efforts should also be made with the states of Southeast Asia's "southern tier," including Indonesia and Malaysia. This will provide greater strategic depth and more options for U.S. forces.
- The United States should make a concerted effort to engage China on strategic stability and escalation issues. The deployment of new classes of conventional and nuclear weapons will likely complicate arms-control challenges in the coming years, but discussions could nevertheless serve to sensitize Chinese and U.S. policymakers to emerging dangers and, perhaps, lead to mutual restraint in some areas.

Although trends in the military balance are running against the United States, there are many actions that the United States could take to reinforce deterrence and continue to serve as the ultimate force for stability in the Western Pacific.

Acknowledgments

The authors are indebted to many for contributions ranging from substantive and methodological suggestions to administrative support. While we lack the space to thank them all, we do wish to acknowledge the contributions of some of those who have been particularly instrumental. The authors wish to thank Michael Chase, Roger Cliff, Cortez Cooper, David Orletsky, Lara Schmidt, and Christopher Twomey for reviewing this report and offering a number of valuable suggestions that have since been incorporated into the text. Burgess Laird provided a final in-house review, assisted with important content, and helped update key data prior to publication. Members of the RAND Board of Trustees and the Defense Science Board provided invaluable feedback during briefings to them on the study's results.

We also wish to thank the U.S. Air Force for its embrace of independent research on Asian military issues. RAND leadership provided critical backing and direction throughout the process. Andrew Hoehn, whose desire to see the RAND Corporation advance the public debate on Chinese and U.S. military capabilities, played the central role in initiating the study. Kari Thyne was invaluable in keeping the project on track and meeting deadlines. Sarah Hauer provided critical administrative support and preparation of the document, including countless hours working with the bibliography and footnotes. And Lauren Skrabala's careful copyediting made a significant contribution to clarity of expression.

Finally, thanks beyond measure are due to Paula Thornhill for her thoughtful, patient, and insightful leadership in ensuring the successful completion of the work. She read the document numerous times, improving it in ways too numerous to mention and navigating it through a daunting publications process. Needless to say, whatever mistakes in fact or interpretation remain are the responsibility of the authors alone.

Abbreviations

AB	air base
AESA	active electronically scanned array
AEW	airborne early warning
AFB	air force base
AIP	air independent propulsion
ALCM	air-launched cruise missile
AMRAAM	Advanced Medium Range Air-to-Air Missile
ASAT	anti-satellite
ASBM	anti-ship ballistic missile
ASCM	anti-ship cruise missile
AWACS	airborne warning and control system
BMD	ballistic missile defense
C4ISR	command, control, communication, computers, intelligence, surveillance, and reconnaissance
CALCM	conventional air-launched cruise missile
CCS	Counter Communications System
CEP	circular error probable
CG	hull classification for cruiser
CONUS	continental United States
CPGS	conventional prompt global strike
CSG	carrier strike group
CSRS	Counter Surveillance Reconnaissance System

CVN	hull classification for nuclear aircraft carrier
CZ	convergence zone
DCA	defensive counter-air
DD	hull classification for destroyer
DDG	hull classification for guided missile destroyer
DMSP	Defense Meteorological Satellite Program
DoD	U.S. Department of Defense
ECM	electronic countermeasure
ELINT	electronic intelligence
EO	electro-optical
FFG	hull classification for guided missile frigate
FGA	ground attack fighter aircraft
FTR	air supremacy fighter aircraft
FY	fiscal year
GBI	ground-based interceptor
GDP	gross domestic product
GEO	geostationary earth orbit
GLCM	ground-launched cruise missile
GMD	Ground-Based Midcourse Defense
GOES	Geostationary Orbiting Environmental System
GPS	Global Positioning System
GSD	General Staff Department (China)
HARM	high-speed anti-radiation missile
HE	high explosive
HEO	highly elliptical orbit
HTS	HARM Targeting System
IADS	integrated air defense system
ICBM	intercontinental ballistic missile
IISS	International Institute for Strategic Studies

INS	inertial navigation system
IOC	initial operating capability
IR	infrared
IRBM	intermediate-range ballistic missile
ISIS	Islamic State of Iraq and Syria
ISR	intelligence, surveillance, and reconnaissance
JASSM	Joint Air-to-Surface Standoff Missile
JASSM-ER	Joint Air-to-Surface Standoff Missile–Extended Range
JDAM	Joint Direct Attack Munition
LCS	littoral combat ship
LEO	low earth orbit
LPAR	Large phased array radar
LPD	hull classification for landing platform dock
LRASM	long-range anti-ship missile
LRS-B	long-range strike bomber
LST	hull classification for landing ship, tank
MCAS	Marine Corps air station
MEO	medium earth orbit
MILSATCOM	military satellite communication
MIRACL	Mid-Infrared Advanced Chemical Laser
MIRV	multiple independently targetable reentry vehicle
MMA	multimission maritime aircraft
MOS	minimum operating surface
MPA	maritime patrol aircraft
MPS	Ministry of Public Security
MRBM	medium-range ballistic missile
MSS	Ministry of State Security (China)
NATO	North Atlantic Treaty Organization
NIPRNet	Non-Secure Internet Protocol Router Network

NOSS	naval ocean surveillance system
NSA	National Security Agency
OTH	over the horizon
PGM	precision-guided munition
Pk	probability of kill
PLA	People's Liberation Army
PLAAF	People's Liberation Army Air Force
PLAN	People's Liberation Army Navy
PNT	position, navigation, and timing
POES	Polar Orbiting Environmental System
RCS	radar cross-section
RF	radio frequency
SAM	surface-to-air missile
SAR	synthetic aperture radar
SATCOM	satellite communications
SBIRS	Space-Based Infrared System
SCADA	supervisory control and data acquisition
SDB	small-diameter bomb
SEAD	suppression of enemy air defenses
SIGINT	signals intelligence
SIPRNet	Secure Internet Protocol Router Network
SLAM	Standoff Land Attack Missile
SLAM-ER	Standoff Land Attack Missile–Extended Range
SLBM	submarine-launched ballistic missile
SM	Standard Missile
SRBM	short-range ballistic missile
SSBN	hull classification for ballistic missile submarine
SSGN	hull classification for cruise missile submarine
SSK	hull identification for attack submarine ("hunter-killer")

SSN	hull classification for nuclear attack submarine
START	Strategic Arms Reduction Treaty
SURTASS	Surveillance Towed Array Sensor System
T-AGOS	tactical auxiliary general ocean surveillance
TBM	theater ballistic missile
TDOA	time-distance of arrival
TEL	transporter-erector-launcher
THAAD	Terminal High-Altitude Area Defense
THEL	Tactical High-Energy Laser
TLAM	Tomahawk Land Attack Missile
UAV	unmanned aerial vehicle
UHF	ultra high frequency
UN	United Nations
USCYBERCOM	U.S. Cyber Command
VHF	very high frequency
VLS	vertical launch system
WGS	Wideband Global Satellite Communication

CHAPTER ONE

Introduction

As late as the mid-1990s, China's military was large but antiquated in both equipment and operational practice.[1] Since then, it has made great strides in virtually all areas. It has replaced much of its old stock of equipment, updated its doctrine, improved the quality of its personnel, and increased the realism of its training. New systems and capabilities challenge the ability of the United States to operate in areas close to China. The reach of the People's Liberation Army (PLA) has grown dramatically, and it has undertaken a wide variety of deployments for United Nations (UN) peacekeeping assignments and training missions. Since December 2008 it has kept naval forces on station to conduct counter-piracy operations in the Gulf of Aden.

These improvements are partly a function of significantly expanded resources. The sinews of national power grow from the civilian economy, and, in China's case, the economy has grown in real (inflation-adjusted) terms by 442 percent between 1996 and (estimated) 2015. Defense spending, which grew by 620 percent in real terms over the same period (for an annual growth rate of 11 percent), has outstripped growth in gross domestic product (GDP).[2] Just as important, improvements to Chinese military capabilities have also been a function of the reform of organizational structures, pro-

[1] Declassified documents reveal that, in the mid-1980s, the U.S. Central Intelligence Agency judged Chinese forces to be outmoded and suggested that the Soviet Union possessed the capacity to take offensive action in northern China, as well as the ability to take Beijing with reinforcement (Central Intelligence Agency, "Soviet Forces in the Far East," National Intelligence Estimate 11-14/40-81, October 1985). A decade later, a 1995 RAND Corporation study of the PLA Air Force (PLAAF) concluded that it "does not constitute a credible offensive threat against the United States or its Asian allies today" (Kenneth W. Allen, Glenn Krumel, and Jonathan D. Pollack, *China's Air Force Enters the 21st Century*, Santa Monica, Calif.: RAND Corporation, MR-580-AF, 1995).

[2] We note, however, that Chinese government expenditures have risen even more rapidly, and defense spending has fallen as a percentage of the overall government budget. GDP (national currency) and GDP deflator data are from the International Monetary Fund's World Economic Outlook Database as of May 2015. Official Chinese defense budgets can be found in a variety of places, including reports following each year's National People's Congress and in the State Statistics Bureau, *China Statistical Yearbook*, various years. China's official defense expenditure is believed to underrepresent China's overall defense-related spending, but the *growth rate* of China's overall defense spending is believed to be approximately the same as that of the official defense expenditure. For more on Chinese defense budgets and changes to them, see Keith Crane, Roger Cliff, Evan Medeiros, James Mulvenon, and William Overholt, *Modernizing China's Military: Opportunities and Constraints*, Santa Monica, Calif.:

curement, and operational concepts.[3] The PLA is not without important weaknesses, but it continues to improve and narrow the qualitative lead once enjoyed by the United States as it moves toward a leaner but higher-quality fighting force.[4]

Although China is narrowing the gap, the United States maintains a substantial overall lead in defense capabilities. It continues to outspend China. At $560 billion in 2015 (including supplemental spending for ongoing operations), the U.S. defense budget was roughly four times the official Chinese defense budget of $142 billion (converted at the market exchange rate).[5] Not all defense-related expenditures are included in China's official defense budget. According to one estimate, China's 2010 defense spending was roughly 45 percent larger than its official defense budget numbers would suggest, and that percentage probably holds roughly true today.[6] However, the U.S. defense budget also fails to capture much of its security-related spending; veterans' affairs and veterans' pensions alone would add 19 percent to the official budget (not including most intelligence, nuclear weapons, and homeland security expenditures). Given the likelihood that China's defense budget growth will continue to outpace that of the United States, the resource gap will continue to narrow over time.

The United States enjoys a stockpile of equipment that is, overall, more formidable than China's. China is only now readying its first aircraft carrier, while the United States has ten full-sized carriers and nine amphibious assault ships capable of supporting fixed-wing aircraft. The U.S. Air Force's first purpose-built stealth aircraft,

---

RAND Corporation, MG-260-AF, 2005, and George J. Gilboy and Eric Heginbotham, *Chinese and Indian Strategic Behavior: Growing Power and Alarm*, Cambridge, UK: Cambridge University Press, 2012.

[3] For two general works that address all three issues, see Anthony H. Cordesman and Martin Kleiber, *Chinese Military Modernization: Force Development and Strategic Capabilities*, Washington, D.C.: Center for Strategic and International Studies, 2007; and David Shambaugh, *Modernizing China's Military: Progress, Problems, and Prospects*, Berkeley, Calif.: University of California Press, 2002. On the reform of the procurement system and industry, see Crane et al., *Modernizing China's Military*, 2005, pp. 135–190. On the development of doctrine, see James Mulvenon and David M. Finkelstein, eds., *China's Revolution in Doctrinal Affairs: Emerging Trends in the Operational Art of the Chinese People's Liberation Army*, Alexandria, Va.: Center for Naval Analyses, 2005.

[4] On continuing PLA weaknesses, see Michael S. Chase, Jeffrey Engstrom, Tai Ming Cheung, Kristen A. Gunness, Scott Warren Harold, Susan Puska, and Samuel K. Berkowitz, *China's Incomplete Military Transformation: Assessing the Weaknesses of the People's Liberation Army (PLA)*, Santa Monica, Calif.: RAND Corporation, RR-893-USCC, February 2015.

[5] Office of the Secretary of Defense, Chief Financial Officer, *United States Department of Defense Fiscal Year 2016 Budget Request: Overview*, Washington, D.C.: U.S. Department of Defense, February 2015, pp. 1–5.

[6] Off-budget items—those that contribute to defense capabilities but are not included in official accounting of "defense spending"—are a feature of virtually all states, but Chinese budgeting is less transparent than that of the United States. Nevertheless, even in the Chinese case, many of the most important categories can either be found as line items in other parts of the budget or estimated using a variety of means. To arrive at the 45-percent figure cited here, spending or estimated spending on defense pensions and benefits, paramilitaries, research and development, subsidies to dual-use industries, arms imports, and arms sales profits was added to the Chinese official defense budget figure. Budgets for intelligence functions and nuclear programs were not included. For additional details, see Gilboy and Heginbotham, *Chinese and Indian Strategic Behavior*, 2012.

the F-117, achieved initial operating capability in 1983. The Air Force currently operates ten squadrons of 185 F-22 aircraft, the world's first operational fifth-generation fighter.[7] As of June 2014, the Air Force and the Marine Corps had taken delivery of 78 F-35 Joint Strike Fighters, the world's second operational fifth-generation aircraft currently in low-rate initial production.[8] The PLA, for its part, appears highly motivated to develop its own stealth capabilities. By July 2014, it had tested four prototypes of the J-20 and had conducted flight-testing of a second fighter with stealth characteristics, the J-31, but none of these aircraft had yet entered production.[9] The United States also maintains far more (and far more capable) support aircraft, such as tankers and airborne warning and control system (AWACS)–equipped aircraft, and it enjoys a similar lead in deployed satellite capabilities, attack submarines, and anti-submarine platforms.

Nonetheless, a number of factors complicate the U.S. military position vis-à-vis China. The U.S. military is pulled in several directions by a range of world-wide demands, while the PLA enjoys the ability to focus more narrowly. Most obviously, the U.S. military continues to operate in both Afghanistan and Iraq, periodically engages in other contingencies (e.g., Libya in 2011 and Syria from 2014 to the present), and maintains a substantial forward presence in Europe and Asia. The U.S.-led wars in Iraq and Afghanistan not only have consumed enormous resources—with the direct combined costs totaling approximately $1.5 trillion through the end of 2014—but have also led the U.S. military to reorganize and re-equip many of its forces in ways that maximize counterinsurgency rather than conventional capabilities.[10]

Historical legacy and national interests have bequeathed the United States other global tasks, missions, and considerations that have little to do with China but consume large portions of the defense budget. In evaluating its strategic nuclear posture, for example, the United States has focused largely on sufficiency relative to the Russian arsenal, which dwarfs that of China or any other second-tier nuclear power. Moving forward, the U.S. military will be constrained by growing budgetary pressures and increased political emphasis on domestic priorities.

China's narrower focus on a range of primarily East Asian missions, especially Taiwan contingencies, allows it to optimize its forces for those tasks. Geography—the "bones of strategy"—vastly complicates the challenges faced by the United States. Taiwan, toward which the United States has complex historical obligations but which

---

[7] United States Air Force, *Fiscal Year 2016 Budget Overview*, Washington, D.C.: U.S. Department of Defense, February 2015, p. 37.

[8] U.S. Government Accountability Office, *F-35 Sustainment: Need for Affordable Strategy, Greater Attention to Risks, and Improved Cost Estimates*, GAO-14-778, Washington, D.C., September 2014, p. 7.

[9] U.S.-China Economic and Security Review Commission, *2014 Report to Congress*, Washington, D.C.: U.S. Government Printing Office, 2014, p. 311.

[10] Neta C. Crawford, *U.S. Costs of Wars Through 2014: $4.4 Trillion and Counting*, 2014.

is claimed as a renegade province by China, lies less than 100 miles off the Chinese coast. In addition, three formal U.S. allies in Asia (Japan, South Korea, and the Philippines) have territorial or maritime claims that overlap with those of China, as do several non-treaty U.S. partners. While China's proximity to many of these allies can be measured in dozens or hundreds of miles, these areas are thousands of miles from the continental United States.

As China prepares for conflicts close to its periphery, the mainland provides large and relatively secure staging areas for operations. This enables the PLA to focus largely on "tooth" (combat forces) as opposed to "tail" (support assets). While the United States must maintain an extensive sea and air logistical capacity, along with largely space-based communication systems, China can utilize local logistical facilities and land-based communication capabilities. These assets are inherently more robust, more secure, and less expensive than the ship-, air-, or space-based platforms upon which U.S. forces must depend.

Further capitalizing on geographic asymmetry, the PLA has developed a number of relatively inexpensive, high-leverage capabilities designed to keep U.S. forces at arm's length. Although they have come to be known as anti-access/area-denial (A2/AD) capabilities, the original Chinese is better translated as *anti-intervention*.[11] These capabilities include a large number of highly accurate ballistic missiles, high-quality anti-ship cruise missiles, submarines, sophisticated long-range air defense systems, and counter-C4ISR (command, control, communication, computer, intelligence, surveillance, and reconnaissance) capabilities, including counterspace, electronic warfare, cyberwarfare, and anti-radiation systems.

The degree to which the United States and China can marshal their unique advantages to address the situation at hand depends on the scenario considered and the operational conditions associated with it. In general, scenarios close to the Chinese mainland offer important operational advantages to the PLA, while those more distant play to U.S. strengths. The objectives associated with military action would have an equally large impact. A Chinese campaign to take and occupy Taiwan would present different opportunities and challenges for both sides than would, for example, an effort to achieve concessions through a submarine or mine-based blockade of Taiwanese ports.

---

[11] The Chinese term 反介入 (*fanjieru*) is better translated as "anti-intervention." The term that more directly translates to anti-access, 反进入 (*fanjinru*), is generally employed only in reference to the U.S. discussion of Chinese strategy. As Christopher Twomey and Taylor Fravel observe, Chinese analysts do not employ either term to describe Chinese strategy. Nevertheless, the term can be a useful shorthand to describe a range of Chinese capabilities that are more useful for China in denying enemy control of air or maritime spaces than to actively control them for itself. See Christopher P. Twomey and Taylor M. Fravel, "Projecting Strategy: The Myth of Chinese Counter-Intervention," *Washington Quarterly*, Vol. 37, No. 4, 2015. For sources that discuss anti-access strategies, see Roger Cliff, Mark Burles, Michael S. Chase, Derek Eaton, and Kevin L. Pollpeter, *Entering the Dragon's Lair: Chinese Antiaccess Strategies and Their Implications for the United States*, Santa Monica, Calif.: RAND Corporation, MG-524-AF, 2007; and Jonathan Greenert and Mark Welsh, "Breaking the Kill Chain: How to Keep America in the Game When Our Enemies Are Trying to Shut Us Out," *Foreign Policy*, May 16, 2013.

Any realistic and effective U.S. political-military strategy in Asia must account for relative military capabilities. In the context of the U.S. democratic system, much of the process by which such strategies are decided unfolds in public. Each administration regularly produces documents that outline the broad contours of U.S. strategy and provide a basis for discussing and debating the country's international direction. Whatever course the executive branch sets, congressional support, in the form its oversight role and power of the purse, is indispensable to implementing effective policy. Ultimately, the public must also be engaged.

The speed and magnitude of change in the East Asian balance of power will inevitably prompt a vigorous debate on U.S. political and military strategy in the region. The ability to produce policy that will secure U.S. interests in the region depends on a solid analytical foundation. Yet, despite the fact that Asian defense will be central to overall U.S. strategy and military force planning, the available literature on these issues is inconsistent in its coverage.

The State of the Literature

China is becoming an increasingly important driver of U.S. military requirements. Given the importance of East Asian military issues, one might expect both a vigorous public debate on the subject and a substantial analytical literature to support the discussion. To be sure, there is a rich body of work on trends in Chinese military capabilities, much of which offers assessments of Chinese source material.[12] This literature includes specialized works on individual services, especially maritime, air, and missile capabilities.[13] Feeding into this literature, particularly in the past five years, has been a

---

[12] The Office of the Secretary of Defense publishes an annual document on Chinese military developments. The latest, as of this writing, is *Military and Security Developments Involving the People's Republic of China 2015*, Washington, D.C., 2015. For a general discussion, see Ashley J. Tellis and Travis Tanner, eds., *Strategic Asia 2012–13: China's Military Challenge*, Seattle, Wash.: National Bureau of Asian Research, 2012; Cordesman and Klieber, *Chinese Military Modernization*, 2007; Shambaugh, *Modernizing China's Military*, 2002; and Roy Kamphausen and Andrew Scobell, eds., *Right-Sizing the People's Liberation Army: Exploring the Contours of China's Military*, Carlisle, Pa.: Strategic Studies Institute, U.S. Army War College, September 2007.

[13] The volume of this literature permits only a selective citation of sources. On PLA ground forces, see Dennis J. Blasko, *The Chinese Army Today: Tradition and Transformation for the 21st Century*, New York: Routledge, 2012. On maritime capabilities, see Phillip C. Saunders, Christopher Yung, Michael Swaine, and Andrew Nien-Dzu Yang, eds., *The Chinese Navy: Expanding Capabilities, Evolving Roles*, Washington, D.C.: National Defense University, 2011; Bernard D. Cole, *The Great Wall at Sea: China's Navy in the Twenty-First Century*, 2nd ed., Annapolis, Md.: Naval Institute Press, 2010; James C. Bussert and Bruce Eleman, *People's Liberation Army Navy: Combat Systems Technology, 1949–2010*, Annapolis, Md.: Naval Institute Press, 2011; and Office of Naval Intelligence, *People's Liberation Army Navy: A Modern Navy with Chinese Characteristics*, Suitland, Md., August 2009. On the PLAAF, see Richard P. Hallion, Roger Cliff, and Phillip Saunders, eds., *The Chinese Air Force: Evolving Concepts, Roles, and Capabilities*, Washington, D.C.: National Defense University Press, 2012; Kenneth Allen, *The Ten Pillars of the People's Liberation Army Air Force: An Assessment*, Washington, D.C.: Jamestown Foundation, April 2011; and Roger Cliff, John Fei, Jeff Hagen, Elizabeth Hague, Eric Heginbotham, and John Stillion, *Shak-*

wealth of news, reports, and studies on the development of notable new Chinese military capabilities.[14]

Yet, despite a large and growing volume of work on the development of Chinese military capabilities, the literature remains limited in several important respects. Perhaps most important in the context of the public discussion of strategy and priorities, the literature on *relative* capabilities in an operational setting is sparse and underdeveloped. Obviously, military conflict does not take place in an abstract void, and some understanding of the interaction of capabilities, geography, distance, speed, and relevant operational dynamics is necessary to reach meaningful conclusions. A number of well-reasoned and important qualitative assessments of military dynamics have addressed these contextual factors.[15] But with a small number of notable exceptions, there has been little work that attempts to quantify the dynamics in ways that would enable a more structured evaluation—or offer a baseline for discussions and debates about the role or parameters of particular variables.[16]

This analytical gap is particularly notable when contrasted with the Cold War period, during which a large body of operationally informed, open-source analysis was produced and debated. Some of this analysis, particularly early in the Cold War,

---

ing the Heavens and Splitting the Earth: Chinese Air Force Employment Concepts in the 21st Century, Santa Monica, Calif.: RAND Corporation, MG-915-AF, 2011. On China's strategic nuclear forces, see Hans M. Kristensen and Robert S. Norris, "Chinese Nuclear Forces, 2013," *Bulletin of the Atomic Scientists*, Vol. 69, No. 6, November/December 2013; M. Taylor Fravel and Evan S. Medeiros, "China's Search for Assured Retaliation: The Evolution of Chinese Nuclear Strategy and Force Structure," *International Security*, Vol. 35, No. 2, Fall 2010; and John Wilson Lewis and Xue Litai, *Imagined Enemies: China Prepares for Uncertain War*, Stanford, Calif.: Stanford University Press, 2006.

[14] See, for example, Dennis M. Gormley, Andrew S. Erickson, and Jingdong Yuan, *A Low-Visibility Force Multiplier: Assessing China's Cruise Missile Ambitions*, Washington, D.C.: National Defense University Press, 2014; Andrew S. Erickson, *Chinese Anti-Ship Ballistic Missile Development: Drivers, Trajectories, and Strategic Implications*, Washington, D.C.: Jamestown Foundation, May 2013; Andrew S. Erickson, Abraham M. Denmark, and Gabriel Collins, "Beijing's 'Starter Carrier' and Future Steps," *Naval War College Review*, Vol. 65, No. 1, Winter 2012; John A. Tirpak, "Here Comes Adversary Stealth," *Air Force Magazine*, Vol. 95, No. 12, December 2012; and Ian Easton, "China's Evolving Reconnaissance Strike Capabilities," Project 2049 Institute, February 2014.

[15] See, for example, Andrew F. Krepinevich, *Why Air-Sea Battle?* Washington, D.C.: Center for Strategic and Budgetary Assessment, 2010; and William S. Murray, "Revisiting Taiwan's Defense Strategy," *Naval War College Review*, Vol. 61, No. 3, Summer 2008.

[16] For exceptions, see Michael A. Glosny, "Strangulation from the Sea? A PRC Blockade of Taiwan," *International Security*, Vol. 28, No. 4, Spring 2004; Owen R. Cote, Jr., *Assessing the Undersea Balance Between the U.S. and China*, MIT Security Studies Working Paper, Cambridge, Mass.: Massachusetts Institute of Technology, February 2011; David A. Shlapak, David T. Orletsky, and Barry A. Wilson, *Dire Strait? Military Aspects of the China-Taiwan Confrontation and Options for U.S. Policy*, Santa Monica, Calif.: RAND Corporation, MR-1217-SRF, 2000; and David A. Shlapak, David T. Orletsky, Toy I. Reid, Murray Scot Tanner, and Barry Wilson, *A Question of Balance: Political Context and Military Aspects of the China-Taiwan Dispute*, RAND Corporation, MR-888-SRF, 2009. Other quantitative work provides analysis that could easily be applied to U.S.-China military dynamics, but is not framed explicitly as such. See, for example, Dean A. Wilkening, "A Simple Model for Calculating Ballistic Missile Defense Effectiveness," *Science and Global Security*, Vol. 8, No. 2, 1999.

focused on nuclear issues.[17] Subsequently, the focus shifted largely to questions of sufficiency in the context of specific conventional military campaigns, with a particular emphasis on a possible campaign in Central Europe.[18] More specialized work looked at individual parts of a Central European campaign, such as armored warfare or air combat.[19] Other work modeled ground combat in the Middle East and U.S.-Soviet submarine combat in the Pacific.[20] This analytical work informed public discussions of critical defense issues and set the stage for important shifts in operational concepts.

It would be wrong to exaggerate the intellectual achievements from the Cold War, which produced outstanding analytical failures in addition to successes, or to disparage contemporary work on Chinese military issues. With a core of dedicated China-watchers and much better access to source material in China than was possible in the Soviet Union, Western analysts have a richer and more complete understanding of many aspects of the Chinese military system than they ever did of the Soviet military. Nevertheless, it is fair to say that the current discussion of China-related military issues suffers from a relative deficit of rigorous analysis of comparative U.S. and Chinese military capabilities in operationally relevant circumstances. Perhaps the paucity of work on this subject derives, at least in part, from an unwillingness to imagine war between states that are not (and need not be) enemies. While this is a laudable sentiment, the United States has security interests in Asia that require imagining the worst, even as it strives for a far superior, peaceful outcome.

---

[17] For an early Cold War case, see the seminal RAND study led by Albert Wohlstetter that assessed the survivability of U.S. bomber bases and recommended strategies for enhancing deterrence. Albert J. Wohlstetter, Fred S. Hoffman, R. J. Lutz, and Henry S. Rowen, *Selection and Use of Strategic Air Bases*, Santa Monica, Calif.: RAND Corporation, R-266, 1954. (A classified version of the report was published the preceding year.) For work from the late Cold War, see Glenn A. Kent and David E. Thaler, *First-Strike Stability: A Methodology for Evaluating Strategic Forces*, Santa Monica, Calif.: RAND Corporation, R-3765, 1989.

[18] Barry R. Posen, "Is NATO Decisively Outnumbered?" *International Security*, Vol. 12, No. 4, Spring 1988; Eliot A. Cohen, "Toward Better Net Assessment: Rethinking the European Conventional Balance," *International Security*, Vol. 13, No. 1, Summer 1988; Barry R. Posen, "Measuring the European Conventional Balance," *International Security*, Vol. 9, No. 3, Winter 1984–1985; John J. Mearsheimer, "Numbers, Strategy, and the European Balance," *International Security*, Vol. 12, No. 4, Spring 1988.

[19] Joshua M. Epstein, *Measuring Military Power: The Soviet Air Threat to Europe*, Princeton, N.J.: Princeton University Press, 1984; Malcolm Chalmers and Lutz Unterseher, "Is There a Tank Gap? Comparing NATO and Warsaw Pact Tank Fleets," *International Security*, Vol. 13, No. 1, Summer 1988.

[20] Joshua M. Epstein, *Strategy and Force Planning: The Case of the Persian Gulf*, Washington, D.C.: Brookings Institution Press, 1987; Barry R. Posen, *Inadvertent Escalation: Conventional War and Nuclear Risks*, Ithaca, N.Y.: Cornell University Press, 1992.

Purpose and Scope of This Study

No single work can fill the gap in the literature discussed above in its entirely, and this study represents more of a beginning than an end to the discussion. Nevertheless, we hope to advance the discussion by offering the broadest array of structured analysis produced to date in the open literature.

This research addressed several key questions: What are the trends in military competition between China and the United States? Are there areas where particular weaknesses or vulnerabilities are emerging? Are there areas where the United States is maintaining (or even improving) its relative position? Are there specific parts of the overall competition that are particularly important in terms of their spillover impact into other areas? What are the effects of geography and distance on the relative capabilities of different types of Chinese and U.S. forces? In the remainder of this section, we address the levels of analysis undertaken and the methods and key features of the analysis. No study can do all things, so we also discuss the limitations on the scope of activities undertaken.

Levels of Analysis

Ten scorecards represent the core of this work. Each looks at relative capabilities in an individual mission area at roughly the operational level. Below that, each scorecard examines the component parts of the competition at the tactical, or element, level. At one level above the scorecards are two campaigns, or scenarios, that provide the context and parameters within which we evaluated each scorecard. Finally, in Chapter Fourteen of this report, we address the strategic implications of the findings for such issues as deterrence and strategic stability.[21]

Operational-Level Analysis: The Scorecards

The report addresses military trends in a number of operational-level mission areas (see Table 1.1), with the assessment of each representing a single scorecard. The scorecards are designed to evaluate critical dynamics in different types of air, maritime, space, and nuclear combat. Each represents a mission area or type of warfare that is relevant to possible military conflict scenarios between the United States and China—and, in particular, to the two campaign-level scenarios outlined later in this chapter. Each is intended to reflect a type or category of conflict that will be recognizable to professional military officers and analysts. (U.S. naval officers will, for example, recognize and understand the term *anti-surface warfare*.)

---

[21] While these terms correspond to U.S. Department of Defense (DoD) terms for levels of conflict, their conception and use have been adapted for this analytical effort. For official DoD definitions of these terms, see U.S. Joint Chiefs of Staff, *DoD Dictionary of Military Terms*, Joint Publication 1-02, Washington, D.C., as amended through March 15, 2015. Strictly speaking, only the strategic, operational, and tactical are officially defined as "levels of war," but the DoD definition of a "campaign" bears many hallmarks of having its own level: "A series of related major operations aimed at achieving strategic and operational objectives within a given time and space."

Table 1.1
Scorecards Analyzed

Number	Description	Details
Air and missile scorecards		
1	Chinese attacks on air bases	China's capability to deny U.S. use of forward air bases
2	U.S. vs. Chinese air superiority	Combat involving U.S. and Chinese air forces attempting to gain air superiority
3	U.S. airspace penetration	U.S. capability to penetrate or destroy Chinese air defenses
4	U.S. attacks on air bases	U.S. capability to deny Chinese use of forward air bases
Maritime scorecards		
5	Chinese anti-surface warfare	Chinese capability to put U.S. surface combatants at risk using missiles, submarines, and aircraft
6	U.S. anti-surface warfare	U.S. capability to destroy Chinese amphibious ships and escorts using submarines and air attack
Space, cyber, and nuclear scorecards		
7	U.S. counterspace	U.S. capability to deny China's use of satellites
8	Chinese counterspace	Chinese capability to deny U.S. use of satellites
9	U.S. vs. China cyberwar	U.S./Chinese capability to gain military advantage from cyber operations
10	Nuclear stability	Capability of both sides to survive and retaliate against a nuclear attack

The intent was not to address all possible types of combat that might occur as part of a conflict between the United States and China but, rather, only those categories likely to be most important.[22] In selecting the scorecards, we considered both Chinese and U.S. operational concepts. In the Chinese case, there is a substantial body of primary and secondary literature on joint and single-service campaigns.[23] PLA writings

[22] The scorecards do not cover all possible aspects of a broader conflict, and they address some only within the context of the ten categories. For example, they do not address ground combat between U.S. and Chinese forces. Although it is quite possible that such combat could occur, its probability and operational importance would likely be secondary to other aspects of the conflict. The fight for information dominance is not treated as a discreet operational arena, but the scorecards on counterspace capabilities and cyberwarfare speak to the problem (as do sections of other scorecards). Certain aspects of anti-submarine warfare are similarly addressed in Chapters Seven and Eight.

[23] Comparing military doctrines is complicated by differences in language and concepts. There is no single word in Chinese that directly corresponds to *doctrine* in the way the term is used in Western militaries. However, there are numerous documents on warfare at the campaign level that, taken together, do offer a comprehensive theory of operational efficacy. Chinese sources consulted include 张玉良 [Zhang Yuliang], ed.,《战役学》 [*The Science of Military Campaigns*], 2nd ed., Beijing: National Defense University Press, 2006; 薛兴林 [Xue Xinglin], ed.,《战役理论学习指南》 [*Campaign Theory Study Guide*], Beijing: National Defense University Press, 2001;《中

on joint "island landing campaigns" and the Chinese Navy's "offensive campaigns against coral island reefs" most directly outline the operational measures and intermediate objectives that would be associated with landings on Taiwan or the islands of the South China Sea. But a "joint anti-air raid campaign" would be a critical subcampaign conducted in conjunction with either of these operations to reduce sortie generation capacity and defend against U.S. air strikes on the mainland. Descriptions of other types of campaigns and activities informed the scorecards' framing.[24] For example, Chinese writings specify that information operations would be conducted before, during, and after most campaign types and that space and nuclear deterrent operations would also be considered.

The operational concepts of both sides suggest that, regardless of strategic circumstances, both will seek to gain the initiative through offensive action. This report therefore includes some scorecards that are framed in terms of PLA offensive objectives and some that see U.S. forces seeking decisive outcomes. All assume active opposition by the adversary.

The Chinese literature suggests that the PLA would seek to achieve "three superiorities" (三权): information dominance, air superiority, and sea superiority. It would aim to disrupt U.S. forward operating bases near the conflict zone, primarily through the use of missile strikes (scorecard 1), while attempting to sink U.S. aircraft carriers or push back their areas of operation using submarines, missiles, and, possibly, air attacks (scorecard 5). The PLAAF would seek to gain at least temporary air superiority over critical areas (scorecard 2). Information warfare would be an integral part of all operations and would include efforts to blind or disable (temporarily or permanently) U.S. satellites (scorecard 8) and to disrupt or commandeer U.S. computer systems through cyber attack (scorecard 9).

The United States, for its part, would seek to gain air superiority through both air-to-air battles (scorecard 2) and by penetrating Chinese airspace to strike air defense targets and command-and-control facilities (scorecards 3 and 4). Air and missile strikes might also be undertaken on radar installations and ballistic missile sites. The United States would also seek to destroy Chinese surface assets, including forces dedicated

---

国人民解放军第二炮兵部队》 [PLA Second Artillery],《第二炮兵战役学》 [*Science of Second Artillery Campaigns*], Beijing: People's Liberation Army Press, 2004; 中国空军百科全书编审委员会 [Editorial Committee of the People's Liberation Army Air Force Encyclopedia],《中国空军百科全书》 [*China Air Force Encyclopedia*], Beijing: Aviation Industry Press, 2005; and 寿晓松 [Shou Xiaosong], ed., 《战略学》 [The Science of Military Strategy], 3rd ed., Beijing: Military Science Press, 2013. For secondary sources, see Mulvenon and Finkelstein, *China's Revolution in Doctrinal Affairs*, 2005; and David M. Finkelstein, "China's National Military Strategy Revisited: An Overview of the 'Military Strategic Guidelines,'" in Andrew Scobell and Roy Kamphausen, eds., *Right-Sizing the People's Liberation Army: Exploring the Contours of China's Military*, Carlisle, Pa.: U.S. Army War College, 2007.

[24] For example, the literature on PLAAF "air offensive campaigns" and "air defensive campaigns" and on the PLA Second Artillery "conventional missile assault campaigns" provides important details on force disposition, the sequencing of operations, and campaign priorities.

to landing operations and surface action groups operating in an air defense or anti-submarine capacity (scorecard 6). It would likely also undertake limited counterspace and cyber operations (scorecards 7 and 9), especially if it were attacked first in those domains.

The nuclear scorecard (scorecard 10) falls into a different category from the others and relates to the security of second-strike capabilities and incentives for vertical escalation, rather than the "balance of capabilities" per se. Neither side would look to use nuclear weapons at the start of hostilities. But the security of nuclear forces would weigh heavily on leaders' minds during a conflict, and, under some circumstances, pressures could build to cross the nuclear threshold, if only for demonstration purposes. In general, when both sides' nuclear forces have a secure second-strike capability, both sides' incentives to launch a first strike are reduced. The nuclear scorecard therefore focuses on the degree of confidence each side would have in the security of its own nuclear forces while acknowledging that other military and political factors will also be important in determining overall escalation stability.

Tactical (Element)–Level Analysis

While the specific structure of each scorecard chapter in this report differs, most include an assessment of several different components (or elements) of the larger competition. The scorecards generally evaluate the ability to detect, target, and attack enemy systems or objectives. To take a single example, the Chinese anti-surface warfare (scorecard 5) analysis looks at four capabilities:

1. Chinese over-the-horizon (OTH) intelligence, surveillance, and reconnaissance (ISR), including radar and satellites, and the ability to detect U.S. surface combatants and pass targeting information on to relevant offensive elements
2. the submarine threat to U.S. surface ships (with and without ISR cueing)
3. hostile aircraft and air-launched cruise missile (ALCM) threats
4. the potential future anti-ship ballistic missile (ASBM) threat to U.S. surface ships (see Figure 1.1).

In several cases, we modeled small engagements (often between individual or small numbers of systems) in addition to larger operational dynamics.

As in the selection of the scorecards themselves, the assessment of element-level competition is intended to provide indicators of trends in key areas, rather than address all possible aspects of each operational area. For example, it is likely that, in the event of a conflict, PLA Navy (PLAN) surface action groups would focus on air defense or anti-submarine operations relatively close to the Chinese coast. It is possible, but not likely, that Chinese surface ships could engage the U.S. fleet, but these engagements are not considered in the Chinese anti-surface warfare scorecard (scorecard 5). A variety of metrics related to ISR problems and three different methods of attack (subma-

Figure 1.1
Scorecard Example and Associated Element-Level Competitions

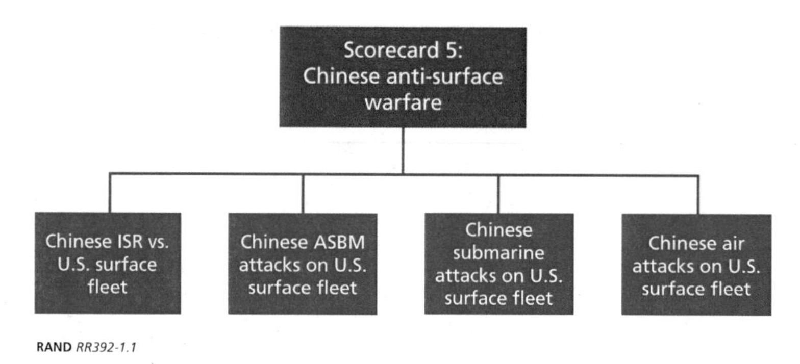

RAND *RR392-1.1*

rine, ALCM, and ballistic missile) do, however, provide a wealth of data on the changing threat level faced by U.S. surface craft.

Campaign (Scenario)–Level Analysis

Each of the scorecards addresses relative capabilities in the context of two campaign-level scenarios: a Taiwan invasion scenario and a Spratly Islands conflict scenario (see Figure 1.2). As two potential flashpoints in the U.S.-China relationship, these scenarios are often treated interchangeably. Yet they differ not only in the political context of the stakes but also in terms of geography. Taiwan lies roughly 160 km off the Chinese

Figure 1.2
Scorecards Assessed in the Context of Two Scenarios

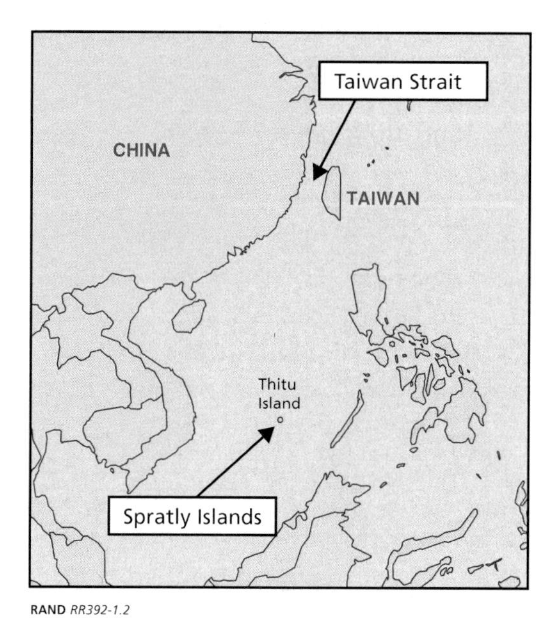

RAND *RR392-1.2*

mainland, while the Spratly Islands range between roughly 900 and 1,300 km from China's Hainan Island. Operations in support of even a local war in these areas could be conducted over vast areas, with certain types of space and ISR operations potentially spanning the globe. Nevertheless, the difference in the center of gravity in these campaigns can be of critical importance.

The Spratlys are beyond the range of most Chinese terrestrially based communication systems, so PLA forces would be forced to rely more heavily on satellites. Relatively more U.S. operating bases and areas would be out of range of PLA Second Artillery conventional ballistic missiles in a Spratly scenario. Transit times for Chinese surface ships and submarines would be longer than in the Taiwan case (a major factor for diesel submarines, in particular), while participating PLAAF and PLAN aircraft would be limited by the range and availability of tankers, which are presently few in number. For the U.S. side, which would be heavily reliant on satellite ISR and refueling tankers in both scenarios, the difference is less pronounced, though limitations on U.S. basing infrastructure near the South China Sea would inhibit operations.

It should be noted that both scenarios, by design, share certain similarities. They both postulate a Chinese occupation of real estate that Beijing claims. Because the two scenarios share operational similarities but are set at different distances from China, they isolate and highlight the effects of distance on relative capabilities. The range of distances involved—roughly 100 miles from the mainland in the case of Taiwan and 700 miles in the case of the Spratly Islands—effectively bookends a wide variety of plausible Asian scenarios. Geography and distance are key factors in any realistic net assessment, and their impact should play a large role in determining plans and strategy, as well as broader security and diplomatic policy.

If we had examined other types of conflicts, the scorecards might be defined somewhat differently. For example, if we were evaluating a Chinese blockade of Taiwan, a separate scorecard for U.S. anti-submarine warfare capabilities might be justified. Furthermore, the fact that these scenarios assume relatively high-intensity combat does not necessarily mean that they would be the most challenging for U.S. commanders. More limited conflicts could pose different types of challenges, since U.S. political leaders might disallow some types of U.S. military operations.

Strategic-Level Analysis

The scorecard analysis supports the assessment of broad strategic questions: What broad trends are evident in relative U.S.-China military capabilities? What is the likely impact of these trends on regional stability and deterrence? What are their implications for optimizing U.S. regional force posture? How and to what extent might military trends shape U.S. thinking about relations with allies and partners in the Asia-Pacific region? In light of the evolving balance of forces, should the United States adjust its engagement with China? Are there implications for crisis management or possible arms-control efforts?

Taiwan and Spratly Islands Scenarios

The operational-level scorecards are the focus of this report; the campaign-level scenarios are provided primarily as vehicles for assessing the effects of distance on military capabilities. Here, we discuss some important assumptions associated with each scenario. (Additional, mission-specific details are provided in Chapters Three through Twelve.) The assumptions provide a framework for analysis and are not intended to represent the most likely paths toward conflict or to predict real-world developments.

Taiwan Invasion

A more assertive China moves to isolate Taiwan further on the world stage, inadvertently pushing Taipei toward *de jure* independence. When diplomatic pressure fails to dissuade Taipei from changing course, Chinese leaders decide to occupy the island by force. In the lead-up to war, Taiwan appeals for U.S. assistance, and, given the ambiguous circumstances of conflict, Washington decides to use military force to protect the island.

The scenario assumes that, as tensions mount, both sides prepare militarily. The PLA deploys additional combat and support aircraft to the Nanjing Military Region, sorties its most advanced submarines, and deploys its forces out of garrison to forward staging areas. The United States moves additional aircraft and ships to the region and raises alert levels. Politically, the scenario assumes that the United States is allowed to operate freely from bases in Japan, the PLA is permitted to strike U.S. bases in Japan, and U.S. forces are allowed to attack nonstrategic targets in mainland China.

Spratly Islands Occupation

The Philippines moves toward exploiting oil and gas resources in the South China Sea. China denounces Manila's "provocative behavior" and begins to harass Philippine ships and platforms. The Philippines reinforces its position in the Spratly Islands, especially on Thitu Island, and China responds by dispatching naval forces and occupying the island. With nationalist demonstrations becoming more unruly in both countries, the United States dispatches a carrier battle group to the area to reassure allies and tamp down tensions. A U.S. surveillance aircraft is destroyed by Chinese naval surface-to-air missiles (SAMs), and the United States decides to eject Chinese forces from Thitu.

The scenario assumes that both sides have deployed additional naval and air assets to bases within range of the Spratly Islands and have sortied surface and submarine elements to positions around the islands. Because of the origins of the conflict, the United States is allowed to use bases in the Philippines and Japan. Despite a desire to keep the conflict limited, both sides see their credibility at stake and are willing to strike bases that are being actively used by the other.

Summary

Each scorecard addresses a different mission area at roughly the operational level. The scorecards are assessed in the context of two larger campaign scenarios: a Taiwan invasion scenario and a Spratly Islands scenario. Based on the results of the scorecard analysis, larger strategic questions about the impact of evolving capabilities are also addressed. For a graphical representation of this structure, see Figure 1.3.

Methods and Features of the Analysis

The scorecard analyses are characterized by a number of features: they address relative capabilities, they look at trends over time, they assess the impact of geography and distance, and they are designed to maximize analytical transparency and replicability.

Assessing Relative Capabilities

Although most of the scorecards are expressed in terms of the ability of one side or the other to achieve a particular objective, all of them examine relative capabilities in a contested environment. What is the capability of each side to accomplish relevant operational tasks in the face of opposition by the other side? The analysis of Chinese anti-surface warfare capabilities (scorecard 5), for example, examines the ability of Chinese forces to attack U.S. carriers *in the face of opposition by the United States*. Therefore, in considering the threat posed by China's growing fleet of modern diesel and nuclear

Figure 1.3
Graphical Representation of the Four Levels of Analysis

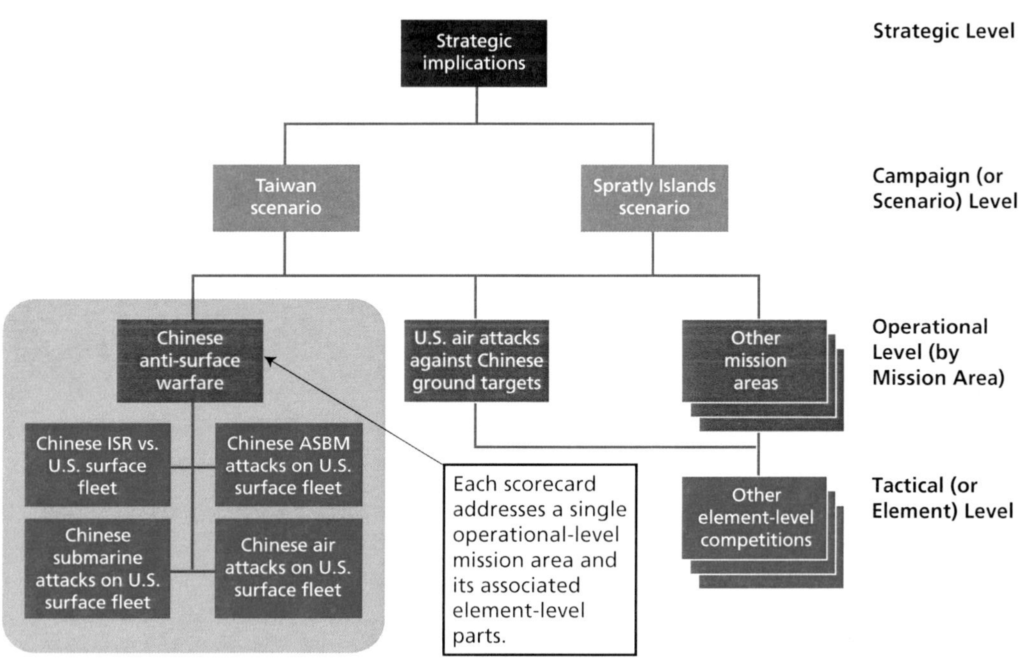

RAND *RR392-1.3*

attack submarines, we also consider U.S. anti-submarine warfare capabilities and tactics, as well as the ability of U.S. carriers to deny Chinese submarines opportunities to attack through high-speed maneuver. In the case of Chinese attacks on U.S. air bases, we consider the availability of hardened shelters for aircraft, along with the ability to reconstitute runways.

Both U.S. and Chinese capabilities have improved and continue to improve so that, in assessing trends over time, the question becomes the degree of relative change. In some cases, answers derived from a static examination of force structures are less than obvious. In air-to-air combat, for example, is the PLAAF's replacement of second- and third-generation fighters with fourth-generation fighters more important than the relatively slower U.S. transition toward fifth-generation fighters? In tackling such questions, dynamic modeling—which considers both qualitative and quantitative factors—is necessary. The differing nature of warfare in each area requires a somewhat different approach but, in each scorecard, we sought to bring a dynamic approach to bear.

Evaluating Trends Over Time (1996–2017)

To capture trends over time, the analysis for each scorecard looks at relative capabilities at four points in time, evenly spaced in seven-year increments: 1996, 2003, 2010, and 2017. While trend lines can clearly change—and there is in fact great variation even over this period—the historical analysis provides an empirical grounding that is sometimes missing from future projections or judgments.

In addition to their function in providing evenly spaced increments useful for tracking change, the specific snapshot years are significant in terms of changes to the Asian military situation. The 1996 starting point establishes a baseline for relative capabilities at a time when the Taiwan Strait crisis marked an acceleration in Chinese defense spending and a shift in China's procurement and training priorities toward Taiwan-relevant capabilities.[25] By 2003, the United States was fully engaged in both Afghanistan and Iraq while the PLA was adapting to new "operational guidelines" (*gangyao*) that emphasized the role of technology and, especially, information in modern warfare.[26] By 2010, with tensions between Beijing and Taipei in remission and an increasing level of conflict over maritime issues, China began to place relatively greater emphasis on power projection. And the 2017 analysis provides a glimpse of plausible future capabilities in a time frame that is close enough to the present to enable informed judgments about both sides' force structure.

---

[25] In 1995 and 1996, China sought to pressure Taiwan's President Lee Teng-hui, whom it accused of seeking to undermine the one-China policy, by "testing" ballistic missiles off Taiwan's coast. The United States responded by dispatching a U.S. aircraft carrier, the USS *Abraham Lincoln*, and escorting ships to the area.

[26] The guidelines were adopted in 1999. See David M. Finkelstein, "Thinking About the PLA's Revolution in Doctrinal Affairs," in Mulvenon and Finkelstein, *The Revolution in Doctrinal Affairs*, 2005, p. 10.

Evaluating the Impact of Distance on Capabilities

As mentioned earlier, the scorecards assess the effects of distance on relative capabilities by looking at two scenarios at different distances from China. Analysis of the two scenarios, centered roughly 160 km (Taiwan) and 900–1,300 km (Spratly Islands) from the Chinese coast, will also be useful in thinking about other scenarios arising at or between those distances. Needless to say, all scenarios (or actual conflicts) are affected by parameters other than distance (e.g., specific location relative to bases, objectives, and forces committed), but much of the scorecard analysis, especially at the tactical level, will remain relevant to a broad range of potential conflicts—from Japan scenarios to air or naval combat off Vietnam and Taiwan blockade scenarios.

Flexibility and Transparency

In contemplating research design, a major objective was flexibility and transparency. We employed a variety of combat models and simulations to assess operational-level dynamics, but we also made a strong effort to keep the modeling effort relatively streamlined. The inputs are described in some detail, as are most key features of the models themselves. In some cases, the modeling work and Monte Carlo simulations are fully replicable, using the information provided in this report. In other cases, we provide citations to guide readers to additional information about modeling dynamics. Importantly, each scorecard includes a robust qualitative discussion of factors that loom large in that particular mission area.[27]

Scope, Boundaries, and Parameters

The analytical effort presented here was ambitious by almost any standard. It involved assessing relative capabilities in ten mission areas, in the context of two scenarios, across four snapshot years, and in ways that are relatively transparent. It is therefore especially important to be clear about the limits and boundaries of the analysis. No study can do all things—or all things equally. We have therefore made judgments about which avenues to pursue and what to leave for others (or for later) based on the larger objectives of the study.

No Overarching Scenario Model

The most significant choice concerned the level of conflict that would be assessed and modeled. To maximize the applicability of the work to a range of scenarios, the analysis focuses on the mission area (for example, air-to-air or anti-surface warfare). While the balance of forces at the larger campaign (or scenario) level is addressed in some detail and informed by the operational-level work, we did not try to create a larger model that would link the mission-level analysis together in a comprehensive way. In other words,

---

[27] For more on transparency and modeling, see Richard J. Hillestad, Bart E. Bennett, and Louis Moore, *Modeling for Campaign Analysis: Lessons for the Next Generation of Models, Executive Summary*, Santa Monica, Calif.: RAND Corporation, MR-710-AF, 1996.

to maximize the applicability of the work to a range of scenarios, the analysis focuses on the analysis of more fundamental, mission-level, events.

Any conflict in East Asia would be an immensely complex affair, with surface, air, missile, and subsurface elements supported by space and electronic elements that would, themselves, be contested. A single, unified model that would account for all these elements could be useful for some purposes, particularly for comparing the performance of different force structures. But especially for those without access to the model or its coding, we concluded that such a model would be extraordinarily difficult to present in ways that would be comprehensible and transparent to readers. Those who wish to employ the modeling concepts presented here are welcome to develop a more comprehensive framework, and we may pursue such an effort in the future.

In the meantime, the mission-area focus employed here allows us to present important parts of the equation with greater clarity and transparency. Within the context of each mission area, this report systematically introduces intermediate and overall objectives, force structure, time and distance factors, and other elements, together with the mechanics and results of the modeling effort. At the same time, although we did not create a single overarching model to integrate the results of all scorecards, we also did not analyze each scorecard in isolation.

This report discusses the scenarios in the context of each scorecard and then revisits them in the final two chapters, which address overall performance and trends. Those chapters evaluate the relative importance of each scorecard in the context of the relevant political objectives and geography outlined in the scenario. The scorecard "weighting" is partly based on our understanding of operational and strategic issues, but it also draws heavily on the scorecard analysis. Moreover, both the individual scorecard analyses and the scenario summaries in Chapter Thirteen also consider the impact of the other scorecards on one another. For example, in looking at air-to-air combat and efforts to gain air superiority (scorecard 2), we consider the impact of PRC attacks on U.S. air bases (scorecard 1).

No Assessment of "Red" Versus "Green"

The scorecards do not assess potential combat between Chinese forces ("red") and third-party (i.e., non-U.S., or "green") forces that might be aligned with U.S. forces. To the extent that there is a de facto division of labor between the United States and its military allies and partners in Asia, the impact on the scorecards examined is limited. In attempts to penetrate Chinese airspace (scorecard 3), for example, Taiwanese forces would have limited impact. Where necessary, we have parameterized critical variables that would be heavily affected by green forces and their capabilities. We consider, for example, the differing impacts on air and naval force requirements if a PLA ground campaign in Taiwan required seven versus 21 days.[28]

---

[28] This is not to say that we assume that the PLA would win such a campaign. Rather, if a Taiwan or Spratly Islands campaign were to continue beyond that time frame, the United States would have ample time to mobilize

Selectivity in Metrics

The analysis presented here necessarily simplifies a great number of conflict characteristics. The emphasis throughout is on developing and assessing metrics in each area that provide a sense of the level of difficulty faced by each side in achieving its objectives. Apart from practical limitations, selectivity is driven largely by the desire to make the work transparent and replicable. Moreover, given the complexities and uncertainties in modern warfare, one could make the case that it is better to capture a handful of important dynamics than to present the illusion of comprehensiveness and precision. All that said, the analysis is grounded in recognized conclusions from a variety of historical sources on modern warfare, from the air war over Korea and Vietnam to the naval conflict in the Falklands and SAM hunting in Kosovo and Iraq.

Dealing with Uncertainty

In many cases, important variables may be unknown. This may be partly a function of the limited source material available in the public domain. Equally important are the uncertainties of modern warfare. (The effectiveness of systems or programs can often not be precisely known until they are tested in combat.) In still other cases, the question may be one of relative effort. We cannot predict with certainty, for example, how many ballistic missiles the Chinese high command would decide to allocate to attacks on air bases as opposed to command-and-control facilities. To the extent possible, then, we have used a range of values to examine the boundaries (or parameters) of given problems.

Personnel Quality and Training

One aspect that is only partially captured in this analysis is the quality of personnel training and performance. In some cases, such as assumptions about fighter pilot quality, we have applied a modest discount rate (explained later) to the performance of Chinese personnel. This is justified by differentials, documented in a variety of open-source material, in annual hours of flight times by the pilots of the two countries, as well as the realism and scale of training opportunities more generally. In other cases, such as the noise levels associated with different submarines, we assume that the impact of training is already accounted for in publicly released government assessments of submarine performance. In general, we have erred on the side of caution in building personnel-related qualitative differences into our modeling of combat dynamics.

This caution may serve to "advantage" China in parts of the analysis (i.e., to make its capabilities appear greater than they actually are). The impact, however, is much larger on the analysis of the earlier periods (and especially 1996), when the

---

its military resources and would, in all likelihood, emerge the victor. The outcome of the other contests within that span of time is therefore critical.

difference in training quality was greatest.[29] The United States still enjoys significant advantages in most areas of personnel quality, training, and experience. But the PLA has made great strides with the introduction of a professional noncommissioned officer corps, increasingly realistic training, more interaction with foreign militaries, and some limited out-of-area deployments. As late as the 1990s, mixed groups of Chinese warships almost never set out from port, much less ventured beyond the first island chain. Today, such mixed deployments for training are routine, and the PLA has initiated long-range operational deployments (including the evacuation of Chinese citizens from Libya and for ongoing counter-piracy operations in the Gulf of Aden).

Exclusively Open-Source Analysis

Our research team worked strictly with publicly available reference works. To ensure consistency, we employed standardized data sources. For equipment inventories, we relied on the appropriate years of *The Military Balance*, published by the International Institute for Strategic Studies (IISS). Where knowledge about equipment subsystems and capabilities was required (e.g., the circular error probabilities of various ballistic missile systems), the team referenced various Jane's databases (e.g., *Jane's Fighting Ships*). In some cases not covered by standardized works, we conducted additional research. For example, we analyzed Google Earth imagery (and online groups that exploit Google Earth) to determine how many Chinese bases might be available to support PLAAF operations off Taiwan or around the Spratly Islands.

In no cases did we "scrub" data found in the standardized sources to make them conform to nonpublic government sources. This approach will inevitably result in force structure numbers and capability parameters that are inexact. Given this, and other uncertainties (e.g., the inherent friction of war and the unknown political-military context of hypothetical future wars), the analysis presented here should not be taken as predictive of a precise course of future conflict. Nevertheless, the analysis, which accounts for the dynamic interactions between geographic, temporal, and material factors, is intended to capture the general magnitude of the challenges facing U.S. commanders in each area. Moreover, given that the same methodology and metrics are employed for each snapshot year, with only the inventories held by each side at that time varied according to the best available information, the approach is particularly well suited to capturing the direction and speed of change in the balance of power.

---

[29] Because the impact is significantly greater in the earlier periods than in later ones, our caution in inserting judgments about personnel quality and training may also understate the magnitude and rate of change over time. The Chinese military shows every indication of appreciating the importance of personnel quality. Looking forward, we can expect it to continue to improve its training methods.

Summary of Findings

An examination of trends (from 1996 to 2017) across all of the scorecards suggests a number of broad conclusions. First, trend lines are moving against the United States across a broad spectrum of mission areas. In some cases, the change is extraordinarily rapid. Improvements to the range and capability of Chinese cruise missiles and, especially, ballistic missiles have dramatically outstripped improvements to the survivability of U.S. forward air bases. Similarly, new generations of more capable Chinese aircraft, missiles, and, especially, submarines pose a major threat to U.S. aircraft carrier operations within 1,000 miles (and possibly farther) from China's coast. The threats to both land bases and carriers would be particularly serious at the outset of a conflict.

Second, although the overall trends are poor, trends vary by mission area, and, in some areas, U.S. relative capabilities remain robust or even dominant. With only marginal improvements to Chinese anti-submarine warfare capabilities, the U.S. submarine fleet remains capable of doing substantial damage to China's surface fleet. Similarly, although penetrating Chinese airspace has become increasingly risky, the deployment of stealth aircraft and standoff strike weapons has partly mitigated the problem.

Other conclusions emerge from looking at the scorecards in the context of individual scenarios. In this regard, the third conclusion is that the challenges for the U.S. military are particularly large and pressing in the Taiwan case, in which most of the scorecards move from U.S. advantage to rough parity or disadvantage between 1996 and 2017. Proximity to the Chinese mainland makes Chinese anti-access capabilities—particularly threats to land-based airpower and U.S. surface ships—especially daunting. And with impaired access by land- and sea-based airpower, all of the other U.S. tasks become more difficult.

Fourth, despite emergent difficulties for the U.S. side in the Taiwan case, the difficulties and vagaries of amphibious assault suggest that the PLA should also take little comfort from the results. Failure modes are numerous, and any one could spell disaster on the beachhead. This is, in short, a conflict that both sides should wish to avoid and one that would likely entail high losses to both. Nevertheless, the possibility of miscalculation or missteps leading to war cannot be discounted.

Fifth, the Spratly Islands scenario analysis suggests that Chinese power diminishes rapidly across even relatively modest distances. The U.S. basing structure is also not optimized for operations in the South China Sea, so both sides would be able to bring less to the fight. But Chinese capabilities suffer much more, as the PLA lacks the support structure necessary to sustain significant combat forces at a distance, and fewer of the PLA's ground-based ballistic and cruise missile forces would be in range of relevant targets. As in the case of the Taiwan scenario, the trend lines are moving in a negative direction for the United States. Further improvements in relative Chinese capabilities can be expected, but, with less asymmetry in the geographic dimension, they are likely to come at a higher cost to China.

Finally, the PLA can pose problems—and potentially win wars—without catching up to the United States in terms of overall quality, sophistication, or system numbers.[30] By many standards, the Chinese military continues to lag far behind that of the United States. However, the scorecard analysis shows that it is necessary to consider the operational circumstances of specific regional scenarios in evaluating the balance of power in any tangible or meaningful way. Available forces, basing, the objectives of the two sides, the available time for mobilization, the distance between various operationally relevant points, and the movement speeds of the assets involved are critical. The scorecards do not show that China has "caught up," but they do indicate that it does not have to catch up in order to jeopardize the U.S. ability to achieve operational objectives in several key conflict scenarios, particularly those in China's immediate front yard.

---

[30] The language used here is borrowed from Thomas Christensen's essay title "Posing Problems Without Catching Up," though his discussion is focused more on the potential for political and military conflict rather than the military balance itself. See Thomas J. Christensen, "Posing Problems Without Catching Up: China's Rise and Challenge for U.S. Security Policy," *International Security,* Vol. 25, No. 4, Spring 2001.

CHAPTER TWO

Different Paths: Chinese and U.S. Military Development, 1996–2017

The Chinese and U.S. militaries have taken two different paths of development since the mid-1990s. The U.S. military, despite downsizing after the Cold War, maintains global reach, missions, tasks, and considerations. The Chinese military, which has modernized in recent years and now regularly conducts counter-piracy operations in the Gulf of Aden, nevertheless remains focused primarily on missions in East Asia.[1]

In this chapter, we discuss in more detail how the Chinese and U.S. militaries have evolved in recent years. We first discuss the development of the Chinese military through 1996 and its standing just after the crisis in Taiwanese relations, when China began rapidly increasing its military spending. We then discuss how the Chinese military has evolved since 1996. Finally, we examine how the U.S. military has evolved since 1996, including capabilities that it is likely to develop in the next several years.

The Development of China's Military

China's military, the Chinese People's Liberation Army (中国人民解放军), was established in 1927 as the military arm of the Communist Party of China. Even today it remains a "Party army" whose first loyalty is to the Party, not the Chinese nation.

The PLA was initially a guerrilla army. By the late 1940s, however, it had evolved into a more conventional force, with training provided by the Soviet Union and equipment supplied by the Soviets or captured from the army of the Republic of China government (whose equipment was supplied by the United States).[2] After the Communist Party came to power in 1949, and particularly after the Korean War began in 1950, the Soviet Union supplied training and equipment on a large scale. The Soviets also built factories in China to manufacture many of the same weapons that equipped their own military. In the summer of 1960, however, the Soviets withdrew their eco-

[1] In addition to the Gulf of Aden mission, in 2013 and 2014, the PLA Navy conducted at least three nuclear and diesel submarine deployments in the Indian Ocean.

[2] Lucien Bianco, *Origins of the Chinese Revolution, 1915–1949*, Stanford, Calif.: Stanford University Press, 1971, pp. 167–198.

nomic and military assistance to China.[3] This withdrawal coincided with the beginning of a severe economic downturn that resulted from the policies of the Great Leap Forward of 1958–1960. China's economy began recovering in 1962, but by 1966 had only just returned to pre–Great Leap Forward levels when Chinese leader Mao Zedong launched the Great Proletarian Cultural Revolution, resulting in several more years of economic disruption.

Between 1964 and 1971, the Chinese government moved the locus of its industry, including its weapon factories, from urban coastal areas to remote interior locations. This "third-line" program was intended to protect China's industry from possible invasion by the United States or the Soviet Union and involved not only new investment in China's interior but also the relocation and dispersal of existing factories.[4] At the same time, from the Korean War until the early 1970s, China was subject to economic and military sanctions imposed by Western countries. The combination of economic and social disruptions and China's isolation from both the Soviet Union and the West meant that its conventional military doctrine and technology progressed little between the 1950s and 1970s. China did make major advances in strategic weapons, testing its first nuclear weapon in 1964 and developing a range of land- and submarine-based weapons. Its first land-based intermediate-range ballistic missile (IRBM) achieved operational status in 1971.[5]

The 1950s, 1960s, and 1970s were a period of frequent conflict and confrontation for China. After the armistice on the Korean peninsula in 1953, China remained engaged in confrontations with the United States and Taiwan; fought a border war with India in 1962; engaged in a confrontation with the Soviet Union that peaked with 1969 clashes on the Manchurian border; supported North Vietnam in its wars with France, the United States, and South Vietnam; and fought its own war with Vietnam in 1979, which resulted in a standoff that periodically boiled over into border clashes until the mid-1980s.

By 1978, two years after Mao Zedong's death, Deng Xiaoping had taken control of China's government and gradually introduced a program of economic liberalization and improved relations with the West, including the establishment of official diplomatic relations with the United States on January 1, 1979. This gave China access to Western commercial markets, investment, technology, and military assistance. At the same time, China deemphasized military spending, with official defense expenditures falling from 4.6 percent of GDP in 1978 to 1.5 percent in 1989 and 1.0 percent in 1996.[6] China's armed forces were downsized as well, with numbers of active-duty per-

---

[3] Allen, Krumel, and Pollack, *China's Air Force Enters the 21st Century*, 1995.

[4] See Barry Naughton, "The Third Front: Defence Industrialization in the Chinese Interior," *China Quarterly*, Vol. 115, September 1988.

[5] John Wilson Lewis and Xue Litai, *China Builds the Bomb*, Stanford, Calif.: Stanford University Press, 1988.

[6] National Bureau of Statistics of China, *China Statistics Yearbook*, Beijing, 2006.

sonnel falling from an estimated 4.3 million in 1978 to 3 million by 1989.[7] The military confrontation with Taiwan eased, as did tensions with the Soviet Union. Other than small-scale clashes with Vietnam and continuing tensions with the Soviet Union, the 1980s were a time of relative peace and security for China.

At the end of the 1980s and throughout the 1990s, the Chinese military received several significant shocks. The collapse of the Eastern Bloc and the dissolution of the Soviet Union placed China's Communist Party on the political defensive. In response to the Chinese government's use of the PLA to violently suppress popular demonstrations on June 4, 1989, the United States and the European Communities (later the European Union) imposed arms embargoes and economic sanctions, halting the military and technological infusions China had been receiving since the 1970s. In early 1991, the United States overwhelmed the Iraqi military, which had been organized and equipped much like China's.

Subsequent developments in Taiwan also increased the salience of military affairs. In 1996, China fired short-range ballistic missiles (SRBMs) into the ocean near Taiwan in an apparent effort to deter the election of a pro-independence government in Taiwan. The United States signaled its intent to defend Taiwan against a Chinese use of force by dispatching two aircraft carrier battle groups to the waters around Taiwan. The Chinese military's inability to locate—much less attack—these aircraft carriers demonstrated its inability to successfully use force against Taiwan should the United States intervene.

In 1999, the U.S.-led North Atlantic Treaty Organization (NATO) bombing campaign against Yugoslavia (Operation Allied Force) once again demonstrated the increasing power of U.S. precision-strike capabilities. The war also highlighted the erosion of international norms of absolute sovereignty in domestic affairs, and many Chinese viewed the bombing of the Chinese Embassy in Belgrade as an intentional hostile act.

In response to these events, the PLA incrementally adjusted its operational concepts in an effort to provide a more realistic template for the modernization of its equipment, training, and operational practice. In 1993, it introduced the concept of "local wars under high-technology conditions." By the late 1990s, the concept was modified to "local wars under informationized conditions," indicating the decisive role that data-intensive systems and processes play in modern war.[8] In 1999, PLA commanders issued new classified campaign guidance documents (纲要), fleshing out the concept, for each service. In the same year, the PLA also adopted its first-ever joint campaign guidance

---

[7] International Institute for Strategic Studies, *The Military Balance*, London: Routledge, 1978, p. 56, and 1989, p. 146.

[8] Originally seen in PLA military commentary, the concept was incorporated into the 2002 Military Strategic Guidelines for the New Period. On the evolution of these concepts, see Blasko, *The Chinese Army Today*, 2012, pp. 15–17; and David M. Finkelstein, "China's National Military Strategy Revisited," 2006, p. 96.

document and joint logistics campaign guidance document.[9] These were accompanied by specific combat-doctrine manuals (战斗条令) for individual service arms.[10]

China began importing advanced weaponry from the former Soviet Union and specifically sought to counter key aspects of the U.S. military's force projection capabilities in the 1990s. Having fallen to 1 percent of GDP by 1996, Chinese military spending began to increase rapidly. Between 1996 and 2015, China's official military expenditure increased by 620 percent in real terms, growing at an average annual rate of roughly 11 percent—faster than China's robust rate of economic growth. Official military spending, which does not include all military-related expenditures, accounts for roughly 1.3 percent of GDP today. Increased budgets and doctrinal changes have been complemented by efforts to improve and modernize PLA equipment, training, and personnel.

China's Military in 1996

In 1996, China's military was vast but not very well equipped, especially by Western standards. It had approximately 2.9 million personnel: 2 million in the Army, 300,000 in the Navy, 500,000 in the Air Force, and 100,000 in the strategic missile forces (known as the Second Artillery Force). Despite its huge size, the PLA ground forces were primary light infantry with around 8,000 main battle tanks, just two-thirds as many as the U.S. Army (which had only 500,000 soldiers). The PLA had only about 4,500 armored infantry fighting vehicles and armored personnel carriers, compared with more than 30,000 in the U.S. Army. PLA ground forces did have a large artillery force, with more than 14,000 howitzers, compared with about 6,000 in the U.S. Army. The vast majority were towed pieces, however, while most howitzers in the U.S. Army were self-propelled. The PLA's other ground-based equipment was also outdated. The vast majority of its tanks, for example, were locally built versions of the Soviet T-54.[11]

The equipment of the PLAN was similarly outdated. It had about 80 attack submarines, some of which may not have been fully serviceable, and all but five were conventionally powered diesel electric boats. Three-quarters of its attack submarines were locally built versions of the Soviet *Romeo* class, which entered service in the 1950s. All but two of the remainder of its diesel boats were of the only slightly more capable *Ming* class. China's five nuclear attack submarines were extremely noisy. Its only nuclear ballistic missile submarine (SSBN), also noisy, had never conducted an operational patrol and carried missiles with a range of only about 1,000 nm. PLAN surface ships

[9] Office of Naval Intelligence, *China's Navy 2007*, Suitland, Md., 2007, p. 28.

[10] For example, see 中国人民解放军空军 [People's Liberation Army Air Force], 《中国空军百科全书》 [*China Air Force Encyclopedia*], 航空工业出版社 [Aviation Industry Press], 2005, pp. 328–330.

[11] IISS, *The Military Balance*, 1996.

consisted of 57 destroyers and frigates, only three of which carried even short-range SAMs, rendering China's surface ships virtually defenseless against modern anti-ship cruise missiles (ASCMs).[12]

Between them, the PLAN and PLAAF operated more than 5,000 fighters and attack aircraft. Virtually all were based on 1950s-vintage Soviet MiG-17s, MiG-19s, and MiG-21s. The PLAN and PLAAF also operated some 430 light and 145 medium bombers (also based on 1950s Soviet designs), but, with the exception of some bombers that carried torpedoes or ASCMs, China's bombers and attack aircraft were equipped only with unguided gravity bombs. Similarly, nearly all of the long-range land-based SAMs operated by the PLAAF were based on the Soviet SA-2, which first entered service in the 1950s.[13]

Aside from the single nuclear ballistic missile submarine, China's nuclear forces were primarily embodied in land-based missiles operated by the Second Artillery. These weapons included about a half-dozen intercontinental ballistic missiles (ICBMs) capable of reaching the 48 contiguous states, perhaps a dozen shorter-range ICBMs capable of reaching Europe or Alaska, and about 70 IRBMs (with a range of 3,000–5,000 km) and medium-range ballistic missiles (MRBMs, with a range of 1,000–3,000 km) capable only of reaching targets in Asia. The Second Artillery also had a modest number of conventionally armed, mobile SRBMs with ranges of less than 1,000 km, some of which were test-fired into the waters near Taiwan in 1995 and 1996.[14]

In 1996, China had only a rudimentary capability to support forces beyond its borders. It acquired its first jet-powered heavy airlift aircraft—14 Russian-made Il-76MDs—in the 1990s, supplementing its aging fleet of 1950s designs. It had no aerial refueling tankers, AWACS, or electronic countermeasure (ECM)–equipped aircraft in its inventory. China's maritime support was only slightly more developed, with limited amphibious lift and at-sea replenishment capabilities. And despite a push during the 1980s by PLAN commander and later vice chairman of the Central Military Commission Liu Huaqing for a Chinese aircraft carrier, the country remained almost two decades from having an operational carrier in service.

China's military training in 1996 was considered poor. Exercises were highly scripted, with predetermined outcomes. Virtually all officers, whose quality was ques-

---

[12] IISS, *The Military Balance*, 1996. On the submarines discussed in this paragraph, see *Jane's Strategic Weapon Systems*, "Romeo Class (Project 633)," December 12, 2014; *Jane's Fighting Ships*, "Ming Class (Type 035)," February 13, 2015; *Jane's Fighting Ships*, "Han Class (Type 091/091G)," February 13, 2015; *Jane's Fighting Ships*, "Xia Class (Type 092)," February 13, 2015; *Jane's Sentinel Security Assessment*, "China and Northeast Asia Procurement," February 16, 2012.

[13] IISS, *The Military Balance*, 1996; *Jane's Sentinel Security Assessment*, "China and Northeast Asia Procurement," 2012.

[14] IISS, *The Military Balance*, 1996; *Jane's Strategic Weapons Systems*, "DF-3 (CSS-2)," May 7, 2014; *Jane's Strategic Weapons Systems*, "DF-11 (CSS-7/M-11)," November 30, 2012; *Jane's Strategic Weapons Systems*, "DF-15," June 5, 2014; *Jane's Sentinel Security Assessment*, "China and Northeast Asia Procurement," 2012.

tionable, were graduates of PLA military academies or had been directly promoted from the enlisted ranks without receiving a higher education.[15]

Modernization Since 1996

China's increased military expenditures since 1996 have emphasized the modernization of equipment and improvement of personnel recruitment, training, and preparation.

The expansion of the Second Artillery Force has arguably been the most prominent aspect of the PLA's modernization since 1996. Originally, the Second Artillery Force was responsible only for China's nuclear missiles. In the 1990s, it began acquiring conventionally armed SRBMs. The numbers of these systems have steadily increased so that, as of 2015, the U.S. Department of Defense (DoD) estimated that China has "at least" 1,200 SRBMs in its inventory.[16] China has also acquired conventionally armed MRBMs, with an estimated 36 DF-21C launchers in its inventory by 2015 (with an unknown number of missiles).[17] The more recent versions of these missiles are believed to be quite accurate, capable of hitting such targets as airfields and ports with a variety of warheads.[18] As of 2012, the Second Artillery Force also fielded between 200 and 500 highly accurate conventional ground-launched, land-attack cruise missiles, with a range of at least 2,000 km.[19]

The Second Artillery and PLAN have also modernized China's nuclear forces. Throughout the 1990s, China's nuclear deterrent against the United States was based on liquid-fuel missiles at fixed locations and was theoretically vulnerable to a preemptive first strike. With the deployment of the DF-31A, however, it now has road-mobile, solid-fuel missiles that can strike anywhere in the United States. China has

---

[15] David E. Johnson, Jennifer D. P. Moroney, Roger Cliff, M. Wade Markel, Laurence Smallman, and Michael Spirtas, *Preparing and Training for the Full Spectrum of Military Challenges: Insights from the Experiences of China, France, the United Kingdom, India, and Israel*, Santa Monica, Calif.: RAND Corporation, MR-836-OSD, 2009, pp. 29–37, 50–57.

[16] Office of the Secretary of Defense, *Military and Security Developments Involving the People's Republic of China 2015*, April 2015, p. 8.

[17] The number of MRBMs in service is not entirely clear, in part because of uncertainty in the literature about how many DF-21s are conventionally armed and how many are nuclear-armed. According to DoD, China had between 75 and 100 DF-21s of all types (including nuclear) in service as of 2012 (Office of the Secretary of Defense, *Military and Security Developments Involving the People's Republic of China*, Washington, D.C., 2012, p. 29). According to a 2014 report in *Jane's Strategic Weapons Systems*, "It seems more likely that the number in service is around 300" (*Jane's Strategic Weapons Systems*, "DF-21," June 24, 2014). The 2015 edition of *The Military Balance* reported that China had 134 MRBM launchers, including 36 conventionally armed DF-21C launchers (IISS, *The Military Balance*, 2015, p. 237).

[18] *Jane's Strategic Weapon Systems*, "DF-21," June 24, 2014.

[19] Office of the Secretary of Defense, *Military and Security Developments Involving the People's Republic of China 2012*, May 2012, p. 29.

also deployed a new generation of SSBNs, the *Jin* class (Type 094), armed with the new JL-2 submarine-launched ballistic missile (SLBM), giving China its first credible sea-based deterrent.[20] Given these improvements, China's potential vulnerability to a hypothetical preemptive first strike has been greatly reduced.

PLAN equipment modernization has emphasized the acquisition of modern diesel-powered submarines; larger, more modern surface vessels; and an array of anti-ship missiles, including the world's first ASBM. Conventionally powered submarine acquisitions since that time have included 12 *Kilo*-class submarines purchased from Russia (including ten very quiet *Kilo 636*s), along with 25 of the relatively modern, domestically built *Song* and *Yuan* classes. Unlike the PLAN's older diesel submarines, armed only with torpedoes, all newer PLAN submarines (including the *Kilo 636N* and the *Song* and *Yuan* classes) are capable of launching ASCMs. Eight of the *Kilo*-class submarines have long-range (200-km), supersonic ASCMs. The *Yuan* class, which entered service in 2006, boasts an air independent propulsion (AIP) system that gives it greater underwater endurance and makes it less vulnerable to detection than submarines forced to spend more time on or near the surface.[21]

In addition, China has commissioned two modern *Shang*-class nuclear-attack submarines. Four improved *Shang*-class boats are expected, and the class will ultimately replace the technically problematic *Han* class.[22] DoD also reports that, within the next decade, China likely will also construct the Type 095 guided-missile nuclear-attack submarine (SSGN), which will incorporate advanced quieting technologies and may provide anti-ship roles with both torpedoes and ASCMs, as well as a submarine-based land attack capability.[23]

China has also comprehensively modernized its surface fleet. Historically weak in air defense, China commissioned eight modern destroyers equipped with SAMs with ranges of 100 km or more between 2004 and 2015.[24] These ships include the *Luyang II* (Type 52C) and *Luyang III* (Type 52D), both equipped with the HQ-9 SAM,

---

[20] China had operated a single *Xia*-class ballistic-missile submarine since the late 1980s, but it is not clear whether this vessel was ever considered operationally capable. Four of the new *Jin*-class ships have reportedly been launched, and a fifth is under construction. Three boats had been commissioned by 2012, with the fourth expected by the end of 2015, and the last in 2017. *Jane's Fighting Ships*, "Jin Class" (Type 094), February 13, 2015.

[21] Jesse L. Karotkin, Senior Intelligence Officer (SIO) for China at the Office of Naval Intelligence, "Trends in China's Naval Modernization," Testimony to the U.S.-China Economic and Security Review Commission, January 30, 2014.

[22] Office of the Secretary of Defense, *Military and Security Developments Involving the People's Republic of China 2015*, Washington, D.C., April 2015, p. 9.

[23] Office of the Secretary of Defense, *Military and Security Developments Involving the People's Republic of China 2015*, 2015, p. 9.

[24] The number includes all ships built, including some that were not commissioned. They include the *Luyang II* (equipped with the HQ-9), the *Luzhou* (equipped with S-300 Rif), and the *Luyang III* (equipped with the HQ-9 SAM) (*Jane's Fighting Ships*, "Luyang II [Type 052C] Class," February 16, 2015; *Jane's Fighting Ships*, "Luzhou Class [Type 051C]," February 13, 2015).

and the *Luzhou* (Type 51C), equipped with the S-300 Rif. Equipped with advanced SAM capabilities, the newer destroyers thus provide the PLAN an increasingly robust fleet air defense capability, able to defend surface combatants from aircraft and long-range ASCMs, even when away from the cover of land-based fighters and SAMs.

All of China's major surface warships are now equipped with advanced ASCMs. The five *Luyang II*–class destroyers have the very long-range (280-km) YJ-62 system while the *Luyang III* class is fitted with the new vertically launched YJ-18 supersonic ASCM with a reported range of 178 km.[25] The *Luyang III* features both a new universal vertical launch system (VLS) capable of housing SAMs, SSMs, and anti-submarine missiles as well as a streamlined hull and superstructure.[26] China's four *Sovremenny*-class destroyers, purchased from Russia between 1999 and 2006, have been fitted with modern, long-range supersonic ASCMs, with a range of between 160 and 240 km.[27] PLAN aircraft also can carry long-range ASCMs including the YJ-83K with an estimated range greater than 100 nm.[28] China has begun deploying coastal defense batteries equipped with a shore-based version of the 280-km-range YJ-62 ASCM.[29] And China has deployed a land-based MRBM—the DF-21D ASBM—with a maneuverable warhead and a range of more than 810 nm.[30] As a senior intelligence officer for the Office of Naval Intelligence testified, this system "will allow China to significantly expand its 'counter-intervention' capability further into the Philippine Sea and South China Sea."[31]

Clearly, the PLA Navy's surface fleet has made remarkable strides. As late as 2003, only about 14 percent of its destroyers and 24 percent of its frigates might have been considered modern—capable of defensive and offensive operations against a capable enemy.[32] By 2015, those figures had risen to 65 percent and 69 percent, respectively.

---

[25] The Office of Naval Intelligence reports that a "similar capability" will be extended to the *Song*-, *Yuan*-, and *Shang*-class submarines. See Office of Naval Intelligence, *The PLA Navy: New Capabilities and Missions for the 21st Century*, Washington, D.C., March 2015, pp. 16, 19.

[26] *Jane's Fighting Ships*, "Luyang III (Type 052D) Class," February 16, 2015.

[27] Submarine and surface ship weapon information is from *Jane's Fighting Ships*, "Spreadsheet: World Naval Ship Fleets," February 12, 2015.

[28] Office of Naval Intelligence, *The PLA Navy*, 2015, p. 16.

[29] *Jane's Air-Launched Weapons*, "YJ-8K (C-801K), YJ-82 (C-802AK/KD) and YJ-83 (C-803)," October 15, 2009; *Jane's Air-Launched Weapons*, "YJ-8K (C-801K), YJ-82 (C-802AK/KD) and YJ-83 (C-803)," December 31, 2012; and Office of the Secretary of Defense, *Military and Security Developments Involving the People's Republic of China 2013*, May 2013, p. 7.

[30] Office of Naval Intelligence, *The PLA Navy*, 2015, p. 24, and Office of the Secretary of Defense, *Military and Security Developments Involving the People's Republic of China 2014*, June 2014, p. 7.

[31] Karotkin, "Trends in China's Naval Modernization," January 30, 2014.

[32] With the understanding that military technology constantly changes and that, therefore, the definition of "modern" will change, our own categorization is based on sensors, defenses, and weapons, as well as performance characteristics of the ship (e.g., speed and ease of detection). For the purposes here, we define *Sovremenny*-,

Unlike the case with fighter aircraft, where degree of capability is widely compared by "generation," there is, however, no commonly accepted definition of "modern" with regard to warships, and some of those included in the above figures would not meet all potential criteria. Even the *Luyang III*, China's most modern destroyer (one in service, five others launched, and an additional five building), is inferior to the Burke (with 62 in service) in a number of respects.[33] Nevertheless, Chinese surface ships might challenge the U.S. Navy under the right operational circumstances and as part of a larger "system of systems."

PLAAF and PLAN aviation equipment modernization has focused on the acquisition of modern fighter aircraft with advanced air-to-air missiles, glass cockpits, long-range SAMs, and precision air-to-ground munitions. China's modern fighter aircraft include the Russian-designed Su-27 and Su-30 "Flanker," roughly comparable to the U.S. F-15 in size and flight performance, and the indigenously designed J-10 "Firebird," roughly comparable to the U.S. F-16. As of 2015, China had imported approximately 175 Flankers from Russia and had produced another 105 licensed models (designated the J-11).[34] It also currently operates around 170 copies of an improved variant (the unlicensed J-11B), built primarily from domestically produced components.[35] Although the J-10 is a Chinese design, Israeli Aircraft Industries is suspected of having provided design assistance. Some 294 J-10s had been built by the end of 2014, with production continuing at a rate of about 30 aircraft per year.[36] As of 2015, roughly half (736 of 1,432) of China's fighters and fighter-bombers were modern, fourth-generation aircraft, while the remainder were legacy aircraft based on the 1950s-era MiG-21 (see Table 4.1 in Chapter Four).

Whether legacy or modern, Chinese fighter and fighter-bombers all carry modern air-to-air missiles, including, in most cases, advanced radar-guided beyond-visual-range missiles. Chinese precision air-to-ground munitions include domestically developed laser-guided and satellite-guided bombs, high-speed anti-radar missiles, and ALCMs

---

Luhai-, *Luyang I-*, *Luyang II-*, *Luyang III-*, and *Luzhou*-class destroyers as modern. We also define *Jiangwei-* and *Jiangkai*-class frigates as modern.

[33] This is particularly true in comparison to the *Arleigh Burke* Flight IIA, which is 25 percent larger than the *Luyang III*. It has 96 VLS tubes to the Luyang III's 64, accommodates two helicopters against the *Luyang III*'s one, and has a visibly smaller radar. *Jane's Fighting Ships*, "Luyang III (Type 52D) Class," February 16, 2015; and *Jane's Fighting Ships*, "Arleigh Burke (Flight IIA) Class," April 2, 2015.

[34] SIPRI Arms Transfer Database, as of May 15, 2015. IISS, *The Military Balance*, 2015, lists 172 Flankers (including Su-27SK, Su-27 UBK, and Su-30 MKK), in addition to the J-11s, currently in service.

[35] IISS, *The Military Balance*, 2015, pp. 241–242.

[36] IISS, *The Military Balance*, 2015; *Janes's World Air Forces*, "China," March 27, 2015; *Jane's All the World's Aircraft*, "SAC (Sukhoi Su-27SK) J-11A," June 30, 2014; *Jane's All the World's Aircraft*, "Sukhoi Su-30M," August 26, 2014; *Jane's All the World's Aircraft*, "SAC (Sukhoi Su-27) J-11," June 30, 2014; *Jane's Sentinel Security Assessment*, "Air Force, China," March 5, 2015. Numbers include both PLAAF and PLAN aviation assets.

with ranges of 100 km or more. In addition, the Su-30 aircraft China imported from Russia have a variety of air-to-ground precision munitions.[37]

China's acquisition of long-range SAMs has included the import from Russia of a reported 40 batteries of long-range systems with ranges of 100–200 km. China has also developed and, as of 2015, deployed another 16 batteries of domestically produced long-range systems.[38] Each battery has four missile-launcher vehicles, each of which carries four missiles, for a total of at least 900 long-range surface-to-air missiles, not including spares.[39]

The PLA has also made great strides since 1996 in developing C4ISR and counter-C4ISR capabilities. It has invested heavily in a variety of means to increase the capacity and resiliency of its communication and command capabilities, which include fiber-optic, wireless, and satellite communication systems.[40] It has constructed at least one skywave OTH radar system and deployed imaging, synthetic aperture radar (SAR), and electronic intelligence satellites to increase its ability to locate targets beyond the horizon. The PLA has also developed a variety of kinetic and nonkinetic weapons and jammers aimed at countering an adversary's ability to use space-based platforms.

For many years, the modernization of PLA ground force proceeded more slowly than that of the air, naval, and missile forces. Since roughly 2010, however, the introduction of new equipment, to include modern tanks, armored personnel carriers, and artillery appears to have accelerated. New amphibious and air-transportable vehicles have also been developed for the PLAN's marines and PLAAF's airborne forces. The PLA has developed and fielded main battle tanks, the Type-98 and Type-99, which are comparable in capability to the U.S. M1A1. By 2015, these tanks accounted for about 640 of its total of 6,540 tanks, the remainder being incremental developments on the

---

[37] *Jane's Air-Launched Weapons*, "Chinese Laser-Guided Bombs (LGBs)," January 22, 2010; *Jane's Air-Launched Weapons*, "LT-2 Laser-Guided Bomb (LS-500J)," February 26, 2015; *Jane's Air-Launched Weapons*, "Fei Teng Guided Bombs (FT-1, FT-2, FT-3, FT-5, FT-6)," January 16, 2015; *Jane's Air-Launched Weapons*, "YJ-91, KR-1 (Kh-31)," March 31, 2015; *Jane's Air-Launched Weapons*, "KD-63 (YJ-63), K/AKD-63," January 28, 2014; *Jane's All the World's Aircraft*, "Sukhoi Su-30M," August 26, 2014; *Jane's Strategic Weapons Systems*, "Multirole ASM (KD-88)," March 6, 2014.

[38] The 2009 *Military Power of the People's Republic of China* (p. 66) shows a total of 16 batteries of HQ-9s, with another 15 batteries of the domestically produced medium-range, mobile HQ-12. Subsequent versions of the DoD report do not list system numbers. The 2015 version of *The Military Balance*, like the editions in several preceding years, shows only around eight batteries, indicating some uncertainty.

[39] *Jane's Land-Based Air Defence*, "S-300P," February 19, 2014; *Jane's Land Warfare Platforms*, "HQ-9/FT-2000," February 20, 2015; Office of the Secretary of Defense, *Military and Security Developments Involving the People's Republic of China 2015*, April 2015, p. 36.

[40] Kevin Pollpeter, "Towards an Integrative C4ISR System: Informationization and Joint Operations in the People's Liberation Army," in Roy Kamphausen, David Lai, and Andrew Scobell, eds., *The PLA at Home and Abroad: Assessing the Operational Capabilities of China's Military*, Carlisle, Pa.: Strategic Studies Institute, U.S. Army War College, 2010.

Soviet T-54.[41] The introduction of new classes of armored infantry fighting vehicles (AIFVs) and armored personnel carriers (APCs) into what was once an overwhelmingly infantry force has also proceeded rapidly—raising the total number of AIFVs and APCs from 4,540 in 2010 to 8,870 by 2015.[42] With a ground force 1.6 million strong, the PLA remains, on balance, less mechanized than the 1.1 million–strong U.S. Army (including National Guard and Reserve), which operates close to 30,000 AIFVs and APCs (including mine-resistant, ambush-protected [MRAP] patrol vehicles).[43]

China maintains a large artillery force of more than 13,000 pieces. As of 2015, only 2,280 of these systems were self-propelled models, the remainder being towed howitzers. China has made a significant investment in self-propelled multiple-rocket launchers, fielding more than 1,800 such systems (in addition to 54 towed launchers) as of 2015.[44]

Chinese efforts to develop long-range power projection forces have lagged behind the development of its anti-access capabilities, though there have been several breakthroughs in recent years. In January 2013, the PLAAF began test flights of a domestically produced heavy lift aircraft, the Y-20. Currently, the largest and most capable operational military transport aircraft in the inventory remains the Il-76MDs imported during the early 1990s. A 2005 agreement to purchase 30 additional Il-76s has remained unfulfilled due to production issues in Russia. China has also fielded an air-deployable armored vehicle for the PLA's airborne forces.[45] The PLAAF added its first aerial refueling tankers in the late 1990s, but the capability remains limited, consisting of ten modest-sized aircraft based on the H-6 bomber design. In 2014, it took delivery of the first of three IL-78M tanker aircraft from Ukraine.

In a major milestone reflective of the breadth of depth of the modernization of the PLA, in September 2012, the PLAN introduced its first aircraft carrier into service, and only two months later, Chinese J-15 "Flying Shark" aircraft successfully conducted

---

[41] IISS, *The Military Balance*, 2015, p. 238; and *Jane's Armour and Artillery*, "NORINCO Type 98/Type 99 (ZTZ-98/ZTZ-99) MBT," April 1, 2014.

[42] IISS, *The Military Balance*, 2010, p. 400; IISS, *The Military Balance*, 2015, p. 239.

[43] IISS, *The Military Balance*, 2015, p. 42.

[44] IISS, *The Military Balance*, 2015, p. 239; Office of the Secretary of Defense, *Military and Security Developments Involving the People's Republic of China 2013*, May 2013, p. 75; *Jane's Land Warfare Platforms*, "NORINCO AR2 300mm (12-Round) Multiple Launch Rocket System," July 22, 2013; *Jane's Land Warfare Platforms*, "China Precision Machinery Import and Export Corporation (CPMIEC) 302mm WS-1B (4-Round) Artillery Rocket System," July 22, 2013; *Jane's Land Warfare Platforms*, "China Precision Machinery Import and Export Corporation (CPMIEC) 300mm (10-Round) A100 Multiple Rocket System," March 5, 2012; *Jane's Land Warfare Platforms*, "China Precision Machinery Import and Export Corporation (CPMIEC) 320mm (4-Round) WS-1 Artillery Rocket System," July 22, 2013.

[45] *Jane's Land Warfare Platforms*, "NORINCO Type 63A Light Amphibious Tank," October 15, 2014; *Jane's Land Warfare Platforms*, "NORINCO Type 77 Armoured Personnel Carrier," March 13, 2015; *Jane's Land Warfare Platforms*, "Chinese Amphibious Assault Vehicle ZBD-2000 (ZBD-05)," March 13, 2015; *Jane's Land Warfare Platforms*, "NORINCO Airborne Assault Vehicle (AAV) ZLC2000 (ZBD-03)," March 13, 2015.

their first-ever carrier takeoffs and landings.[46] In doing this, China became, as a 2015 report from the U.S. Office of Naval Intelligence observed, "only the fifth country in the world to possess conventional takeoff and landing fighters aboard an aircraft carrier."[47] While the *Liaoning* is smaller than the U.S. *Nimitz*-class and *Ford*-class carriers and not as well suited to conducting long-range power projection as U.S. carriers, it will be able to augment the PLAN's fleet air defense capabilities. It should also be understood as China's initial training investment into the realm of modern aircraft carriers and carrier-based aviation. The Chinese navy appears committed to developing a force of several carriers in the future, but will not have a fully developed mission-capable carrier within the time frame covered by this report.

China has roughly doubled its amphibious lift capacity since 1996, and its fleet now includes four large, domestically built *Yuzhao* (Type 071)–class amphibious transport docks (LPDs) in addition to older craft. The *Yuzhao* LPDs can carry up to four of the new *Yuyi* air cushion landing craft plus four or more helicopters, armored vehicles, and troops on long-distance deployments.[48] The Office of Naval Intelligence expects additional *Yuzhao* construction in the near term as well as a follow-on amphibious assault ship that is larger and has a full flight deck for helicopters.[49] Together, these enhancements in amphibious capabilities since 1996 herald China's development of a modern expeditionary warfare and OTH amphibious assault capability.

In addition to modernizing its doctrine and equipment, the PLA has also sought to improve the quality of its personnel and training. It has increased educational requirements for new officers and enlisted personnel. Recruits are now expected to have at least graduated from middle school. Since 2001, the PLA has targeted college graduates, and, in 2009, some 100,000 college graduates volunteered.[50] Approximately half of the PLA's officers are now recruited from civilian universities, which are thought to provide a higher-quality education than the PLA's academies.[51] Military service reforms in 1999 established the basis for a professional noncommissioned officer corps. All noncommissioned officers must have at least a high-school education or—for those who are not high-school graduates—a "certificate of professional qualification" obtained from a PLA academy, civilian college, research institute, or industrial college where they have received the requisite training.[52]

---

[46] Office of Naval Intelligence, *The PLA Navy*, 2015, p. 13.

[47] Office of Naval Intelligence, *The PLA Navy*, 2015, p. 23.

[48] Office of Naval Intelligence, *The PLA Navy*, 2015, p. 15.

[49] Office of Naval Intelligence, *The PLA Navy*, 2015, p. 15.

[50] Blasko, *The Chinese Army Today*, 2012, p. 56.

[51] Johnson et al., *Preparing and Training for the Full Spectrum of Military Challenges*, 2009, pp. 32–33.

[52] Mark K. Snakenberg, "Junior Leader PME in the PLA: Implications for the Future," *Joint Force Quarterly*, Vol. 62, No. 3, 3rd Quarter 2011; Blasko, *The Chinese Army Today*, 2012, p. 60; Johnson et al., *Preparing and Training for the Full Spectrum of Military Challenges*, 2009, pp. 30–32.

The PLA is also improving the quality of its training and exercises by increasing their realism, complexity, and "jointness." Traditionally, training was conducted with small units belonging to a single branch (e.g., infantry, frigates, fighter aircraft) and in benign conditions that included familiar terrain, daylight, good weather, and either with no opposing force or with opposing forces whose actions were predetermined and briefed to each other. Now, training is routinely conducted on unfamiliar terrain, at night or in bad weather, and against opposing forces whose actions are not predetermined. The frequency of combined-arms (different branches within a single service) and joint (units from different services) training has also increased, as has the scale of the exercises. Some training areas have dedicated opposition forces that simulate the tactics of potential adversaries and are allowed to defeat the visiting unit. Finally, rigorous evaluation and critique have become an integral part of PLA training, with units required to meet standardized performance benchmarks or else undergo remedial training.[53]

To be sure, the PLA continues to suffer weaknesses in its system of education and training. Some of these are institutional. The PLA Air Force, for example, lacks a true equivalent to the U.S. Air Weapons School, which trains officers who then return to U.S. units to organize and lead tactical training at the unit level. Other problems are related to organizational culture and the highly centralized, top-down nature of Chinese military decisionmaking. Institutional and cultural factors interact in ways that make improvements to the overall operational capability slower than the advances seen in Chinese hardware.[54] Nevertheless, China has made great strides in terms of the quality of its personnel and training overall. And in some areas, such as the regular inclusion of electronic warfare operations in training exercises, its practices may now equal those of other leading powers.

Development of the U.S. Military Since 1996

The fundamental forces shaping the U.S. military since 1996 have differed markedly from those shaping the Chinese military. The differences stem primarily from the respective reach, missions, and tasks of each military, with the Chinese military

---

[53] On Chinese military training, see Roy Kamphausen, David Lai, and Travis Tanner, eds., *Learning by Doing: The PLA Trains at Home and Abroad*, Carlisle, Pa.: Strategic Studies Institute, U.S. Army War College, 2012, and Blasko, *The Chinese Army Today*, 2012. On PLAAF training, see Kevin Lanzit, "Education and Training in the PLAAF," in Richard P. Hallion, Roger Cliff, and Phillip C. Saunders, eds., *The Chinese Air Force: Evolving Concepts, Roles, and Capabilities*, Washington, D.C.: National Defense University Press, 2012, and Kenneth Allen, *The Ten Pillars of the People's Liberation Army Air Force*, Washington, D.C.: Jamestown Foundation, 2011. On "jointness" training, see Kevin McCauley, "The PLA's Three-Pronged Approach to Achieving Jointness in Command and Control," *China Brief*, Vol. 12, No. 6, March 2012.

[54] For a comprehensive treatment of persistent weaknesses in the PLA's system of training and education as well as other areas, see Chase et al., *China's Incomplete Military Transformation*, 2015.

focused largely on Asia and the U.S. military responsible for a global array of missions. At the same time, the U.S. military faced reduced defense budgets at the end of the Cold War, followed by the new challenges of extended operations in Iraq and Afghanistan. Today, it faces new budgetary pressures under a political focus on deficit reduction and domestic priorities.

The U.S. Military in 1996

As the Warsaw Pact dissolved, the Soviet Union collapsed, and the Cold War ended with a whimper, the U.S. military found itself unchallenged as the world's preeminent armed force. Its strength was vividly demonstrated in 1991, when U.S. forces that had been organized, trained, and equipped to fight World War III in Central Europe instead threw their weight against Saddam Hussein's "battle-hardened" but ultimately overmatched army in the deserts of Saudi Arabia, Kuwait, and Iraq.

Operation Desert Storm only temporarily shifted the Pentagon's attention from its core policy challenge of the 1990s: defining a credible construct by which to size and shape the armed forces. The Base Force study (1990), the Bottom-Up Review (1993), the Commission on the Roles and Missions of the Armed Forces (1995), the first Quadrennial Defense Review (1997), and the report of the National Defense Panel (1997) each sought to establish a clear and durable rationale for the post–Cold War force structure.[55] The "two major theater wars" metric—sizing the force to enable it to fight and win two simultaneous wars against regional powers (usually Iraq and North Korea)—prevailed. However, this construct faced challenges from those on the political left who questioned its ambition and from those on the right who questioned its parsimony.

Regardless of the metric used to assess the adequacy of the force, the U.S. force structure was substantially reduced between 1990 and 1996 (see Table 2.1). The military shed roughly 571,000 active-duty personnel, a reduction of more than 25 percent. Other cutbacks included

- active U.S. Army divisions, from 18 to 12 (33-percent reduction)
- aircraft carrier battle groups, from 14 from 13 (7-percent reduction)
- Navy surface combatants, from 206 to 116 (44-percent reduction)[56]
- attack submarines, from 136 to 82 (40-percent reduction)

---

[55] See Colin Powell, "Building the Base Force: National Security for the 1990s and Beyond," annotated briefing, September 1990; Les Aspin, *Report on the Bottom-Up Review*, Washington, D.C.: Office of the Secretary of Defense, October 1993; Commission on the Roles and Missions of the Armed Forces, *Directions for Defense*, Washington, D.C., May 1995; Office of the Secretary of Defense, *Report of the Quadrennial Defense Review*, Washington, D.C., May 1997; and National Defense Panel, *Transforming Defense: National Security in the 21st Century*, Washington, D.C., December 1997.

[56] Surface combatants are large warships other than aircraft carriers and submarines: frigates, destroyers, cruisers, battleships (for 1990 only), and littoral combat ships (for 2010 and 2015).

Table 2.1
Selected U.S. Force Elements, 1990–2015

Force Element	1990	1996	2003	2010	2015
Active personnel (millions)	2.118	1.547	1.427	1.580	1.433
Army divisions (active)	18	12	10	10	10
Aircraft carriers	14	13	12	11	10
Surface combatants	206	116	106	100	105
SSNs	136	82	56	57	59
Large amphibious vessels	63	39	38	31	31
U.S. Air Force FTR/FGA	3,444	2,485	2,413	2,158	1,570
U.S. Air Force bombers	301	195	199	154	139

SOURCES: IISS, *The Military Balance*, 1991, 1996, 2003, 2010, and 2015 editions.
NOTES: FTR = air supremacy fighter aircraft, whose primary mission is air-to-air combat. FGA = ground attack fighter aircraft, whose primary mission is air-to-surface strike.

- large amphibious vessels, from 63 to 39 (38-percent reduction)
- fighter-bomber aircraft, from 3,444 to 2,485 (28-percent reduction)
- bomber aircraft, from 301 to 195 (35-percent reduction).

U.S. Force Development, 1996–2015

Most U.S. force components were reduced further between 1996 and 2015. These cuts were particularly deep for submarines (20 percent), heavy bombers (29 percent), and fighter aircraft (37 percent). The demands of the wars in Iraq and Afghanistan resulted in some growth in the total number of personnel by 2010, but that number has again declined. While the number of U.S. systems may have decreased, their quality has improved. There have been improvements at both the system level, such as avionics upgrades to aircraft, and the platform level, including the entry of F-22 aircraft and the replacement of DD-956 *Spruance*-class destroyers by Aegis-equipped DDG-51 *Arleigh Burke*–class vessels.

There were three notable changes in U.S. forces between 1996 and 2015.[57] First, by 2015, precision munitions, especially air-delivered ones, had become the norm rather than the exception. Precision-guided munitions constituted only 7 percent of

[57] This is not to say that these were the only important changes during that time. The rapid fielding of the MRAP vehicle could certainly be considered an important change, given the number of lives it has saved in Iraq and Afghanistan. The changes we discuss are the most germane to the Sino-U.S. military balance, however.

the air-to-surface weapons employed in the 1991 Gulf War but 70 percent of those used during the "major combat" phase of Operation Iraqi Freedom in 2003.[58]

Second, the Air Force and Navy have benefited from improvements in key platforms. F-22 "Raptor" procurement is now complete, and the fifth-generation capability represents a substantial step forward from the "legacy" F-15, though it will also interface with legacy aircraft in ways that improve their battle-space awareness and overall effectiveness. Especially important, the F-22 should be able to operate in heavily defended airspace that would be prohibitively dangerous to older jets.

As of early 2015, the Navy operates 62 DDG-51 *Arleigh Burke*–class guided missile destroyers.[59] Equipped with the Aegis air and missile defense system, these vessels have supplanted older *Spruance* destroyers, which offered minimal air defense capabilities. Newer *Arleigh Burke*–class ships—called "Flight IIA"—have further improvements over earlier ships in the class, including the ability to carry and operate two helicopters. The Navy has stabilized the size of its nuclear attack submarine fleet and is gradually fielding the new SSN-774-class submarine. Eleven *Virginia*-class submarines had been commissioned as of early 2015.[60] The *Virginia* class incorporates enhancements first seen in the cost-prohibitive *Seawolf* class and represents a substantial improvement over the longtime mainstay of the SSN fleet, the *Los Angeles* class.[61]

A third change has been the wide introduction and integration of unmanned aerial vehicles (UAVs) into the arsenal. In 1996, neither the Air Force nor the Navy had any operational UAVs, while the Army and Marine Corps each fielded a handful.[62] By 2012 the U.S. military operated 7,500 UAVs, the large majority of which were small UAVs such as the RQ-11 Raven, Wasp, and RQ-7 Shadow.[63] Armed UAVs, especially the MQ-1 Predator and MQ-9 Reaper, have been heavily used in Afghanistan and Iraq, and the RQ-4 Global Hawk and MQ-4C Triton are increasingly important to U.S. ISR capabilities.[64]

The most dramatic and unexpected influences on the size and development of the U.S. military during this period were the wars in Afghanistan and Iraq. Direct military

---

[58] Andrew F. Krepinevich, *Operation Iraqi Freedom: A First-Blush Assessment*, Washington, D.C.: Center for Strategic and Budgetary Analysis, 2003, p. 14.

[59] IISS, *The Military Balance*, 2015, p. 43.

[60] *Jane's Fighting Ships*, "Virginia Class," March 24, 2015; IISS, *The Military Balance*, 2015, p. 55.

[61] Norman Polmar, *The Naval Guide to the Ships and Aircraft of the U.S. Fleet*, Annapolis, Md.: Naval Institute Press, 2005, pp. 75–77.

[62] IISS, *The Military Balance*, 1996, pp. 26–29.

[63] Jeremiah Gertler, *U.S. Unmanned Aerial Systems*, Washington, D.C.: Congressional Research Service, January 3, 2012, p. 8.

[64] See, e.g., Christopher Drew, "Drones Are Weapons of Choice in Fighting Al-Qaeda," *New York Times*, March 16, 2009; and Gertler, "U.S. Unmanned Aerial Systems," January 3, 2012.

costs alone amounted to some $1.5 trillion through the end of fiscal year (FY) 2014.[65] The war against the Islamic State of Iraq and Syria (ISIS) and other smaller conflicts have, thus far, seen operations on a significantly smaller scale with correspondingly lower costs.[66] Nevertheless, these operations too have taken a toll on U.S. forces and readiness. Future costs will include additional billions needed to "reset" the force—that is, to replace or repair equipment destroyed, damaged, or worn out in the course of these prolonged conflicts. Estimates for the price of this reset vary.[67] Then–Under Secretary of Defense for Acquisition, Technology, and Logistics Ashton Carter suggested in 2009 that something larger (and more expensive) than a reset may be necessary:

> [A]fter six years of war it makes no sense to restore forces to the state they were in before the war. So "reset" becomes "modernization" as a practical matter, and modernized forces cost more than the older forces they replace.[68]

Through 2017, the U.S. military will continue to evolve, but the likely changes will be evolutionary rather than revolutionary. F-35 schedules continue to slide, and although the U.S. Government Accountability Office predicts that more than 350 aircraft will be delivered to the U.S. Air Force, Navy, and Marine Corps by 2017, it does not expect that the aircraft will be ready for initial operational testing until 2017.[69] At the same time, further shrinkage in the Air Force fighter inventory can be expected.[70] Meanwhile, although the Air Force has begun a program to develop a long-range strike bomber (LRS-B), the anticipated launch has slipped from 2018 to some indefinite future date.[71]

---

[65] Crawford, "U.S. Costs of Wars Through 2014: $4.4 Trillion and Counting," 2014.

[66] Based on statements by DoD officials about the daily costs of war and other anecdotal evidence, the National Priorities Project estimated the costs as of May 2015 at about $2.5 billion. For more on the methodology, see National Priorities Project, "Costs of National Security," 2015.

[67] In June 2006, for example, the Army Chief of Staff and the Marine Corps Commandant testified before Congress that reset costs could amount to $15 billion and $5 billion a year for the respective services and, at least in the Army's case, these costs will continue a "minimum of two to three years" after the two wars end (General Peter Schoomaker, cited in Belasco, *The Cost of Iraq, Afghanistan, and Other Global War on Terror Operations Since 9/11*, 2011, p. 51).

[68] Ashton B. Carter, "Defense Management Challenges for the Next American President," *Orbis*, Winter 2009, p. 43, fn. 4.

[69] U.S. Government Accountability Office, *F-35 Joint Strike Fighter*, 2013, pp. 6, 23.

[70] David Axe, "China's Fighters Won't Match US," *The Diplomat*, March 10, 2011.

[71] For a reference to the 2018 operational date, see, e.g., Office of Management and Budget, *Terminations, Reductions, and Savings, Budget of the U.S. Government, FY 2010*, Washington, D.C., 2010. On the program start, see Office of the Under Secretary of Defense (Comptroller) Chief Financial Officer, *United States Department of Defense Fiscal Year 2014 Budget Request, Overview*, April 2013, pp. 7–20.

In FY 2015, the Air Force took delivery of its first seven KC-46A tankers, with another 12 systems scheduled for FY 2016.[72] Although the KC-46A incorporates some new capabilities (e.g., boom and drogue delivery on the same sortie), it is essentially a direct replacement for the aging KC-135 and is designed primarily to keep the fleet flying rather than to transform it in any fundamental way. The heavy and medium airlift fleet, for its part, may face greater challenges. The total number in service declined by 85 aircraft between 2010 and 2015 (from 720 to 635), the C-17 line was halted at 224 aircraft in 2013, and the aging C-5A (8 aircraft in 2015) is being retired.[73] Nevertheless, despite some decline in the fleet, the U.S. military maintains an impressive refueling capability compared to China's ten low-capacity H-6U tanker aircraft.

The Navy expects to take delivery of the *Gerald Ford* (*CVN-78*), the first in a new class of aircraft carriers, in March 2016, but faces more questions about the future of its surface combat fleet.[74] In October 2013, it took delivery of the first of its *Zumwalt*-class destroyers, a stealthy platform with small crew size and a number of new technologies (as well as 50 percent greater displacement).[75] However, although originally slated to replace the *Arleigh Burke* class, cost and technology challenges stopped procurement at three ships. The Navy will instead revert to procurement of an improved *Arleigh Burke*, the "Flight III," which will begin delivery in 2016. The Navy has also commissioned four littoral combat ships (LCSs) of a planned 52-ship force. A small frigate with modular packages that can be added or removed for different missions, the LCS is considered by many to be too weak to survive in high-intensity warfare but may fill a variety of important niches, such as mine warfare.[76]

The development of new technologies, such as long-range precision strike, challenge a range of existing U.S. platforms and capabilities, and the U.S. Air Force and Navy continue to debate appropriate responses. Partly for this reason, many U.S. next-generation systems are hotly debated, with the likely number and shape of key systems uncertain as of this writing.

---

[72] Office of the Secretary of Defense, "United States Department of Defense Fiscal Year 2016 Budget Request," February 2015, p. 5-2.

[73] IISS, *The Military Balance*, 2010, 2015; *Jane's Sentinel Security Assessment*, "Procurement, United States," April 2, 2015.

[74] The Navy retired the USS *Enterprise* in December 2012, temporarily leaving the nation with a ten-carrier Navy until the USS *Gerald Ford* is delivered in March 2016. Ronald O'Rourke, *Navy Ford (CVN-78) Class Aircraft Carrier Program: Background and Issues for Congress*, Washington, D.C.: Congressional Research Service, March 24, 2015, p. 4.

[75] Ronald O'Rourke, *Navy DDG-51 and DDG-1000 Destroyer Programs: Background and Issues for Congress*, Washington, D.C.: Congressional Research Service, June 12, 2015.

[76] On the shipbuilding plan, see O'Rourke, *Navy Force Structure and Shipbuilding Plans*, 2012, p. 4. For a critical view, see William D. Hartling, "It's Time to Sink the Littoral Combat Ship," *Defense One*, August 25, 2014.

China in U.S. Defense Policy

For many years, there was asymmetry in the attention paid to Asian missions and scenarios by China and the United States. While the PLA optimized its forces for these tasks, and particularly for capabilities most relevant to Taiwan scenarios, U.S. military attention was spread across a much broader array of global missions. This asymmetry persists, but it is narrowing somewhat as China builds capabilities for more distant tasks and the United States focuses more heavily—though certainly not exclusively—on Asia.

In 2004, Chinese premier Hu Jintao directed the PLA to prepare for "new historic missions." The guidance associated with these missions, including the safeguarding of China's expanding national interests and helping to maintain world peace, was framed less narrowly than comparable guidance in the past. It opened the door to more types of missions farther from home, including humanitarian assistance and disaster relief, noncombatant evacuation operations, the protection of sea lines of communication, peacekeeping operations, and counter-piracy activities. China has stepped up its activities in some of these areas. For example, the PLAN has participated in international counter-piracy escort operations in the Gulf of Aden since 2008; and the Chinese military evacuated Chinese nationals from Libya in February 2011 and from Yemen in April 2015.

At the same time, Washington has begun to focus more heavily on U.S. deterrence capabilities in Asia—particularly the challenge posed by Chinese power—in shaping policy, plans, and procurement. The clearest and most comprehensive public statement to this effect can be found in a January 2012 DoD document, *Sustaining U.S. Global Leadership: Priorities for 21st Century Defense*. The document highlights the "mix of evolving challenges and opportunities" in Asia, and states, "while the U.S. military will continue to contribute to global security, *we will of necessity rebalance toward the Asia-Pacific region*."[77]

While such broadly framed policy guidance emphasizing presence in Asia is new, the general policy direction is not. Post–Cold War cuts came primarily in Europe and in the continental United States, with far fewer in Asia. More recently, additional assets have flowed toward the Pacific. In 2003, the United States established a continuous bomber presence mission at Andersen Air Force Base (AFB) in Guam, with a regular rotation of B-1, B-52, and B-2 squadrons. The 2006 *Quadrennial Defense Review Report* mandated that, by the end of 2010, 60 percent of the U.S. attack submarine fleet should be home-ported in the Pacific—an objective that was achieved with time to spare.[78]

---

[77] U.S. Department of Defense, *Sustaining U.S. Global Leadership*, 2012.

[78] U.S. Department of Defense, *Quadrennial Defense Review Report*, Washington, D.C., February 6, 2006.

In September 2009, the U.S. Navy Chief of Naval Operations and the U.S. Air Force Chief of Staff signed a memorandum of agreement to begin work on the so-called Air-Sea Battle Concept.[79] The Air-Sea Battle Concept has since been folded into a larger framework, which also includes the U.S. Army and U.S. Marine Corps, called the Joint Concept for Access and Maneuver in the Global Commons (JAM-GC), but air and sea operations remain the focus of interservice integration, however. The central idea of these new initiatives is that "future joint [service] forces will leverage cross-domain synergy—the complementary vice merely additive employment of capabilities in different domains."[80] The few official public documents on these concepts do not mention particular countries or threats. Nevertheless, the primary purpose of these concepts is to counter "emerging anti-access and area-denial security challenges," making it clear that China is a chief concern.[81]

Because of the long development times associated with most new military systems, procurement often evolves much more slowly than changes to operational concepts. After the end of the Cold War, the U.S. military adopted a new generation of military systems, largely optimized for low-intensity conflict, and some capabilities relevant primarily to high-intensity conflict were allowed to atrophy. While many of these adjustments proved beneficial to operations in Afghanistan and Iraq, as operations in both those nations wound down, there was a growing sense that meeting new challenges in Asia and on NATO's eastern flank, and possibly from Iran, would require rebalancing the force.

In November 2014 this sentiment was given renewed emphasis and concrete top-level support in then–Secretary of Defense Chuck Hagel's unveiling of the Defense Innovation Initiative, more widely known as the Third Offset Strategy. The strategy is focused on "sustaining and advancing U.S. superiority against potential adversaries" who "have been modernizing their militaries" and "developing and proliferating disruptive technologies." It aims to achieve its goal through a multipronged effort that includes long-range research and development focused on identifying, developing, and fielding breakthrough technologies; reinvigorated wargaming to develop and test alternative ways of achieving objectives; and the development of new innovative operational concepts to meet emerging threats.[82]

---

[79] Christopher P. Cavas, "USAF, U.S. Navy to Expand Cooperation: Air-Sea Battle Will Close Gaps, Boost Strength," *Defense News*, November 9, 2009.

[80] U.S. Department of Defense, *Joint Operational Access Concept (JOAC)*, version 1.0, Washington, D.C., January 17, 2012, p. ii.

[81] U.S. Department of Defense, *Joint Operational Access Concept (JOAC)*, 2012, foreword.

[82] See Sydney J. Freedberg, "Hagel Lists Key Technologies for US Military; Launches Offset Strategy," *Breaking Defense*, November 16, 2014; and Robert O. Work, Deputy Secretary of Defense, "The Third U.S. Offset Strategy and Its Implications for Partners and Allies," speech at the Willard Hotel, January 28, 2015.

In an era of constrained budgets, adjustments to deployment patterns, operational concepts, and procurement will be critical to the United States' ability to deter Chinese adventure and, if that fails, prevail in a conflict. Those adjustments will, in other words, have a significant impact on how the results of the following scorecards change over time.

CHAPTER THREE

Scorecard 1: Chinese Capability to Attack Air Bases

This chapter examines the Chinese capability to strike U.S. air bases with ballistic and cruise missiles to impede U.S. air operations from those locations. It briefly discusses Chinese concepts for the employment of precision missile forces against air bases, surveys the development of China's ballistic and cruise missile force structure, and examines the ability of these forces to suppress U.S. air base operations in both the Taiwan and Spratly Islands scenarios in four snapshot years: 1996, 2003, 2010, and 2017 (projected). Finally, it offers conclusions about recent trends and future possibilities.

The analysis shows that China's conventional missile forces have expanded their capabilities over the past 15 years to the point that the PLA can now contest U.S. air base operations within roughly 1,500 km of Chinese territory. This capability will indirectly impinge on a much larger range of U.S. capabilities, complicating the air superiority battle and, by extension, making the use of enabling assets, such as AWACS, tankers, and electronic warfare enabling aircraft, near the battle area both risky and difficult.

Chinese Military Thought on the Use of Precision Missiles

The PLA has the most active ballistic missile program in the world. The Second Artillery, the branch of the PLA devoted to nuclear and conventional ballistic missile forces, currently has missiles in production ranging from short-range conventionally armed SRBMs to nuclear-armed intercontinental-range weapons.[1] Over the past decade and a half, the PLA has made dramatic improvements in the number, quality, and range of its conventionally armed ballistic and cruise missiles.

These improvements are part of a broader portfolio of counter-intervention or anti-access capabilities that China is developing to challenge the ability of the United

[1] While there are no strict definitions, SRBMs are generally classified as having ranges of less than 1,000 km, MRBMs have ranges of 1,000–3,000 km, IRBMs have ranges of 3,000–5,500 km, and ICBMs have ranges in excess of 5,500 km.

States to project power into the Western Pacific.[2] PLA doctrine does not use the term *anti-access*, but it does discuss using precision ballistic and cruise missiles for waging "counter-air strike campaigns" to deny a more capable adversary, like the United States, the ability to generate combat sorties.[3] Precision missiles are a particularly useful capability for the weaker party because they offer the potential to pin a larger and more advanced air force on the ground long enough for the PLAAF to overfly targets and deliver a large volume of precision-guided weapons.[4] In this manner, PLA analysts view missile strikes on air bases as an integral part of how the PLAAF would gain at least temporary or limited air superiority during the opening phase of a war.[5]

Large numbers of accurate ballistic and cruise missiles allow China to threaten key aspects of U.S. air base operations, such as runway surfaces, unprotected aircraft, hardened aircraft shelters, fuel supplies, and logistics facilities.[6] Many Chinese observers suggest that missile strikes on air bases would be part of the opening salvos of a war.[7]

There are, of course, an array of countermeasures available to the United States, so the conventional missiles of the Second Artillery do not necessarily provide the PLA with an assured war-winning capability. However, countermeasures will come at a cost, and China's growing force of sophisticated ballistic and cruise missiles increasingly place U.S. military forces on the wrong end of a cost-imposition calculus. The following sections explore some of the dynamics of this emerging competition.

Force Structure

This section reviews China's inventory of conventionally armed ballistic and cruise missiles. It begins with a summary of developments through 2015 and then estimates

[2] On the PLA's anti-access strategy, see Thomas G. Mahnken, "China's Anti-Access Strategy in Historical and Theoretical Perspective," *Journal of Strategic Studies*, Vol. 34, No. 3, June 2011; Cliff et al., *Entering the Dragon's Lair*, 2007; and Mark A. Stokes, *China's Evolving Conventional Strategic Strike Capability: The Anti-Ship Ballistic Missile Challenge to U.S. Maritime Operations in the Western Pacific and Beyond*, Arlington, Va.: Project 2049, September 2009.

[3] Cliff et al., *Entering the Dragon's Lair*, 2007, pp. 62–64, 81–83.

[4] Numerous studies discuss the synergetic effects of using ballistic and cruise missiles to open windows of opportunity for a less capable air force. See, for example, Shlapak et al., *A Question of Balance*, 2009, pp. 36–37, 42–43, 51.

[5] Cliff et al., *Entering the Dragon's Lair*, 2007, pp. 62–71.

[6] In addition to doctrinal writings, there is further evidence that the Second Artillery intends to use its forces to target airfields. Open-source imagery analysis has identified mock airfields in China that have been used as targets for Second Artillery theater ballistic missile (TBM) tests. These airfields are located in the extreme western edge of Gansu Province and can be seen in Google Earth imagery at 40.476625 degrees, 93.497609 degrees. See Sean O'Connor, "Dragon's Fire: The PLA's 2nd Artillery Corps," *IMINT and Analysis Blog*, June 26, 2010.

[7] Cliff et al., *Entering the Dragon's Lair*, 2007, pp. 31–34.

the country's holdings by 2017. There are, however, a number of important uncertainties with regard to both current and future inventory numbers. Where uncertainties loom particularly large, they have been noted in the text, and a range of possible force structures are included in future projections.

Development of the Second Artillery, 1996–2015

The Second Artillery has dramatically expanded the number and improved the quality of its conventional ballistic and cruise missile forces over the past 15 years. Table 3.1 summarizes this development.

In 1996, the Second Artillery's SRBM force consisted of a handful of DF-15 and DF-11 missiles.[8] The Second Artillery began developing these single-stage, solid-fueled, road-mobile missile designs during the 1980s, originally for export. The first generation of the DF-11 had an inertial guidance system with a 600-meter circular error probable (CEP) and the first-generation DF-15 had some form of terminal guidance that led to 300-meter CEP accuracy.[9] By 2003, a second generation of the DF-15 had been developed and was beginning to be deployed. A new variant of the DF-11—the DF-11A—added Global Positioning System (GPS) midcourse updates and may have added an optical correlation terminal guidance system to improve accuracy to the 20– to 30-meter CEP level. Without the optical correlation terminal guidance, the DF-11A would have a 200-meter CEP. The DF-15A added GPS updates and a radar terminal correlation system to reduce its CEP to 30–45 meters. By 2009, another variant, the DF-15B—reportedly added an active radar seeker and a laser range finder to reach a CEP of five to ten meters.[10]

While the Second Artillery was making these qualitative improvements, the overall size of the conventional SRBM force expanded. In 1996, IISS assessed that China had a small but indeterminate number of surface-to-surface missiles in its inventory. By 2010, that inventory had expanded to roughly 350–400 DF-15s and 700–750 DF-11s.[11] In 2015, the U.S. Department of Defense reported that, in total, China had "at least 1,200 short-range ballistic missiles."[12] All of China's SRBM-equipped brigades are based within range of Taiwan, and in recent years, China has also introduced

---

[8] This report uses Chinese designations for missile systems (e.g., DF-15), but includes the Western designation (e.g., CSS-6) in some cases.

[9] CEP describes the radius of a circle centered on a target within which 50 percent of missiles land. More accurate systems have lower CEPs. See *Jane's Strategic Weapon Systems*, "DF-11 (CSS-7/M-11)," December 11, 2014, and *Jane's Strategic Weapon Systems*, "DF-15," June 23, 2015.

[10] *Jane's Strategic Weapon Systems*, "DF-15," June 23, 2015.

[11] Office of the Secretary of Defense, *Military Power of the People's Republic of China*, August 2010, p. 66.

[12] Office of the Secretary of Defense, *Military and Security Developments Involving the People's Republic of China 2015*, April 2015 p. 8.

Table 3.1
Chinese Conventionally Armed Theater Ballistic and Cruise Missiles

Missile Type	Range (km)	Warhead (kg)	CEP (m)	Number in Inventory			
				1996	2003	2010	2017
SRBMs							
DF-11	280–350	500–800	500–600	Small number	175	700–750	~1,200
DF-11A	350	500	20–30				
DF-15	600	500	300	Small number	160	350–400	
DF-15A	600	600	30				
DF-15B	600–800	600	5				
MRBMs							
DF-21C	2,500	500	50	0	0	36–72	108–274
DF-16[a]	800–1,000	?	?	0	0	0	
IRBMs							
IRBM	5,000	500	30–300	0	0	0	Possible
Cruise missiles							
DH-10	1,500–2,000	400	5–20	0	0	200–500	450–1,250
ALCM	3,300	400	5–20	0	0	In inventory	

SOURCES: *Jane's Strategic Weapons Systems* data; IISS, *The Military Balance*, 1996, 2003, 2010 and 2015; and Office of the Secretary of Defense, *Annual Report to Congress: Military and Security Developments Involving the People's Republic of China*, Washington, D.C., 2010 and 2014.

[a] Although the DF-16's range technically makes it an SRBM and it is classified as in some publications, the DF-16 straddles the boundary between SRBM and MRBM. We list it as an MRBM, both because of its ability to hit targets beyond Taiwan (e.g., Okinawa) and because of its two-stage construction. IISS, *The Military Balance*, 2015, lists the system as an MRBM.

new SRBMs able to hit more distant targets.[13] These include the DF-15B (range 600–800 km), as well as a new missile, the DF-16 (range 800–1,000 km).

The DF-16 is a two-stage missile that straddles the boundary between SRBMs and MRBMs. Given its reported range, the DF-16 is capable of striking targets on Okinawa and, given its smaller size, it is probably able to do so at a smaller financial cost than employing DF-21Cs (discussed below) for the same task.[14] Both because the DF-16 can attack targets beyond Taiwan (and may be especially well suited to strike U.S. bases on Okinawa) and because of its two-stage construction, we list the missile as

[13] Mark A. Stokes and Ian Easton, *Evolving Aerospace Trends in the Asia-Pacific Region: Implications for Stability in the Taiwan Strait and Beyond*, Arlington, Va.: Project 2049 Institute, May 2010, pp. 10–11.

[14] *Jane's Strategic Weapon Systems*, "DF-16," December 11, 2014.

an MRBM in Table 3.1 and discuss future build rates in that context (below).[15] *Jane's Strategic Weapons Systems* suggests that the DF-16 will replace the DF-15 and perhaps also the DF-11.[16] The 2015 edition of *The Military Balance* credits China with having approximately 12 launchers in service which, depending on the number of reloads per launcher, could suggest anywhere between 24 and 48 missiles.[17]

The Second Artillery's conventionally armed MRBM capabilities are a direct extension of its nuclear-armed ballistic missile programs. Growing out of development work on an SLBM during the 1960s, the two-stage, solid-fueled, road-mobile DF-21 was first designed to replace the liquid-fueled DF-2 tactical nuclear missile. The first DF-21s were tested in the 1980s, carried nuclear payloads and had inertial guidance packages that permitted an accuracy of 700-meter CEP.[18] During the late 2000s, the Second Artillery began to develop a conventionally armed variant with GPS updates and a radar correlation terminal guidance system.[19] There is some confusion in the literature over the designations of the DF-21 variants. Here, we refer to the conventional land-attack variant as the DF-21C.[20]

Assessing the growth of the DF-21C force over time is difficult because many sources do not differentiate between different DF-21 variants. Nuclear-armed variants of the DF-21 appear in assessments as early as 1995.[21] The conventional DF-21C does not appear in estimates until 2010, when IISS listed 36 DF-21C transporter-erector-launchers (TELs) in the Chinese inventory, the same number of DF-21C TELs reflected in IISS in its 2015 report.[22] Assuming one to four missiles per TEL, we estimate that there are between 36 and 144 DF-21Cs in the Second Artillery's inventory as of early 2015. Because the Second Artillery has as many as five missiles per TEL for its SRBM force and one missile per TEL for its ICBM force, missile-to-TEL ratios between one and five seem appropriate for the DF-21C (with the most probable range being between two and four).

Finally, in addition to conventionally armed ballistic missiles, the Second Artillery has developed precision ground-launched cruise missiles (GLCMs) that have a

---

[15] IISS, *The Military Balance*, 2015, p. 237.

[16] *Jane's Strategic Weapon Systems*, "DF-16," December 11, 2014.

[17] IISS, *The Military Balance*, 2015, p. 237.

[18] *Jane's Strategic Weapon Systems*, "DF-21 (CSS-5)," June 24, 2014.

[19] *Jane's Strategic Weapon Systems*, "DF-21 (CSS-5)," June 24, 2014.

[20] Jane's calls the conventional land-attack variant the DF-21A and the anti-ship variant the DF-21B, but recent expert testimony to Congress has identified the conventional land-attack variant as the DF-21C and the anti-ship variant as the DF-21D. Since the preponderance of expert commentary uses these later designations, this analysis will use them as well while referring to the performance characteristics as assessed by Jane's.

[21] IISS, *The Military Balance*, 1995, p. 176.

[22] IISS, *The Military Balance*, 2010, p. 399, and, in the 2015 edition, p. 237.

range of between 1,500 and 2,000 km.[23] The DH-10 features a combination of guidance systems, resulting in a CEP of less than 20 meters.[24] There is some confusion with regard to Chinese cruise missile nomenclature; to simplify, we use "DH-10" in discussing all missiles with the performance characteristics mentioned above (such as the CJ-10 and CJ-20), whether they are ground- or air-launched. The DH-10 entered service around 2007, and by 2010, the Second Artillery possesses an estimated 200–500 of these systems.[25] No estimated numbers have been provided in standard DoD reports since that time. There is also an air-launched variant that can be launched from the PLAAF's H-6 bombers. H-6Hs can carry two DH-10 ALCMs to a combat radius of 1,800 km, meaning that targets as far away as 3,300 km (potentially including Andersen AFB on Guam) could be attacked with cruise missiles.[26] The newest H-6 variant, the H-6K, can carry six DH-10s and has new engines for increased range. As of 2014, 36 of the PLAAF's 106 bombers were the newer H-6K and the other 70 bombers were the older H-6A/H/M versions.[27] There are no estimates regarding the number of air-launched DH-10s in the PLAAF's inventory.

The net result of these advances is illustrated in Figure 3.1, which uses colored bands to depict the number of weapons that can reach a given range. In 1996, the small Chinese missile inventory (range shown in green) could reach only as far as Taiwan and U.S. bases in Korea. By 2010, the Second Artillery's 1,000 conventional SRBMs (shown in pink) could reach U.S. bases in South Korea, while the yellow region reflects the hundreds of DF-21Cs and DH-10s that could reach U.S. bases in Japan. The green region reflects that there is some number of ALCM variants of the DH-10 in the PLAAF's inventory, which could threaten Guam. By 2017, the number of missiles capable of striking bases in Japan and Guam will have increased further.

Estimating the 2017 Inventory

Estimating a 2017 inventory for China's conventionally armed ballistic and cruise missiles is complicated by the inherent difficulties of discerning intentions and estimating build rates, as well as the uncertainties about current numbers. For several key Chinese systems, the U.S. government has not updated its published estimates of inven-

---

[23] Office of the Secretary of Defense, *Military Power of the People's Republic of China 2010*, August 2010; *Jane's Strategic Weapons Systems*, "C-602 (HN-1/-2/-3/YJ-62/X-600/DH-10/CJ-10/HN-2000)," May 12, 2015.

[24] Guidance includes inertial navigation, terrain contour matching, and global navigation satellite systems. *Jane's Strategic Weapons Systems*, "C-602 (HN-1/-2/-3/YJ-62/X-600/DH-10/CJ-10/HN-2000)," May 12, 2015.

[25] The first mention of the DH-10 in a DoD report to Congress was in 2008. For the 2010 estimate, see Office of the Secretary of Defense, *Military Power of the People's Republic of China 2010*, August 2010, p. 66.

[26] Office of the Secretary of Defense, *Military and Security Developments Involving the People's Republic of China*, 2010, p. 32.

[27] IISS, *The Military Balance*, 2015, p. 242. For a report on the H-6K, see *Jane's Sentinel Security Assessment*, "China, Procurement," October 15, 2012. On the DH-10, see Easton, *The Assassin Under the Radar*, 2009, p. 3; and Gormley et al., *A Low-Visibility Force Multiplier*, 2014.

Figure 3.1
Second Artillery Missile Threats to Bases in the Western Pacific, 1996–2017

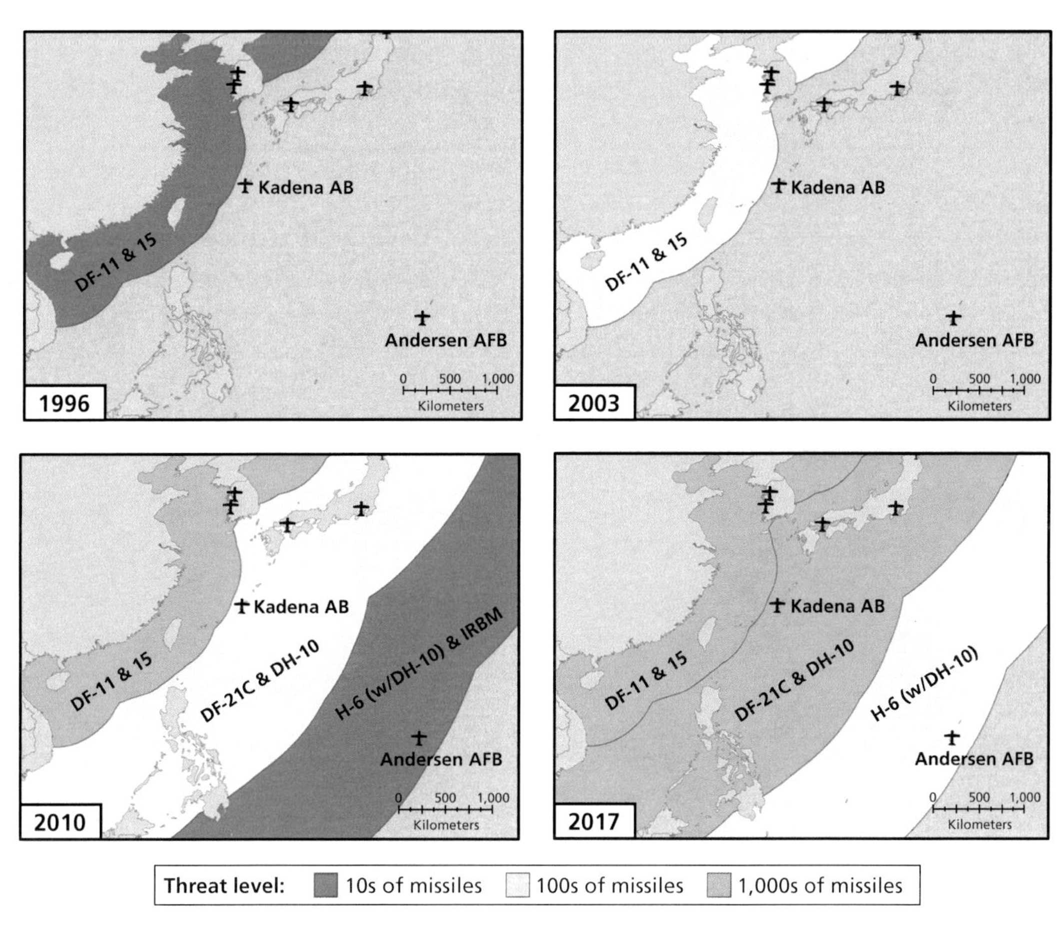

RAND *RR392-3.1*

tory numbers since 2010, and it is unclear how much production has occurred since that time. Also, past build rates and inventory change are not sure indicators of future developments. China could, for example, decide that it has sufficient SRBM forces and shift its efforts to focus on longer-range systems. And the expansion of Chinese inventories may be constrained, to a degree, by unit force structures and manpower limitations. The PLA continues to downsize, even as it modernizes, retiring older systems from the inventory and adding new and more effective ones. The projections in Table 3.1 provide only a possible range of inventory figures.

SRBMs

Based on U.S. Department of Defense reporting, the Second Artillery appears to have increased the number of SRBMs deployed by 50 and 100 per year between 2002 and

2009.[28] Since that time, reported numbers have held roughly steady, with no substantial growth in the number of SRBMs deployed.[29] One possibility is that the numbers in both periods are affected by lags in detection and reporting and may not be accurate. Alternatively, the Second Artillery may have decided that it had an adequate SRBM force and shifted its primary focus toward other systems. Given the substantial manpower and infrastructure requirements of missile units, the number of brigades available may be a limiting factor in total missiles deployed. Furthermore, the replacement of older missiles with more capable ones and the use of missiles for testing may result in some new production being absorbed without a corresponding increase in overall inventory. For these reasons, we believe that the SRBM inventory will remain roughly 1,200, with an increasingly large portion of that force comprised of missiles with ranges that, like those of the DF-16, approach those of MRBMs.

MRBMs

Uncertainty also holds in the case of China's MRBM production and inventory. Based on the historical production rates of analogous systems, China could conceivably produce between ten and 68 MRBMs per year.[30] This range would pertain independently to both DF-21 and DF-16 production (including nuclear and conventional variants). The upper-bound production rate would represent a significant change in resource allocation, though a figure of roughly half that might be imagined in the case of DF-16 production, especially if China employed that system as a replacement for its DF-15s. As a lower bound, we estimate that China might have 108 MRBMs by 2017, based on lower-bound estimates for DF-21Cs (36 missiles) and DF-16s (24 missiles) in 2015, plus a production rate of 24 missiles per year for the next two years. For the higher-bound, we estimate that China might reach 274 MRBMs by 2017, based on the current higher-bound estimates for DF-21Cs (144 missiles) and DF-16s (48 missiles), plus a production rate of roughly 40 missiles per year over each of the next two years.

Cruise Missiles

The DH-10 entered service around 2007, and the U.S. Department of Defense's *Military and Security Developments Involving the People's Republic of China 2010* report

---

[28] Office of the Secretary of Defense, *Military Power of the People's Republic of China 2002*, Washington, D.C., July 2002, p. 16; Office of the Secretary of Defense, *Military Power of the People's Republic of China*, Washington, D.C., 2009, p. 22.

[29] Office of the Secretary of Defense, *Military Power of the People's Republic of China 2009*, March 2009, p. 22, and 2012, p. 29, and *Military and Security Developments Involving the People's Republic of China 2014*, June 2014, p. 40.

[30] For the purposes of bounding potential production rates we examined U.S. production of comparable MRBMs during the Cold War. The maximum rate of U.S. MRBM (in this case Pershing II) production was approximately 68 missiles per year. The lower production rate estimate, ten missiles per year, was derived from our estimate of all types of DF-21s. See Rep. Joseph P. Addabbo, "B-210596 L/M," legal decision, Washington, D.C.: U.S. Government Accountability Office, February 1, 1983.

to Congress on China's military and security developments estimated an inventory of 200–500.[31] If accurate, this implies that Chinese production averaged anywhere between 50 and 150 missiles over this three-year period. DoD reports on Chinese military developments in 2011 and 2012 did not show any increase in the GLCM inventory, and its reports since then have provided no specific figures. But given that the PLA is deploying an air-launched version and developing a naval version of the DH-10, it is unlikely that production lines have been shut down.[32] Using the range of missiles provided in 2012 (200–500) and respective low and high production estimates of 50 and 150 per year, respectively, we derived a low 2017 estimate of 450 land-attack cruise missiles and a high estimate of 1,250 (including ground, air, and sea variants).[33] The Chinese could probably increase production rates, and the high estimate is not necessarily an absolute upper bound, though it seems a reasonable figure.[34]

IRBMs

The only U.S. base in the Western Pacific not currently threatened by conventional ballistic missiles is Andersen AFB on Guam. The only Chinese IRBM that can reach Andersen AFB is the liquid-fueled, single-stage, nuclear-armed DF-3.[35] Given this situation, analysts have argued for years that China would be interested in developing a conventionally armed IRBM to threaten Andersen in the next five to ten years.[36] China validated this speculation when it announced in early 2011 that it was developing a 4,000-km, conventionally armed IRBM.[37] Subsequent publications have projected that the new, still undesignated, IRBM will enter into service sometime between 2016 and 2019.[38]

---

[31] Office of the Secretary of Defense, *Military Power of the People's Republic of China 2010*, August 2010, p. 66.

[32] Office of the Secretary of Defense, *Military and Security Developments Involving the People's Republic of China*, 2012, 2013, 2014, and 2015 versions.

[33] With more data, one could imagine conducting a portfolio analysis of Chinese missile investments that examines the trade-offs between expanding one class of weapons (such as the GLCM DH-10) at the expense of other classes (such as the ALCM DH-10).

[34] For comparative purposes, Raytheon produced 496 BGM-109 Tomahawk cruise missiles (a missile close in size and payload to the DH-10) using a single factory in FY 2009—a year in which the United States was not engaged in a high-intensity conflict. U.S. Department of Defense, *Fiscal Year 2009 Budget Request: Program Acquisition Costs by Weapon System*, Washington, D.C., May 2009.

[35] *Jane's Strategic Weapon Systems*, "DF-3 (CSS-2)," May 7, 2014.

[36] See, for example, Stokes, *China's Evolving Conventional Strategic Strike Capability*, 2009, p. 32.

[37] Zhang Han and Huang Jingjing, "New Missile Ready by 2015," *People's Daily Online*, February 18, 2011; Doug Richardson, "China Plans 4,000 km–Range Conventional Ballistic Missile," *Jane's Missiles and Rockets*, March 1, 2011.

[38] *Jane's Sentinel Security Assessment*, "Strategic Weapons Systems," April 2015; and U.S.-China Economic and Security Review Commission, *2014 Report to Congress*, Washington, D.C.: U.S. Government Printing Office, November 2014, p. 316.

If the Second Artillery develops this weapon, it might look something like a more accurate version of the Russian RSD-10 Pioneer (SS-20) IRBM developed during the Cold War. The RSD-10 was a solid-fueled, road-mobile, two-stage IRBM with a 5,000-km range and a 450-meter CEP. Its 1,360- to 1,580-kg payload would be sufficient to add terminal guidance and a maneuvering reentry vehicle to increase accuracy while still having 500 kg left over to deliver a militarily significant warhead on the order of the DF-11, DF-15, or DF-21.[39] At the same time, the PLAN is "improving its over the horizon (OTH) targeting capability with sky wave and surface wave OTH radars, which can be used in conjunction with reconnaissance satellites to locate targets at great distances from China, thereby supporting long range precision strikes."[40]

In the next section, we discuss the force size required to execute a useful mission against targets, such as Andersen AFB, in the Second Island Chain.

U.S. Forces' Geographic Challenge

The growing number, range, and accuracy of Chinese ballistic and cruise missiles constitute a serious threat to forward U.S. air bases. The challenge to U.S. air efforts is compounded by geography and the scarcity and limited capacity of air bases within reasonably easy flying distances of key scenarios, including the Taiwan and Spratly Islands scenarios considered in this report. Table 3.2 lists the distances of U.S. air bases to China, as well as the distances to a central location relevant to the two scenarios.

The table shows that only two U.S. air bases are located within U.S. fighters' unrefueled combat radius (less than 1,000 km) of the Taiwan Strait.[41] (For comparative purposes, China has 39 air bases within 800 km of Taipei.) The scarcity of U.S. operating areas close to Taiwan means that Chinese commanders could force U.S. aircraft to operate at much greater distances by focusing fires on and denying the use of just a few locations. The next closest bases, Kunsan and Osan, might be denied to combat operations by their South Korean hosts and are, in any case, well within SRBM range of Chinese territory.[42]

To be sure, U.S. fighters can and probably would operate from more distant locations using in-flight refueling, but this would substantially reduce the sortie generation rate and on-station time of the aircraft employed. It would also add to the number of

---

[39] GlobalSecurity.org, "RT-21M / SS-20 SABER Specifications," web page, last updated July 24, 2011.

[40] Lee Fuell, Technical Director for Force Modernization and Employment, National Air and Space Intelligence Center, testimony before the U.S.-China Economic and Security Review Commission, January 30, 2014.

[41] Of these, one (Futenma) is a Marine Corps air station (MCAS) with a single 9,000-foot runway, limited infrastructure, and a mission (largely oriented toward supporting U.S. Marine Corps ground forces) that would limit its role in the air superiority battle.

[42] Chinese SRBM attacks on bases in Korea would require long-distance movement to position assets in northeastern China, within range of Korea.

Table 3.2
U.S. Bases in the Western Pacific and Key Distances

Air Base	Distance to Nearest Chinese Territory (km)	Distance to Center of Taiwan Strait (km)	Distance to Thitu Island (km)
Kadena AB	650	770	2,200
Futenma MCAS	650	770	2,200
Kunsan AB	390	1,360	3,000
Osan AB	400	1,500	3,170
Iwakuni MCAS	940	1,560	3,130
Yokota AB	1,100	2,200	3,700
Misawa AB	850	2,630	4,200
Andersen AB	2,950	2,870	3,330

tankers for which basing would have to be found and security provided in the air, as well as on the ground. Moreover, as indicated earlier, the even more distant bases in Northeast Asia, such as Andersen and Misawa, are now coming within range of Chinese attack, albeit at a lower level of threat.

The distances between U.S. bases and the Spratly Islands scenario (centered on Thitu Island) are significantly greater than in the Taiwan scenario. As discussed later, however, Chinese aircraft would also be flying greater distances and from a smaller number of airfields.

Operational Analysis: Taiwan Scenario

This section examines the effectiveness of Chinese missile forces in shutting down U.S. air base operations during a Taiwan scenario. We use Kadena AB and Andersen AFB to show how Chinese missile forces could threaten three critical aspects of air base operations: runways, aircraft on the ground, and fixed structures (such as fuel tanks or hangars). The results are intended to illustrate the dynamics of the competition created by the Second Artillery's modernization, as well as how these dynamics could respond to potential changes in missile inventory and adaptations on the part of U.S. air forces.

The modeling employs open-source estimates of important parameters and cannot predict the precise outcomes of ballistic and cruise missile attacks on U.S. air bases. Not only might open-source estimates be incorrect, but Chinese planners might also view specific aspects of the problem differently and might therefore allocate attacks differently. Nevertheless, by employing the same methodology and same background assumptions for each of the snapshot years—and varying only the equipment inventories and capabilities to reflect our assessment of what each side actually had at the time—the results should provide a good sense of trends over time, as well as the general magnitude of the problem.

In this analysis, the Second Artillery attacks runways using ballistic missiles armed with runway-penetrating submunitions. Aircraft can be parked in the open, in hangars, or in hardened shelters, and different weapons are used to attack in each of the different cases. Aircraft parked in the open are attacked with ballistic missiles armed with small submunitions.[43] Aircraft parked in hangars and hardened shelters are attacked with cruise missiles armed with unitary warheads.

During a war between China and Taiwan or between the United States and China in the South China Sea, Kadena AB and Andersen AFB would play key roles. Kadena AB is the U.S. base with permanently stationed fighters nearest to both the scenarios in question, and Andersen AFB is the nearest base on U.S. territory. Therefore, if the Second Artillery were able to suppress air operations at these bases for a significant amount of time, it could have severe consequences for the outcome of the campaign.

In 1996 and 2003, China did not have sufficient numbers of conventional ballistic missiles or cruise missiles to threaten Kadena AB or Andersen AFB. During this period, the United States could have effectively operated at will from these bases. By 2010, Kadena AB could have been reached by a significant number of DH-10 GLCMs and DF-21Cs. Andersen AFB remained outside the threat ring of conventional ballistic missile systems, though it is within range of ALCMs carried by H-6H bombers. By 2017, Kadena could be within range of hundreds of MRBMs and perhaps more than 1,000 DH-10s. Andersen AFB could, depending on the evolution of the threat, face anything between a moderately increased threat from ALCMs to an entirely new class of conventional IRBMs and a much more substantial force of cruise missiles.

Runway Attacks

Destroying portions of a runway to deny its use by enemy aircraft requires that the attacker "crater" it such that no minimum operation surface (MOS) is left. MOS requirements vary by aircraft size and performance characteristics and indicate the minimum length and width of unbroken runway that a fighter, tanker, or bomber requires to take off and land. To illustrate the runway targeting problem, we take the case of Kadena AB. Kadena AB has two runways, each roughly 12,000 feet long. One is 200 feet wide and the other 300 feet wide.

First, consider the case of fighter operations from Kadena. The nominal MOS for a fighter is 5,000 feet long and 50 feet wide.[44] Thus, if an attacker can destroy enough

[43] The efficiency of using ballistic missiles armed with small submunitions to attack aircraft parked in the open has been explored in detail by John Stillion and David T. Orletsky, *Airbase Vulnerability to Conventional Cruise-Missile and Ballistic-Missile Attacks: Technology, Scenarios, and U.S. Air Force Responses*, Santa Monica, Calif.: RAND Corporation, MR-1028-AF, 1999.

[44] Note that, here, we are using publicly available U.S. military standards. Chinese planners may employ different assessments about minimum operating surfaces and arrive at significantly different conclusions. A

of the runway that there is no undamaged 5,000- by 50-foot section, then the attacker has shut the air base to fighter aircraft operation until repair crews can reopen the runway.[45] Given that Kadena's has 12,000-foot-long runways, the attacker would need to cut each runway in two places, for a total of four cuts. (An illustrative overlay of these cuts is depicted in Figure 3.2.) Further, given the width of the runways (200 and 300 feet), each lengthwise cut would require several craters across the runway, with the exact number depending on their size and exact position.

Large aircraft, such as tankers, required a larger MOS. As with fighters, the precise MOS required for a tanker depends on a host of factors, including the fuel load of the tanker and weather conditions. A commonly used planning MOS for tankers is 7,000 feet by 147 feet.[46] This larger MOS yields fewer potential operating surfaces,

Figure 3.2
Kadena Air Base

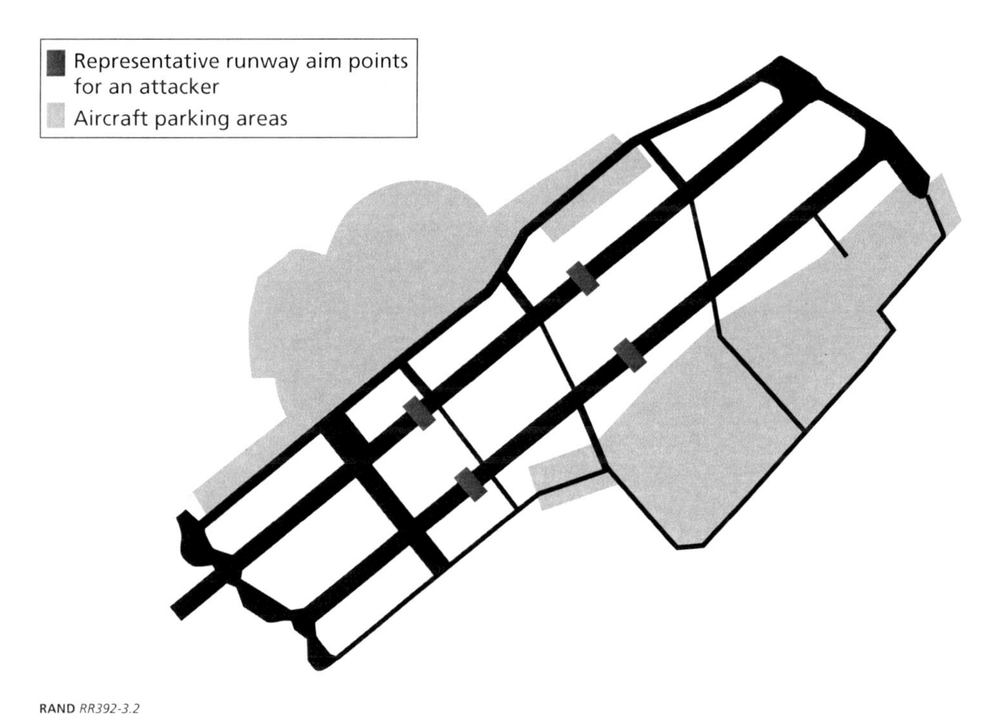

RAND *RR392-3.2*

similar caveat would apply to a range of other factors, such as required confidence levels in a certain level of destruction. MOS figures presented here are from U.S. Air Force, *Airfield Damage Repair Operations*, Air Force Pamphlet 10-219, Vol. 4, May 28, 2008.

[45] Other surfaces, such as taxiways, might be large enough to enable fighter operations, but for the purposes of this illustrative analysis, we assume that they are not usable.

[46] This MOS requirement is the same for KC-135s and KC-10s, the two tankers currently in the Air Force inventory. See Headquarters Air Mobility Command, *Airfield Suitability and Restrictions Report*, Scott AFB, Ill., December 4, 2007, p. xii.

requiring fewer aim points for an attacker to destroy to shut down the base. In this case, a single cut point on each runway could, if properly placed, close Kadena to tankers.

Given these parameters, the question then becomes whether a given inventory of Chinese missiles can destroy two to four sections of runway, and, if so, how many times they can re-attack sections of runway after they are repaired. The effectiveness of a Chinese attack on a runway depends primarily on five variables: the number of missiles, the accuracy of the missiles, the time it takes U.S. forces to repair a destroyed section of runway, the effectiveness of U.S. missile defenses, and the quality of Chinese intelligence.

There is little open-source information about MRBM warheads, but given the PLA's emphasis on attacking air bases and the reported development of high-explosive submunitions for the DF-21C, it is likely that the missile has runway attack capabilities.[47] Indeed, as a ballistic missile with high terminal velocity, it would be well suited to this task. For modeling purposes, we assume that the submunition characteristics are similar to those used in the U.S. Air Force's BLU-67 anti-runway bomb—specifically, a 4.5-kg penetrating submunition carrying 2.75 kg of high-explosives.[48] According to Air Force planning documents, a warhead with these characteristics would create a five-foot-diameter crater in a runway.[49]

A ballistic missile capable of carrying 82 of these anti-runway submunitions could dispense them at a height and velocity such that they would land within a circle with a radius 300 feet distributed around the missile impact point.[50] The resulting density of craters inside this 600-foot-diameter area is sufficient to damage the runway to the point that no 50-foot-wide section remains clear, so long as the submunition footprint covers the entire width of the runway.

Methodology

We simulated attacks on an air base's runways as follows: First, we calculated the number of aim points that the attacker needs to hit on the runways to deny an aircraft an MOS with the necessary length. As discussed earlier, in the case of Kadena, four aim points must be hit to deny operations to fighters and two aim points must be hit to deny operations to tankers and other support aircraft.

Next, we calculated how many missiles the attacker would have to fire to have a 90-percent chance of destroying both aim points (in the case of tanker operations) or

---

[47] On DF-21C characteristics, including warhead types, see *Jane's Strategic Weapon Systems*, "DF-21 (CSS-5)," June 24, 2014.

[48] For details, see *Jane's Air-Launched Weapons*, "Penetrating and Area Denial Bombs," March 18, 2005.

[49] For munitions effects, see U.S. Air Force, *Airfield Damage Repair Operations*, 2008, p. 51.

[50] While the precise mechanism used to dispense the submunitions can create uneven dispersion in this 600-foot-diameter footprint, for the purposes of this analysis, we assume that there is a uniform probability of a submunition landing on any point in the dispersion footprint.

all four aim points (in the case of fighter operations). The number of missiles required depends, in part, on the reliability of the missile and the effectiveness of missile defenses, with the number of missiles that arrive and dispense submunitions determined through Monte Carlo simulations.[51]

We then considered the accuracy of the missile. Having arrived successfully in the target area, does it hit its target? This depends on the footprint over which the submunitions are dispensed (discussed earlier) and the CEP of the missile. For missile accuracies of less than or equal to a 50-meter CEP (such as that of the DF-21C), a single missile has an extremely high likelihood of hitting near enough to its aim point to do sufficient damage and cut the runway. As missile CEP increases beyond 50 meters, multiple missiles must be fired at a single aim point to have a 90-percent chance of successfully cutting the runway.

For illustrative purposes, Table 3.3 shows the probability of cutting a single 200-foot-wide runway with between one and four missiles for a case assuming a missile CEP of 40 meters, a 50-foot-wide fighter MOS, and an 85-percent reliable missile.[52] In this case, an attacker who wishes to achieve at least a 90-percent chance of shutting down a runway with a single aim point must fire at least two missiles at each runway aim point.[53]

Table 3.3
Probability of a Single Runway Cut Using a 40-Meter CEP Missile

Number of Missiles Fired	1	2	3	4
Probability of destroying a runway aim point	0.84	0.98	0.99	1

[51] Failures to reach its release point could be due to mechanical defects or missile defenses, such as land-based Patriot batteries or Standard Missile 3 (SM-3) weapons fired from Aegis ballistic missile defense destroyers or cruisers. For a given missile reliability (on a scale of 0 to 1), we've assign a random value that identifies whether the missile successfully reaches its target.

[52] For more detail on the methodology, the Monte Carlo simulation proceeds as follows. A random draw, based on the missile's reliability, determines whether a missile fails. If the missile does not fail, then its impact point is randomly chosen from a normal distribution corresponding to the CEP of the missile. Next, for each of the 82 submunitions, a random draw determines whether the submunition fails (based on its reliability). If the submunition does not fail, then its impact point is randomly chosen by making two random draws from uniform distributions (one for the radius from the missile impact point and one for the angle) such that all points within the submunition dispersal pattern are equally likely to be struck. An algorithm then searches through the impact points to determine whether there remains an undamaged width of runway greater than or equal to the specified MOS width. If no such MOS exists, then the case is marked as a successful cut.

[53] Achieving a 0.9 probability of cutting all aim points on a base with multiple aim points is harder than achieving a 0.9 probability against any single point. In the example in Table 3.3, the 0.98 probability of making a single cut with two missiles would translate to a 0.92 probability of cutting the runways at all four aim points if two missiles were fired at each ($0.98^4 = 0.92$). There are other cases in which an attacker would have to fire additional missiles at each aim point to ensure a sufficiently high combined probability of cutting all the aim points on a base.

Due to a range of uncertainties about Chinese and U.S. capabilities, we explored the impact of different parameter values. We considered CEP values of between five and 150 meters, missile reliabilities of between 50 and 100 percent, U.S. runway repair times of between four and eight hours, submunition dispersion patterns of between 20 and 350 meters in radius, different required MOS dimensions, and the quality of Chinese intelligence and battle damage assessment.[54]

Results

Table 3.4 presents results of a Chinese ballistic missile attack on the Kadena AB using baseline assumptions, including 75-percent missile reliability (with the other 25 percent malfunctioning or destroyed by defenses), a CEP of 50 feet, perfect Chinese battle damage assessment, and eight-hour repair times for a single runway cut. We show results for a range of missile inventories that might be available and used in each of our snapshot years. We note that the PLA may elect to hold some of its missiles in reserve or divide their use among several targets, so the different possibilities in the table can also be taken to reflect different choices related to usage, as well as uncertainty about the total available number of missiles.[55] The analysis shows that if the PLA employed 36 missiles against Kadena (under the assumptions discussed earlier), it could shut Kadena to fighter operations for four days or to tanker operations for more than 11 days.

Table 3.4
Days of Runway Closure at Kadena Air Base Employing Baseline Assumptions

	MRBM Inventory Employed and Closure Times (days)			
	2010		2017	
Operation Type	**36 missiles**	**72 missiles**	**108 missiles**	**274 missiles**
Fighter	4	10	16	43
Tanker	11.3	23.3	35.3	90.6

NOTES: In the baseline case, we assume that it takes eight hours to repair a single runway cut, 75 percent of missiles reach their targets, the missiles have a 50-meter CEP, the Chinese have perfect intelligence about the results of attacks, and the entire missile inventory is used in runway attacks against Kadena AB. Because the number of missiles available in 2010 and the projected number in 2017 are uncertain, we provide a range of estimates for each.

[54] If the Chinese had perfect battle damage assessment, then they would only have to re-attack those portions of runway that had not been damaged by a missile salvo or that had been repaired by rapid runway repair crews. At the other extreme, if the Chinese have no battle damage assessment, then they would have to blindly re-attack all runway aim points according to their estimation of how long it would take a repair crew to open an MOS.

[55] The Chinese might, for example, use some of these missiles to attack Yokota AB or Misawa AB if tanker sorties or other critical missions were being run from these locations. Also, Chinese leaders might want to hold missiles in reserve to counter later U.S. deployments (or for political leverage). Hence, even if the Second Artillery had 274 missiles available in 2017, it might, depending on the circumstances, use closer to 108.

To assess the sensitivity of the results to changes in individual variables, we ran a series of excursion cases to determine closure times for fighter operations (see Table 3.5). The excursions show high sensitivity to changes in at least some variables. By adjusting just two variables—reducing runway repair times to four hours and arrival rates to 50-percent closure time for fighter operations drops from four days to 0.6 days (an 85-percent reduction). Such brief windows of opportunity would not provide a Chinese planner with sufficient latitude to conduct operations free of U.S. Air Force interference.

In addition to demonstrating the sensitivity of results to uncertainty, the data in Table 3.5 also illustrate the importance to U.S. air operations of taking both active and passive measures to improve survivability and recovery. Against a series of attacks employing 108 missiles, for example, a combination of better missile defenses and improved runway repair could ensure that Kadena would be closed to operations for fewer than four days. Against an attack by 274 MRBMs, however, the base might be closed for ten days, even with improved runway repair capabilities and missile defenses. Denying Chinese battle damage assessment, however, would again reduce Kadena AB closure to three days. And if an additional runway were added to Kadena (or if an additional runway on Okinawa were used), then the closure time might be less than two days.[56]

Table 3.5
Days of Runway Closure to Fighter Operations at Kadena Air Base, Baseline and Excursion Cases

	MRBM Inventory Employed and Closure Times (days)			
	2010		2017	
Scenario	36 missiles	72 missiles	108 missiles	274 missiles
Baseline Fighter Operations Closure Time	4.0	10.0	16.0	43.0
Excursion 1: 50% missile arrival rate; 4-hour runway repair time	0.6	2.2	3.6	10.6
Excursion 2: Assumptions above, plus lack of Chinese battle damage assessment	0.3	0.6	0.9	2.3
Excursion 3: All assumptions above plus an additional runway at Kadena AB	0.2	0.4	0.6	1.5

NOTES: In the baseline case, we assume eight hours to repair a single runway cut, that 75 percent of missiles reach their targets, a 50-meter CEP, and perfect Chinese intelligence. The excursion cases include the listed changes off this baseline.

[56] It is worth noting here that the maximum range of the DF-15B (800 km) is sufficient to reach Kadena AB from a small slice of Eastern China. Therefore, these cases could alternatively be interpreted as situations in which a mix of DF-15Bs, DF-16s, and DF-21Cs are fired at Kadena AB.

Ensuring that the potential closure time of key bases remains manageable requires sustained attention. This is a major departure from the past two decades, during which U.S. planners could establish air bases with modest thought to defense, survivability, and recovery.

Parking Area Attacks

There are many similarities between attacking aircraft parked in the open and attacking runways. Both cases would involve the use of ballistic missiles armed with submunitions. However, the submunitions used to attack aircraft in the open can be smaller—on the order of one pound.[57] A DF-21C-class missile could carry hundreds of these submunitions, blanketing hundreds of square feet so that every aircraft parked in the area would have a high probability of being damaged. Depending on the payload and accuracy of the missile, the size of the parking area to be attacked, and the effectiveness of active defenses, a given inventory of missiles may be able to make multiple sweeps of a base's parking areas.[58]

For example, with an inventory of 108 MRBMs with CEPs of 50 meters, attacking parking areas at Kadena AB with submunition warheads optimized to hit fighter-sized aircraft, and facing no active defenses, the Second Artillery would be able to conduct four full sweeps of the parking area. One could, of course, mix this with a runway attack, using three-quarters of the inventory to shut the base for over a week while sweeping the parking area once.

Cruise Missile Attacks on Infrastructure

The ability of a cruise missile to successfully attack infrastructure targets, such as hangars, hardened aircraft shelters, or fuel tanks, depends on the size of the structure being attacked, the accuracy of the cruise missile attacking those targets, and the lethal radius of the cruise missile. There are 27 hangars and hardened shelters at Kadena AB.[59] Assuming that China has cruise missiles with CEPs of 10 meters, its low 2010 estimated inventory of 200 would suffice to make more than three full attacks on all hangars and hardened shelters at Kadena. (Two cruise missiles per target would be

---

[57] The U.S. BLU-42 submunition is an example. See Andreas Parsch, "BLU-42/B," *Designations of U.S. Aeronautical and Support Equipment*, last updated October 23, 2012.

[58] Jane's assesses that China has equipped the DF-15 and DF-21 with submunition warheads but does not specify the types of submunitions it has tested. Therefore, this analysis uses a representative submunition warhead design that would be well within Chinese capabilities. A 500-kg payload with 75 percent of its weight devoted to submunitions could carry 825 one-pound bomblets that could blanket a 900-foot-diameter circle such that every fighter-sized target would be very likely to sustain damage. Spreads customized for larger aircraft could blanket an even larger area. See Stillion and Orletsky, *Airbase Vulnerability to Conventional Cruise-Missile and Ballistic-Missile Attacks*, 1999, pp. 5–28.

[59] The 15 hardened aircraft shelters at Kadena were not designed to defeat a modern precision-guided weapon. While they would keep a fighter safe from the small submunition attack targeting aircraft in the open, they likely could not stop a cruise missile flying at Mach 0.8 carrying 400 kg of high explosives.

necessary to have a greater than 90-percent chance of destroying the shelters.)[60] This calculation does not account for cruise missile defenses, and cruise missiles are generally easier to shoot down than ballistic missiles because of their slower speed. Nevertheless, the prospects are grim. Defenses might be struck with ballistic missiles as part of an integrated attack. And even with the smallest estimate of China's current cruise missile inventory, there are more than seven cruise missiles for every fixed hangar and shelter on Kadena.[61]

Cruise missiles could also be used to attack fuel stores. Open-source imagery analysis identifies six partially buried circular fuel tanks at Kadena, each 50 meters in diameter. A single cruise missile with ten-meter CEP has a greater than 90-percent chance of hitting a target of this size. A salvo of 60 DH-10s could target every hangar, hardened aircraft shelter, and identified large fuel tank, such that each (individually) would have a greater than 90-percent chance of being hit. Needless to say, if the United States has (or could) bury or otherwise hide fuel tanks or other parts of the critical logistical infrastructure, that would limit the damage somewhat.

Shifting focus from Kadena AB to Andersen AFB, the only current conventional missile threat to Guam is ALCMs fired from H-6 bombers. Using cruise missiles to deliver small submunitions against aircraft in the open is less effective than using ballistic missiles, but it can be done.[62] An attacker would require 33 arriving DH-10s to blanket the parking areas at Andersen AFB. Open-source imagery also suggests that there are six hangars on Guam that could be targeted, and a salvo of 20 missiles would provide a high assurance of destroying them all, even if 33 percent of the cruise missiles failed or were shot down.

The 53 missiles required to attack both hangars and open-air aircraft parking ramps could be delivered by 27 H-6H or as few as nine H-6K bombers. Such an attack would, however, force the bombers to fly beyond the unrefueled combat radius of escort fighters, which would potentially make the H-6s highly vulnerable to intercept by U.S. aircraft, depending on the course and status of the larger air superiority battle. In the case of Andersen AFB, U.S. fighters might also be put on standby to scramble and destroy cruise missiles in flight. Finally, as in the Kadena case, additional

---

[60] These are straightforward calculations using probability-of-kill equations based on weapon CEP and target radius. See Irving Lachow, *The Global Positioning System and Cruise Missile Proliferation: Assessing the Threat*, Discussion Paper 94-04, Cambridge, Mass.: School of Government, Harvard University, June 1994.

[61] Cruise missile defenses include SAMs (such as the surface-launched advanced medium-range air-to-air missile [SLAMRAAM] or Patriot), rapid-firing gun systems (such as the Phalanx close-in weapon system), or airborne fighter aircraft.

[62] Cruise missiles are less effective submunition-delivery vehicles due to their generally smaller payloads. The U.S. BGM-109 Tomahawk Land Attack Missile (TLAM) D variant carries 166 submunitions and can blanket a roughly 200- by 400-meter area. Jane's assesses that there are submunition variants of the DH-10, and we use the TLAM-D as a proxy for the DH-10 submunition variant. *Jane's Strategic Weapons Systems*, "C-602 (HN-1/-2/-3/YJ-62/X-600/DH-10/CJ-10/HN-2000)," May 12, 2015.

ground-based defenses might be established to attempt to destroy cruise missiles as they approach.

If the Second Artillery were to develop a conventional IRBM, however, the threat to Andersen AFB would significantly increase. Andersen AFB has generally been used to support larger aircraft, such as bombers or tankers. As discussed earlier, the MOS requirements for these aircraft are longer and wider than those for fighters, and ballistic missiles armed with appropriate warheads are well suited to cutting runways. Thus, even though Andersen AFB has two runways of more than 10,000 feet, only two sections of runway would need to be destroyed to deny a large aircraft its MOS (see Figure 3.3).

With an inventory of just 50 IRBMs, China could keep Andersen AFB closed to large aircraft for more than eight days (assuming missile reliability of 75 percent and eight-hour repair times), even if the PLA is denied battle damage assessment of the

Figure 3.3
Andersen Air Force Base

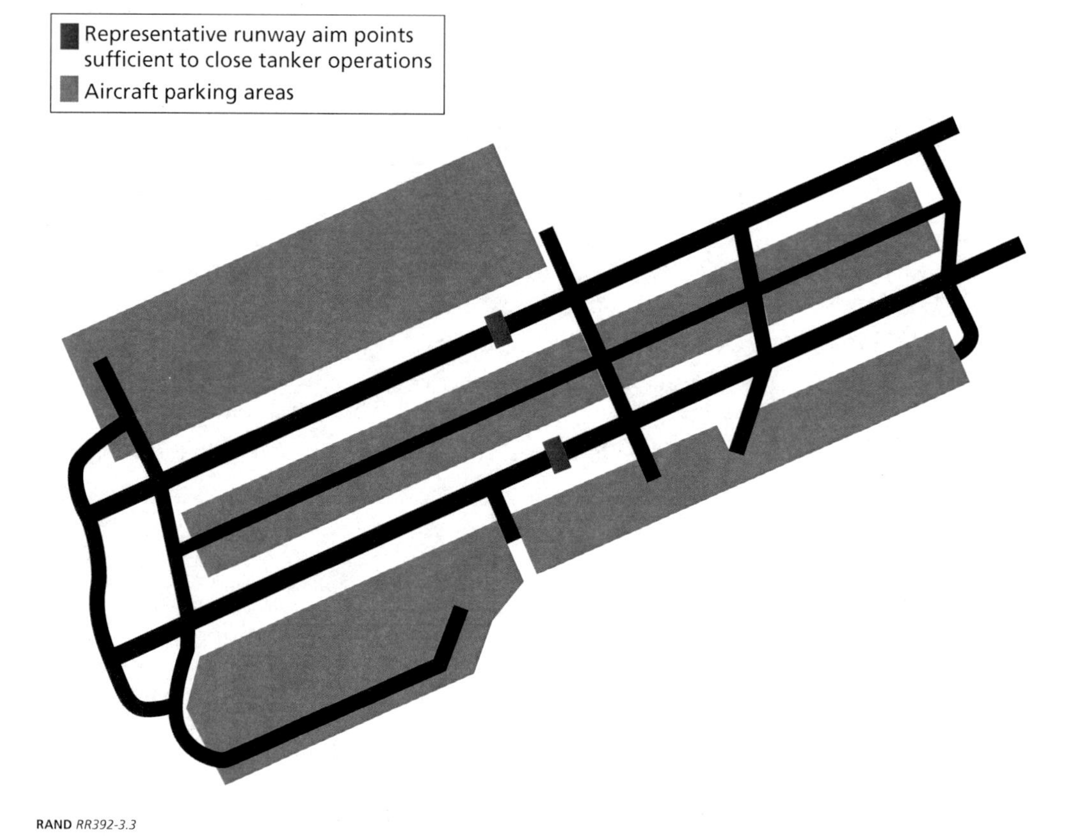

RAND *RR392-3.3*

success of its strikes.[63] Parking-area attacks would be even more effective than in the Kadena case; each missile could distribute its submunitions in a wider pattern because the target aircraft are larger.[64] Warning times would be shorter for ballistic missiles than for cruise missiles, increasing the chance of catching aircraft on the ground. With 100 IRBMs, the PLA could make a full sweep of all unsheltered aircraft parking areas and then use the rest of its inventory to keep Andersen shut to large aircraft for 11 days.

A Note on Mixed Attacks

Although the previous sections considered runway, parking-area, and infrastructure attacks in isolation, in an actual conflict, an intelligent adversary would structure an attack in whatever manner seems to offer the best chance of shutting down the air base. Because the U.S. Air Force needs an aircraft, fuel, and a piece of runway that meets the MOS requirement of the available aircraft to conduct a sortie, destroying at least one of these elements would prevent the aircraft from flying. A conservative attacker might attack all elements in parallel, allocating, for example, 108 of a 274-MRBM inventory for runway attacks, using another 108 to conduct four sweeps of the parking areas, and keeping the remaining 58 missiles in reserve. Another option available to an attacker would be to allocate some TBMs to attack missile defense assets to increase the ability of follow-on salvos (including cruise missiles) to reach their targets.

Operational Implications

Kadena AB is the closest U.S. Air Force base to the Taiwan Strait (see Table 3.3, earlier in this chapter). It is a sizable air base, with two 12,000-foot runways, and the 18th Wing, 5th Air Force, is permanently stationed there. These factors would make it an attractive site from which to base during a war over Taiwan. China's ability to shut down operations at the base for two weeks at the outset of a conflict would potentially disrupt a host of other key activities. For example, Navy fighters flying from carriers positioned outside the DF-21D ASBM range could be denied the tankers they need to reach the Taiwan Strait if these aircraft are destroyed on the ground by submunitions or if anti-runway attacks deny them their MOS. An air-to-air war over Taiwan would also be complicated if the fighters that the U.S. Air Force planned to use were destroyed on the ground or unable to take off due to damaged runways (see scorecard 2).

---

[63] These results assume a missile CEP of 50 meters, a 500-kg warhead armed with the same 82 anti-runway submunitions as discussed in the Kadena AB case.

[64] Each missile could distribute submunitions over a circle with a 1,148-foot radius, whereas they would need a 710-foot radius to target fighters.

Operational Analysis: Spratly Islands Scenario

The operational conditions relevant to the Spratly Islands scenario are quite different from those in the Taiwan scenario. In addition to the large, permanent bases discussed earlier, we postulated that, during this scenario, U.S. forces could operate from two additional bases in the Philippines. While these bases would be central to the U.S. strategy in a Spratly Islands conflict, our analysis suggests that they also have serious limitations.

Tambler AB and Antonio Bautista AB have a combined parking area of roughly 1,200,000 square feet, less than 4 percent the size of the parking area at Kadena (which is more than 27,700,000 square feet). Furthermore, Antonio Bautista AB has only one approximately 8,000-foot-long runway and Tambler AB has only one approximately 10,000-foot-long runway. Hence, should the Chinese attempt to attack the runways at these bases, three aim points would be sufficient to deny both fighter and tanker operations at both. Tambler AB lies outside of DF-21C range, meaning that China would have to use air-launched DH-10s to attack it. An inventory of 72 DF-21Cs with 50-meter CEPs could keep Antonio Bautista AB shut for more than 11 days.[65]

Given that the Philippine bases are so small and the number of fighters at each base in this scenario is correspondingly low (12 F/A-18s at Tambler AB and five assorted aircraft at Antonio Bautista AB), the PLA could well decide that it makes more sense to use DH-10 attacks to destroy the aircraft on the ground rather than cutting the runways. Five submunition-armed DH-10s would be sufficient to blanket all the parking areas at the two bases, which would enable the PLA to preserve its DF-21C force as a hedge against escalation. Allocating 15 DH-10s for each attack (three per target), the PLA could attack all parking areas three times and using only a small portion of its likely DH-10 inventory, which could be much more appealing than using its scarcer DF-21C missiles.

To attack Tambler AB, however, the DH-10s would need to be air-launched from H-6 bombers. The H-6 bombers might be vulnerable to U.S. Navy and Air Force combat air patrols, depending on the escort package the PLAAF could muster to defend its H-6s. And the longer flight time of the DH-10s relative to that of DF-21Cs could give aircraft more warning and potentially allow them to scramble and escape before the missiles arrived. Finally, the slower speed of cruise missiles also could increase the effectiveness of U.S. missile defenses, to include defensive fires by whatever fighters were airborne at the time the missiles arrived.

The larger balance of capabilities would of course be a critical factor in deciding the ultimate success impact of a Chinese missile campaign. We have just discussed the growing Chinese capability to attack U.S. bases with ballistic and cruise missiles. But

[65] These runway attack results are based on assumptions similar to those in the baseline case in the Kadena AB attack results, discussed earlier.

as scorecard 4 suggests, U.S. offensive capabilities against China's basing infrastructure in the Spratly scenario are also very robust (see Chapter Six) and the air balance is significantly better in the Spratly scenario than in the Taiwan one. H-6 bombers might therefore face more daunting challenges reaching launch points for ALCMs without being engaged by U.S. aircraft.

Chinese land reclamation in the South China Sea, the construction of small air bases there, and the positioning of SAM systems on those islands are unlikely to provide decisive military advantage for China, though they could conceivably provide a (likely temporary) buffer against U.S. airpower that might make the operation of Chinese bombers less risky.

Conclusions

In its operations over the past 20 years, the U.S. military has had the luxury of setting up secure main operating bases close to the site of conflict from which it could generate a large number of aircraft sorties. Indeed, although the United States planned to operate from air bases that could be subject to attack during the Cold War, it has not actually done so since World War II. However, the increasing quantity and quality of Second Artillery missiles means that the United States must again think about operating while under attack. Moreover, the challenge posed by modern, accurate ballistic and cruise missiles is fundamentally different from the threat posed by the Soviets in the 1980s. The threat then was from unguided bombs and inaccurate Scud missiles. Against this threat, a combination of point defenses, offensive counter-air activities, dispersal, camouflage, and hardening offered the hope of survival.

Precision weapons pose a different challenge, however. In the face of increasing Chinese missile inventories, the United States will want to pursue a portfolio approach to build a robust basing posture. An intelligent attack would attempt to identify and target the weakest link in the system of an air base, so a balanced portfolio of investments should include combinations of dispersal, base hardening, missile defenses, and new operating techniques to limit the impact of attacks on flight operations.

Dispersal would create more runways that the Chinese would have to attack and, if combined with camouflage and efforts to impede Chinese surveillance capabilities, would make it more difficult for the PLA to target aircraft on the ground. Hardening would force the Chinese to individually target aircraft parked inside shelters rather than wholesale on the tarmac. Aircraft in shelters are not visible from the sky or space, so they might also create uncertainty about the location of U.S. assets. Even missile defenses that could destroy only a fraction of incoming missiles could force the Chinese to allocate more missiles per salvo, depleting the Second Artillery's inventory faster. Preparing to use such operating techniques as arresting cables to shorten the MOS required for an aircraft to land could increase the number of aim points that would have to be hit to effectively close U.S. runways. These changes would require a concerted effort, and they would take time to implement. By the time an actual conflict starts, it will be too late to make these adjustments.

In addition to base resiliency measures, it is possible that strong performance in the areas addressed by the other scorecards could improve the ability of U.S. air bases to weather attacks. If the Navy can successfully cope with threats from Chinese submarines and DF-21D ASBMs, and if it could operate close to China (see the scorecard 5 discussion in Chapter Seven), then carrier-based fighters could prevent H-6 bombers from reaching the release arcs necessary to attack Andersen AFB with ALCMs—even if Chinese ballistic missiles had suppressed sorties from Kadena AB. If U.S. airpower could successfully penetrate Chinese airspace (scorecard 3, Chapter Five) to attack Chinese air bases (scorecard 4, Chapter Six), then H-6s might be neutralized on the ground. However, it should be noted that achieving both of these conditions would be easier if U.S. aircraft were able to operate from forward air bases, and that ability would depend, in part, on the survivability of air bases.

The primary takeaway from this analysis is that China is increasing its ability to threaten U.S. forward air bases, challenging the way the United States has conducted military operations for the past 20 years. If China develops a conventional IRBM able to reach Andersen AFB—as it says it intends to—then every U.S. base in the Western Pacific would be within range of a conventional ballistic missile. This does not mean that China can pursue its operations at will or that it will dominate the battlespace. Rather, the improvements made by the Second Artillery mean that the U.S. military will no longer be able operate from secure bastions during crises and conflict. U.S. military leaders must think creatively about how to maintain the ability to conduct air operations in a high-threat environment.

Scorecard Coding

Figure 3.4 provides our summary coding of the results of scorecard 1, "Chinese Attacks on Air Bases." As is the case for the coding of the other scorecard results, the assessment is based on selective criteria. We urge readers to consider the results in the context of the broader analysis presented in this chapter and in the context of the other scorecards. Advantage in this scorecard is evaluated in terms of the Chinese ability to disrupt flight operations at relevant U.S. bases during the early phases of each conflict to a sufficient degree that it calls into question the ability of the United States to prevail in an air campaign.

In 1996 and 2003, the Chinese lacked significant capability to attack or close U.S. bases outside of Korea, and we coded both scenarios as major U.S. advantages. Given the importance of Kadena AB in a Taiwan contingency and the Chinese ability to cause at least a temporary closure of that facility by 2010, we coded the Taiwan

Figure 3.4
Scorecard 1 Summary Coding

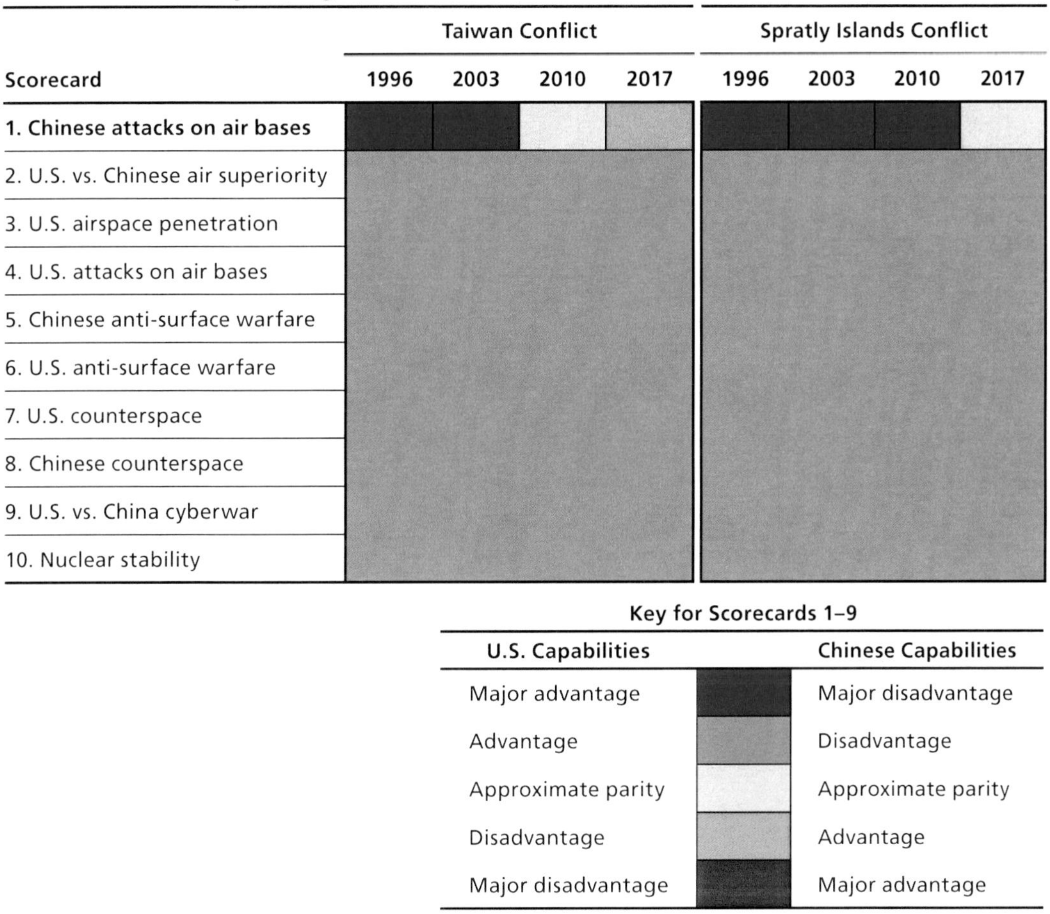

	Taiwan Conflict				Spratly Islands Conflict			
Scorecard	**1996**	**2003**	**2010**	**2017**	**1996**	**2003**	**2010**	**2017**
1. Chinese attacks on air bases								
2. U.S. vs. Chinese air superiority								
3. U.S. airspace penetration								
4. U.S. attacks on air bases								
5. Chinese anti-surface warfare								
6. U.S. anti-surface warfare								
7. U.S. counterspace								
8. Chinese counterspace								
9. U.S. vs. China cyberwar								
10. Nuclear stability								

Key for Scorecards 1–9

U.S. Capabilities		**Chinese Capabilities**
Major advantage		Major disadvantage
Advantage		Disadvantage
Approximate parity		Approximate parity
Disadvantage		Advantage
Major disadvantage		Major advantage

scenario as *approximate parity* in that year. In 2017, Kadena might be closed for two weeks or more while Guam's Andersen AFB will also begin to face significant threat, shifting the Taiwan scenario coding to *Chinese advantage*.[66]

In the Spratly scenario, U.S. air forces can operate from bases farther from Chinese offensive systems. The difference in distance between, for example, Kadena AB and Andersen AFB to the Spratly Islands is proportionately less than the difference in their respective distances to Taiwan. Given that Andersen is under a significantly lower threat than Kadena, the proportionate loss of sortie generation would be less significant than in the Taiwan case. Moreover, the United States could base its air assets in the Philippines itself, and at least some of the possible basing locations would not

[66] The likelihood that Andersen AFB would not be closed for the duration and that U.S. air operations would continue, if at a degraded rate, from multiple air bases spread throughout the Asia-Pacific prevents this scorecard from being coded as a major Chinese advantage.

be within range of Chinese DF-21Cs. Tankers operating out of Australia could, given the somewhat reduced distances and (more importantly) smaller number of fighters involved, play a relatively larger role in the Spratly Islands scenario. Hence, we coded the scenario as U.S. advantage through 2010 and approximate parity in 2017 as more U.S. basing locations begin to fall within range of Chinese systems.

CHAPTER FOUR

Scorecard 2: Air Campaigns Over Taiwan and the Spratly Islands

One of the key threats facing Taiwan in any conflict with China is the possibility of sustained air attacks by a survivable and effective Chinese air force. The PLAAF has not historically posed much of a threat to neighboring countries. However, as the force upgrades to aircraft with air-to-ground sensors and precision munitions, it may become a greater threat to Taiwan than the ballistic missile inventory that is typically the focus in discussions of such a scenario. The shift in Chinese procurement priorities has been supported by a transformation in Chinese operational thinking and writing about airpower. The PLAAF is moving away from its traditionally heavy emphasis on territorial air defense and toward a more independent strategic role that combines offensive and defensive action.[1]

To be sure, ballistic missiles will continue to play a key role in, for example, attacking U.S. and Taiwanese SAM sites and air bases. But a U.S.-style strategic bombing campaign by China using hundreds of sorties could deliver more munitions each day against targets in Taiwan than the total SRBM inventory. Given the size of the PLAAF's aircraft inventory and the magnitude of the ballistic and cruise missile threat to Taiwan's air bases, the Taiwanese air force may find itself heavily reliant on the U.S. Air Force and Navy to defend its airspace.

Similarly, in a conflict in the Spratly Islands, China would likely attempt to use its increasingly capable long-range aviation assets to strike targets or to protect its naval and ground forces engaged in the local area. This chapter examines trends in relative U.S. and Chinese air-to-air capabilities and the ability of U.S. forces to prevail in different types of air superiority campaigns, disrupting and defeating enemy air-to-ground missions and protecting U.S. and coalition forces from air attack.

We begin with a discussion of the methodology employed. We then assess the dynamics and possible outcomes of an air war associated with a Taiwan scenario, outlining the evolution of China's force structure, evaluating the capabilities of PLA and U.S. aircraft, and assessing the outcome of an air-to-air campaign run in conjunc-

---

[1] For general background on the PLA Air Force, see Hallion, Cliff, and Saunders, *The Chinese Air Force*, 2012. On PLA airpower employment concepts, see Cliff et al., *Shaking the Heavens and Splitting the Earth*, 2011.

tion with a Chinese invasion of Taiwan. We then conduct a similar assessment of the Spratly Islands scenario.

We conclude that gaining air superiority, especially in the Taiwan case, has become far more challenging, and gaining full air superiority in a short war, or in the first few weeks of a longer one, is far from certain. This is not to say that Chinese aircraft and pilots will best U.S. air forces in combat or achieve favorable kill ratios (though expected losses to the U.S. side would be higher in 2017 than in 2003 or 2010), but, rather, that prevailing quickly enough to influence the naval and ground battle in and around Taiwan will be difficult. The air balance in a Spratly Islands scenario remains robust, but even here, the task is becoming somewhat more difficult.

We do not attempt to link the results of modeling of missile attacks against U.S. air bases from Chapter Three (scorecard 1) to the analysis of air-to-air combat.[2] However, even without explicitly considering ballistic and cruise missile attacks, which could have a major impact on U.S. air base operations, gaining air superiority over Taiwan would likely not be easy or quick. Put together, these two scorecards suggest that, at least in the early stages of a conflict near the Chinese mainland, air superiority would be hotly contested and the ability of U.S. aircraft to assist with the ground fight might be limited.

Methodology

For this analysis, we used the air-to-air module of a large campaign model developed by RAND, including robust algorithms for adjudicating air-to-air combat.[3] We developed the inputs to the model to produce plausible kills per sortie, using exchange ratios based on open-source weapon performance and historical experience.

As Figure 4.1 illustrates, China can bring a large force structure to bear in this scenario. Later, we discuss in more detail how PLAAF and PLAN aviation aircraft can operate from about 40 air bases within unrefueled fighter range of Taiwan (about 800 km). The PLA maintains 18 regiments of fighters and bombers, typically with 24 aircraft each in the case of fighters, in the Nanjing Military Region alone, and would almost certainly call on reinforcements from other regions in the event of conflict.[4] Even if China holds back a substantial number of fighters for defensive counter-air (DCA)

[2] As noted in Chapter One, we do not attempt to create a single model that integrates the scorecards into a single campaign model. Our emphasis is on presenting discreet and transparent pieces of analysis that inform an understanding of the larger problem.

[3] We employed the Simplified Tool for Analysis of Regional Threats, which is a large campaign model. In this case, we used only the air-to-air module, which incorporates modified Lanchester algorithms.

[4] This figure includes PLAAF inventory in the Nanjing region and PLAN air units assigned to the East Sea Fleet. IISS, *The Military Balance*, 2015, pp. 244–245.

Figure 4.1
Taiwan Scenario Air Campaign

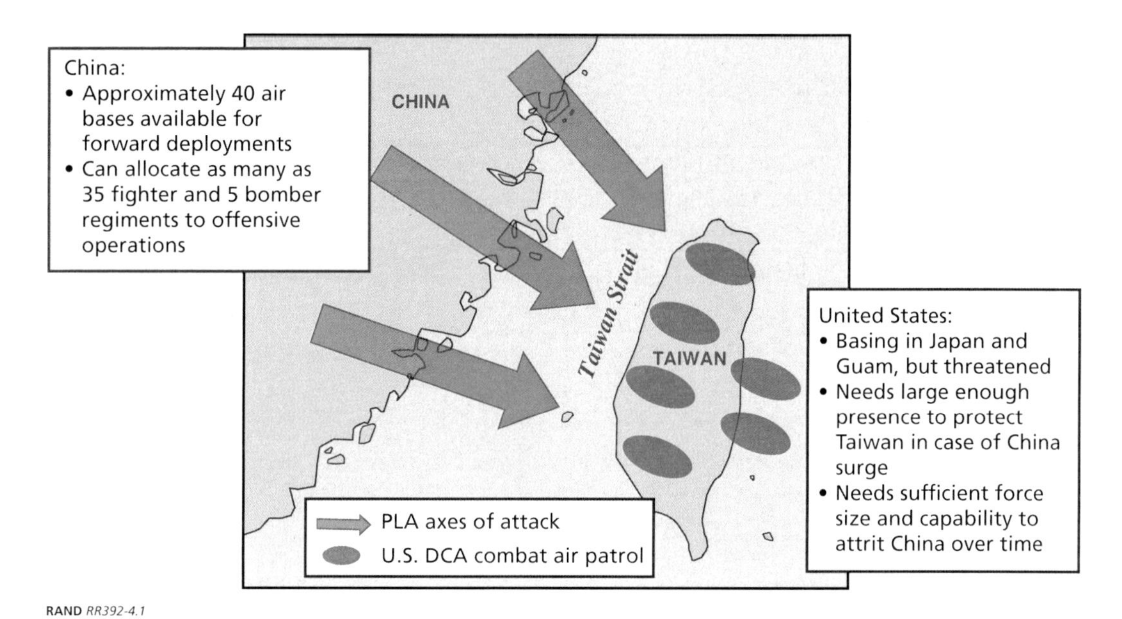

RAND *RR392-4.1*

missions, as we postulate in our scorecard 3 analysis (see Chapter Five), a force structure of up to 800 aircraft could be dedicated to a Taiwan campaign.[5]

In contemplating deterrence or warfighting in a Taiwan scenario, U.S. commanders would want airpower to perform two types of tasks. First, U.S. Navy and Air Force aircraft (hereafter simplified as *air forces*) would want to be able to maintain a large enough continuous presence in Taiwan's airspace to successfully disrupt a large "surge" of strikes across the country. This would take the PLA air forces out of the short-term equation in the ground and naval battle. Second, U.S. commanders would want to destroy enough Chinese aircraft over time to compel or convince the Chinese to abandon their air campaign before Taiwan capitulates. This capability will be a function of the size of the respective air forces and the kill ratios achieved.

Ideally, U.S. air forces would be able to perform both tasks, but, at a minimum, they would want to perform the second task. Prevailing in a campaign, however short, will generally be less demanding than maintaining the forces necessary to defeat a surge. Assuming that attrition objectives can be achieved quickly enough, a favorable strategic outcome will nevertheless be achieved, though admittedly at higher cost to Taiwanese forces battling on the ground. All things considered, in the context of a

[5] Most likely several hundred of these aircraft would, at least initially, be held in reserve. However, the majority are likely to be shorter-range systems, leaving the most modern fourth-generation aircraft available to attack Taiwan and the South China Sea.

fight over Taiwan, the second task will be a more realistic objective, though the question of whether U.S. air forces can defeat a surge remains relevant.

Our analysis examines the size and capability of U.S. forces necessary to accomplish these two tasks against Chinese forces present in 1996, 2003, 2010, and 2017. In an air campaign in which both sides could bring all their forces to bear and basing is unlimited, the United States would—given the number and capability of its aircraft and pilots—prevail quickly and easily. In the real world of the Pacific, however, only a fraction of U.S. air forces are deployed at any given time, basing is limited to a handful of locations, and all of the bases near enough to Taiwan to provide easy access to the battle area are under threat from missile attack. Hence, assessing the size of the force structure required for the United States to prevail and matching that force structure against available basing (or a potential range of available basing), provides a more meaningful sense of the air-to-air balance than merely looking at exchange rates or total losses.

The Spratly Islands scenario (shown in Figure 4.2) is similar in terms of overall objectives. However, the geography of the South China Sea makes the challenges for

Figure 4.2
Spratly Islands Scenario Air Campaign

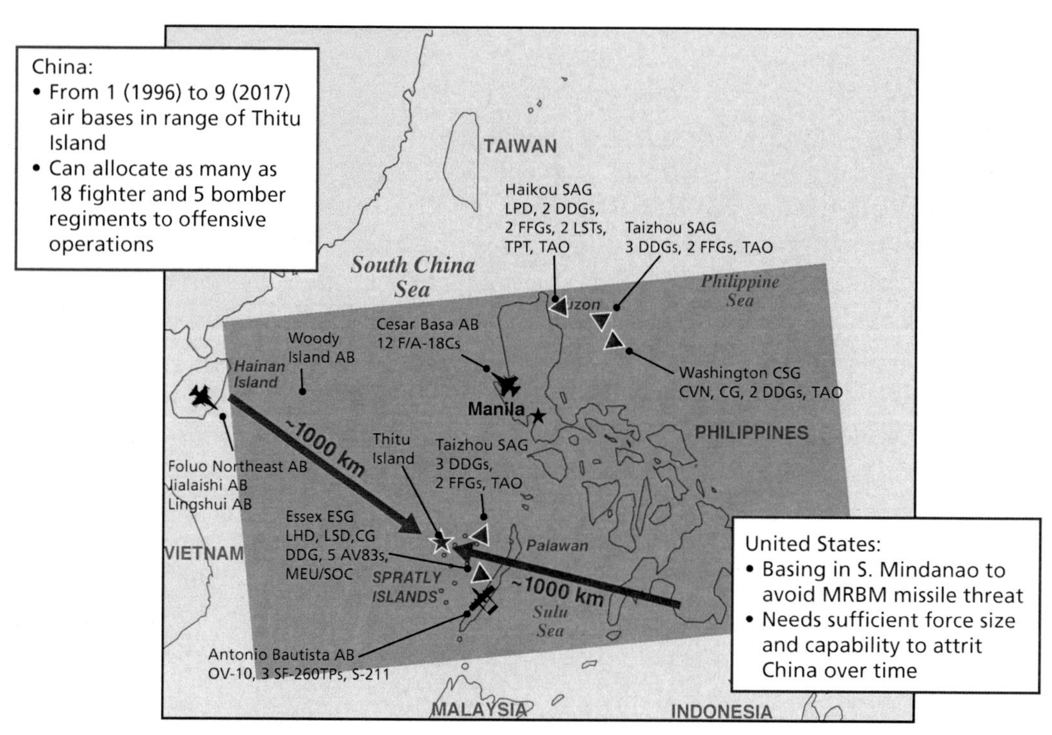

NOTES: Shaded area represents the air/missile exclusion zone declared by China. CG = cruiser. CSG = carrier strike group. ESG = expeditionary strike group. LST = landing ship, tank. MEU/SOC = Marine expeditionary unit, special operations–capable. SAG = surface action group. TAO = oiler. TPT = transport ship.

RAND *RR392-4.2*

each side different. Although the scenario we postulate is centered on Thitu Island in the Spratly chain, the island itself is quite small, and it would require only a few sorties by China to totally destroy the infrastructure on the island (a poorly surfaced 4,000-foot runway and a few dozen buildings). However, a landing of military forces onto the island would require significant air cover, particularly if opposed by a reinforced Philippine garrison or U.S. naval forces. These factors imply that, in this scenario, defending against a short-term surge operation may be more relevant than a protracted attrition campaign.

Taiwan Scenario

From an operational air-to-air perspective, the Taiwan scenario is the more demanding of the two scenarios. The challenge lies more in geography than in the number or quality of equipment or personnel, though rapid improvements in China's force structure enable it to exploit geographic advantage. While the United States is limited to a handful of bases, most of which are distant from the conflict area, China has roughly 40 bases from which it can conduct unrefueled operations over Taiwan. In this section, we discuss both sides' force structure, evaluate the capabilities of the forces, and present the results of our modeling of the Taiwan air-to-air campaign.

Force Structure Inputs

Table 4.1 summarizes the total inventory of PLA (air force and navy) fighter and strike aircraft in each snapshot year, as well as current figures for early 2015. Whereas China had just taken delivery of its first handful of fourth-generation fighters in 1996, by 2010, modern fourth-generation variants accounted for almost 30 percent of the force. By 2015, the figure was 51 percent, and by 2017, we estimate it will reach roughly 62 percent. Between 2010 and 2015, China's inventory of fourth-generation fighters increased from 383 to 736, yielding an impressive average of 70 modern fighters added per year.

In modeling the Taiwan air campaign, we calculated that there is sufficient basing within proximity of Taiwan for China to employ as many as 1,000 fighters, or roughly 80 percent of the inventory available since 2003.[6] (In 1996, China possessed thousands of obsolete fighters, and a smaller percentage of these would have been able to engage in the fight.) Of this total, we assume that 400 aircraft are dedicated to defensive roles over Chinese territory. As we discuss in Chapter Five (scorecard 3), we determined the size of this defensive force by calculating the number of combat air patrols required to

---

[6] China frequently redeploys air units from one air base to another for training exercises, so we assume that they are able to redeploy, as appropriate, for the two scenarios, with the primary limiting factor being total available capacity at Chinese military air bases. While this may exaggerate Chinese flexibility, we do limit deployment to military air bases (as opposed to civilian air fields), an assumption that might understate Chinese flexibility.

Table 4.1
PLAAF and PLAN Combat Aircraft Inventory, 1996–2017

Aircraft	1996	2003	2010	2015 (current)	2017
Air superiority aircraft					
2nd generation					
J-5 (FTR)	400	—	—	—	—
J-6 (FTR)	3,300	550	—	—	—
3rd generation					
J-7 (FTR)	570	700	588	528	450
J-8 (FTR)	130	232	360	168	100
4th generation					
J-10 (FTR)	—	—	150	294	350
Su-27/J-11 (FTR)	24	100	136	340	400
Su-30 MKK/J-16 (FGA)	—	58	97	97	121
J-15	—	—	—	5	30
Strike and bomber aircraft					
H-5 (bomber)	430	90	20	—	—
H-6 (bomber)	145	138	132	136	140
Q-5 (strike)	500	330	150	120	100
JH-7 (fighter-bomber)	—	18	156	240	240

SOURCES: IISS, *The Military Balance*, 1996, 2003, 2010, and 2015.

cover major cities and fill gaps in SAM coverage. Chinese writings on air campaigns stress that even in an "offensive air campaign," defensive tasks, including air intercept, remain important.[7]

Of the remaining 600 or so aircraft that could potentially be used in an offensive role, we attempted to keep a roughly 2-to-1 ratio between air-to-air–equipped escort sorties and air-to-ground attack sorties. This ratio generally reflects the numeric breakdown in capabilities by PLA aircraft type and by mission-specific regiments. It should also provide fairly robust protection for strike aircraft against U.S. DCA operations over Taiwan. Table 4.2 summarizes the number of PLAAF and PLAN aircraft employed in the model for offensive (strike and escort) purposes. Note that the aircraft totals by mission do not necessarily have a 2-to-1 ratio, since the calculation is

[7] Cliff et al., *Shaking the Heavens and Splitting the Earth*, 2011.

Table 4.2
PLAAF and PLAN Aircraft Quantities and Missions Used to Model the Taiwan Scenario

Aircraft	1996	2003	2010	2017
Air-to-air escort fighters				
2nd generation				
J-5 (FTR)	—	—	—	—
J-6 (FTR)	312	144	—	—
3rd generation				
J-7 (FTR)	288	336	216	—
J-8 (FTR)	30	58	134	96
4th generation				
J-10 (FTR)	—	—	—	192
Su-27/J-11 (FTR)	4	72	120	264
Su-30 MKK/J-16 (FGA)	—	48	72	72
Total	**634**	**658**	**542**	**624**
Strike and bomber aircraft				
H-5 (bomber)	20	20	—	—
H-6 (bomber)	100	100	120	130
Q-5 (strike)	192	144	48	—
J-8 (strike variant)	90	134	34	24
JH-7 (fighter-bomber)	—	—	120	192
Total	402	398	322	356

based on the number of sorties, and some aircraft types and basing locations provide a higher sortie rate than others.

We used the U.S. force structure required to prevail in the scenario as an output rather than as an input. In constructing the required force, we first built forces using carrier-based air—up to a maximum of 72 aircraft. This equates to approximately two CSGs of aircraft, although with force protection and other missions, a third CSG may be required to provide 72 aircraft over Taiwan.

Next, we used fifth-generation fighters in the 2010 and 2017 time frames. For 2010, we assumed that one wing (with the canonical 72 combat-coded aircraft) would be available to deploy to the theater, and, for 2017, we assumed that two wings could deploy. Although we did not attempt to differentiate the specific capabilities of the

F-22 and F-35 for this study, our 2017 deployable force would likely consist of both types of aircraft.[8]

Finally, we added fourth-generation fighters until the victory condition was achieved. Because we are examining a defensive counter-air mission, the primary U.S. aircraft types to consider are the U.S. Navy F/A-18 (C/D or E/F models, depending on the year) and the U.S. Air Force F-15C. As the results show, in some cases more aircraft are required than the planned quantities of these types; hence, additional aircraft types (most likely the U.S. Air Force F-16C) may be necessary.

Assessing Sufficiency

Having arrived at a necessary force structure to achieve U.S. objectives, we measure that force against the U.S. capability to maintain those forces in the operational area. Sufficiency may be measured against two standards. The first and most liberal measure would be the total size of the U.S. Air Force fighter inventory, which numbered approximately 17 wings in 2010. (We assume 72 aircraft per wing throughout this analysis.) The second and more restrictive measure is the capacity of U.S. bases within practical flight distance of Taiwan. This figure is more difficult to ascertain, given that fighters would compete for space with a variety of other U.S. aircraft, including bombers, tankers, mobility aircraft, AWACS- and Joint Surveillance Target Attack Radar System–equipped aircraft, and electronic warfare assets. Some of these requirements are addressed further in Chapters Five, Six, and Seven (scorecards 3, 4, and 5).

One assessment suggests that although there are sufficient parking areas at Andersen AFB, Guam, for 250 aircraft, space available to fighters would not likely exceed four to five squadrons, or roughly one and a half wings (roughly 100–125 aircraft).[9] Given its size and location, Andersen would likely base a disproportionate number of tankers and other support aircraft. A relatively larger proportion of space at bases in Japan, which tend to be smaller, might be reserved for fighters. Assuming a single wing at Kadena AB, one at Misawa AB, and one and a half at Andersen AFB, the United States might base roughly 3.5 U.S. Air Force fighter wings in the area without undue stress on basing capacity, plus an additional contingent at sea.

By squeezing additional fighters onto U.S. bases where fighter wings are now located (Andersen AFB, Kadena AB, and Misawa AB) and stationing additional squadrons on other U.S. air facilities (e.g., Yokota AB and the two MCASs in Japan), the total upper limit might conceivably be raised to six wings of ground-based fighters plus tanker support. With allowance for an additional 72 U.S. Navy fighters (or slightly

---

[8] Although the F-35 may not have reached full initial operating capability (IOC) by 2017, wartime exigency might see it deployed to the theater for operational purposes.

[9] Note that the flight operations of five squadrons of F-22s would require the support of some 68 tanker aircraft, each taking up roughly twice the space of a fighter. Five squadrons would be sustainable only if no bombers were based at Andersen. Eric Stephen Gons, *Access Challenges and Implications for Airpower in the Western Pacific*, dissertation, Santa Monica, Calif.: Pardee RAND Graduate School, RGSD-267, 2011, pp. 81–82.

more than the equivalent of a single U.S. Air Force fighter wing) on board two aircraft carriers, the United States might, depending on circumstances, be able to maintain between 4.5 and seven fighter wings within range of Taiwan.

Other factors might increase or decrease total U.S. basing capacity. Gaining the ability to use Japanese Air Self-Defense Force bases or civilian airfields (for support aircraft if not fighters) could further expand the total number of deployable wings. The Guidelines for U.S.-Japan Defense Cooperation, released in April 2015, stipulate that "in order to expand interoperability and improve flexibility and resiliency," U.S. and Japanese armed forces will "enhance joint/shared use" of facilities.[10] On the other hand, as Chinese ballistic and cruise missile capabilities continue to develop, PLA attacks might succeed in at least temporarily inhibiting flight operations at some of these locations. Also, in the face of China's missile threat, the United States may choose to mitigate vulnerability by dispersing aircraft, limiting the inventory that could be absorbed even in the event that new locations become available.[11]

Force Capability Inputs

In addition to quantities of aircraft, the model also requires sortie rate and air-to-air capability inputs for each side. To calculate sortie rates, we used a simple model that accounts for mission distance, loiter time, aircraft turn time, crew duty day, and crew ratio. Allowable crew flight time is set by U.S. Air Force regulations that specify maximum hours of flight time and duty days for one-, seven-, 30-, and 90-day campaigns.[12] Our analysis accounts for the likely high intensity of the early stages by using the weekly limit for the first seven days of the campaign and then switching to the monthly limit.

Figure 4.3 plots sortie rate as a function of distance (base to loiter location) assuming a 1.25 crew ratio with weekly and monthly restrictions in place for a mission with a two-hour on-station time. Flying from Kadena AB on Okinawa (roughly 770 km from the center line between Taiwan and the mainland), U.S. aircraft could fly 1.6 sorties per day for the first seven days and 0.9 sorties per day thereafter. From Andersen AFB on Guam (roughly 2,870 km from the center line), aircraft could achieve rates of 0.8 sorties per day for the first seven days and 0.5 thereafter.[13]

---

[10] The Guidelines for U.S.-Japan Defense Cooperation, April 27, 2015, Section III.7.

[11] Some of the facilities located in in Japan are situated in particularly disadvantageous positions. Misawa AB, for example, is more than 2,600 km from Taiwan, or almost as far from the fight as Andersen AFB, but it lies just 850 km from the nearest Chinese territory—well within range of Chinese MRBMs. Some types of support operations might be conducted from U.S. bases in South Korea, but in addition to even more severe geographic disadvantages, the South Korean government would be extremely reluctant to permit the use of these bases for combat missions in a war over Taiwan.

[12] These standards are established in Air Force Instruction 11-202, Vol. 3, *General Flight Rules*, October 22, 2010.

[13] Note that fighter missions beyond 3,100 km are not possible because the duty day would exceed Air Force Instruction 11-202 restrictions.

Figure 4.3
Sortie Rate as a Function of Distance

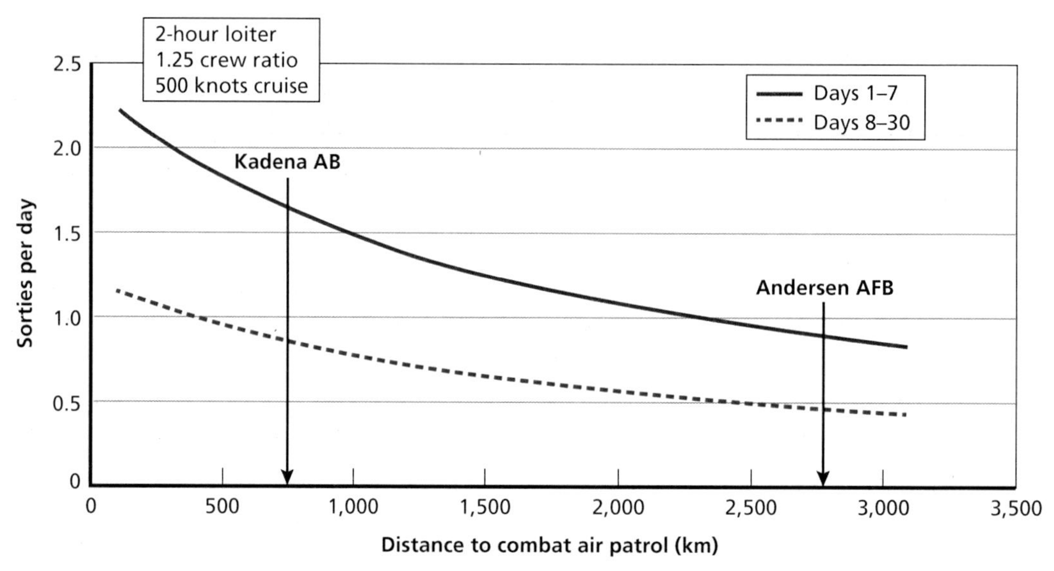

SOURCE: Air Force Instruction 11-202, Vol. 3, 2010.
RAND *RR392-4.3*

The figure highlights the dropoff in sortie generation capability when moving beyond the first week of a campaign, as well as the degradation with increasing range. Sortie rates could potentially be improved by increasing the crew-to-aircraft ratio, adding maintenance personnel and spare parts, and using waivers for flight hours, but this model appears to predict real-world sortie generation rates fairly accurately.[14] Since Chinese sortie-rate-capabilities are poorly understood, we simply used the same model for those sortie rates.[15] However, Chinese basing is typically between 200 km and 600 km from Taiwan, so overall sortie rates will be higher for the PLAAF and PLAN.

The final and most difficult inputs to the model involve air-to-air capability. The model requires two general types of capability inputs: overall "intensity" (typically, kills per sortie for air-to-air combat) and relative lethality and vulnerability for each aircraft type. For this analysis, we set overall Chinese capability at 70 percent of U.S. capability to account for differences in training and experience. Chinese pilots reportedly fly fewer hours than their U.S. counterparts, though they have narrowed the gap.[16] Simi-

---

[14] James A. Winnefeld, Preston Niblack, and Dana J. Johnson, *A League of Airmen: U.S. Air Power in the Gulf War*, Santa Monica, Calif.: RAND Corporation, MR-343-AF, 1994.

[15] China's maintenance capabilities are almost certainly not as good as those of the United States, but it would enjoy advantages operating from home fields close to resupply for spare parts and depot maintenance.

[16] IISS reports the difference at 100–150 flying hours for Chinese fighter pilots (up from just 24 hours per year during the height of the Cultural Revolution), compared with 160 for U.S. pilots. Much more significant is the U.S. Air Force's greater access to and use of sophisticated simulators and realistic air combat training. We note

larly, Chinese pilot training lacks the level of realism and complexity associated with pilot training in the United States, though PLAAF training has become more realistic over time.[17] Related to training, experience, and China's relatively hierarchical military philosophy are differences in operational practice. The PLAAF penchant for ground-controlled intercepts, for example, provides less opportunity for Chinese pilots to exercise initiative when encountering unexpected tactical situations. While the percentage value adjustment to effectiveness is subjective, we felt that a degree of difference was nevertheless justified by differences in the two systems.

The "kills per sortie" estimate is limited to fewer than two, based on a typical loadout of six to eight air-to-air missiles and historic missile probabilities of kill below 50 percent.[18] Note that this intensity can still produce large daily losses. For instance, if a wing of 72 U.S. aircraft is flying 1.5 sorties per day, the wing could potentially kill as many as 216 Chinese aircraft each day in the target-rich environment over Taiwan. This can be compared to the experience of the 8th Air Force in World War II, which lost up to 69 bombers on a single day.[19]

Since open-source data on aircraft performance are limited to a small number of parameters, we grouped aircraft types into classes with similar levels of capability and used an open-source tactical air combat model to examine the relative capabilities of each class.[20] Aircraft and weapon characteristics—such as radar and missile range, semi-active versus active missiles, and the use or nonuse of radar cross-section (RCS) reduction (captured by simply dividing the standard vulnerability score by 10)—defined the classes. We used tactical scenarios, which were simple engagements between opposing four-ships of aircraft approaching head-on at the same altitude, to generate relative lethality and vulnerability scores for the various classes. Table 4.3 summarizes the scores for each class with the 70-percent factor included. Again, note that we did not attempt to make fine distinctions within classes. We have kept the scores of U.S. aircraft constant over our four snapshot years, an approach that obviously does not capture the full effect of upgrades to aircraft and missile capabil-

that here, too, the PLAAF is narrowing the gap. For example, it has established three squadrons of "aggressor" aircraft (flown by some of China's best pilots) and introduced unscripted air combat exercises. IISS, *The Military Balance*, 2015, pp. 48, 242; and National Air and Space Intelligence Center, *People's Liberation Army Air Force 2010*, Wright-Patterson AFB, Ohio, August 1, 2010.

[17] The complexity of PLAAF "Red Flag"–type exercises held at Dingxin is increasing, and it has introduced annual competitions for air-to-air combat ("golden helmet" competitions) and air-to-ground attack ("golden dart" competitions). Feng, "2012 in Review," *Information Dissemination*, December 28, 2012.

[18] Stillion and Orletsky, *Airbase Vulnerability to Conventional Cruise-Missile and Ballistic-Missile Attacks*, 1999.

[19] Eighth Air Force Historical Society, "WWII 8thAAF Combat Chronology, January 1944 Through June 1944," web page, undated.

[20] The model is the Tac Brawler model. For information on the model, see R. M. Kerchner et al., *The TAC Brawler Air Combat Simulation Analyst Manual*, rev. 3.0, Decision Science Applications Report No. 668, 1985.

Table 4.3
Aircraft Lethality and Vulnerability Scores (with modifier for differences in pilot training applied)

Generation	Relative Lethality	Relative Vulnerability
2nd–3rd	0.1	1.7
4th	1.4	1.0
U.S. 4th	2.0	0.7
U.S. 5th	2.0	0.1

ity over the period considered. However, given our desire to capture broad trends, this omission should not greatly affect our overall findings.

Taiwan Scenario Results

We evaluated two different standards against which U.S. force structure requirements could be judged. The first and more demanding standard is the capability to blunt a major Chinese surge of aircraft, preventing the PLA air forces from influencing other aspects of the battle even when they commit maximum air resources. The second is prevailing through attrition, or the capability to destroy enough PLA aircraft to convince or compel Chinese commanders to abandon or scale back air operations before the end of a Taiwan conflict. In the following sections, we address each of these measures individually.

Defeating a Surge

An air campaign over Taiwan could take many forms. In the analysis presented here, we envisage China flying air-to-air and air-to-ground sorties from 39 air bases in the Guangzhou and Nanjing Military Regions. These sorties would most likely be organized into separate strike packages, with a single air-to-ground (or suppression of enemy air defenses [SEAD]) regiment supported by two regiments of air-to-air escorts.[21] With the number of aircraft available, the PLAAF and PLAN could generate around 15 three-regiment packages. Theoretically, they could be airborne simultaneously, but given both a likely desire to phase air attacks and, more importantly, the difficulties of airspace management over Taiwan, half this number is a more realistic surge by Chinese forces.

Defending Taiwan against such a surge was the first metric we used to judge relative capability. Since the objective is to disrupt the air-to-ground aircraft and minimize successful attacks, U.S. aircraft must first defeat the air-to-air escorts. Thus, we assessed the minimum force necessary to match the combat power (essentially, the

[21] These latter two could be employed with a regiment forward playing a sweep role and the other providing a close escort to the strike aircraft themselves. Cliff et al., *Shaking the Heavens and Splitting the Earth*, 2011.

number of sorties multiplied by the capability mix) of the escorts. Although this may not initially seem like a challenging objective, recall that this combat capability must be kept over or very near Taiwan continuously to respond to what could be a very short-warning Chinese attack.[22] This metric is also a useful surrogate for measuring the U.S. ability to achieve air superiority and air dominance continuously throughout a campaign. Success on this measure implies that the United States can match any large-scale Chinese attack and prevent most interference by Chinese airpower in events over and on Taiwan.

The resulting size of the required U.S. force structure is presented in Table 4.4. The top line specifies the size of the force required to be airborne on-station at any given time. The bottom line shows the total force required in theater to keep those aircraft airborne and on station 24 hours a day, seven days a week. This line depicts a range of values, depending on where the force is based. The smaller figure represents the required force structure if all aircraft can be based on Okinawa (or at an equivalent distance from Taiwan). The larger figure represents the force structure if all aircraft are based in Guam (or at an equivalent distance from Taiwan). In most cases, U.S. forces will be based out of a variety of bases and the actual requirement would fall somewhere in between these values. We note, however, that as the threat from Chinese ballistic and cruise missiles to nearby bases increases (see the scorecard 1 analysis in Chapter Three), a relatively greater share of U.S. aircraft may operate from more distant bases.

The forces needed on station would have been quite modest in 1996 and could, for example, have consisted of four continuous DCA patrols of four aircraft each. However, the requirements would have grown rapidly as China modernized its air forces. By 2010, somewhere between nine and 20 wings would have been required in

Table 4.4
Force Structure Requirements to Match a PLAAF and PLAN Surge Over Taiwan (measured in U.S. fighter wings of 72 aircraft each)

Requirement	1996	2003	2010	2017
Air wings required on-station over Taiwan to match PLA air surge	0.2	0.8	1.4	2.0
Air wings required in theater to maintain on-station force	1.6–2.1	5.3–10.6	9.3–19.6	13.8–29.9

NOTES: In each snapshot year, the U.S. force is assumed to include 72 U.S. Navy fighters (two CSGs). The 2010 force includes 72 fifth-generation fighters, and the 2017 case includes 144 fifth-generation fighters. All remaining aircraft are fourth-generation fighters. The range of values in the table reflects different assumptions about where most U.S. fighters are based, with the smaller figure indicating closer basing and the larger more distant.

[22] Flight times to Taiwan from some of the closer Chinese bases would be 15–30 minutes, though indicators provided signals intelligence (SIGINT) or other sources could provide greater warning times.

theater. Given the limited capacity of basing on Okinawa and the threat from ballistic missiles to those bases (see the scorecard 1 analysis in Chapter Three), the actual requirement would probably be near the higher end of this range. Given the overall regional (Guam plus Japan) basing of roughly six wings, this requirement would have been unsustainable. Indeed, the higher end of this range exceeds the total number of fighter wings in the U.S. inventory. Even the 2003 requirement of between roughly five and 11 wings would probably have stressed the limits of U.S. basing capacity. Needless to say, sustaining the requirement for the 2017 case is well beyond reach.

This analysis suggests that the United States could have difficulty obtaining the 24/7 air dominance desired, and it may be forced into its own surge operations in which it seeks to gain air superiority for a limited amount of time over a limited area, perhaps in support of Taiwanese ground force operations. At one level, the inability to dominate the airspace over a contested area completely and without interruption against a major world power is a natural condition and consonant with major power wars of the past. For the United States in the post–Vietnam War period, however, this is a new condition.

Attrition Victory

As an alternative to this 24/7-air dominance strategy, we next examined the requirement to simply inflict unsustainable attrition on the Chinese forces. In the steady state, U.S. and Chinese aircraft fly at a sortie rate determined by their range from Taiwan—seven days at a higher rate, followed by a lower, more sustainable rate. For each 24-hour time step in the model, we calculated attrition for both sides based on the mix of aircraft capabilities and the force ratio between them. The force ratio is important not only for calculating the loss rate but also for identifying the fraction of the two forces that can actually engage each other. If one side greatly outnumbers the other, the larger force cannot employ all of its sorties against the adversary.

Although we cannot predict with any certainty when the PLA air forces might "break," the loss of 50 percent of China's air-to-ground capability would almost certainly lead its commanders to scale back their efforts with an eye to preserving the remainder of the force. Within the model, we adopted this standard—50-percent loss of PLA air-to-ground aircraft—as the effective end of the campaign. We assessed the U.S. force structure that would have to be maintained in theater to achieve this objective within a seven-day time frame and a 21-day time frame. Given that the Chinese side will be conducting air-to-ground missions, attacking targets on Taiwan and potentially elsewhere, until the U.S. side can achieve air dominance, the shorter time frame will be preferred. It will, however, also require more resources.

Table 4.5 presents the results of the attrition cases. The U.S. force structure, measured by the number of fighter wings (standardized to 72 aircraft), that would be required to prevail in seven days is shown in the first line, while the 21-day case is presented in the second. As in the surge cases discussed earlier, we include a range for

Table 4.5
U.S. Force Requirements for Attrition Victory, Taiwan Scenario (number of fighter wings of 72 fighters each)

	Wings Required to Win by End of Last Day of Campaign			
Campaign Type	**1996**	**2003**	**2010**	**2017**
7-day campaign	0.8	2.1–2.6	3.1–4.6	4.1–7.0
21-day campaign	0.3	0.8	1.7–2.2	2.8–4.1

NOTES: In each time period, the force is assumed to include 72 U.S. Navy fighters (2 CSGs). The 2010 force includes 72 fifth-generation fighters, and the 2017 case includes 144 fifth-generation fighters. All remaining aircraft are fourth-generation fighters. The range of values provided reflects different assumptions about where the bulk of fighters are based, with the smaller figure indicating that all aircraft can be based near Taiwan and the larger indicating that they are all based at distant locations.

each year, depending on where the bulk of U.S. forces are based. The smaller number within each range holds if all aircraft can be based at Kadena AB (or at other locations of approximately equal distance to Taiwan), while the larger number pertains to cases in which all U.S. aircraft are based in Guam (or at other locations at an approximately equal distance from Taiwan). Again, most cases would, in reality, see mixed basing and a required force structure somewhere within the range presented.

The results suggest that an air campaign over Taiwan in 1996 would have posed little challenge to the United States, regardless of basing. In fact, U.S. Navy aircraft alone would have been sufficient to achieve steady-state objectives in either a seven- or 21-day campaign. In 2003, approximately two and a half wings (around 200 aircraft) would have been needed to defeat a Chinese air offensive within a seven-day time frame, even if nearby (i.e., Okinawa) basing were unavailable. Yet, as the scorecard 1 analysis showed, basing on Okinawa was not yet under significant threat in 2003. With secure bases in Okinawa, or if the seven-day time requirement could be relaxed by a day or two, the requirement could drop to just over two wings.

As China's wholesale replacement of obsolete aircraft with modern ones hits its stride, the U.S. task of prevailing quickly in an attrition contest becomes more difficult. By 2010, between three and 4.5 U.S. fighter wings would have been required to win in a seven-day campaign, with the higher end of the spectrum more likely if nearby basing on Okinawa were denied. Relaxing the time requirement from seven to 21 days would reduce requirements to between roughly 1.5 to two wings. It would have been possible to find basing for this force, though not without difficulty in a partly denied environment.

By 2017, even relaxing the time frame by which air superiority is to be achieved leaves the United States with force requirements that would be somewhat problematic from a basing perspective. Assuming that Kadena AB is largely denied by 2017 and that most fighters would be based primarily on Guam or an equivalent distance, roughly four fighter wings would be required to win in a 21-day campaign. With the equivalent of one wing supplied by naval air, the U.S. military would likely find bases on Guam or at other relatively secure locations for the remaining 200 fighters and the roughly equivalent number of tankers that would support them. But those bases would almost certainly be crowded, making U.S. aircraft highly vulnerable to missile attack.

As discussed in Chapter Three (scorecard 1), even basing on Guam is not entirely secure from attack. And as the scorecard 5 analysis in Chapter Seven highlights, U.S. carriers operating within unrefueled range of Taiwan would also be under substantial risk of detection and attack.[23] This would further increase the demand for U.S. Air Force tanker aircraft, and basing for those assets would also have to be found. Time requirements could be relaxed beyond 21 days to bring down the force requirement—an outcome that might be forced on U.S. commanders—but it should be noted that until air superiority is achieved, the PLA air forces would have opportunities to attack targets on Taiwan. Stretching the air campaign well beyond 21 days risks making the air battle irrelevant to the larger conflict.

Attacks on Chinese Air Bases

Given the challenges faced by U.S. forces by 2017, other measures might be necessary to improve the prospects for success. Here, we examine one such possibility. Specifically, we examine whether U.S. attacks on Chinese air bases could significantly reduce Chinese sortie generation and, if so, whether that could reduce the level of U.S. forces required to win in an attrition campaign. As discussed in Chapter Five (scorecard 3) and Chapter Six (scorecard 4), a significant fraction of the Chinese air bases within unrefueled aircraft range of Taiwan are vulnerable to attack by standoff cruise missiles and penetrating bombers.

We examined the effect of temporary base closures by degrading Chinese sortie rates according to the runway attack results in our scorecard 3 and 4 analyses. U.S. attacks on Chinese air bases would carry costs—in the form of demand for additional fighters to serve as strike escorts—and benefits. These escorts would protect bombers from Chinese DCA en route to their targets. We therefore add the required fighter escorts to larger U.S. force structure requirements, even as the requirement for fighters over Taiwan is reduced.[24] The additional requirement for escort fighters ranges from a high of 45 aircraft in 2003 to a low of 22 fighters in 2017.[25]

Since the U.S. ability to attack Chinese runways diminishes over the duration of a given campaign as the available stock of standoff munitions is expended, the effects should be more pronounced in the short seven-day case than in the more extended

[23] We account for some of this effect by moving back the CSG operating locations, and hence reducing the sortie rate each year, from 500 km in 1996 to 2,000 km in 2017.

[24] We assumed that escort fighters would be based in Okinawa in the 1996 and 2003 cases and in Guam in the 2010 and 2017 cases.

[25] The decrease after 2003 results from a growing number of Chinese bases becoming inaccessible to U.S. bombers, which drives escort requirements, and increased U.S. reliance on standoff missiles to attack those bases.

21-day case. Also, since fighting from distant bases places U.S. aircraft defending Taiwan at a larger numerical disadvantage in individual engagements than fighting from closer bases, we expect attacks on Chinese air bases to have a larger proportional effect on later years, when the missile threat makes distant basing more necessary, than earlier. Figure 4.4 presents the results for the seven-day case without U.S. attacks on Chinese air bases (represented by the bar on the left in each year) and with U.S. attacks (represented by the bar on the right for each year).

As can be seen, for the 1996 and 2003 cases, attacking air bases in China does not reduce the required U.S. force structure. In those years, the additional fighters required to escort bombers on those attacks more than counteract the effects of reduced PLAAF sortie rates on U.S. requirements. This should not be surprising, given that the PLA aircraft in those years are easily killed, whether engaged in the air or on the ground. The 2010 and 2017 cases, when the PLA has more modern aircraft at its disposal, do show some benefit to air base attack. The 2010 case shows a 26-percent reduction in required U.S. force structure, and the 2017 case shows a 27-percent reduction. Although the 2010 and 2017 force requirements are still quite large, the one- and two-wing reductions as a result of attacking air bases produce a demand for forces that would be easier for the U.S. basing structure in Asia to accommodate.

We also assessed other cases with and without U.S. attacks on Chinese air bases. In a seven-day campaign in which U.S. aircraft were based in closer proximity to Taiwan, attacks on Chinese bases again produced gains for the U.S. side, though they

Figure 4.4
U.S. Force Requirements for Attrition Victory With and Without Attacks on PLA Air Bases (measured in number of fighter wings)

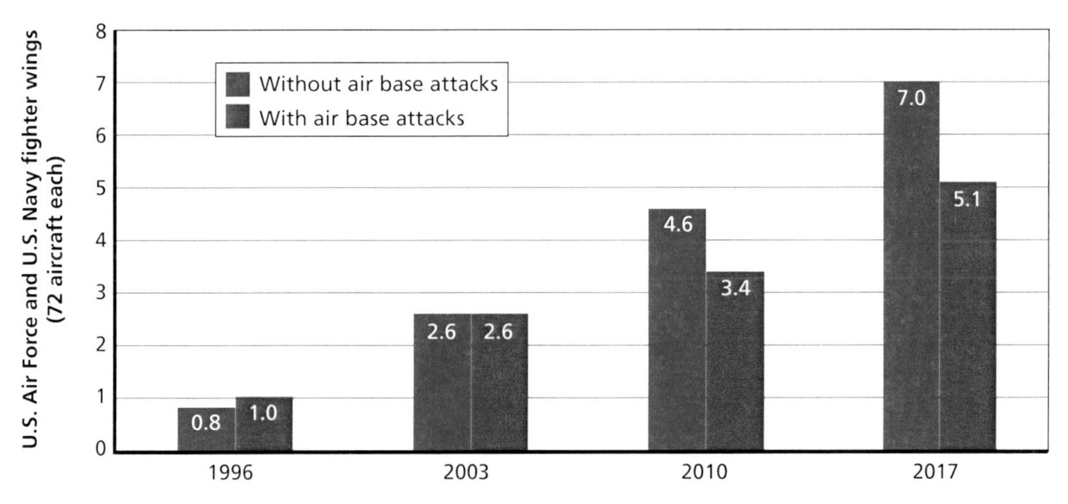

NOTE: These results pertain to a seven-day campaign in which U.S. aircraft are based primarily on Guam (or at a roughly equal distance from Taiwan). Wings are standardized to 72 aircraft. Some 72 U.S. Navy fighters flying from CSGs are included in the counts.

RAND *RR392-4.4*

were more modest (with a 10-percent reduction in the required force structure in 2010 and a 20-percent reduction in 2017). And we examined cases involving the impact in a longer, 21-day campaign and found benefits, but only in the 2017 case. In 2017, conducting attacks against Chinese air bases reduced U.S. force requirements by between 11 and 17 percent, depending on assumptions about where the bulk of U.S. aircraft are based.

Spratly Islands Scenario

The forces available to China in a Spratly Islands conflict are limited by the smaller number of bases and the greater distance to the area and, therefore, are significantly smaller in number (see Table 4.6). With only a handful of tankers in the PLA force structure, we limited participating aircraft to those that could conduct unrefueled operations over the Spratly area. The few tankers available would probably be used to "top off" China's longer-range aircraft before they departed for contested airspace. With the exception of bombers, there are few strike aircraft that could range the Spratly Islands until the 2010 time frame. The number of air-to-air escorts would also be limited. By 2017, however, the PLAN and PLAAF could field a substantial force against targets in

Table 4.6
PLAAF and PLAN Aircraft Quantities and Missions Used Model the Spratly Islands Scenario

Aircraft	1996	2003	2010	2017
Strike aircraft				
H-6	80	100	100	100
J-8	—	18	—	—
JH-7	—	—	48	120
Total	80	118	148	220
Air-to-air escort aircraft				
J-6	—	—	—	—
J-7	—	—	—	—
J-8	—	—	—	—
J-10	—	—	—	—
Su-27 / J-11	24	72	96	216
Su-30MKK/J-16	—	24	96	92
Total	24	96	192	308

this area, though Chinese aircraft would still be operating near the limit of their range and the total number would still be constrained by the small (but growing) number of bases within range.

When China's first aircraft carrier becomes operational, probably sometime after 2017, an additional dozen carrier aircraft might be available locally, though the carrier itself could be vulnerable to submarine attack and air-launched ASCMs. Also, as of this writing in June 2015, China is undertaking significant land reclamation operations at a half a dozen locations in the South China Sea and building an airstrip large enough to take military aircraft at Fiery Cross Reef.[26] These facilities could host a handful of SAMs and fighter aircraft. While such systems could play a role in peacetime operations and so-called "gray zone conflicts" (which involve the non-lethal use of military force), they are unlikely to be a significant factor in high-intensity military operations against U.S. forces beyond the first hours of a conflict.[27] China could improve its position more significantly by adding a larger number of tanker aircraft, AWACS, and a more robust basing infrastructure in the southern portions of the Guangzhou Military Region.[28] However, these developments are unlikely to occur before the 2017 period considered by this report, if at all.

Other modeling assumptions, such as aircraft capabilities, sortie rates, and U.S. force flows, are all consistent with those used in the Taiwan analysis. Note, however, that as the distance Chinese aircraft must fly in this scenario increases, the sortie disparity diminishes. With U.S. basing in the southern portion of the Philippines, both sides would be operating approximately 1,000 km from Thitu Island, the focus of the scenario.[29] This compares with 400 km for the Chinese and up to 2,500 km for the U.S. side in the Taiwan scenario. Even without bases in the Philippines, U.S. aircraft could fly from bases in Japan or Guam that are between 1,300 and 2,100 km from

[26] James Hardy and Sean O'Connor, "China Building Airstrip-Capable Island on Fiery Cross Reef," *Jane's Defence Weekly*, November 20, 2014; "China's Land Reclamation in Disputed Waters Stokes Fears of Military Ambitions," *The Guardian*, May 8, 2015.

[27] SAM systems have historically caused the greatest challenge when they are able to hide. On South China Sea islands, there would be no space for SAMs to maneuver or conceal themselves, and they would likely be easily destroyed by a small handful of weapons. Airstrips on reclaimed land would, for this reason, be lightly defended and could be easily closed to traffic. For a detailed Administration statement on the new islands, see David Shear, Assistant Secretary of Defense for Asian and Pacific Security Affairs, testimony before the Senate Committee on Foreign Relations at the hearing "Safeguarding American Interests in the South and East China Seas," May 13, 2015.

[28] China is developing a large transport aircraft, the Y-20, which is likely to have both a tanker and AWACS variant. The final resolution of its engine issues, which have delayed its production, will mark a very significant development for Chinese military capabilities. In the Spratly scenario examined here, a larger number of tankers would enable Chinese aircraft to fly from more distant bases and reduce the PLA's current dependence, in this campaign, on a handful of relatively vulnerable bases. *Jane's All the World's Aircraft*, "XAC Y-20 Kunpeng," January 7, 2015.

[29] Closer basing is available in the Philippines, but such air bases as Clark and Antonio Bautista would be in range of Chinese MRBMs.

the Spratly Islands. Although these are long ranges and would require the support of substantial refueling assets, the disparity between the two sides' operating distances would nevertheless be substantially less than in the Taiwan scenario. This, coupled with the smaller force size China is able to use, should significantly reduce U.S. force requirements.

As before, we begin by examining the forces necessary to match a Chinese air surge that might include as much as 50 percent of PLA aircraft assigned to the campaign. We also examine the impact of attacking PLA air bases on this requirement. Figure 4.5 presents the size of the force structure that the U.S. military would have to maintain in theater to keep patrols large enough to defeat a Chinese air surge over the Spratly Islands 24/7, as well as the size of the requirement if strikes on Chinese air bases were also conducted.

In 1996, the United States could almost certainly have maintained enough forces in theater (half an air wing) to defeat a Chinese surge even without striking Chinese air bases. The 2003 case is somewhat more questionable, given the expected modest level of U.S. commitment in this scenario, but it might nevertheless have been possible to support the required force (roughly four wings), especially given the lack of the DF-21C threat to U.S. bases at that time. By 2010, however, it would have become virtually impossible to maintain sufficient forces for a decisive 24/7 presence over the battle area—at least not without also reducing Chinese air sorties by attacking Chinese

Figure 4.5
U.S. Force Structure Required in Theater to Maintain Constant Patrol Over Spratly Islands Capable of Defeating a PLA Air Surge

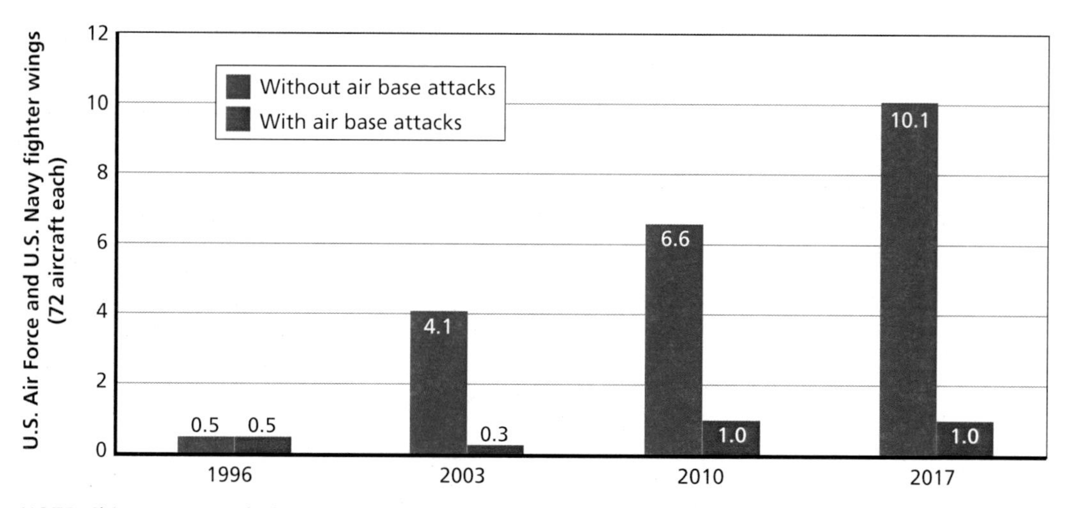

NOTE: China's surge includes 50 percent of Chinese aircraft available for this scenario. In each snapshot year, the U.S. force is assumed to include 72 U.S. Navy fighters (two to three CSGs). The 2010 force includes 72 fifth-generation fighters, and the 2017 case includes 144 fifth-generation fighters. All remaining aircraft are fourth-generation fighters. The figure assumes U.S. basing in southern Mindanao.
RAND *RR392-4.5*

air bases. Keeping even a relatively modest force of aircraft on station around the clock will always be challenging for a force based 1,000 km away.

Attacking Chinese air bases could dramatically reduce the size of the force required to achieve quick, around-the-clock domination of the battle area, however. Given the distance from China to the Spratly Islands, Chinese aircraft would be limited to using a small number of air bases, all of which would be relatively near the Chinese coast. For this analysis, we assumed that bases on Hainan are protected only by fighter aircraft until 2017, at which point we assumed that modern long-range SAM batteries are deployed there. The SAMs are not a hugely important driver, however; with the small number of bases under consideration, the current U.S. inventory of standoff weapons can accomplish much of the mission without exposing manned aircraft to SAM fires.

Given these parameters, attacks on the small set of relatively exposed air bases result in a high rate of sortie suppression (see the scorecard 4 analysis in Chapter Six). The impact on the U.S. ability to defeat Chinese air surges against targets in and around the Spratly Islands is dramatic (see Figure 4.5). With the exception of the 1996 case, in which the need for escort fighters is larger than the savings from an air base attack, the addition of an air base attack can remove most of China's capability. Attacks by standoff weapons and a small number of escorted bomber sorties are sufficient to reduce Chinese sortie rates at the nine bases within range of the Spratly Islands to no more than 25 percent of their original values. The reduction in U.S. aircraft required to defeat a Chinese air surge is large enough that U.S. naval airpower alone would likely be able to meet the demand, though the force would still require bomber and tanker support from the U.S. Air Force.

The most significant uncertainty in the analysis is whether international political considerations might dissuade the United States from conducting attacks against Chinese bases. The Spratly scenario would probably not, depending on the road to war, engage the same level of national U.S. commitment that a war over Taiwan would. Given the potentially escalatory effects of striking locations on the mainland or Hainan Island in what might otherwise be a limited war, there is some chance that these attacks might not be allowed, despite their operational promise. Based on both this fact and the difficulty of maintaining sufficient forces to defeat a Chinese air surge, we tested one final metric, the ability to achieve a 50-percent reduction in air-to-ground sorties (see Table 4.7). If the United States cannot dominate the airspace from the outset against all air threats, can it maintain the forces necessary to defeat the Chinese air threat through attrition within an operationally relevant time frame?

The results suggest that prevailing in a short, seven-day attrition campaign would require roughly one-third the force structure needed to dominate the airspace from the outset and defeat an air surge (in the absence of attacks on Chinese bases). Prevailing within 21 days would, in the 2010 and 2017 cases, require roughly one-fifth the inventory needed to keep forces continuously on station to defeat a surge. The most difficult attrition case examined, the seven-day 2017 case, requires approximately

Table 4.7
U.S. Force Required to Destroy 50 Percent of China's Strike Aircraft Within Seven or 21 Days, Spratly Islands Scenario

	Wings Required to Win by End of Last Day of Campaign			
Campaign Type	**1996**	**2003**	**2010**	**2017**
7-day campaign	0.3	1.1	2.1	3.6
21-day campaign	0.1	0.8	1.2	2.0

NOTES: In each period, the force is assumed to include 72 U.S. Navy fighters (two to three CSGs). The 2010 force includes 72 fifth-generation fighters, and the 2017 case includes 144 fifth-generation fighters. All remaining aircraft are fourth-generation fighters. The table assumes U.S. basing in southern Mindanao.

260 aircraft, including 80 U.S. Navy fourth-generation fighters, 144 U.S. Air Force fifth-generation aircraft, and 36 U.S. Air Force fourth-generation fighters. This number of U.S. Air Force fighters could stretch the basing capacity of the three primary dual-use air bases on Mindanao (located in General Santos City, Zamboanga, and Cagayan de Oro City).

However, some relaxation of the time requirement would lower U.S. force requirements and, thus, likely alleviate the basing problem. Additional basing could be available in Luzon if the TBM threat was not considered critical. Alternatively, aircraft could, depending on political alignments, fly from more distant bases located in third countries, though more aircraft might be required if the longer transit times resulted in reduced time on station. Overall, it would appear that China would have little hope of gaining and maintaining air superiority over and around Thitu Island if the United States committed a substantial force to the conflict.

Conclusions

The United States continues to maintain unparalleled air-to-air capabilities. Even in the most challenging cases examined in this chapter, the United States does not "lose" the war in the air. Among the cases considered, the largest loss to U.S. aircraft occurs in a 2017 Taiwan scenario—one that envisions denied basing close to Taiwan and an extended 21-day campaign. Even in this case, though, the United States loses only 78 aircraft in air-to-air combat (though many more U.S. aircraft could be destroyed on the ground) and achieves a kill ratio of some 13 to one.[30]

[30] As is the case for all the modeling discussed in this report, this number should be treated with circumspection. Advanced Chinese electronic warfare, jammed communications, the loss of U.S. ISR platforms, or shortages of air-to-air weapons could make the campaign messier and more difficult.

However, these observations should not obscure a deeper and equally meaningful set of realities. Specifically, the political and operational context of the conflict is critical, and the analysis shows that while the United States will not soon lose an air war in Asia, achieving its objectives in a politically and operationally relevant time frame is, in some cases, becoming far more challenging. Historically, when China has used force, it has shown a strong inclination to pursue tight and contained operations that can be concluded quickly. It may be relatively more inclined to consider the use of force if it believes that it could achieve its operational objectives before the United States could bring its forces fully to bear. Similarly, the credibility of U.S. military assurances to allies and partners could become weaker if it appears to them that U.S. strategy (and theory of victory) depended on extended operations.

The analysis also highlights the importance of geography. Any plausible conflict in East Asia would put U.S. forces at a disadvantage in terms of relative distance to the battle area. Beyond that, the specific location and parameters of conflict would have a critical impact on outcomes. In a Taiwan conflict, the United States would have only one major air base within 800 km of the focus of conflict, while the Chinese would have 39. This basing mismatch would enable a large portion of China's increasingly modern strike and fighter force to reach the battle area and put U.S. fighter aircraft at a substantial numerical disadvantage.

The Spratly Islands scenario, in contrast, would see Chinese pilots, like their U.S. counterparts, flying from substantial distances but with fewer tankers and command and control aircraft to support them. Moreover, Chinese aircraft would be flying from a much more limited set of bases—probably numbering less than a dozen—and these bases would be highly vulnerable to U.S. air and missile attack. The analysis suggests that the United States would likely be capable of maintaining complete air dominance from the outset if attacks on Chinese air bases were permitted, and it would be capable of maintaining enough aircraft within range of the area to prevail in an attrition contest if they were not (depending on basing assumptions and the duration of the conflict). The addition of one or more small Chinese airfields in the Senkaku Islands and the deployment of SAM systems there are unlikely to change this equation significantly. However, the construction of a more extensive basing infrastructure in southern China, Hainan, and the Paracel Islands, together with more long-range aircraft, tankers, and a conventionally armed IRBM force might shift the balance at some point after 2017.

Scorecard Coding

Figure 4.6 provides our summary coding of the results of scorecard 2. Advantage in this scorecard is evaluated in terms of whether the United States could gain general air

Figure 4.6
Scorecard 2 Summary Coding

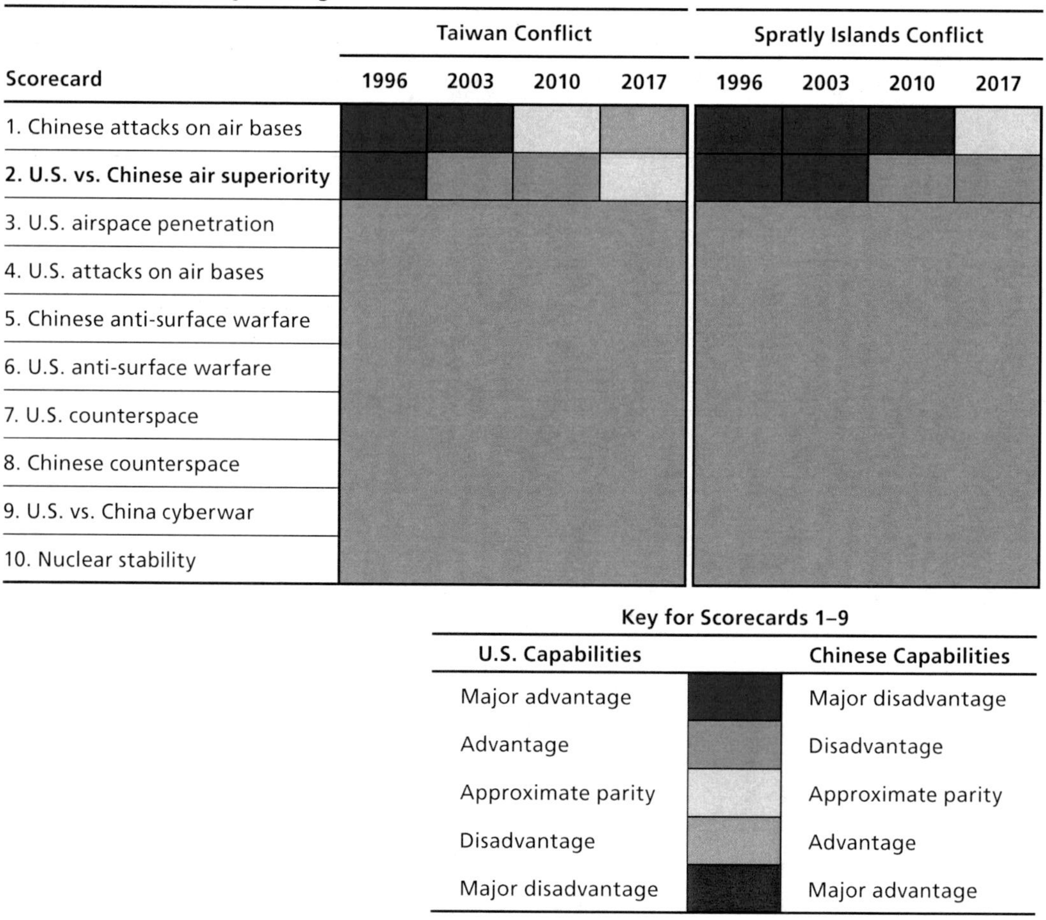

	Taiwan Conflict				Spratly Islands Conflict			
Scorecard	1996	2003	2010	2017	1996	2003	2010	2017
1. Chinese attacks on air bases								
2. U.S. vs. Chinese air superiority								
3. U.S. airspace penetration								
4. U.S. attacks on air bases								
5. Chinese anti-surface warfare								
6. U.S. anti-surface warfare								
7. U.S. counterspace								
8. Chinese counterspace								
9. U.S. vs. China cyberwar								
10. Nuclear stability								

Key for Scorecards 1–9

U.S. Capabilities		Chinese Capabilities
Major advantage		Major disadvantage
Advantage		Disadvantage
Approximate parity		Approximate parity
Disadvantage		Advantage
Major disadvantage		Major advantage

superiority in the first weeks of a conflict (*U.S. advantage*) or whether airspace would remain contested (*Chinese advantage*).[31]

For the Taiwan scenario, we coded the scorecard as *U.S. advantage* through 2010. As late as 2010, between 1.7 and 2.2 U.S. air wing equivalents would be sufficient to achieve air superiority in a 21-day Taiwan campaign. These numbers would not strain the inventory, and it would be within the capacity of U.S. bases to absorb the force. In the 2017 case, we coded the scorecard as *approximate parity*, since the number of air wings required in a 21-day period (between 2.8 and 4.1) begins to stretch the limits of local basing capacity.[32] In the Spratly Islands case, the entire period is characterized by

---

[31] While the latter would not constitute an outright Chinese aerial victory, it would largely deny allied forces air cover and, therefore, constitute Chinese advantage.

[32] In the interest of maximizing the clarity of our analytical methodology, we examine capabilities in individual types of operational areas without explicitly modeling the interactions between them. (We do discuss interac-

U.S. advantage (though the margin of advantage declines over time), since the number of air wings required to prevail can easily be accommodated. As noted earlier, this analysis does not account for the problem of operating air bases under attack and may, therefore, understate the degree to which China could challenge U.S. air superiority, especially at the outset of a conflict near the Chinese mainland. Equally importantly, this analysis does not address adequacy in more protracted conflicts, which are likely to see greater U.S. advantage, though that advantage is diminishing over time.

---

tions throughout the report, however.) Our coding of the air superiority scorecard does not consider the effects of missile attacks on U.S. air bases, the effects of submarine and missiles threats to aircraft carriers, or attacks on Chinese air bases by U.S. airpower. Considered in the context of other scorecards, the air superiority campaign would be challenging by 2017.

CHAPTER FIVE

Scorecard 3: U.S. Penetration of Chinese Airspace

In recent wars, the U.S. ability to penetrate hostile airspace and attack large numbers of targets has provided it with important advantages. Following the display of U.S. airpower in the Gulf War, PLA planners recognized that contesting U.S. access to Chinese airspace was increasingly critical to the larger outcome of a possible conflict. China has pursued two simultaneous paths to mitigate the threat of U.S. airpower. The first is the development and deployment of assets that threaten U.S. air bases and aircraft carriers, including improved ISR capabilities, ballistic and cruise missiles, submarines, and strike aircraft (see the analysis of scorecards 1 and 5 in Chapters Three and Seven, respectively). The second path, and the topic of this chapter, is the steady improvement of China's integrated air defense system (IADS).

Since the early 1990s, China has invested in a combination of foreign purchased and domestically produced long-range SAM batteries, modern interceptor aircraft, and early-warning radars to strengthen its air defense capability. The United States has not, for the most part, specifically tailored the development of its strike capabilities to counter Chinese air defense modernization, but strides in tactics, stealthy aircraft, and standoff cruise missile development would all play a role.

This chapter assesses the ability of U.S. aircraft to penetrate Chinese airspace or use lethal SEAD to neutralize PLA air defenses. First, we present an overview of China's air defense modernization and U.S. advances in airpower projection capabilities since 1996. Second, we model the degree of risk faced by U.S. bomber aircraft in penetrating air defenses (including SAMs and interceptor aircraft) to strike at a notional target set within China. (We do not address the ability of U.S. aircraft to locate or destroy specific targets after reaching those positions, though a portion of this problem is discussed in Chapter Six, scorecard 4.) Third, we explain how we employed a simple model to assess the capability of U.S. aircraft to destroy or neutralize Chinese SAMs in one-on-one engagements and discuss those results in the context of other operational considerations. Finally, we provide a scorecard of trends in the U.S. capability to penetrate Chinese airspace in the Taiwan and Spratly Islands scenarios.

Balance of Forces: Chinese IADS Modernization

Following the U.S. demonstrations of air dominance in the 1990s, China embarked on an ambitious program to modernize its obsolete air defense network, bolstering the three core components of an IADS—early warning, SAM systems, and airborne interceptors—while at the same time improving connectivity and doctrine.

Early Warning

Since 1996, China has not only built advanced radars that improve overall early-warning coverage through ground-based sensors and the introduction of airborne early-warning aircraft, but it has also designed and deployed radar systems that are reportedly optimized to detect stealthy aircraft.[1] In 1999, China sought to acquire the highly capable Phalcon Airborne Early Warning (AEW) system from Israel, but it was prevented from doing so by U.S. diplomatic efforts.[2] After this setback, the PLAAF implemented indigenous AEW development programs, producing three viable airframes within a decade:

- KJ-2000: A-50 airframe with Chinese-designed electronically steered phased-array radar (four commissioned between 2006 and 2007; and four in service as of 2015)
- KJ-200: Y-8 airframe with a "balance-beam" active electronically scanned array radar (four in service since the late 2000s)
- Y-8 AEW: Y-8 airframe with rotodome (probably for export only).

Long-Range Surface-to-Air Missiles

China's sole long-range SAM in the early 1990s was the obsolete HQ-2 (SA-2) system. With a range of 35 km, these relatively immobile batteries provided little coverage and could be easily neutralized by modern SEAD operations. Lacking indigenous expertise, China's initial modernization efforts centered on importing modern SAM systems from Russia. To date, Russia has delivered 160 S-300 PMU (SA-10C) and S-300 PMU-1 and -2 (SA-20A and B) launchers with attendant radars, munitions, and support equipment.[3] By the mid-2000s, China had successfully reversed-engineered Russian and Western technologies and incorporated elements of both the S-300 and Patriot

---

[1] Carlo Kopp, *Russian/PLA Low Band Surveillance Radars*, Air Power Australia, Technical Report APA-TR-2007-0901, updated April 2012. Recent reporting describes stealth detecting X-band radar deployed on a large UAV. See "Divine Eagle, China's Enormous Stealth Hunting Drone, Takes Shape," *Popular Science*, May 28, 2015.

[2] Steven Lee Myers, "U.S. Seeks to Curb Israeli Arms Sales to China," *New York Times*, November 11, 1999.

[3] SinoDefence.com, "S-300PMU (SA-10) Air Defence Missile System," web page, last updated May 8, 2008. Jane's puts the potential number of launchers somewhat higher, claiming up to 240 launchers and 1,000 missiles (all acquired by the mid-2000s). See *Jane's World Air Forces*, "China—Air Force, Procurement," March 1, 2013.

batteries into its HQ-9 system, which has a range of approximately 200 km.[4] Around the same time, the PLA also introduced the 50-km-range HQ-12 SAM. Unlike older HQ-2 (SA-2) systems, the new generation of SAMs is more mobile and incorporates jamming-resistant technology.[5] Since the development of the HQ-9, China has stopped acquiring S-300 (SA-10 and SA-20) launchers from Russia.

Russian media reports that China signed a contract in September 2014 to purchase six battalions of Russia's newest and most advanced long-range self-propelled SAM system, the S-400 *Triumf* (NATO designation SA-21 *Growler*) for an estimated $3 billion.[6] Subsequent statements on the significance of China's acquisition by researchers with the PLA Academy of Military Science and the China Aerospace Science and Industry Corporation lend credibility to Russian reporting. If accurate, China's purchase would further strengthen the PLA's integrated air defense system. The S-400 reportedly includes an active electronically scanned array (AESA) radar and can target aircraft, cruise missiles, as well as tactical and ballistic missiles at ranges up to 400 km and at speeds of up to 4.8 km per second.[7] If the S-400s are delivered to China, it is unlikely they will reach IOC by 2017, but, given the possibility that they will be available by then, we include them in our analysis.

Air Interceptors

China's air defense fighters were woefully out of date in 1996. Since that time, Beijing has retired its second-generation J-5 (MiG-17) and J-6 (MiG-19) fighters, incorporated modern weapons and avionics into its third-generation aircraft (J-7 and J-8), and purchased fourth-generation aircraft (Su-27 and Su-30) from Russia. China and Russia signed a co-production contract for 200 Su-27 fighters (designated J-11A) in 1996. In the early 2000s, China halted production to redesign the aircraft using indigenous technology. The redesigned aircraft was designated the J-11B and entered service around 2008.[8]

China began production of the J-10, its indigenous fourth-generation fighter in the mid-2000s. Over the next few years, China will continue to replace older aircraft

---

[4] *Jane's Land Warfare Platforms*, "HQ-9/FT-2000," February 20, 2015.

[5] Carlo Kopp, *Surviving the Modern Integrated Air Defence System*, Air Power Australia, Analysis 2009-02, February 3, 2009.

[6] "Russia Confirms Arms Deal to Supply China with S-400 Air Defense Systems," *Sputnik International*, April 13, 2015; "China Signs Contract to Purchase Russian S-400 Missile Systems," *Russia Beyond the Headlines*, April 14, 2015; and Catherine Putz, "Sold: Russian S-400 Missile Defense Systems to China," *The Diplomat*, April 14, 2015.

[7] GlobalSecurity.org, "HQ-19 Anti-Ballistic Missile Interceptor," updated April 20, 2015; and *Jane's Land Warfare Platforms*, "S-400," April 17, 2015.

[8] On the Chinese military aviation industry and its drive to absorb new technologies, see Phillip C. Saunders and Joshua K. Wiseman, "China's Quest for Advanced Aviation Technologies," in Hallion et al., *The Chinese Air Force*, 2012.

with modern J-10 and J-11 fighters.[9] Based on recent production rates, we project that by 2017 more than 60 percent of the PLAAF fighter inventory will be composed of fourth-generation aircraft.[10] China has made progress in developing fifth-generation fighters. It has conducted extensive flight tests of the J-20 (first tested in January 2011), and demonstrated the smaller J-31 at the Zhuhai air show in November 2014.[11] Many questions remain about how stealthy these designs are, whether sufficiently powerful engines can be procured for them, and what other fifth-generation technologies may or may not be incorporated. And because they will not be available in meaningful numbers within our analytical time frame, we did not consider their potential impact as part of this study.[12] Table 5.1 shows the PLA inventory of AEW systems, interceptors, and SAM launchers for each snapshot year. (Note that while the table displays the number of launchers, SAM sites typically have a battery of four to six launchers.)

As it has acquired new equipment, China has also updated its air defense doctrine.[13] Chinese sources highlight the threat posed by stealth, precision, and long-range strike and enumerate several ways in which Chinese air defenses are evolving to cope:

- The importance of "key point defense" is waning while that of "large area defense" is growing. Given the threat from standoff strike, early warning and engagement must occur as far forward as possible.
- Fixed defenses are giving way to "mobile air defense," with mobility considered key to concentrating firepower and plugging holes in air defense coverage.
- Defensive air defense is being replaced by "offensive air defense," with a greater role for counterattack.
- Single-service air defense is being replaced by joint air defense, reflecting the broader PLA trend toward joint approaches to warfare.[14]

---

[9] On the modernization of the PLAAF inventory and the quality of systems, weapons, and equipment, see David Shlapak, "Equipping the PLAAF: The Long March to Modernity," in Richard P. Hallion, Roger Cliff, and Phillip C. Saunders, eds., *The Chinese Air Force: Evolving Concepts, Roles, and Capabilities*, Washington, D.C.: National Defense University Press, 2012.

[10] We input recent production rates by comparing inventories from the 2003 and 2010 editions of IISS, *The Military Balance*.

[11] See *Jane's All the World's Aircraft*, "CAC J-20," February 3, 2015; and *Jane's All the World's Aircraft*, "SAC Shen Fei," January 7, 2015.

[12] The U.S. government estimates that the J-20 will not enter service until at least 2018. *Jane's All the World's Aircraft*, "CAC J-20," February 3, 2015.

[13] For a more complete summary of Chinese thinking on air defense campaigns, see Cliff et al., *Shaking the Heavens and Splitting the Earth*, 2011.

[14] Wang Fengshan, Yang Jianjun, and Chen Jiesheng, 《信息时代的国家防控》 [*National Air Defense in the Information Age*], Beijing: Aviation Industry Press, 2004, pp. 113–122; Xue Xinglin, ed., 《战役理论学习指南》 [*Campaign Theory Study Guide*], 2002; Zhang Yuliang, ed., 《战役学》 [*The Science of Military Campaigns*], 2006.

Table 5.1
Chinese AEW Aircraft, Interceptors, and SAM Launchers, 1996–2017

Asset Type		1996	2003	2010	2015 (current)	2017
AEW aircraft						
KJ-2000		—	—	4	4	4
KJ-200		—	—	4	4+	4–8
Interceptor aircraft	**Generation**					
J-5 (MIG-17)	2nd	400	—	—	—	—
J-6 (MIG-19)	2nd	3,300	550	—	—	—
J-7 (MIG-21)	3rd	570	700	588	528	450
J-8 (Finback)	3rd	130	232	360	168	100
J-10	4th	—	—	150	294	350
Su-27/J-11	4th	24	100	136	340	400
Su-30 MKK/J-16	4th	—	58	97	97	121
J-15	4th	—	—	—	5	30
SAM launchers	**Range (km)**					
HQ-2 (SA-2)	35	500+	500+	300+	300+	200+
S-300 PMU (SA-10C)	100	32	32	32	32	32
S-300 PMU-1 (SA-20A)	150	—	32	64	64	64
S-300 PMU-2 (SA-20B)	200	—	—	64	64	64
HQ-12 (KSA-1)	50	—	—	24	24	48
HQ-9	200	—	—	32	32+	64
S-400 (SA-21)[a]	400	—	—	—	—	16

SOURCES: Missile ranges from Planeman, "Bluffer's Guide: Fortress China: Main Area Defense Systems," discussion forum posts, SinoDefence.com, February 3, 2009; and *Jane's Strategic Weapons Systems*, "S-400 Triumf (SA-21 'Growler')," July 17, 2013. Inventory numbers are from IISS, *The Military Balance*, 1996, 2003, 2010, and 2015.

NOTES: Many of the interceptors listed are multipurpose aircraft and may not be used in the interceptor role. Numbers include PLAN aviation and PLAAF aircraft.

[a] Because of recent reporting on China's purchase of the S-400, we include it in our analysis for 2017, though we note that it may not be available by that time.

Adjusting doctrine requires considerable experimentation and a willingness to work across stovepiped bureaucracies. Effective changes to operational practice will not be as fast as the acquisition of new equipment, yet it is clear that the Chinese air forces are moving toward more flexible and modern modes of operation.

Balance of Forces: United States

In the Gulf War, the United States relied on SAM suppression, stealth, and standoff weapons to penetrate Iraqi air defenses. A SEAD campaign employing a combination of airborne jamming by EF-111 and EA-6B aircraft and kinetic attacks using high-speed anti-radiation missiles (HARMs) eliminated many air defense sites, while stealthy F117s struck well-defended targets. The Gulf War also saw the first use of AGM-86C conventional ALCMs (CALCMs) in a standoff role.[15] In subsequent air campaigns in the Balkans and Middle East, the United States continued to rely on SEAD, stealth, and standoff weapons to penetrate and neutralize enemy air defenses. Since 1996, the Air Force has heavily invested in stealth and standoff capabilities, while the Navy has focused on electronic attack to disrupt enemy air defenses with EA-6B and now EA-18G aircraft.

U.S. SEAD Forces

High attrition inflicted by Soviet-designed SAM systems during the Vietnam War prompted U.S. planners to develop a more effective SEAD doctrine, which was further refined in subsequent years.[16] By the close of the Cold War, the Pentagon had embraced a SEAD concept that involved

- *communication jamming* using Air Force EC-130H and Navy EA-6B electronic attack aircraft
- *radar jamming* by Navy EA-6Bs and Air Force EF-111s
- *physical destruction* of enemy air defense radars with AGM-88 HARMs, Mach 3 weapons, fired from tactical fighters.[17]

---

[15] Some 35 missiles were fired from B-52s. See J. T. Nielson. "CALCM: The Untold Story of the Weapon Used to Start the Gulf War," *Aerospace and Electronic Systems Magazine*, July 1994, Vol. 9, No. 7.

[16] For a history of U.S. SEAD concepts and equipment, see Anthony M. Thornborough and Frank B. Mormillo, *Iron Hand: Smashing the Enemy's Air Defenses*, London: JH Haynes and Co., 2002.

[17] The AGM-88 HARM was introduced in 1986 and remains the principal anti-radiation weapon in the U.S. inventory. The HARM's maximum range exceeds that of early-generation SAMs, including the SA-2, SA-3, and SA-6, and previously allowed SEAD aircraft to attack from beyond the target's engagement envelope. See *Jane's Air-Launched Weapons*, "Raytheon AGM-88 HARM (High-Speed Anti-Radiation Missile)," October 22, 2014.

In recent years, long-range SAM batteries have made U.S. SEAD missions more dangerous by outranging HARMs. At the same time, force reductions and DoD's preference for low-observable systems, such as B-2 and F-22 aircraft, have eroded the priority accorded to dedicated SEAD capabilities, leading to an atrophy of U.S. capabilities.[18] In the future, the Air Force may use fifth-generation fighters to attack targets at closer ranges with smaller munitions that are compatible with the internal weapon carriage of stealthy aircraft.[19] Key developments in the evolution of U.S. SEAD capabilities included the following:

- *Replacement of the F-4G with the F-16CJ:* The Air Force retired the F-4G "Wild Weasel" after the Gulf War. The two-seat F-4G was considered a superior SEAD platform, partly because it carried a dedicated electronic warfare officer to locate and target enemy SAMs and jam incoming missiles.[20] The F-16CJ, currently the U.S. Air Force's sole HARM-firing aircraft, is equipped with the HARM Targeting System (HTS), which improves the range and accuracy of the AGM-88. While U.S. Navy EA-6B and F/A-18 aircraft are HARM-capable, they lack HTS and so may be less effective HARM-launching platforms.[21]
- *Retirement of the EF-111 "Raven":* The retirement of the EF-111 in 1998 left the U.S. Air Force without an airborne radar jamming capability. Consequently, the Air Force has had to rely on Navy and Marine Corps EA-6B and EA-18G aircraft to provide electronic warfare support on SEAD missions. The EA-6B and EA-18G are slower aircraft with shorter ranges, and they require more sorties and tanking than the EF-111.[22]
- *Introduction of fifth-generation fighters:* Although its primary role is air superiority, the F-22 is expected to play a role in SEAD as well. Its stealth capability allows it to approach SAMs undetected, its powerful avionics can detect enemy radar emissions, and its supercruise ability helps it outrun enemy missiles. However, the F-22 is not compatible with the HARM (which does not fit in the F-22's internal weapon bay), and it would have to rely on subsonic glide weapons to attack enemy

[18] U.S. General Accounting Office, *Combat Air Power: Funding Priority for Suppression of Enemy Air Defenses May Be Too Low*, Washington, D.C., April 10, 1996; Michael W. Pietrucha, "The Comanche and the Albatross: About Our Neck Was Hung," *Air and Space Power Journal*, May–June 2014.

[19] Amy Butler, "Bomber in a Pinch," *Aviation Week and Space Technology*, Vol. 172, No. 21, May 31, 2010. U.S. SEAD operations against Chinese SAM batteries and early-warning emitters are analyzed in more detail in the scorecard 4 analysis in Chapter Six.

[20] U.S. General Accounting Office, *Electronic Warfare: Comprehensive Strategy Still Needed for Suppression of Enemy Air Defenses*, Washington, D.C., November 25, 2002.

[21] Federation of American Scientists, "F-16 Fighting Falcon," web page, undated.

[22] Benjamin Lambeth, "Kosovo and the Continuing SEAD Challenge," *Air and Space Power Journal*, Vol. 16, No. 2, 2002. According to Lambeth, the speed difference between the F-16CJ and the EA-6B contributed to friction in SEAD missions during Operation Allied Force.

SAM sites.[23] The F-35 faces similar compatibility issues with the HARM. Both the F-22 and the F-35 will be in high demand for other (non-SEAD) missions.

- *Replacement of the EA-6B with the EA-18G:* The U.S. Navy replaced the EA-6B "Prowlers" with EA-18G "Growlers" between 2010 and November 2014. The Growler is an electronic warfare variant of the F/A-18F and carries the same jamming pods as the Prowler. In addition to improved airborne electronic attack capabilities, the Growler is capable of carrying not only HARMs for SEAD missions but also the Advanced Medium Range Air-to-Air Missile (AMRAAM) for self-defense.[24] The U.S. Marine Corps 27 EA-6Bs are scheduled for replacement by F-35Bs equipped with electronic warfare pods in 2019.[25]
- *Upgrade of EC-130H:* The U.S. Air Force has long relied on the EC-130H "Compass Call" to provide electronic warfare support, including jamming enemy communications. The Air Force plans to upgrade the EC-130H's electronic warfare suite to allow the Compass Call to disrupt enemy search and acquisition radars more effectively.[26]

Stealth

First-generation stealth aircraft were able to successfully penetrate the Iraqi IADS with no losses. Since then, the United States has retired the F-117 and introduced improved stealth designs in the form of B-2 bombers and F-22 fighters. The U.S. Air Force, Navy, and Marine Corps plan to procure large numbers of F-35 fighters in the near future. However, these advances in stealth capability have fallen short of initial plans. The United States had cut production of F-22 fighters from the 648 planned in 1991 to 187 by the time production lines were closed.[27] It also faces many cost, technical problems, and delays in the introduction of F-35 fighters, and it is unclear how many (if any) will be truly operational by 2017.

Standoff Strike Weapons

Since the 1980s, the U.S. military has developed and deployed a range of conventionally armed ALCMs, in addition to the conventional variant of the TLAM (TLAM-C), which is deployed on naval ships and submarines:

---

[23] Butler, "Bomber in a Pinch," 2010.

[24] *Jane's C4ISR and Mission Systems*, "Boeing EA-18G Growler," July 28, 2014.

[25] U.S. Marine Corps, *Marine Aviation Plan 2015*, Washington, D.C., 2014.

[26] U.S. Air Force, "EC-130H Compass Call," fact sheet, May 27, 2005.

[27] On changes to planned F-22 procurement, see *Jane's All the World's Aircraft*, "Lockheed Martin (645) F-22 Raptor," April 8, 2013.

- *CALCM:* In 1986, the U.S. Air force began to convert AGM-86B nuclear ALCMs into conventional weapons, designated AGM-86C CALCMs. These weapons have ranges exceeding 1,000 km and were first used in Operation Desert Storm.[28]
- *SLAM/SLAM-ER:* Beginning in the mid-1980s, the U.S. Navy began developing the AGM-84 Standoff Land Attack Missile (SLAM), based on its Harpoon anti-ship missile. First tested in 1990, the SLAM has a range of roughly 100 km.[29] The upgraded SLAM–Extended Range (SLAM-ER), which became operational in March 2000, boasts a range of 300 km.[30]
- *JASSM/JASSM-ER:* The Joint Air-to-Surface Standoff Missile (JASSM), which can be carried by both fighter and bomber aircraft, entered service with the Air Force in late 2010 and has a range of 400 km. The JASSM–Extended Range (JASSM-ER), with a range of approximately 1,000 km, entered service on Air Force bombers in April 2014.[31] The U.S. military intends to purchase around 2,500 JASSMs and almost 3,000 JASSM-ERs, with production ending in the late 2020s.

Table 5.2 shows how inventories of electronic attack aircraft, lethal SEAD aircraft, bombers, and standoff missiles have evolved in the U.S. Air Force and U.S. Navy across our snapshot years. (Numbers of standoff missiles are estimates.) As can be seen in the table, the Air Force and Navy have taken somewhat different paths in building their airspace penetration capabilities. The Air Force has emphasized stealth

Table 5.2
U.S. Air Defense Suppression and Strike Assets, 1996–2017

Asset Type	1996	2003	2010	2015 (current)	2017
Electronic attack					
EF-111 (Air Force)	40	—	—	—	—
EA-6B (Navy, Marine Corps)	143	120	96	27	27
EC-130 (Air Force)	27	27	14	14	14
EA-18G (Navy)	—	—	7	106	106
Total	210	147	117	147	147

[28] *Jane's Air-Launched Weapons*, "AGM-86 Air-Launched Cruise Missile (ALCM) and CALCM," October 22, 2014.

[29] *Jane's Air-Launched Weapons*, "AGM-84E SLAM, AGM-84H/K SLAM-ER," October 22, 2014.

[30] *Jane's Air-Launched Weapons*, "AGM-84E SLAM, AGM-84H/K SLAM-ER," October 22, 2014.

[31] *Jane's Defence Weekly*, "USAF Approves JASSM-ER," December 15, 2014; U.S. Air Combat Command, "Why COCOMs Continue Calling Upon the B-1B Lancer," March 9, 2015.

Table 5.2—Continued

Asset Type		1996	2003	2010	2015 (current)	2017
Lethal SEAD						
F-4G (Air Force)		54	—	—	—	—
F-16CJ (Air Force)		80	210	210	210	210
F-117 (Air Force)	Multirole	40	40	—	—	—
F-22 (Air Force)	Multirole	—	—	139	177	177
F-35 (Navy, Marine Corps, Air Force)	Multirole	—	—	—	—	?*
Total		174	250	349	387	387+
Long-range bombers						
B-52H		94	90	71	72	44
B-1B		95	60	64	63	65
B-2A		—	21	20	20	16
Total		189	171	155	155	125
Standoff missiles (est.)	**Range (km)**					
CALCM (Air Force)	1,300	130	200	450	450	450
SLAM-ER (Navy)	300	—	500	700	700	700
JASSM (Air Force)	370	—	—	800	1,300	1,500
JASSM-ER (Air Force)	930	—	—	—	555	1,000

SOURCES: Aircraft numbers are from IISS, *The Military Balance*, 1996, 2003, 2010, and 2015, with estimates for 2017 based on U.S. programs of record and IISS, *The Military Balance*, 2013. Missile ranges and inventories are from *Jane's Air-Launched Weapons*, "AGM-86 Air-Launched Cruise Missile (ALCM) and CALCM," October 22, 2014; *Jane's Air-Launched Weapons*, "AGM-84E SLAM, AGM-84H/K SLAM-ER," October 22, 2014; and *Jane's Air-Launched Weapons*, "AGM-158A JASSM (Joint Air-to-Surface Standoff Missile), AGM-158B JASSM-ER and LRASM," August 28, 2014.

NOTES: According to a study by the U.S. Government Accountability Office, the services will have taken delivery of more than 350 F-35s by 2017, but the aircraft will not have been fully tested and will not have reached IOC by that time. See U.S. Government Accountability Office, *F-35 Joint Strike Fighter*, 2013, pp. 6, 23. Many aircraft are multipurpose, and the categories in the table are not necessarily exclusive or definitive. The assets listed under "lethal SEAD" include HTS-equipped legacy aircraft as well as stealthy aircraft that might be used in a SEAD role; they do not include legacy aircraft equipped with HARM but not HTS.

aircraft and striking from beyond enemy air defense range with standoff weapons but has allowed its electronic attack capability to atrophy. The Navy remains committed to disrupting enemy air defenses through jamming and has invested relatively modestly in stealth and standoff capabilities. In actual operations, of course, the two services would be operating together, presenting a complex set of challenges to an adversary.

Penetrating Chinese Defenses

To get a sense of trends in Chinese air defenses and the U.S. capability to penetrate them, we employed an air penetration model that calculates the risk to attacking aircraft attempting to reach an aircraft weapon's range of each target in a notional target set. Here, to gain a general sense of changes to defensive and offensive capability, we make a number of simplifying assumptions in the application of the model. The most important is that U.S. bombers are allowed to take any path to their targets and are not limited by range.[32] (More realistic assumptions are applied in the assessment of U.S. attacks on Chinese airfields in Chapter Six.) To reflect the large variety of technology and tactics that the United States could employ, we parameterize several key offensive variables, including observability,[33] weapon ranges (from 30 to 1,000 km standoff), aircraft altitude (from 500 to 40,000 feet), SEAD support (with and without),[34] and fighter escorts (with and without).[35]

For Chinese forces, we mapped SAM and defensive counter-air missions. To determine current locations of early-warning radars and SAM garrisons in China, we used Google Earth imagery.[36] Since historical imagery and future site information are not available, we assumed that newer SAMs simply replaced older ones at the same

---

[32] We also assume that the attacking U.S. aircraft have exact knowledge of the location of air defense assets and can plot a path that minimizes exposure. In the case of relatively mobile SAM systems, we assume that the attacker knows the location only to within 10 km, forcing the penetrating aircraft to remain at a greater distance. On the other hand, we assume that the defending Chinese command-and-control capability works almost flawlessly. If, for example, early-warning radars can detect aircraft, valid tracks will be formed, passed to command-and-control nodes, and then passed on to the SAM or fighter units best located for an engagement. In both cases, we assume that systems function as designed and that both sides know how to use their respective equipment. Reality, of course, would be far messier.

[33] We examined a wide range of parametric radar cross-section levels, from high observability (with radar cross-sections associated with conventional bombers and large fighters) down to levels at which further reduction makes no further difference in survivability. For a brief explanation of stealth principles, see Rebecca Grant, *The Radar Game: Understanding Stealth and Aircraft Survivability*, Arlington, Va.: Mitchell Institute Press, 2010.

[34] We compared a baseline result with no SAM suppression and cases in which four key SAM sites have been removed and air defenses suffer effectiveness degrades to their probability of kill from jamming.

[35] We examined the results with no fighter protection and compared those with cases in which escorts were able to completely neutralize the air interception threat within a specified distance from the coastline.

[36] In doing this work, we capitalized on a vibrant online community that identifies and updates information on SAM sites in China and elsewhere. An overlay of SAM sites on Google Earth can be downloaded from Sean O'Conner, "Worldwide SAM Site Overview," *IMINT and Analysis*, June 2, 2013. Overhead imagery of the layout and disposition of the site can be viewed by zooming in on individual sites.

locations, using the SAM system numbers and types provided in Table 5.1.[37] Chinese doctrine calls for the employment of some fighters in DCA missions, even during offensive campaigns.[38] We estimate that approximately 16 patrols of four aircraft each could be continuously maintained over Chinese territory, requiring roughly 400 aircraft dedicated to the mission. To simplify the capability assessment, the model uses three notional interceptor types for DCA patrols, each with characteristics of several similar individual aircraft designs. Table 5.3 shows the number of each type of DCA patrol for each year, as well as relevant capabilities of the notional aircraft.

Figure 5.1 highlights the locations of SAM systems and DCA deployments, with estimated coverage areas for each year.

Having established parameters for offensive and defensive forces, we then considered a notional target set within China. We acknowledge that, even in a major conflict with China, the United States might not bomb targets in that country—a decision that would be made at the highest political levels. Our analysis is not an attempt to model an actual bombing campaign, nor is it meant to reflect the likely targets that might be struck in such a campaign. The intent was simply to select a collection of infrastructure and military targets that might cluster roughly in areas where operational targets might be found. In other words, this list of "targets" is intended to facilitate the test of U.S. ability to fly aircraft over operationally relevant areas in China.

To do this, we employ a representative sample of approximately 2,000 target locations to depict a notional distribution of possible targets relative to air defense emplace-

Table 5.3
Defensive Counter-Air Patrols and Aircraft Characteristics

Generation	1996	2003	2010	2017	Radar Range ($1m^2$/km)	Max. Missile Range (nose-on km)	Max. Intercept Radius (km)
2nd	7	—	—	—	20	15	570
3rd	9	16	14	9	53	40	825
4th	—	—	2	7	100	88	1,400

NOTES: Radar ranges, missile ranges, and intercept radius data are composite figures derived from *Jane's All the World's Aircraft* reports on the interceptors listed in Table 5.1. Note that the performance of even different models of single aircraft types (e.g., the J-7) can vary widely, depending on the specific aircraft variant and mission configuration. China is continuously upgrading the radar, avionics, missiles, and other capabilities of its older aircraft. Nevertheless, there are limits to what can be done with the older designs due to weight and other limitations.

[37] We also plotted early-warning systems and their approximate detection ranges against targets of various RCSs. Most early-warning sites were identified by the system. We placed proxy early-warning radars with performance levels similar to YLC-2s (detects 0-dB targets at 250 km) at sites where the radar type could not be determined.

[38] Roger Cliff, *The Development of China's Air Force Capabilities*, testimony before the U.S.-China Economic and Security Review Commission, Santa Monica, Calif.: RAND Corporation, CT-346, May 20, 2010.

Figure 5.1
SAM and Defensive Counter-Air Coverage

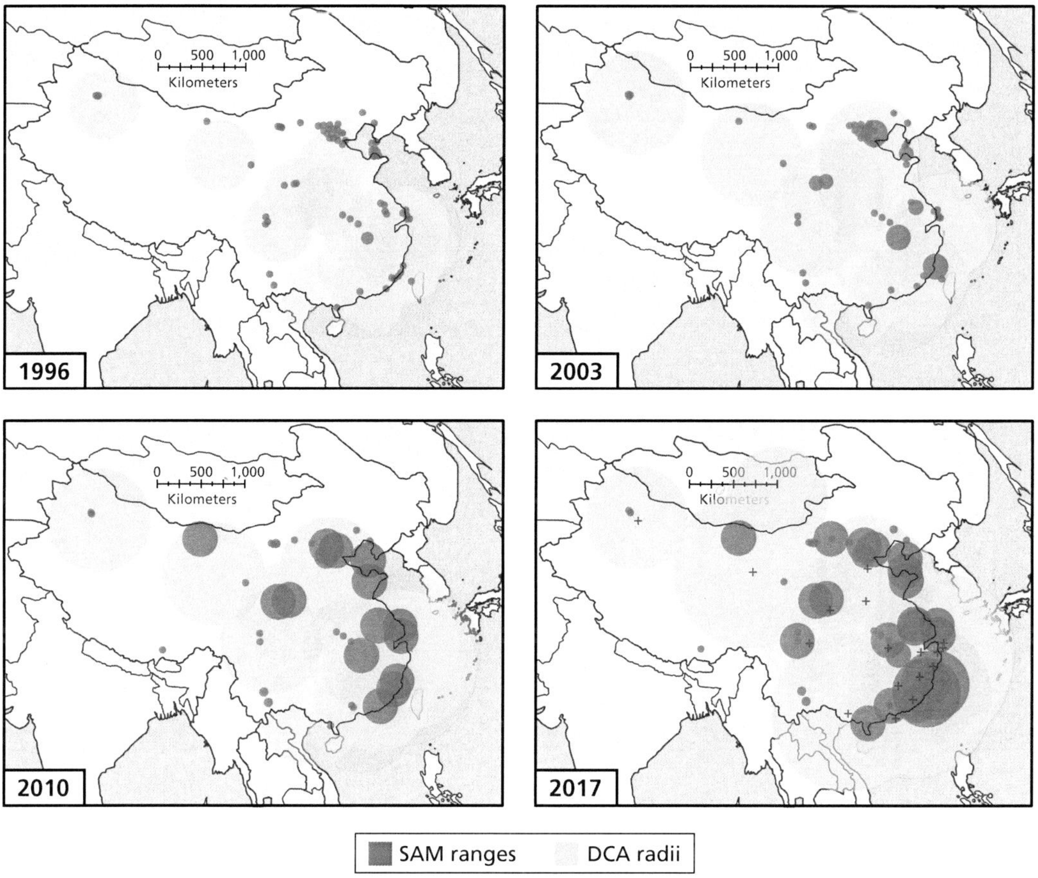

SOURCES: Missile ranges are from Planeman, "Bluffer's Guide: Fortress China: Main Area Defense Systems," 2009. SAM site and air base locations are from O'Conner, "Worldwide SAM Site Overview," *IMINT and Analysis Blog*, June 26, 2009.
RAND *RR392-5.1*

ments.[39] Figure 5.2 plots the locations (i.e., notional targets) used in the analysis. For the Taiwan scenario, we focused on the ability of U.S. aircraft to safely attack a subset of 823 targets (highlighted in red in Figure 5.2) that are located within 1,000 km of Taipei. The analysis of the Spratly Islands scenario focused on 100 target locations within 1,300 km of Thitu Island (green in Figure 5.2).

---

[39] These "targets" were identified via various databases and included air bases, shipyards, refineries, electrical power generation facilities, SAM and early-warning radars, and ballistic missile garrisons—in other words, some military targets and some infrastructure targets. Note that we are not accounting for some specific types of targets, such as hardened command-and-control facilities or mobile SAM and TBM launchers. These targets require specialized surveillance and munitions to successfully attack, so our evaluation did not attempt to address them

Figure 5.2
Targets in China

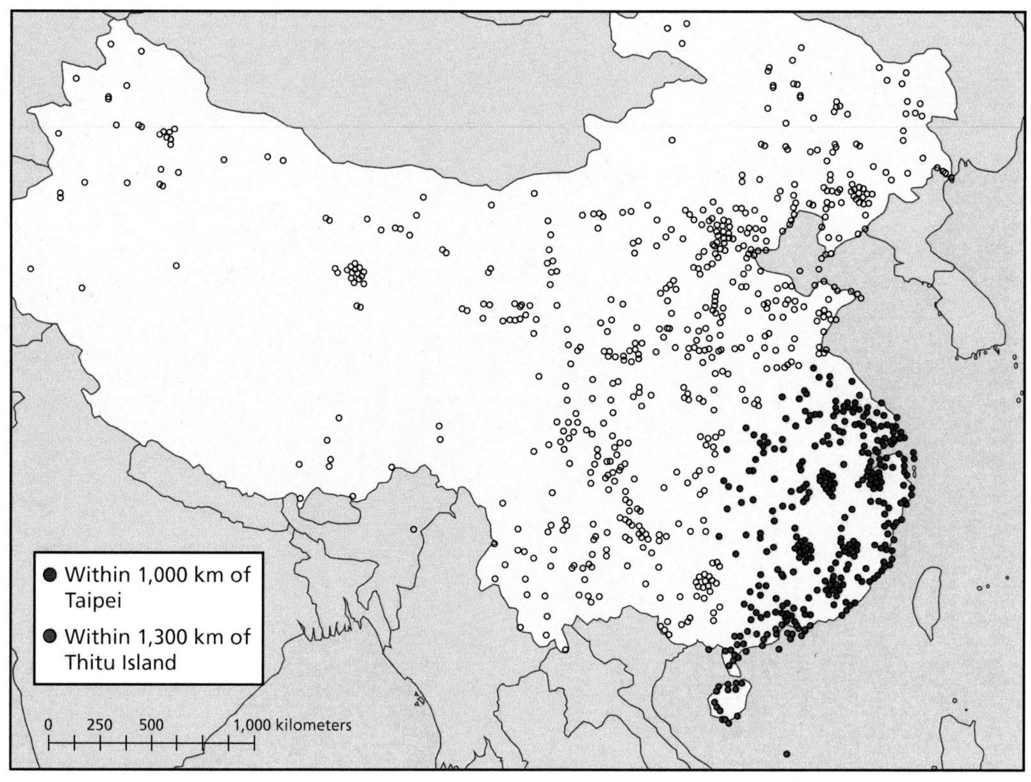

RAND *RR392-5.2*

The air penetration model assesses the degree of risk in attacking each target. The analytical problem is divided into three parts: (1) detection, (2) engagement by SAMs, and (3) engagement by air interceptors. The model begins with a contest between Chinese early-warning radars attempting to detect intruders and U.S. aircraft seeking to avoid detection. We calculated detection ranges for each early-warning and SAM battery against aircraft of different notional RCS levels and altitudes, and we plotted the results for all SAM and early-warning radars on a grid over the area of interest.

If detection occurs, the model assesses the ability of each SAM system to engage the attacking aircraft, given missile range and speed and the characteristics of the attacking aircraft.[40] Early-warning or SAM radar detection of U.S. aircraft also trig-

directly. However, these targets will most likely be located in areas where we already have targets, and so the risk level to reach them should be captured in our analysis.

[40] Having calculated all potential engagements, the model then calculates the lowest-risk path from the start point to the weapon standoff range of the target and back out to the start point. We used the number of engagements and probability of kill (Pk) per engagement to calculate overall risk due to SAMs with a simple $1 - (1 - Pk)^N$ formula. We used 0.75 as the Pk per engagement. When we included defensive ECM, we simply halved this Pk

gers air interception efforts by in-range DCA patrols. The likelihood of a successful interception is determined by early-warning coverage, the respective speeds and ranges of the aircraft, and the interceptor's operational and engagement ranges. Where early-warning coverage is effective, the model assumes that fighter aircraft receive cueing. Where it is not, fighters are assumed to patrol over a given area and move to engage if U.S. aircraft are within range of their onboard sensors (see Table 5.3).[41]

Finally, the model plots the Pks of all SAM engagements and air interceptions that a strike package faces at each point on the grid, and it plots the lowest-risk path from the start point to within weapon standoff range of the target and back out to the start point. Having determined the best route, the model then determines the aggregate risk faced by each strike package (given any particular RCS level and altitude) to produce a risk level between 0 (no chance of being destroyed) and 1 (assured destruction) for an attack on each target. For the purposes of coding, we label a risk of less than or equal to 0.1 as "moderate."[42]

Air Defense Penetration in the Taiwan Scenario

A Chinese conflict with Taiwan serves as the initial reference point for our analysis. We examined the ability of U.S. aircraft to access Chinese airspace across from the Taiwan Strait.

Baseline Case

Table 5.4 shows the percentage of targets within 1,000 km of Taipei that can be struck with moderate risk to aircraft of differing levels of detectability. Detectability is parameterized to reflect different aircraft sizes and RCS reduction.[43] The results correspond

to 0.375. For reference, this Pk is much higher than has been achieved in many historical cases. In Vietnam, for example, the SA-2 achieved a 0.1 Pk per missile fired in 1965 and a 0.025 Pk in 1966. However, since the Vietnamese and Chinese generally fired salvos of two or three missiles per engagement, the Pk per engagement (which we use here as our measure) was closer to 0.2 in 1965 and 0.05 in 1966. Also, the Vietnamese case involved many instances of firing blind or firing outside the engagement envelope, pushing the Pk down, whereas our model only includes engagements undertaken under relatively propitious circumstances. Finally, modern Russian double-digit SAMs have highly automated digital fire-control systems similar to those found in Western systems, with sophisticated seekers and a high degree of resistance to jamming. These characteristics would tend to push Pk up from those historical antecedents. For an estimate of Pks during the Vietnam War, see Steven J. Zaloga, *Red SAM: The SA-2 Guideline Anti-Aircraft Missile*, New York: Osprey Publishing, 2007, p. 19.

[41] When strike aircraft move beyond the range of escorting fighters (and are therefore alone), we assumed that if an adversary DCA fighter were able to close to within engagement range of the attacking bomber and maintain position sufficiently long, the probability of inflicting a mission-kill would be 1.0, halved to 0.5 when ECM aircraft are present.

[42] While a 0.1 probability of suffering a mission-kill will hardly seem "moderate" to a commander (much less the pilot of the aircraft), the single-shot Pks used to derive these probabilities were somewhat high and actual losses would presumably be somewhat lower.

[43] High RCS is 1 decibel per square meter, or roughly the RCS of a legacy bomber or large fighter with no improvements for RCS reduction. At the other end of the spectrum, very low detectability is simply defined as

Table 5.4
Percentage of Targets Accessible to Direct Attack at Moderate Risk to Attacking Aircraft, Taiwan Scenario

Detectability of Attacking Aircraft	% of Targets Accessible at Moderate Risk to Attacker			
	1996	2003	2010	2017
High	29	15	1	0
Medium	38	18	4	2
Low	74	57	51	42
Very low	100	100	100	93

NOTES: The target set for this analysis includes the 823 locations within 1,000 km of Taipei discussed elsewhere in this report. Values are derived from the air penetration model described in the text.

to high-altitude flight with direct attack munitions (i.e., with general-purpose bombs rather than missiles) and no SEAD. We consider the impact of SEAD and standoff weapons later in this chapter.

The results of the base case illustrate important trends. In 1996, obsolete Chinese SAMs and fighters with limited range and weaponry provided only spotty coverage. Consequentially, even U.S. legacy aircraft were able to access large portions of Chinese airspace, corresponding to low-risk access to approximately 29 percent of the Taiwan scenario target set at high RCS on direct-attack missions. By 2010, high-RCS aircraft could no longer attack targets across from Taiwan without standoff weapons, SEAD support, or other means of enhancing survival. These findings also indicate that marginal RCS improvements to existing fourth-generation airframes may not be sufficient to produce significant gains in air penetration capability, though moving to "low" and "very low" RCS levels would have a more meaningful impact.

On the other hand, stealthy aircraft remain capable of penetrating reasonably large portions of Chinese airspace even in later years. However, this finding comes with critical caveats. First and foremost, the small number of fifth-generation fighters—combined with the growing demand for them in the air-to-air war—will limit the number of such aircraft available for striking targets in China or escorting penetrating bombers. As of early 2015, the United States had 264 stealthy F-22 and F-35 fighters in service but more than 2,459 F-15, F-16, and F-18 fighters and fighter-bombers in its inventory, yielding about nine legacy aircraft for every fifth-generation aircraft.[44]

the level with low-risk access to 100 percent of the target set against the 2010 threat. The intervening medium and low levels of detectability are simply evenly spaced in decibels per square meter between these two extremes. There is a large body of literature discussing real-world RCS issues and values. See, for example, Grant, *The Radar Game*, 2010, and Serdar Cadirci, *RF Stealth (Or Low Observable) and Counter-RF Stealth Technologies: Implications of Counter-RF Stealth Solutions for Turkish Air Force*, thesis, Monterey, Calif.: Naval Postgraduate School, March 2009.

[44] Inventory figures are from IISS, *The Military Balance*, 2015.

Improvement in Survivability Using Standoff Weapons

Increasing the range of weapons deployed by aircraft can significantly improve the mission survivability of legacy aircraft by allowing the firing aircraft to remain at greater distances from the target. This could enable standoff aircraft to play a role in air strikes even during the first weeks of combat (i.e., before the Chinese air defense system is degraded). In 2017, the model suggests that medium-RCS aircraft could attack 90 percent of the Taiwan target set with moderate risk to the aircraft using 400-km-range weapons (JASSM) or 100 percent of the Taiwan target set using 1,000-km weapons (CALCM and JASSM-ER), as opposed to 2 percent using 30-km-range direct-attack (or free-fall) munitions.

However, like stealth aircraft, the inventory of standoff weapons is limited. Given this fact, the most logical employment concept is to use standoff missiles against the most heavily guarded targets, leaving lower-risk targets for direct-attack munitions. We applied this tactic to evaluate attacks against the near-Taiwan target set by medium-RCS aircraft (see Table 5.5). We ran the model using half of the total U.S. inventory of CALCMs and JASSMs, assuming that the United States would want to hold back a significant portion of its inventory for other contingencies.[45] (For comparative purposes, the table also shows the percentage of targets that could be attacked using direct-attack munitions only.)

The model results show that, by 2010, larger U.S. inventories of standoff weapons significantly mitigate—but do not negate—the impact of improving Chinese IADS on the ability of U.S. legacy aircraft to penetrate Chinese airspace. The 2017 outcome suggests that growing quantities of cruise missiles can create net gains for the United

Table 5.5
Percentage of Targets Accessible to Attack With and Without the Use of Standoff Weapons, Taiwan Scenario

	% of Targets Accessible to Attack			
Attack Method	**1996**	**2003**	**2010**	**2017**
Standoff and direct attack	46	20	11	17
Direct attack only	38	18	4	2

NOTES: Figures for standoff and direct attack employ the disposable inventory of standoff weapons against the Taiwan scenario target set (within 1,000 km of Taipei). Values were generated using the air penetration model described in the text. The table shows results for medium-RCS aircraft and moderate risk to the attacker.

[45] Full inventory data on standoff missiles can be found in Table 5.2. Where standoff weapons are employed, we estimated that eight missiles would be necessary to achieve a 90-percent probability of destroying each target, assuming that (1) each target includes three aim points, (2) the missiles have a reliability of 90 percent, (3) each missile has a 0.8 probability of kill against an aim point, and (4) there is a 20-percent chance of each missile being shot down.

States. However, although the U.S. inventory of standoff weapons is growing, it will never be unlimited. Hence, in a war lasting more than a few weeks, the overall U.S. ability to attack targets at reasonable risk to the aircraft involved will largely depend on the degree to which Chinese air defenses (and particularly double-digit SAMs) are destroyed or degraded by the time the inventory of standoff weapons reaches critical levels. Both stealth technology and long-range missiles improve U.S. airspace penetration capability. However, neither is currently available in the necessary quantities to offset China's increasing ability to exclude legacy aircraft from its airspace. Though the United States is planning to procure a total of 2,443 stealthy F-35s in the coming years, the program suffered significant delays, and some question how many will ultimately be purchased.[46]

Other Operational Adjustments

Given the limitations on time and space, a full analysis of all of the tactical and operational options available to both sides is impossible. Suffice it to say, however, that both sides have a host of measures that could improve their respective positions, in addition to those discussed here. The U.S. military could seek to improve the odds through the use of SEAD, fighter cover, or low-altitude attack, while Chinese forces could employ mobility, surprise, and selective engagement strategies.

Most air campaigns open with an effort to degrade enemy air defenses through a combination of jamming and kinetic-kill weapons. If U.S. SEAD efforts (e.g., HARM strikes and ECM) succeeded in neutralizing four of the most threatening SAM sites and in reducing the Pk of SAM and air intercept engagements by 50 percent, penetration results would improve significantly.[47] For example, in 2010, medium-RCS aircraft would have been able to reach 21 percent of all Taiwan scenario targets with moderate risk to the attacker (up from 4 percent without SEAD). In 2017, the same aircraft would be able to reach 16 percent of Taiwan scenario targets (up from 2 percent without SEAD). However, because of improvements in the mobility and jam resistance of Chinese SAMs, it may become difficult and costly to execute SEAD missions that would produce the results given here. (Lethal SEAD engagements between U.S. strike aircraft and Chinese SAMs are modeled later in this chapter.)

The United States could provide fighter cover to protect its strike aircraft from interception by enemy fighters. Our analysis of this option suggests that escorts could have reduced the risk to strike aircraft, particularly in the 1996 and 2003 periods, when much of the threat to attacking U.S. aircraft came from defensive combat air rather than SAMs. However, these gains diminish substantially in the 2010 and 2017 cases as Chinese SAMs become more capable and as U.S. refueling tankers are pushed

---

[46] For the current status and prospects, see U.S. Government Accountability Office, *F-35 Joint Strike Fighter: Assessment Needed to Address Affordability Challenges*, Washington, D.C., GAO-15-364, April 2015.

[47] This Pk degrade is an attempt to account for many possible effects from ECM and electronic attack.

farther from the Chinese coast. Moreover, as in the case of SEAD, providing fighter escorts (or offensive combat air sweeps) would require larger force packages and potentially expose additional aircraft to SAM attack.

Another possible U.S. tactic to reduce risk is the use of low-altitude flight. During the later stages of the Cold War, U.S. Air Force pilots trained extensively in low-altitude flight to mitigate the risk from long-range SAMs. However, low-altitude flight makes aircraft vulnerable to fire from short-range SAM and anti-aircraft artillery. Our modeling of low-altitude flight (in which we assumed that penetrating aircraft must deal with ten randomly placed anti-aircraft artillery batteries in each target area in addition to longer-range systems) shows substantial gains in the survivability of legacy aircraft. However, our understanding of the modeling results is tempered by several considerations: The assumption of perfect information about defenses is even more questionable in the case of short-range air defenses than in the case of long-range SAMs, U.S. forces no longer train as extensively in low-altitude flight and attack, and the British abandoned low-level attack for the duration of the Gulf War after their use of the tactic resulted in heavy losses.[48]

The Chinese also have tactical options. Although the PLAAF has limited recent combat experience, it does have a rich post-1949 air defense history on which to draw for concepts and ideas.[49] Chinese doctrine emphasizes mobility, deception, and ambush in the employment of SAMs. It also emphasizes the massing of resources (SAMs or aircraft) to create local and temporary superiorities, even when the overall balance of airpower may be adverse to PLA forces. Recently, Chinese doctrine has also called for the more flexible employment of airpower and the employment of special units to conduct "air hunting" and "air sweeps" within designated areas—a departure from the PLAAF's emphasis on ground-controlled interception.[50] While current Chinese

---

[48] British losses during the Gulf War may not be representative of those associated with low-level airspace penetration, since the former came largely during the strikes themselves and were associated with particularly hazardous conditions. On issues related to training for low-level flight, see U.S. General Accounting Office, *Military Training: Limitations Exist Overseas but Are Not Reflected in Readiness Reporting*, Washington, D.C., GAO-02-525, April 30, 2002. The report states, "Very few of the training needs can be satisfied" for the U.S. Air Force base on Okinawa. See also, Rebecca A. Efroymson, Winifred Hodge Rose, Sarah Nemeth, and Glenn W. Suter, *Ecological Risk Assessment Framework for Low-Altitude Overflight by Fixed-Wing and Rotary-Wing Military Aircraft*, Oak Ridge, Tenn.: Oak Ridge National Laboratory, 2000.

[49] Chinese aircraft and anti-aircraft artillery contested U.S. air operations over North Korea during the Korean War. Chinese SAMs jousted with Taiwanese reconnaissance aircraft over Chinese airspace during the 1960s, destroying several U-2s at high altitude. The PLAAF dispatched large numbers of SAM and anti-aircraft artillery units to North Vietnam and fought a sustained and, by many measures, highly effective campaign against U.S. bombers during the Vietnam War. And Chinese air defenses reportedly engaged Vietnamese air elements in 1979. Although Chinese air defenses did not win all of their battles, they proved dogged and resourceful, complicating the task of the attacker.

[50] On the operational aspects of Chinese air defense thinking, see Cliff et al., *Shaking the Heavens and Splitting the Earth*, 2011, pp. 130–143.

training may not prepare its forces to execute all of these options on a wide scale, it is becoming more extensive and realistic over time.

Summary of the Taiwan Scenario

Despite possessing an impressive array of technological and tactical approaches to improve airpower projection capability, the U.S. ability to penetrate Chinese air defenses in the context of an air campaign over Taiwan has declined. This does not mean that U.S. airpower is by any means impotent against Chinese IADS but, rather, that U.S. defense planners can no longer rely on an ability to access Chinese airspace with the ease to which they have become accustomed in other post–Cold War conflicts. The permissive environment that U.S. airpower would have enjoyed over Chinese territory in 1996 has evolved into a contested one. While the United States can still execute stealthy and standoff attacks against some targets without endangering the penetrating aircraft, a variety of tactics and survivability techniques may have to be employed to enable the large-scale use of nonstealthy aircraft and direct-attack munitions. But given the demands of the air superiority contest and the threat to forward bases, there may be few resources to devote to SEAD and escort missions, limiting U.S. options for air penetration and air strike.

Air Defense Penetration in the Spratly Scenario

How do U.S. prospects in the Spratly Islands scenario compare with those in the Taiwan case? In the Spratly scenario, the majority of PLA combat power will be generated from bases along China's southern coast and on Hainan Island. Our limited target set here includes those within 1,300 km of Thitu Island, the focal point of the scenario. (This distance represents the longest unrefueled range of maritime strike fighters in the PLA arsenal, the Su-30 and JH-7, if fuel reserve for maneuvering is maintained.)[51]

The scenario-relevant area for the Spratly Islands scenario differs from that in the Taiwan scenario in three important respects. First, Beijing has not historically deployed its most modern SAM systems to the southern coast. The longest-range SAM systems guarding the area in 2010 were the HQ-2 (range 35 km) and HQ-12 (range 50 km). Given the increased salience of South China Sea security interests and the possible deployment of China's new *Jin*-class SSBNs to the Yulin naval base on Hainan Island, we credit the area with two HQ-9 SAM batteries in 2017. Second, the smaller area that might support Chinese operations includes a far smaller number of targets (roughly 10 percent as many as in the Taiwan case).[52] The smaller number of targets

---

[51] See *Jane's All the World's Aircraft*, "Sukhoi Su-30M," August 26, 2014, and *Jane's All the World's Aircraft*, "XAC JH-7," January 7, 2015.

[52] The Spratly Islands scenario includes only 85 targets, compared with 823 in the Taiwan case. Once again, this should not be considered the actual set of targets that an air campaign would neutralize. But because they were generated using standardized categories of facilities, the relative number reflects the impact of distance on the magnitude of a possible operational target set.

allows limited inventories of stealth aircraft and standoff weapons to have a much greater impact. Finally, the relevant target set does not extend as deeply into the Chinese mainland, making it easier for tankers (and the fighter escorts they might support) to approach relatively closer to target areas. The following sections provide a detailed account of these findings.

Baseline Outcome

As in the analysis of the Taiwan conflict, we initially assessed risks for high-altitude aircraft with direct-attack munitions. Table 5.6 provides the percentage of targets accessible at "moderate" risk in the Spratly scenario, with comparable figures for the Taiwan scenario.

The Spratly scenario threat environment is significantly more permissive than that in the Taiwan scenario, and the modeling results exhibit a less steep negative trend. Moreover, the same limited inventory of U.S. stealth aircraft can play a larger role in this scenario. In both scenarios, stealth aircraft remain capable of penetrating the airspace surrounding many targets. But in the Taiwan case, the limited number of stealth aircraft, as well as the heavy demand for them in other missions, may prevent them from being deployed against a large percentage of targets. In the Spratly case, the target set is an order of magnitude smaller, and stealth aircraft may therefore be able to attack a correspondingly larger percentage of targets. Moreover, other requirements for U.S. stealth aircraft, such as the air superiority battle, are likely to be more modest, freeing more of them for strike missions.[53]

Needless to say, should China deploy more advanced air defenses than anticipated in Southern China, the penetration of Chinese air space would become more difficult,

Table 5.6
Percentage of Targets Accessible to Direct Attack at Moderate Risk to Attacking Aircraft, Taiwan and Spratly Islands Scenarios

Detectability of Attacking Aircraft	% of Targets Accessible at Moderate Risk to Attacker							
	1996		2003		2010		2017	
	Spratly Islands	Taiwan	Spratly Islands	Taiwan	Spratly Islands	Taiwan	Spratly Islands	Taiwan
High	58	29	38	15	14	1	9	0
Medium	58	38	40	18	21	4	17	2
Low	86	74	72	57	63	51	52	42
Very low	100	100	100	100	100	100	98	93

NOTES: The Spratly Islands scenario target set includes 85 targets on Hainan and the mainland within 1,300 km of Thitu Island. The Taiwan scenario target set includes the 823 targets within 1,000 km of Taipei. These values were generated using the air penetration model described in the text.

[53] See the scorecard 2 analysis in Chapter Four for more detail.

but factors related to geography and the target set would continue to make the U.S. task easier than that associated with the Taiwan scenario.

Standoff Weapons in the Spratly Islands Scenario

As in the Taiwan case, standoff weapons could have a major impact in the Spratly scenario by enabling even legacy aircraft to deliver weapons from beyond the effective range of Chinese air defenses. Table 5.7 presents the percentage of Spratly scenario targets that could be struck with moderate risk to the attacking aircraft with a combination of direct and standoff attack (with comparative figures for the Taiwan case). As in the Taiwan case, the United States is assumed to allocate standoff missiles to well-defended targets and conduct direct attacks against targets that are less well defended.[54]

Compared with the Taiwan case, the same arsenal of U.S. standoff weapons employed against fewer targets leads to even greater improvements in the Spratly results. The U.S. inventory of standoff weapons in 2010 and in 2017 (projected) enables even medium-RCS aircraft to threaten the entire Spratly Islands scenario target set at moderate risk to the attacking aircraft. The 2010 and projected 2017 results show a net improvement for the United States, due largely to the introduction and projected growth in the number of JASSM cruise missiles.

Lethal Suppression of Enemy Air Defenses

We now turn to the lethal SEAD, or the destruction of SAM systems, which would almost certainly figure into any large-scale effort to penetrate and strike within Chinese airspace. Prior to launching large-scale air campaigns over contested airspace, the United States typically executes SEAD operations to degrade and destroy ground-based air defenses. SEAD encompasses a broad array of activities and includes lethal attacks

Table 5.7
Percentage of Targets Accessible to Attack by Standoff and Direct Attack, Taiwan and Spratly Islands Scenarios

Scenario	1996	2003	2010	2017
Spratly Islands	58	55	94	100
Taiwan	46	20	11	17

NOTES: The table shows results for medium-RCS aircraft and moderate risk to the attacker (defined by a probability of destruction of less than or equal to 0.1). Figures include access to standoff and direct attack employ the disposable inventory of standoff weapons against the Spratly scenario target set (within 1,300 km of Thitu Island). These values were generated using the air penetration model described in the text.

[54] The numbers in the table are for medium-RCS aircraft, and, as in the Taiwan case, the United States uses 50 percent of its standoff inventory, allocating eight missiles per target.

(e.g., physical destruction by bombs or missiles) and nonlethal measures (e.g., jamming communications and sensors). Due to the impracticality of eliminating all hostile air defenses, U.S. SEAD concepts often focus first on neutralizing long-range SAMs to establish high-altitude sanctuaries and then expanding to attacking other targets.

In recent conflicts, U.S. SEAD concepts have proven effective in suppressing the early-generation Soviet-designed SAMs that historically made up the bulk of PLA defenses. However, Pentagon planners are concerned that China's procurement of advanced Russian SAMs and comparable indigenous systems could enable the PLA to frustrate U.S. SEAD efforts. We addressed this by evaluating the performance of U.S. SEAD packages against an evolving arsenal of SAM batteries. The results will help us better understand the relative ability of U.S. SEAD missions to facilitate the types of air penetration missions discussed here.

Methodology

As an indicator of relative capabilities over time, we simulated one-on-one duels between the systems available to both sides in 1996, 2003, 2010, and 2017 (projected). Real-world SEAD missions can involve substantially larger packages of forces, as well as complex tactics. On the defender's side, SAMs would typically be arranged into mutually supporting positions. Visual or electronic decoys would be deployed, and fighter aircraft might orbit nearby to distract and engage SEAD aircraft. Attacking forces would attempt to approach from multiple directions, in numbers greater than could be readily sorted and engaged. Some of these numbers would include decoys and electronic warfare assets, and tactics could include feints and other deceptive operations.

With a variety of options available to both sides, actual SEAD missions and defending against them would involve rapid adjustments to tactics. Given these possibilities, the results of the one-on-one engagements modeled here will not predict actual outcomes in the complex air-to-ground combat environment that would characterize a military conflict between the United States and China. Nevertheless, the analysis does provide a first-order indicator of the respective challenges facing the two parties. While the underdog in one-on-one combat may be able to design force packages and tactics to compensate for disadvantage, the side that enjoys advantages in single combat will presumably have a relatively easier time than it would otherwise in more complex engagements.

In our analysis, we focus on the end of a SEAD mission, in which a strike aircraft employs a weapon against a located and active SAM site. The simulated U.S. concept of operations involves flying at the known enemy SAM position, firing a weapon at the SAM radar from maximum range, and then disengaging. The ability of the defending SAM to detect and identify attacking aircraft depends on the RCS of the strike aircraft and the characteristics of the defending surveillance radar. When a detection occurs, the SAM crew can choose to attack the U.S. aircraft or to tear down and relocate. We assume the SAM can detect the attacker's weapon launch and calculate an expected time of arrival. If this is greater than the teardown time, the SAM crew will relocate and escape.[55]

If the SAM is able to detect the incoming SEAD aircraft and launch missiles that can reach the aircraft's position *before* the aircraft can launch a SEAD weapon, then the SAM wins the engagement.[56] If the aircraft can launch a SEAD weapon that hits the SAM's position before the SAM battery can return fire and destroy the aircraft, the SEAD aircraft wins the engagement. If both the SAM and SEAD aircraft are able to launch weapons that reach their targets' locations, the engagement is scored as a mutual kill. Finally, if the SAM tears down and relocates, both platforms survive the encounter, although the SEAD mission may be considered successful.[57]

For each snapshot year examined, we pitted contemporary U.S. SEAD-capable aircraft against the most capable PLA SAM battery deployed in the operational areas most relevant to the scenario. In each year, U.S. fourth-generation aircraft armed with HARM missiles is one primary U.S. option.[58] We also include fifth-generation aircraft armed with subsonic glide munitions as an alternative SEAD platform for the 2010 and 2017 cases. Table 5.8 shows the notional matchups for each year in the Taiwan and Spratly Islands scenarios.

To capture various possibilities, we parameterize several key characteristics of the U.S. forces and tactics: aircraft RCS, weapon range and velocity, the altitude of attacking aircraft, and the use of jamming against SAM radar.[59] The estimated performance characteristics of Chinese SAM batteries are listed in Table 5.9. Note that

---

[55] We also assume that the SEAD weapon is autonomous and so will home in on a target even if the launch aircraft disengages or is destroyed. In contrast, the SAM-launched missile needs to be continuously guided by the engagement radar. If the SAM radar tears down or is destroyed, any missile that it was guiding becomes inert.

[56] Note that the coding for the engagements outlined here produces more decisive results than are perhaps warranted. No weapon is fully reliable. Both aircraft and SAMs have defensive measures that can improve survivability, including jamming, decoys, defensive fire, and point defenses. A more accurate interpretation of a "win" here is gaining first-strike advantage. Thus, a "SEAD win" means that the U.S. aircraft can effectively attack the SAM without fear of retaliatory fire. A "SAM win" indicates that the U.S. aircraft will have to evade one or more surface-to-air missiles before it has the opportunity to launch its weapon, and a "mutual kill" means that both the SAM and the aircraft will endure attacks from the other.

[57] This obviously reflects a choice favoring the SAM's survivability. It could remain in place instead, resulting in one of our other three outcomes.

[58] These would primarily include the F-16CJ, the EA-6, and the F/A-18, all of which are equipped with HARM targeting systems.

[59] RCS levels are parameterized from larger conventional aircraft with no reduction down to the point that it makes no further difference to survivability. We considered SEAD weapon ranges from 0 to 200 km. The two types of weapons that the United States is likely to employ, HARMs and small-diameter bombs (SDBs), both have ranges of approximately 110 km when launched from high altitudes. We examined outcomes for weapon speeds of Mach 3.0 (HARM) and Mach 0.8 (SDB). We ran both high-altitude cases, in which U.S. aircraft fly at 40,000 feet, and low-altitude cases, in which they fly at 500 feet. We examined a base case without electronic warfare disruption and a case in which U.S. jamming degrades PLA radar detection ranges by 50 percent. On U.S. SDBs and HARMs, see *Jane's Air-Launch Weapons*, "GBU-39/B Small Diameter Bomb (SDB I),

Table 5.8
SEAD-Versus-SAM Matchups, by Year

Year	United States	China (Taiwan Scenario)	China (Spratly Islands Scenario)
1996	4th-generation aircraft with HTS	HQ-2 (SA-2; S-75)	HQ-2 (SA-2; S-75)
2003	4th-generation aircraft with HTS	SA-10 (S-300 PMU)	HQ-2 (SA-2; S-75)
2010	4th-generation aircraft with HTS or 5th-generation aircraft	SA-20 (S-300 PMU-2)	HQ-2 (SA-2; S-75)/HQ-12
2017	4th-generation aircraft with HTS or 5th-generation aircraft	SA-21 (S-400)[a]/HQ-9 (S-300 PMU-2)	HQ-9 (S-300 PMU-2)/SA-21 (S-400)[a]

[a] As noted earlier in the text, the S-400 is not currently in the Chinese inventory and may not be by 2017, but reports indicate China and Russia have reached agreement on the sale of 36 launchers and these could potentially be in place by 2017. If not, then HTS-armed fourth-generation aircraft would be matched against the HQ-9s (S-300 PMU-2).

Table 5.9
Selected Chinese SAM Characteristics

SAM Type	Radar Range (km)	Missile Range (km)	Missile Speed (Mach)	Fire Delay (sec)	Teardown (min)
HQ-2 (SA-2)	80+	35	3.5	30	30
S-300 PMU-1 (SA-20A)	200	150	6.0	10	10
S-300 PMU-2 (SA-20B)	250	200	6.0	10	5
S-400 (SA-21)	400	400	6.0	10	5

SOURCES: Carlo Kopp, *Almaz S-300P/PT/PS/PMU/PMU1/PMU2; Almaz-Antey S-400 Triumf; SA-10/20/21 Grumble/Gargoyle*, Air Power Australia, Technical Report APA-TR-2006-1201, updated January 2014; *Jane's Land-Based Air Defence*, "S-75 Family," February 17, 2015.

although SAM radar teardown times can found in specifications for the systems, actually achieving these times would require substantial training and practice on the part of the PLA operators and may be regarded as optimistic.

SEAD Results for the Taiwan Scenario

Analytical results show HARM-armed fourth-generation aircraft to be excellent SEAD platforms against older HQ-2s, even without evasive tactics or offensive electronic warfare support. The long range of the HARM (in excess of 100 km) gives the United States the ability to attack from beyond an HQ-2's detection range. The

GBU-39B/B Laser SDB," January 2, 2015, and *Jane's Air-Launched Weapons*, "AGM-88 (High-Speed Anti-Radiation Missile)," October 22, 2014.

S-300 PMU-1 (from 2003) and S-300 PMU-2 (from 2010) create a more threatening environment and can be engaged only if the attacking aircraft flies at low altitude or if electronic warfare aircraft are able to successfully jam Chinese search and engagement radars. However, both of these options would be problematic. Low-altitude flight would expose the attacker to short-range air defense and man-portable air defense threats. In the case of electronic warfare, modern SAM systems like the S-300 series incorporate jam-resistant technology, and U.S. EA-18G and EC-130H aircraft would have to operate in a very-high-threat environment. The difficulties faced by fourth-generation aircraft in tackling S-400 batteries, would be even more severe. Using fifth-generation aircraft could improve results, but because no fifth-generation aircraft is projected to be HARM-compatible within our time frame, the gains will be less than they might otherwise be.

Baseline Case

Figure 5.3 shows model outcomes for our base cases. In these cases, U.S. fourth-generation aircraft, armed with HARMs and flying at high altitude, attack different SAM systems without the benefit of electronic warfare support. Each diagram represents the performance of a particular SAM system against attacking aircraft and weapons, while the blue dot represents an attacking aircraft and SEAD weapon. (In the base case, the blue dot represents a fourth-generation aircraft armed with HARM.) The RCS of the attacking SEAD aircraft rests on the horizontal axis, while the range of the SEAD weapon is on the vertical axis. The green area represents combinations of parameterized variables in which the SEAD aircraft "wins" against the SAM system represented in the diagram, yellow represents a "mutual kill," and red indicates that the SAM "wins."

As shown in Figure 5.3, U.S. fourth-generation aircraft, such as the F-16CJ, armed with HARMs are able to easily defeat HQ-2 SAMs under the conditions specified. Against the more advanced SAM systems available to China in later years, however, U.S. legacy aircraft are unable to win in a simple one-on-one matchup. As China deploys larger numbers of advanced, long-range SAMs, the risk of conducting SEAD with such aircraft will increase dramatically.

As suggested earlier in this chapter, the results in real life would depend on the maneuver and firing tactics employed by both sides, as well as the technical vagaries of detection ranges and targeting. Outcomes would not be as predictable or clear-cut as the figure suggests. Moreover, the plots present one-on-one duels between the most advanced SAMs in the Chinese inventory at any one point in time against a given type of SEAD system and do not reflect the full range of potential systems that might be engaged. Nevertheless, the plots in Figure 5.3 provide a sense of the relative difficulty of the lethal SEAD mission over time and the degree to which one side or the other would have to employ superior tactics or allocate additional assets to gain the edge or remain competitive.

Figure 5.3
Baseline Results for U.S. Fourth-Generation Aircraft Versus Selected Chinese SAM Systems

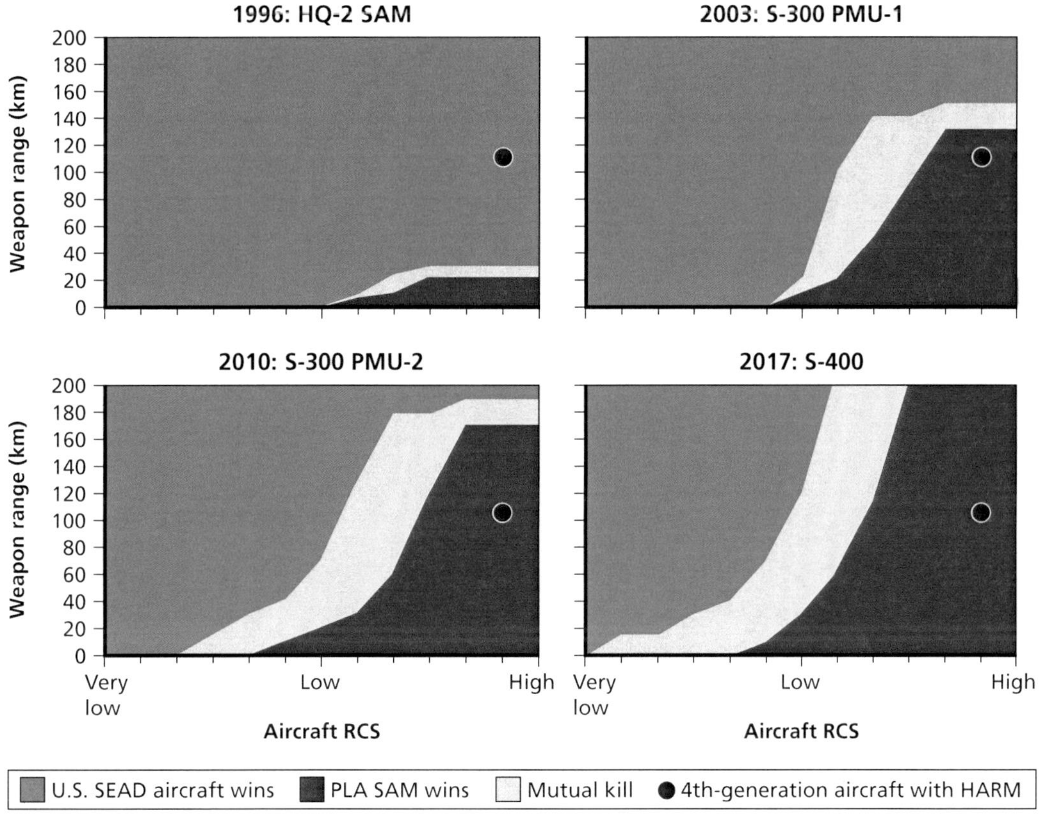

SOURCES: Inputs used to generate the figures include aircraft speed, HARM range and speed, SAM range and speed, and detection ranges. Sources for SAM inputs are *Jane's Strategic Weapons Systems*, "S-300/Favorit (SA-10 'Grumble'/SA-20 'Gargoyle')," July 18, 2013; Kopp, "Almaz S-300P/PT/PS/PMU/PMU1/PMU2; Almaz-Antey S-400 Triumf; SA-10/20/21 Grumble/Gargoyle," 2014; and *Jane's Land-Warfare Platforms*, "S-75 Family," February 2015. Source for HARM data is *Jane's Air-Launched Weapons*, "AGM-88 HARM (High-Speed Anti-Radiation Missile)," October 22, 2014.

RAND *RR392-5.3*

Stealth

The use of fifth-generation aircraft could significantly improve the results of U.S. SEAD operations. Stealth characteristics reduce the effective engagement ranges of enemy SAMs and thus enable SEAD strikes from safe distances. Figure 5.4 shows the results of a U.S. aircraft employing a subsonic glide weapon (such as a small-diameter bomb) against PLA SAMs in 2010 and 2017. Purple represents cases in which the SAM unit is able to tear down and escape prior to weapon impact.[60] A fifth-generation air-

---

[60] In the earlier plots in Figure 5.3, the Mach 3 HARM reaches the target quickly enough that tearing down and moving the SAM is not an option. In the stealth case, U.S. aircraft deploy subsonic weapons, and Chinese SAMs may tear down to escape, depending on the weapon release range.

Figure 5.4
Fifth-Generation Aircraft with Glide Weapons Versus Selected PLA SAM Systems

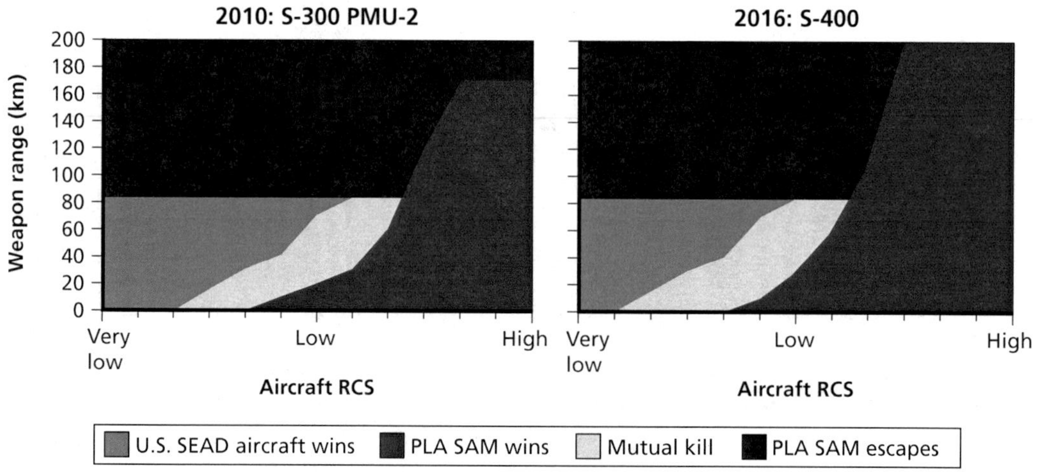

2010: S-300 PMU-2
2016: S-400
Weapon range (km)
Aircraft RCS
Very low
Low
High
U.S. SEAD aircraft wins
PLA SAM wins
Mutual kill
PLA SAM escapes

SOURCES: Inputs used to generate the figures include aircraft speed, HARM range and speed, SAM range and speed, and detection ranges. Sources for SAM inputs are *Jane's Strategic Weapons Systems*, "S-300/Favorit (SA-10 'Grumble'/SA-20 'Gargoyle')," July 18, 2013; Kopp, "Almaz S-300P/PT/PS/PMU/P-MU1/PMU2; Almaz-Antey S-400 Triumf; SA-10/20/21 Grumble/Gargoyle," 2014; and *Jane's Land-Warfare Platforms*, "S-75 Family," February 2015. Source for HARM data is *Jane's Air-Launched Weapons*, "AGM-88 HARM (High-speed Anti-Radiation Missile)," October 22, 2014.
RAND *RR392-5.4*

craft and weapon combination is not plotted on this figure, since the RCS levels of U.S. stealth aircraft types are not publicly available. Presumably, however, that plot would lie considerably to the left (indicating a lower RCS) of the earlier plots for fourth-generation aircraft.

One drawback of fifth-generation SEAD aircraft is that they are incompatible with the HARM and so would have to attack enemy emplacements with subsonic glide weapons. Current U.S. standoff glide weapons have ranges of 100 km or more when employed from high altitudes. A subsonic SDB released from maximum range (~110 km) will reach the target after approximately seven minutes of flight, whereas S-300 and S-400 systems can tear down in as little as five minutes, leaving an empty radar site for the SDBs to attack. Additionally, standoff glide bombs are more vulnerable than HARMs to GPS jamming and point defenses.[61] To achieve higher kill confidence, fifth-generation fighters may have to release their weapons from closer ranges, where they may (depending on their RCS level) be more vulnerable to Chinese SAMs.

The launch of supersonic (Mach 3) anti-radiation weapons from fifth-generation fighters would prevent the possibility of SAMs escaping.[62] In 2006, Boeing

[61] Charles Smith, "China Take's Aim at U.S. GPS," *Newsmax*, November 20, 2007.

[62] A Mach 3 weapon would close 100 km within three minutes, faster than the time required to tear down an S-300 system.

won a contract to begin development of a Joint Dual-Role Air Dominance Missile, an AMRAAM-sized supersonic missile that could engage both ground and air targets.[63] The program was terminated under the FY 2013 budget due to "higher Air Force priorities." The successor program, dubbed the Triple Target Terminator (T3) and sponsored by the Defense Advanced Research Projects Agency, is a long-range missile that could be carried internally by fifth-generation aircraft and would be capable of engaging aircraft, cruise missiles, and land-based air defense targets.[64] A stealthy fighter with such a weapon would be able to attack enemy missile sites, potentially from beyond detection range. However, it is unclear when the weapon might be deployed.

Electronic Jamming

Electronic warfare operations could remain a key enabler of successful SEAD if they possess sufficient capability to affect radar detection and engagement ranges. To assess the possible impact, we modeled a 50-percent reduction in SAM radar range. To be clear, the reduction of SAM radar ranges assumed here is meant to parameterize the problem and demonstrate possible impact; it is not a prediction about effectiveness. Over the past two decades, China has placed a heavy emphasis on ECM and counter-ECM while the United States has gained combat experience against a range of SAM systems in Iraq and elsewhere.

With the assumed effect on SAM system ranges, legacy SEAD aircraft such as the F-16CJ would continue to dominate obsolete HQ-2 batteries and move into the "mutual kill" zone when encountering S-300 PMU-1 and S-300 PMU-2 SAMs—not ideal but clearly an improvement over the unsupported case. Electronic warfare SEAD missions against S-300 PMU-1 and S-300 PMU-2 batteries might be reminiscent of Vietnam-era "Wild Weasel" missions against North Vietnamese SA-2 SAMs, when U.S. pilots often dodged or took other countermeasures against enemy missile launches in their attempt to score kills. The 2017 matchup against the S-400 still leads to unfavorable outcomes for the United States.

Low-Altitude Attack

Finally, approaching enemy air defense sites from low altitude could also be an effective tactic against long-range radar-guided SAM batteries. Even powerful radars are limited by the radar horizon, which is about 50–70 km for objects 500 feet above ground level—significantly less than the 100-km range of a HARM missile. As noted previously, low-level flight makes the aircraft vulnerable to anti-aircraft fire, short-range missiles, and man-portable air defense. According to one analysis, these low-

---

[63] Marc Sklar, "Taking Aim at the Future," *Boeing Frontiers*, August 2008.

[64] On the status of these programs, see *Jane's Air-Launched Weapons*, "Joint Dual Role Air Dominance Missile (JDRADM), T3 and Next Generation Missile (NGM)," August 19, 2014.

altitude threats were responsible for up to 90 percent of worldwide aircraft combat losses between 1984 and 2001.[65]

Taiwan Scenario Summary

The NATO experience against Yugoslavian SAMs during operation Allied Force suggests that the United States may have difficulty locating adversary SAMs that activate their radars intermittently and employ mobility and concealment. As Chinese equipment and training improve, U.S. forces will find it difficult to locate Chinese SAMs. If and when U.S. ISR assets do locate advanced Chinese SAM systems, U.S. traditional SEAD concepts will be less capable of prevailing in a "kinetic contest" with legacy aircraft. Prosecuting SEAD missions with stealth aircraft or through low-altitude ingress may mitigate some challenges. But the effective use of stealth will require the United States to develop new SEAD weapons for fifth-generation aircraft, and low-altitude attack will create risks from low-altitude threats.

SEAD Results for the Spratly Islands Scenario

Reflecting the importance that the PLA accords to Taiwan contingencies, the Nanjing Military Region opposite Taiwan has enjoyed priority SAM allocation. The areas most relevant to a Spratly Islands scenario, Hainan Island and the southern portions of the Guangzhou Military Region, have historically received lower priority. Hence, whereas the Nanjing Military Region has enjoyed significant air defense upgrades, Google Earth imagery suggests that relevant parts of the Guangzhou Military Region were, as of June 2013, protected primarily by HQ-12 SAM (range 50 km) garrisons. There were HQ-9 battery sites in the Guangzhou Military Region, but not within the defined areas of interests to this scenario (southern Guangzhou Military Region, within 1,300 km of Thitu Island).[66]

Because of recent military construction on Hainan Island and the increasing strategic importance of the South China Sea, however, we project improvements to the PLA SAM installations by 2017. The South Sea Fleet, which was once arguably the least important fleet, has recently received priority allocation of modern submarines and surface combatants.[67] For analytical purposes, we assume that two HQ-9 SAM batteries will have been added to defend the area by 2017, and we credit the

---

[65] Michael Puttre, "Facing the Shoulder-Fired Threat," *Journal of Electronic Defense*, Vol. 24, No. 4, April 2001.

[66] Sean O'Connor, "Worldwide SAM Site Overview," *IMINT and Analysis*, data current as of June 2, 2013. The 2013 data indicate two HQ-9 sites in the Guangzhou Military Region, one occupied and the other empty, both along the boundary of the Nanjing Military Region. It may be theoretically feasible for the PLA to move advanced SAMs to this area from the Nanjing Military Region during a crisis. However, SAM regiments, though mobile, are not designed to operate far from fixed garrison sites and so would be of limited use even if they were successfully transported. For example, a regiment operating in this way would lack access to targeting data from underground fiber-optic cables and would be unfamiliar with the local terrain.

[67] David McDonough, "Unveiled: China's New Naval Base in the South China Sea," *The National Interest*, March 20, 2015.

HQ-9 with similar performance specifications to a Russian S-300 PMU-2. Should the S-400 be delivered and emplaced by 2017, there is some chance that a portion could be assigned to the Guangzhou Military Region, though we view it as likely that most will be assigned to the Nanjing Military Region. Overall, compared with the Taiwan scenario, this region is likely to remain defended by somewhat less capable systems—and significantly fewer of them.

With obsolete HQ-2 batteries being the only long-range SAMs stationed in the relevant region in 1996 and 2003, U.S. fourth-generation aircraft armed with HARM missiles and flying at high altitude would have been dominant, even when unsupported by electronic warfare assets. By 2010, most of these batteries were replaced by HQ-12s. Although we did not model SEAD duels against HQ-12s, the system, with a maximum range of 50 km and 20 minute teardown time, bears a much stronger resemblance to the HQ-2 (SA-2) than to double-digit SAMs. U.S. fourth-generation aircraft armed with HARM missiles would again have enjoyed significant advantages.

The addition of HQ-9 batteries in 2017 would greatly increase the risk to U.S. aircraft.[68] U.S. SEAD could no longer win one-on-one engagements, unless operations were undertaken by fifth-generation aircraft, supported by effective electronic warfare, or employed low-altitude attacks. By employing these measures, U.S. forces would have a reasonable chance of suppressing HQ-9 batteries, especially if they are sparsely deployed and not mutually supporting.

U.S. jamming is likely to be more abundant and effective in the Spratly scenario, as the overall air supremacy burden is lower, freeing up sorties for other missions. As in the Taiwan scenario, offensive electronic warfare leads to an environment in which neither the HQ-9 SAM nor attacking fourth-generation aircraft are strictly dominant. Low-altitude flight could enable F-16CJs to win one-on-one duals with HQ-9s, but it would make attacking aircraft vulnerable to short-range air defense threats. Finally, the employment of fifth-generation aircraft is promising, but those aircraft will be in short supply, and the lack of compatible supersonic munitions could allow SAMs to escape unharmed.

U.S. ISR and SAM "Hide" Tactics

The engagements modeled here assume either that Chinese SAM missile radars are turned on, and are therefore easy to find, or that the U.S. military can locate SAMs using other ISR assets. Although this assumption is required to examine the engagement portion of SEAD missions, historical experience suggests that locating SAM sites can be problematic. U.S. aviators have had occasion to prosecute two high-tempo SEAD campaigns in the post–Cold War era. In Operation Desert Storm against Iraq,

---

[68] The schematic diagram of one-on-one competition in the Spratly Islands scenario in 1996, 2003, and 2010 is identical to the 1996 diagram for the Taiwan scenario in Figure 5.3, while the 2017 Spratly scenario schematic is identical to the 2010 Taiwan cases in Figures 5.3 and 5.4.

coalition forces flew 4,326 SEAD sorties (representing 6 percent of all combat sorties), and in Operation Allied Force against Serbian forces, NATO flew 4,538 SEAD sorties (21.5 percent of all combat sorties).[69]

In the Gulf War, Iraq deployed approximately 125 SAM batteries.[70] After detecting nearly 100 air defense radar emissions in the early hours of the war, coalition forces prosecuted a large-scale SEAD campaign in which EA-6B and EF-111 jammed Iraqi radars while tactical fighters launched HARMs at emitters.[71] The campaign successfully reduced air defense radar emission rates by more than 95 percent within a week, crippling Iraq air defenses.[72]

The 1999 SEAD campaign against Serbia produced less satisfactory results. NATO aviators sought to neutralize Serbia's approximately 40 SA-3 and SA-6 area-defense SAM launchers but were able to destroy only three launchers and ten air defense radar emitters after several thousand SEAD sorties and the expenditure of more than 1,000 HARMs.[73] U.S. losses were very low, and, by that standard, the SEAD campaign could be considered a success. Because many of the launchers survived and continued to present a viable threat, however, SEAD missions also drained resources from other tasks.

Dissimilar SEAD outcomes between Desert Storm and Allied Force stemmed largely from differences in Iraqi and Serbian air defense doctrine. Iraqi SAM crews operated from stand-alone sites in fixed locations, making them easy targets for coalition air attacks. Conversely, Serbians improved their survivability through mobility and concealment. SAM crews activated radars intermittently, shutting down and relocating upon warning of approaching NATO SEAD packages. Consequently, NATO forces had substantial difficulty locating Serb air defense emplacements. Admittedly, the terrain also proved advantageous to Serb defenders. However, the difficulty finding Scud missile launchers in Iraq suggests that Iraqi air defenses might have successfully pursued a SAM strategy similar to Serbia's in spite of less propitious terrain.

China's air defense capabilities and doctrine in the 1990s did not differ substantially from those of Iraq. PLA air defense relied heavily on fixed-location SA-2 batteries with little interconnectivity. Although there was some discussion of mobility in Chinese doctrinal writing, the emphasis on the defense of cities and other fixed positions limited room for maneuver, even if the equipment had permitted it. Success in Desert

---

[69] Christopher Bolkcom, *Military Suppression of Enemy Air Defenses (SEAD): Assessing Future Needs*, Washington, D.C.: Congressional Research Service, January 24, 2005.

[70] William A. Hewitt, *Planting the Seeds of SEAD: The Wild Weasel in Vietnam*, thesis, Maxwell AFB, Ala.: Air University, 1993.

[71] Lambeth, "Kosovo and the Continuing SEAD Challenge," 2002.

[72] Lambeth, "Kosovo and the Continuing SEAD Challenge," 2002.

[73] Randy Cunningham, "Suppression of Enemy Air Defenses: Improvements Needed," Washington, D.C.: Electronic Warfare Working Group, Issue Brief No. 7, June 11, 2001.

Storm suggests that contemporary U.S. SEAD tactics might have been highly successful against PLA SAM crews, though Chinese air defenses may not have collapsed as suddenly as Iraq's, given the larger Chinese SAM inventories and better terrain.[74]

The PLA learned from Kosovo and adjusted its doctrine as it modernized its air defense equipment. It began to place more emphasis on mobility and ambush in air defense doctrine, and it has procured systems capable of rapid employment and mobility. By 2003, U.S. SEAD missions against China would have faced many of the ISR challenges that dogged operations in the Kosovo campaign, particularly because the United States would have been unlikely to deploy SEAD forces larger than those used in Operation Allied Force (48 F-16CJs and 30 EA-6Bs). These challenges, more severe by 2010, are likely to continue growing in magnitude through 2017 as China expands the number of mobile SAM launchers in its inventory, trains operators in mobile tactics, and lays underground fiber-optic networks that allow air defense forces to share surveillance and targeting data from a variety of sources.[75]

Lethal SEAD Summary

Since 1996, Chinese SAM forces have dramatically improved as advanced, long-range, mobile systems have replaced weapons of limited mobility and capability. The United States has made considerably fewer investments in upgrading its dedicated SEAD force, even after Operation Allied Force demonstrated the limitations of current approaches. Stealth and standoff weapons may compensate for some decline in dedicated SEAD capability (relative to the threat), and fifth-generation strike fighters may be able to provide their own SEAD. But this scorecard analysis indicates that the battle between Chinese SAMs and U.S. SEAD has grown more competitive, particularly in the Taiwan Strait region.

In 1996, U.S. airpower had the ability to systematically suppress Chinese air defenses in a way that might have proven reminiscent of Operation Desert Storm. A more proper analogy today may be early SEAD missions over North Vietnamese skies, which involved dangerous cat-and-mouse games that led to high losses for both sides. New tactics and technologies, revolving around fifth-generation fighters, may give the United States at least temporary advantages in some number of these contests. But, without the continuous development of SEAD capabilities, these short-term advancements will not offset ongoing Chinese IADS improvements.

---

[74] The United States operated more than 2,000 combat aircraft from six carriers and numerous bases in Saudi Arabia during the Gulf War; it would have, at most, three carriers and a few close-in air bases (Kadena and Futenma) from which to conduct missions against China.

[75] Carlo Kopp, "Advances in PLA C4ISR Capabilities," *China Brief*, Vol. 10, No. 4, February 18, 2010.

Conclusions

China has made remarkable progress toward improving its air defense capabilities. In less than 20 years, the PLA has turned its air defense network from a flimsy distraction into a robust network that can successfully safeguard its airspace against all but the most advanced technology and tactics. The PLA began its defense modernization process by relying heavily on foreign weapons and an aging air fleet. The Chinese defense industry has since evolved to the point that it can indigenously produce many elements of a formidable air defense system, although it will likely continue to rely on foreign sources for the most advanced technologies.

The U.S. military plans to use stealth, jamming, and standoff weapons to surmount modern IADS. U.S. efforts appear to be successful in meeting many of the challenges, particularly from a qualitative standpoint. However, the United States faces a problem of scale. Given the cost of U.S. high-end systems, they are available only in limited numbers, while a conflict with China could require large numbers. Given the severity of the threat that modern Chinese air defenses pose to legacy aircraft, the U.S. SEAD effort will necessarily be more selective, and U.S. forces as a whole will no longer be able to deliver ordnance in the same volume that was previously possible without exposing U.S. aircraft to substantial risk.

The net effect of these developments on the outcome depends largely on the geographic reach and duration of the conflict. In a fight close to China's coast, such as that posited in our Taiwan scenario, the PLA could employ a dense and redundant infrastructure to mobilize combat power over a vast geographic area. Electronic warfare and other support aircraft face severe challenges accompanying lethal SEAD and strike aircraft to many of the relevant targets, placing the latter at higher risk. At the same time, stealth aircraft and standoff weapons alone might well prove insufficient to neutralize a significant portion of the target set. The United States would find itself hard-pressed to attack mainland China with the necessary frequency and intensity without suffering greater air losses than it has in any war in recent memory.

On the other hand, in conflicts farther from the mainland, such as the case posited in our Spratly Islands scenario, Beijing might find itself defending a small set of bases near its own southern coast. Fewer advanced SAM systems are deployed to this area, and defensive combat air patrols would enjoy less space within which to operate. U.S. support aircraft would have an easier time servicing combat aircraft striking targets near China's periphery, especially once adversary defenses and aircraft on the offshore islands are cleared. As important, the scale of the target set for U.S. planners would be smaller, enabling them to allocate more standoff weapons to each.

Scorecard Coding

Figure 5.5 provides our summary coding of the results of scorecard 3. Coding is based on how easy or difficult it would be for U.S. aircraft to penetrate Chinese airspace to attack Chinese targets, as well as the relative difficulty of U.S. lethal SEAD operations against Chinese long-range SAM systems.

In the Taiwan scenario, a highly competitive dynamic is evident, with the introduction of new capabilities by the respective sides providing each with particular advantages. In 1996, even U.S. legacy aircraft were able to penetrate and attack a high proportion of Chinese targets opposite Taiwan, and U.S. SEAD aircraft were able to neutralize Chinese SAMs when they could be located. We therefore code the scenario as *U.S. advantage* in 1996. As double-digit SAMs and new Chinese fighter aircraft limited the proportion of targets that could be safely struck, the scenario turns to *approximate parity* by 2003. Subsequently, the U.S. introduction of larger numbers of

Figure 5.5
Scorecard 3 Summary Coding

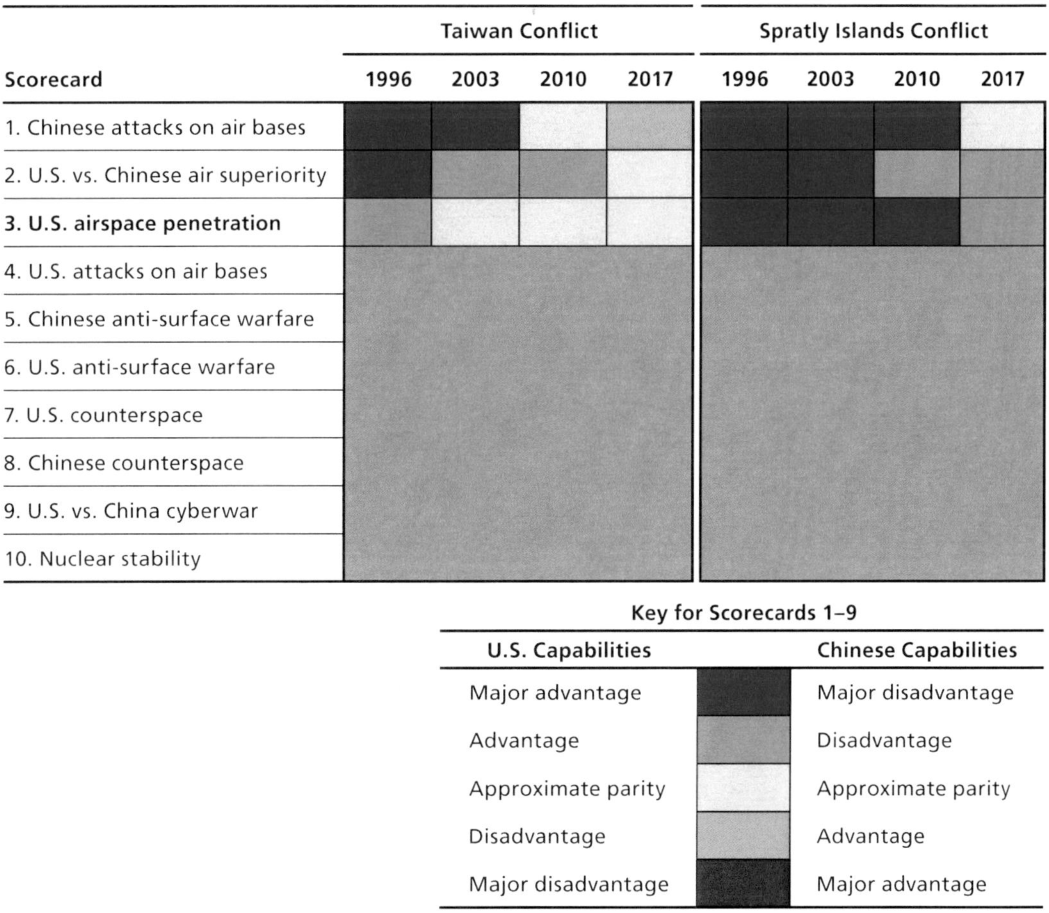

	Taiwan Conflict				Spratly Islands Conflict			
Scorecard	1996	2003	2010	2017	1996	2003	2010	2017
1. Chinese attacks on air bases								
2. U.S. vs. Chinese air superiority								
3. U.S. airspace penetration								
4. U.S. attacks on air bases								
5. Chinese anti-surface warfare								
6. U.S. anti-surface warfare								
7. U.S. counterspace								
8. Chinese counterspace								
9. U.S. vs. China cyberwar								
10. Nuclear stability								

Key for Scorecards 1–9

U.S. Capabilities		Chinese Capabilities
Major advantage		Major disadvantage
Advantage		Disadvantage
Approximate parity		Approximate parity
Disadvantage		Advantage
Major disadvantage		Major advantage

standoff weapons mitigated the impact of further improvements to the Chinese IADS, and stealth provided new U.S. options (albeit represented by a small number of platforms). Given this highly competitive dynamic and the ability of U.S. aircraft to attack some but not all targets with relatively low risk to themselves, we code the scenario as *approximate parity* in both 2010 and 2017.

The results of the Spratly scenario are significantly better for the United States. Fewer Chinese high-end SAM systems have been deployed to southeastern China. This could change as the South China Sea assumes more strategic importance. But other factors will still make the Spratly case easier than the Taiwan one. The relevant targets are not as deep within China in the former case, and are therefore less commonly defended by SAMs with overlapping coverage. As important, with fewer targets to service, the limited supply of U.S. standoff weapons and stealth aircraft can strike a higher percentage of the target set (and conduct re-attacks on individual locations). Hence, we code the scenario as *U.S. advantage* over the entire period, though this advantage holds to a diminished degree by 2017, when some high-end SAM systems and more fourth-generation fighters are likely to be in service there. We can expect that the air penetration problem will grow more difficult beyond 2017, even if the geography of conflict in the South China Sea remains less challenging than in the Taiwan case.

CHAPTER SIX

Scorecard 4: U.S. Capability to Attack Chinese Air Bases

In the context of a conflict with China, the decision about whether to strike targets inside China would be made at the highest political levels and would be based on both operational and other, more strictly political considerations. If the decision were made to strike operational targets, how would U.S. forces fare? In this chapter, we evaluate the U.S. ability to attack Chinese air bases to either close them to air operations or destroy aircraft on the ground. Not only is this an important topic in its own right, but it also serves as a partial proxy for an examination of the more general ability of U.S. aircraft to destroy fixed Chinese ground targets—with allowances for the unique aspects of different target sets.

Since the end of the Cold War, the United States has introduced stealth bombers, GPS-guided precision munitions, and long-range standoff missiles. It has also maintained a substantial fleet of legacy aircraft. All of these systems can threaten enemy air bases. Meanwhile, China has hardened air bases by building protective structures and underground hangers and has strengthened air defenses to make nearby airspace increasingly hostile for potential adversaries. How has the U.S. capability to attack Chinese air bases and the aircraft on them changed over time?

The following analysis of air base attacks is divided into four parts. The first part discusses the number and capability of systems (bombers and missiles) that could be employed by the United States to attack Chinese air bases. The second part assesses the number of Chinese air bases that would be relevant to the Taiwan and Spratly Islands scenarios and the passive defenses associated with each. The third part combines these elements in a model that analyzes the ability of U.S. aircraft to penetrate Chinese defenses, attack runways, and close Chinese bases to flight operations. Finally, the fourth part looks at the ability of U.S. aircraft to strike Chinese aircraft on the ground.

Although the scorecard's results are mixed, they are generally better than those of several others (especially scorecards 1 and 5). Despite the challenge posed by Chinese SAMs, described in the previous chapter, the U.S. ability to attack Chinese air bases and degrade their ability to operate or to destroy the aircraft parked there increased markedly between 1996 and 2003—at least in the context of a short war—and has remained relatively unchanged since then. Standoff missiles and stealthy aircraft significantly mitigate the threat posed by Chinese double-digit SAMs, and, perhaps more importantly, the proliferation of precision weapons would make U.S. attacks far more effective today than they likely would have been 20 years ago.

The advent of precision strike and accurate standoff capabilities have, in other words, put new arrows in the U.S. quiver, just as they have enabled the Chinese ballistic and cruise missile threat to U.S. air bases. The limited inventory of standoff weapons hampers the U.S. ability to sustain attacks in a long war and therefore colors our assessment of the scorecard results, but this scorecard reminds us that not all trend lines in the military competition are relentlessly negative.

U.S. Bombers and Missiles

We begin with an assessment of U.S. strike systems that might be used in attacks on Chinese bases, which provided the first set of inputs for our modeling of air base attacks. The U.S. bomber fleet and much of the current cruise missile inventory were originally designed to penetrate Soviet air defenses and deliver nuclear weapons against strategic targets. Over the past two decades, U.S. Air Force bombers have undertaken a conventional strike role using direct-attack precision-guided munitions (PGMs) and standoff missiles. The Air Force flies three types of long-range bombers: the B-52, B-1, and B-2. Each can be categorized as either a legacy or stealth system.

Legacy Bombers

The B-52 is a subsonic long-range bomber that served as the sole U.S. platform for launching cruise missiles from the air until 2008 and has remained active since.[1] The B-1 Lancer is a supersonic bomber originally designed to penetrate Soviet air defenses. Given the similar weapon loads, our analysis treats both the B-52 and B-1 bombers as generic "legacy bombers" that are nonstealthy and can deliver 20 PGMs or cruise missiles or 30 cluster bombs per sortie.[2] In 1996, legacy bombers lacked GPS-guided weapons and thus carry 60 unguided 500-lb bombs or 30 cluster weapons instead of PGMs. Note that in 2008, Secretary of Defense Robert Gates announced that after 2018, the U.S. military would only employ cluster weapons with submunitions that result in no more than one percent unexploded ordnance.[3] For this analysis, we assume that in 2017, U.S. forces are allowed to employ existing cluster weapons (and we antici-

[1] Air Force Global Strike Command, "B-52 Stratofortress," fact sheet, April 23, 2010.

[2] The B-52 can deliver 20 2,000-lb PGMs or cruise missiles per sortie. The B-1 Lancer is a supersonic bomber that can deliver up to 24 precision weapons per mission. It is expected to be compatible with the JASSM-ER cruise missile (with a range of 1,000 km), which should be available by 2016. Both the B-52 and B-1 can be armed with 30 cluster bombs to attack dispersed targets or with dozens of unguided Mk 82 500-lb bombs. See Federation of American Scientists, "B-52 Stratofortress," web page, undated; and Federation of American Scientists, "B-1 Lancer," web page, undated.

[3] Harold Hongju Koh, Legal Advisor U.S. Department of State, "U.S. Position on Conventional Weapons Negotiations on Cluster Munitions Protocol," Special Briefing, Washington, D.C., November 16, 2011.

pate that suitable replacements will be found for years after 2018, though there could be some period when there are few such weapons in the inventory).

Stealth Bombers

Having reached IOC in 1997, B-2 stealth bombers use a combination of composite materials, a special coating, and stealth shaping to escape detection.[4] The aircraft was able to carry 16 2,000-lb or 32 1,000-lb Joint Direct Attack Munitions (JDAMs) in 2003. In the 2010 and 2017 cases, we gave each aircraft the option of carrying 80 smaller (500-lb) JDAMs. Alternatively, the B-2 can be armed with up to 34 cluster bombs for use against unhardened targets. Due to extensive maintenance needs, B-2s have low readiness rates. At any given time, only between 40 and 50 percent of the 20 aircraft in the U.S. inventory are fully combat-ready.[5]

For analytical purposes, we assumed that the Air Force would fly 24 bombers against Chinese air bases. At the height of the Vietnam War, the United States deployed more than 200 bombers in theater. However, this massive force was built up over a long period, and abundant basing options throughout the region meant that bombers did not have to share space with tactical aircraft or tankers. In a short but intense conflict with China, bombers would compete with other assets for space at forward air bases, and the United States would likely dispatch fewer bombers. By holding the number of bombers constant in each of our snapshot years, we are able to assess the impact of changes to equipment and base defenses, though it should be kept in mind that actual bomber numbers would depend on a range of circumstances.

While we kept the number of bombers engaged constant in each of our snapshot years, we varied the composition of the force depending on the contemporary U.S. force structure. In years in which the B-2 was available, we designated eight of the U.S. bombers in theater as stealthy and the rest as legacy. From 1996 to 2010, legacy bombers are based in Guam, where each aircraft can achieve 0.9 sorties per day against Chinese bases. In 2017, due to the anticipated cruise missile and IRBM threat to Guam (discussed in Chapter Three, scorecard 1), legacy bombers are relocated to Australia. There, sortie rates fall to 0.7 per day. In 2003, all stealth bombers are based in the continental United States (CONUS) and can achieve 0.4 sorties per day. In 2010, stealth bombers can be deployed forward to Guam (sortie rate of 0.9), but they are pushed back to Alaska (sortie rate of 0.7) by 2017 because of the threat to Guam. Table 6.1 summarizes the force levels, sortie rates, and payload of U.S. direct-attack bombers.

---

[4] Federation of American Scientists, "B-2 Spirit," web page, undated.

[5] In part, the readiness rate depends on the definition of the phrase "combat ready." In 2001, the readiness rate, defined by a requirement for low observability, was 31 percent. If stealth is not a factor, the aircraft could be ready roughly 80 percent of the time. In a conflict with China, the B-2's employment would be predicated on its RCS reduction. GlobalSecurity.org, "B-2 Operations," web page, last updated July 24, 2011; and "Readiness Declines in Aging, Overworked Fleet," *Air Force Times*, October 2, 2013.

Table 6.1
Model Input: U.S. Strike Capability, by Year

Bomber Type	1996	2003	2010	2017
Legacy				
Number	24	16	16	16
Sortie rate	0.9 (Guam)	0.9 (Guam)	0.9 (Guam)	0.7 (Aus)
Payload	60 Mk82s or 30 CBUs	20 PGMs or 30 CBUs	20 PGMs or 30 CBUs	20 PGMs or 30 CBUs
Stealth				
Number	—	8	8	8
Sortie rate	—	0.4 (CONUS)	0.9 (Guam)	0.7 (Alaska)
Payload	—	16-32 PGMs or 34 CBUs	80 PGMs or 34 CBUs	80 PGMs or 34 CBUs

SOURCES: Payload data are from Federation of American Scientists, "B-52 Stratofortress," undated; Federation of American Scientists, "B-1 Lancer," undated; and Federation of American Scientists, "B-2 Spirit," undated. B-2 data are from *Jane's Aircraft Upgrades*, "Northrop Grumman (Northrop) B-2A Spirit," September 16, 2014; and *Jane's Missiles and Rockets*, "B-2 Spirit Releases 80 JDAMs in Test," October 17, 2003.

NOTE: CBU = cluster bomb unit.

Long-Range Missiles

In addition to conducting direct-attack bombing sorties, the U.S. Air Force could employ numerous standoff missiles, including CALCMs, JASSMs, and JASSM-ERs.[6] Given their range, these highly accurate weapons can be launched by legacy bombers with minimal risk to the aircraft. They can be used to destroy structures or to crater runways, although they are not optimized for the latter. Our analysis allows the United States to expend half of its total CALCM and JASSM inventories against PLA air bases.

U.S. Navy surface combatants and submarines can supplement bombers by launching TLAMs with a range of 1,200 km or more (depending on the model). Our analysis assumes that U.S. forces will have 200 TLAMs available for use against Chinese air bases.[7] This number is based on a short-war scenario and would be larger

[6] *Jane's Air-Launched Weapons*, "AGM-158A JASSM (Joint Air-to-Surface Standoff Missile), AGM-158B JASSM-ER and LRASM," August 28, 2014.

[7] A typical carrier battle group carries 300 TLAMs. See Ronald O'Rourke, *Cruise Missile Inventories and NATO Attacks on Yugoslavia: Background Information*, Washington, D.C.: Congressional Research Service, 1999. While it may be difficult for U.S. surface warships to approach within range of many of the Chinese bases assessed in this scenario, the deployment of *Ohio*-class SSGNs (cruise missile submarines), each armed with more than 150 missiles, could compensate.

if the war continued beyond a period of several weeks.[8] Of the Tomahawk missiles employed, we assume that half are armed with a 1,000-pound unitary warhead that could be used against aircraft shelters and hangars, while the other half are equipped with a submunition variant (166 bomblets) appropriate for attacking parking ramps.[9] Table 6.2 summarizes availability and performance data on U.S. cruise missiles.

U.S. Bomber Access to Chinese Air Bases

In this section, we identify the Chinese air bases that would be most relevant to the Taiwan and Spratly Islands scenarios and therefore most likely to be targeted by U.S. air attack. We enumerate characteristics of the bases, including runway dimensions

Table 6.2
Model Input: Available U.S. Weapons for Air Base Attack

Weapon	Platform	Range (km)	CEP (m)	Payload	Target	1996	2003	2010	2017
TLAM-C	Naval	1,000+	10	450 kg HE	Hangars	100	100	100	100
TLAM-D	Naval	1,000+	10	166 bomblets	Parking ramp	100	100	100	100
CALCM	Legacy bomber	1,300+	10	450 kg HE	Hangars/ runway	65	100	225	225
JASSM	Legacy bomber	400	5	450 kg HE	Shelters/ runway	—	—	400	750
JASSM-ER	Legacy bomber	930	5	450 kg HE	Shelters/ runway	—	—	—	500

SOURCES: Missile capabilities are from *Jane's Air Launched Weapons*, "AGM-158A JASSM (Joint Air-to-Surface Standoff Missile); AGM-158B JASSM-ER," August 28, 2014; *Jane's Air Launched Weapons*, "AGM-86 Air-Launched Cruise Missile (ALCM) and CALCM," October 22, 2014; *Jane's Strategic Weapons Systems*, "RGM/UGM-109 Tomahawk," May 6, 2014; GlobalSecurity.org, "BGM-109 Tomahawk," undated; and Federation of American Scientists, "U.S. Smart Munitions," undated. Inventory data are from GlobalSecurity.org, "AGM-86C/D Conventional Air-Launched Cruise Missile," web page, last updated July 7, 2011; Ronald O'Rourke, *Cruise Missile Inventories and NATO Attacks on Yugoslavia*, 1999; and GlobalSecurity.org, "AGM-158 JASSM Program Developments," web page, last updated July 7, 2011.

NOTES: CALCM and JASSM numbers are 50 percent of the total U.S. inventory. TLAM numbers reflect naval vessel loadout and a short-war scenario.

---

[8] During the 1991 Gulf War, U.S. forces launched 288 Tomahawk missiles. During Operation Allied Force, they launched roughly 150 Tomahawks against Serbian positions. And during the 2003 invasion of Iraq, the United States launched 725 Tomahawk missiles. See O'Rourke, *Cruise Missile Inventories and NATO Attacks on Yugoslavia*, 1999, and GlobalSecurity.org, "BGM-109 Tomahawk: Tomahawk Operational Use," web page, last updated July 7, 2011.

[9] *Jane's Strategic Weapons Systems*, "RGM/UGM-109 Tomahawk," May 6, 2014. The submunition-dispensing TLAM-D is currently withdrawn from active service. We allow them to be redeployed during a major military crisis with China.

and the number with hangers, shelters, and underground facilities. Finally, we evaluate the degree to which each is accessible to U.S. direct attack with moderate risk to attacking U.S. aircraft.[10]

Identifying Air Bases

Using Google Earth, it is possible to identify more than 200 military air bases in China, including 39 within 800 km of Taiwan and up to nine that could potentially support unrefueled fighter operations in the South China Sea.[11] Figure 6.1 shows the geographic distribution of the bases most relevant to the Taiwan scenario (in red) and the Spratly Islands scenario (in green). Over time, the PLA has improved its ability to protect these air bases by building hangars, hardened shelters, and underground facilities, as well as by strengthening the surrounding IADS.

Figure 6.1
PLA Air Bases Relevant to the Taiwan and Spratly Islands Scenarios

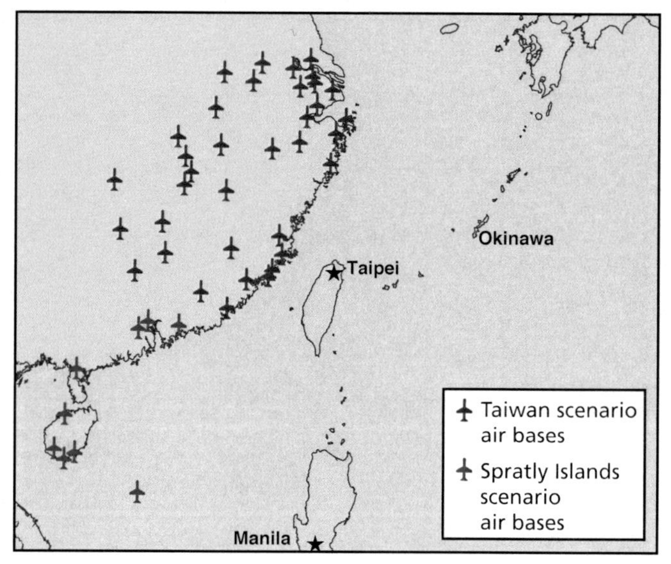

RAND *RR392-6.1*

[10] Here, we employ the same definition of moderate loss as in Chapter Five (scorecard 3)—less than a 10-percent probability of loss—with the same caveats to that definition.

[11] In addition to the air bases considered here, a number of training academies, ground-force air bases, and civil airports and airfields may be modified to accommodate combat aircraft operations, but we do not address the possible use of these airfields in this analysis. The nine South China Sea bases include PLAAF and PLAN air bases within 1,300 km of Thitu Island, a distance representing the longest unrefueled operational range of maritime strike fighters in China's arsenal (the Su-30 and JH-7).

Taiwan Scenario Air Bases

The PLAAF operates 33 air bases and the PLAN operates six air bases within 800 km of Taiwan.[12] Table 6.3 details the number of bases with runways longer than 2,500 meters, as well as the number equipped with hangers, above-ground shelters, and underground shelters. We have made several assumptions for the purposes of this analysis. All bases that have hangars or shelters are treated as if there are enough such structures to accommodate one fighter regiment (~24 aircraft). Additionally, because Google Earth has satellite imagery for only the period since 2003, we estimated the number of hangars and shelters available in 1996 and 2017 using the build rate for protective structures seen in recent years. Finally, we assume that China could disperse aircraft such that only one regiment is deployed at each base.[13]

We evaluated the ability of U.S. aircraft to penetrate Chinese air defenses and conduct direct attacks against each of these Chinese air bases at moderate risk to the attacking aircraft (see Table 6.4).[14] We define *moderate risk* as less than a 10-percent Pk.[15] To assess risk, we employed the same modeling methodology used in Chapter Five, but with a wider range of operational considerations. In this case, we restricted the target set to air bases and assumed that a SEAD campaign has neutralized four key Chinese SAM sites (each equipped with a single battery of launchers),

Table 6.3
PLA Air Bases Within 800 km of Taiwan, 1996–2017

Year	Total Air Bases	Number with Runways Longer Than 2,500 m	Number with Hangars	Number with Shelters	Number with Underground Facilities
1996	39	32	5	2	7
2003	39	32	7	3	7
2010	39	32	9	4	7
2017	39	32	11	5	7

SOURCES: Authors' analysis of Google Earth imagery; Kopp, *People's Liberation Army Air Force and Naval Air Arm Air Base Infrastructure*, 2012.

---

[12] These figures are derived from our analysis of Google Earth imagery and Carlo Kopp, *People's Liberation Army Air Force and Naval Air Arm Air Base Infrastructure*, Air Power Australia, Technical Report APA-TR-2007-0103, updated April 3, 2012.

[13] As noted earlier, we assume no dispersion to airfields that do not serve as air bases during peacetime. Although we make certain assumptions for the purposes of this analysis, we have attempted to be evenhanded: One of the assumptions about dispersion is optimistic while the other is pessimistic.

[14] Using 30-km standoff munitions.

[15] Although we cannot speak to the degree of risk that U.S. commanders might tolerate in planning operations, we note that the definition of *moderate risk* is liberal. Many would regard a 10-percent probability of loss as intolerably high.

Table 6.4
PLA Air Bases in the Taiwan Scenario Accessible to Direct Attack with Moderate Risk to Attacking U.S. Aircraft, by U.S. RCS Level

		Number (%) of Chinese Bases Accessible with Moderate Risk to Attacking Aircraft			
Year	Total PLA Air Bases	U.S. High-RCS Aircraft	U.S. Medium-RCS Aircraft	U.S. Low-RCS Aircraft	U.S. Very-Low-RCS Aircraft
1996	39	27 (69%)	29 (74%)	32 (82%)	33 (85%)
2003	39	23 (59%)	25 (64%)	29 (74%)	33 (85%)
2010	39	7 (18%)	9 (23%)	21 (54%)	26 (67%)
2017	39	2 (5%)	5 (13%)	11 (28%)	14 (36%)

bombers have ECMs that reduce SAM Pks by 50 percent, and escort fighters are able to fully protect bombers from enemy interceptors, though both bombers and fighters are subject to range limitations.[16]

These assumptions reflect the post–Cold War experience of U.S. airpower. The U.S. military has lost no combat aircraft to action by enemy aircraft since 1991, and U.S. airpower has had significant success minimizing attrition from ground-based air defenses using both active and passive measures.[17] As the other scorecard analyses indicate, the conditions that produced those results may not hold when facing China in the Western Pacific. Erosion in the environment since 1996 is accounted for in the modeling by changes to assumptions about the location of tanker support and basing locations. More generally, the assumptions employed here are predicated on some measure of success in other arenas, including the protection of U.S. air bases (scorecard 1), air-to-air combat (scorecard 2), and SEAD (scorecard 3).

Bases accessible by high-RCS aircraft represent those that can be attacked by legacy bombers with direct-attack munitions, while stealth bombers are parameterized by the moderate, low, and very low detectability values defined in Chapter Five. The ability of legacy bombers to reach near-Taiwan air bases falls precipitously between 2003 and 2010 and becomes virtually nonexistent by 2017. If stealth aircraft were able to achieve low or very low detectability levels, they would have continued to have access to a majority of relevant air bases as late as 2010. We assume air- and sea-launched long-range standoff missiles are not restricted in terms of their ability to access air bases

[16] We allow fighters to fly escort missions up to 750 km from the point of last refueling, which is over Taiwan in 1996 and 2003. However, in 2010 and 2017, the PLA air threat is assumed to push tankers back by an additional 100 km and 200 km, respectively. Many bomber sorties to inland bases would have to go unescorted in those years, resulting in unacceptable risk from interceptors.

[17] Bolkcom, *Military Suppression of Enemy Air Defenses (SEAD)*, 2005.

and that 80 percent of them reach their targets, with the other 20 percent falling prey to either malfunction or Chinese defenses.

Spratly Islands Scenario Air Bases

China's basing options are more limited in a South China Sea clash. The epicenter of the Spratly Islands scenario, Thithu Island, lies more than 1,000 km from China's southern coast. Only a handful of bases on Hainan Island and in southern Guangdong Province are close enough to support unrefueled operations by even China's longest-range fighters. Table 6.5 shows the number and characteristics of Chinese air bases that are within range (1,300 km) of Thitu Island in the Spratly Islands group and therefore possible targets for U.S. air attack in the scenario.[18] (As in several other possible events considered in this report, such attacks might be ruled out for political reasons, but we consider their possibility because they represent an operationally important option.)

Having identified the air bases for the Spratly Islands scenario, we performed the same analysis of the U.S. ability to attack these bases as we did in the case of Taiwan. The results, detailing the number (and percentage) of bases accessible at moderate risk to attacking bombers, are presented in Table 6.6. These facilities are much more vulnerable to U.S. attack than the Chinese bases near Taiwan, a condition that remains true throughout the time frame considered. Local deployment of HQ-9s by 2017 prevents legacy and moderate-RCS stealth bombers from conducting direct attacks against a number of bases without substantial risk. But these bases would remain vulnerable to attacks by stealthier aircraft and cruise missiles.

Having outlined U.S. offensive systems and capabilities, the Chinese target air bases and their characteristics, and the ability of U.S. aircraft to penetrate and attack those air bases, we next examined two different types of concepts of operation that

Table 6.5
PLA Air Bases Within 1,300 km of the Spratly Islands

Year	Total Air Bases	Number with Runways Longer Than 2,500 m	Number with Hangars	Number with Shelters	Number with Underground Facilities
1996	0[1]	0	0	0	0
2003	6	6	1	1	0
2010	8	8	2	2	1
2017	9	8	3	3	1

SOURCES: Authors' analysis of Google Earth imagery; Kopp, *People's Liberation Army Air Force and Naval Air Arm Air Base Infrastructure*, 2012.

[18] Extrapolations for base numbers and characteristics for 1996 and 2017 were conducted in the same manner as those for near-Taiwan bases.

Table 6.6
PLA Air Bases in Spratly Islands Scenario Accessible with Moderate Risk to Attacking U.S. Aircraft, by U.S. RCS Level

		Number (%) of Chinese Bases Accessible with Moderate Risk			
Year	Total PLA Air Bases	U.S. High-RCS Aircraft	U.S. Medium-RCS Aircraft	U.S. Low-RCS Aircraft	U.S. Very-Low-RCS Aircraft
1996	0	N/A	N/A	N/A	N/A
2003	6	4 (67%)	6 (100%)	6 (100%)	6 (100%)
2010	8	6 (75%)	7 (88%)	8 (100%)	8 (100%)
2017	9	2 (22%)	3 (33%)	6 (67%)	9 (100%)

the United States might employ: (1) attacks on runways to shut Chinese bases to air operations and (2) attacks on parking areas designed to destroy Chinese aircraft on the ground.

Attacking Runways

The first type of air base attack examined was that designed to shut down flight operations by cutting runways. An air base is considered incapable of sustaining flight operations if runways are cut to such a degree that the undamaged area fails to meet MOS requirements. Air Force planning documents define an MOS required for fighter operations as a 1,500-by-15-meter strip of undamaged concrete.[19] To allow for a reasonable margin of error, we required that runways over 2,500 meters be cut twice to be rendered inoperable, while runways of shorter length need only be cut once. Figure 6.2 depicts notional cuts that would shut down the 2,800-meter runway at the Suzhou air base. Given the numbers of air bases with runways shorter and longer than 2,500 meters, each PLA air base requires, on average, 1.8 cuts to shut down fighter operations. Each cut, in turn, requires that a pattern of craters running perpendicular across the runway be created to ensure that no 15-meter stretch of concrete remains available for the passage of departing or incoming aircraft.[20]

Having established these parameters, we employed Monte Carlo simulations to estimate the number of weapons required to achieve a high confidence (90 percent)

[19] U.S. Department of Defense, *Airfield Damage Repair*, Unified Facilities Criteria 3-270-07, draft, Washington, D.C., June 30, 2003.

[20] PLA runways tend to be approximately 50–60 meters wide. Assuming craters of 5–10 meters in width, three craters would be required to ensure that a single cut is achieved. We estimate that a 2,000-lb bomb makes a ten-meter crater, a smaller unitary warhead makes a five-meter crater, and a bomblet with 1.5 kg of high explosives makes a one-meter-diameter crater. See Shlapak et al., *A Question of Balance*, 2009.

Figure 6.2
Suzhou Air Base

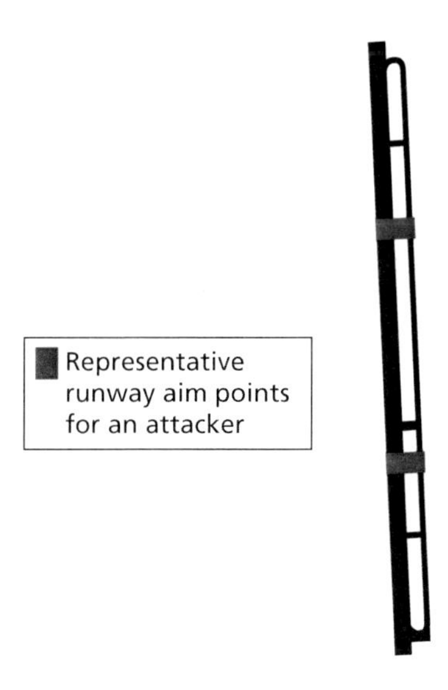

RAND RR392-6.2

of closing PLA air bases.[21] According to these simulations, suppressing an "average" Chinese air base requires 11–16 weapons (each with a CEP of ten meters), depending on warhead size. This number is higher than the roughly six to nine Chinese DF-21C missiles required to shut Kadena AB to U.S. operations (described in Chapter Three, scorecard 1), despite Kadena's longer total runway area and the DF-21Cs' lesser accuracy. The difference is explained by a combination of available warhead types and the inherent advantages of ballistic missiles over the bombs and missiles used by the U.S. side.[22] We assume that the duration of closure is eight hours for runways with ten-meter craters and four hours for runways with five-meter craters, after which the air base must be struck again.

---

[21] We assume an arrival rate of 80 percent for cruise missiles and 90 percent for bombs and PGMs (with the remainder malfunctioning or destroyed by defenses) and weapon CEPs of ten meters (with excursions run for CEPs of 30 meters to capture the effects of jamming).

[22] The Chinese have reportedly acquired runway attack cluster warheads for their DF-21Cs. This would allow a single missile to cover a wider area and create a larger number of craters. Bomblets dispersed by a ballistic missile will travel fast enough to penetrate concrete without the assistance of rockets. In contrast, the slower speed of free-fall bombs and cruise missiles means that purpose-built runway-attack munitions are generally unitary (except in the case of the SG-357 carried by the British JP-233 system, which, at 26 kg, is nevertheless of substantial size), and they are equipped with rockets to facilitate penetration. The U.S. Air Force purchased the rocket-assisted GLU-107B runway-attack munition from France in the late 1980s, but it currently has no dedicated runway-attack weapons. We note that although we have no data on Chinese cost structures, MRBMs can run many times the cost of cruise missiles or free-fall bombs.

Base-Days of Closure

Using the calculations for the systems and weapons needed to close individual average Chinese bases and the types and number of systems available to the United States (listed in Tables 6.5 and 6.6), we then assessed the number of base-days that could be suppressed during the first week of a conflict. A suppressed base-day is equivalent to preventing one air base from conducting flight operations for a 24-hour period. If the results yield 50-percent closure, then Chinese bases are closed, on average, for half (3.5 days) of the first seven days of the conflict. In practice, few bases will suffer average closure. In the case of 50-percent base closure, some bases will likely be shut for the entire period, while others will remain largely untouched. In all cases, the modeling results assume that bases will be struck with direct attack if those attacks entail only moderate risk to attacking aircraft.

Taiwan Scenario Outcomes

Table 6.7 shows the results of Taiwan scenario runway attacks. The last row in the table shows the total percentage of base days closed by U.S. air attacks. The three rows above show which systems account for what proportion of base closure: (1) direct attack by nonstealthy aircraft (where air defense conditions permit), (2) standoff attacks using cruise missiles, and (3) direct attack by stealth bombers. The base closure by stealth bomber attack includes a range of values, depending on differing assumptions about the RCS levels of these aircraft (from moderate to very stealthy). All weapons are assumed to have ten-meter CEP, except for iron bombs dropped in 1996, which have 60-meter CEP. Bombs dropped after 1996 are assumed to be guided, using JDAM kits with GPS guidance.[23]

Table 6.7
Percentage of Time Chinese Bases Could Be Closed by U.S. Air Attack, Taiwan Scenario, First Seven Days

	% of Time Closed			
Sortie Type	**1996**	**2003**	**2010**	**2017**
Legacy bomber	3	23	18	5
Long-range missile	2	2	8	18
Stealth bomber	N/A	3–6	3–18	8–20
Total base closure	5	28–31	29–44	31–43

NOTE: The table shows the average closure rate for the 39 Chinese air bases within 800 km of Taipei.

[23] Michael Russell Rip and James Hasik, *The Precision Revolution: GPS and the Future of Aerial Warfare*, Annapolis, Md.: U.S. Naval Institute Press, 2002. Rip and Hasik estimate that unguided bombs dropped in the fall of 1990 achieved an average CEP of 61 meters.

As the last row in Table 6.7 shows, the U.S. ability to close Chinese air bases improved dramatically from 1996 to 2003. In 1996, U.S. attacks could close the 39 Chinese air bases within 800 km of Taipei for, on average, only 5 percent of the first week. (Given differences in the ease of access to U.S. bombers, air bases near the coast would have been closed for a larger percentage of time than this average, while air bases in the interior would be closed for less time.) Average base closure increased to between 28 and 31 percent of the first week of the war by 2003, with most of the improvement coming with the introduction of precision-guided bombs (specifically, GPS-guided JDAM kits).[24]

Despite improvements to Chinese IADS, the U.S. capability to attack and close Chinese air bases near Taiwan improved further by 2010 (averaging 29–44 percent of the first week). While the ability of legacy aircraft to penetrate and attack diminished, the increased inventory of standoff missiles and the ability to fly stealth bombers from Guam (as opposed to CONUS) more than compensated for this loss. Between 2010 and 2017, the average base closure achieved remains largely unchanged, with more additions and improvements to the U.S. standoff missile inventory compensating for the further diminished role of legacy aircraft.[25] As a reminder, all of these results (as well as those for the Spratly Islands scenario) assume some success in SEAD and ECM operations, as well as escort fighters for legacy bombers, as discussed earlier in the chapter.

Spratly Islands Scenario Outcomes

Using the same U.S. force laydown and methodology, our analysis shows that the United States possesses the capability to comprehensively shut down the small number of Chinese air bases that could support unrefueled fighter operations in the Spratly Islands. Not only are there fewer bases for U.S. forces to target, but the lack of modern SAM batteries in three out of the four snapshot years means that each base is less well-defended against bomber sorties. Table 6.8 shows the percentage of Chinese base-days that could be suppressed by U.S. runway attacks. Because the area lacked strong air defenses in 2003 and 2010, legacy bombers and cruise missiles could have crippled air bases in those years. While the addition of HQ-9 batteries severely reduces the effectiveness of legacy bombers in 2017, U.S. forces can still effectively shut down PLA flight operations by attacking runways with stealth bombers and with the expanded force of cruise missiles.

---

[24] We note that, in 1996, F-15E Strike Eagles were armed with PGMs and could have conducted anti–air base operations. However, they carried limited payloads (two to four PGMs per sortie). Because they would have made a greater impact by flying air superiority sorties, we do not include them in this analysis.

[25] The impact of stealth bombers on base suppression is uncertain and depends on how visible the aircraft are to air defense radars. In 2010, stealth bombers accounted for between 10 and 41 percent of the total closure achieved by U.S. attacks, while they account for between 26 and 47 percent of the total in 2017.

Table 6.8
Percentage of Time Chinese Bases Could Be Closed by U.S. Air Attack, Spratly Islands Scenario, First Seven Days

	% of Time Closed		
Sortie Type	**2003**	**2010**	**2017**
Legacy bomber	85	65	11
Long-range missile	15	35	62
Stealth bomber	0	0	27
Total base closure	100	100	100

NOTES: The table shows average closure rates for Chinese air bases within 1,300 km of Thitu Island. We assessed closure as a total of zero bases in 1996, six bases in 2003, eight bases in 2010, and nine bases in 2017.

Long-Duration Conflicts

We did not model sustained base attacks. However, it is worth considering the possible impact of the conflict's duration on U.S. offensive capability. In a long war, direct attacks by legacy and stealth bombers would be limited only by attrition to the air fleet and the availability of guided bombs, which are both cheap and plentiful. In the case of long-range missiles, however, half of the entire stock of U.S. munitions (with the exception of TLAMs) would be expended to achieve the seven-day results listed in Table 6.8.[26] Viewed from that perspective, the 2003 or 2010 cases may represent the peak of relative U.S. capabilities in an extended war, since the reliance on standoff systems after that suggests that the ability to prosecute more sustained attacks would be limited. On the other hand, to the extent that air bases are closed during the first week, significant damage might be inflicted on aircraft trapped on the ground during that period.

Summary of Runway Attack

Over the past decade and into the future, the United States maintains the means to substantially degrade PLA flight operations against Taiwan. When looking at scenarios near the Spratly Islands, we find that the U.S. Air Force has the ability to comprehensively shut down relevant air bases—assuming permission is given to attack those bases. After adding dramatic new capabilities with the precision guidance revolution between 1996 and 2003, legacy bombers suffered a more gradual but nevertheless significant and continuing erosion in their effectiveness as China deployed more effective and longer-range SAM systems. The addition of new types of long-range standoff missiles and stealth aircraft compensated for the diminished contribution of legacy air-

[26] For long-range missile inventories and sourcing, see Table 6.2.

craft, but standoff missiles and stealth bombers suffer from other types of limitations (particularly from high cost and limited numbers).

Attacking Parked Aircraft

Instead of—or in addition to—attacking runways and closing PLA air bases to operations, U.S. forces could attempt to destroy Chinese aircraft on the ground. In this section, we consider the U.S. ability to cover open parking areas with cluster weapons and to hit hangars and shelters with PGMs and cruise missiles.[27] For this analysis, we again employed the same force structure described earlier (24 bombers of various types and roughly half of the standoff missile force). While this enabled us to compare the potential impact of the two force employment concepts, in an actual conflict, sorties would likely be divided between the two types of missions. We again restrict the time frame to a one-week period.

Aircraft can be parked in different types of locations: parking ramps, hangars, hardened shelters, or underground shelters. The characteristics of Chinese air bases and whether they have hangars, shelters, or underground facilities were presented in Table 6.3 (for the Taiwan scenario) and Table 6.5 (Spratly Islands scenario) earlier in this chapter. For analytical purposes, we assumed that, at any given air base, all aircraft are parked in the open, in hangars, in hardened shelters, or in underground shelters. We allocated U.S. offensive assets to attack particular targets based on the suitability of the system in question to the mission; the number of different types of targets to be struck; and the risk to attacking bombers of striking at individual air bases—an allocation process that involved a variety of excursions to determine a reasonably effective distribution but one that did not necessarily arrive at an optimal solution.[28]

Attacking Aircraft on the Ground

Having allocated weapons to targets, we then modeled the effectiveness of each type of attack. Table 6.9 shows the percentage of bases at which 90 percent of targets (parking-area aim points in the case of open bases and individual hangars or shelters in the other cases) are destroyed. Because RCS levels for stealth bombers are unknown, a range of

[27] It is important to note that we did not directly assess the U.S. ability to destroy parked aircraft. Rather, we examined the ability to successfully attack locations where aircraft can be parked as a surrogate measure.

[28] We allocated 50 percent of total bomber sorties (armed with cluster bombs) and 50 percent of total TLAM missiles (armed with submunition dispensers) to attacks against open parking areas. We allocated 25 percent of total bomber sorties, 50 percent of total TLAM missiles, and all CALCM missiles (all using unitary warheads) to attacks against hangers. We allocated 25 percent of total bomber sorties and all JASSMs, equipped with penetrating warheads, to attacks against hardened shelters. Underground shelters were regarded as invulnerable to attack, though in reality, the U.S. military does maintain some "bunker busters" that might be employed successfully against some of these sites.

Table 6.9
Percentage of PLA Air Bases in Which Aircraft Parking Areas Are More Than 90-Percent Destroyed, Seven-Day Campaign

	1996		2003		2010		2017	
Scenario	% of Bases	Total Bases	% of Bases	Total Bases	% of Bases	Total Bases	% of Bases	Total Bases
Taiwan	28	39	57–69	39	46–64	39	41–56	39
Spratly Islands	N/A	0	100	6	88	8	78–89	9

values is shown for cases in which stealthy aircraft participate in attacks, reflecting the number of bases that can be accessed at moderate risk, depending on the high- and low-end possibilities for the RCS of the aircraft.

The results are broadly consistent with the outcomes for runway attacks discussed earlier in this chapter. In the Taiwan scenario, the U.S. ability to attack aircraft parking and storage areas was limited in 1996, with 28 percent of the relevant base areas destroyed. By 2003, the revolution in precision guidance had netted large gains for the United States, and the percentage of base parking areas that could be destroyed had risen to 57–69 percent (depending on assumptions about the RCS level of U.S. stealth aircraft). The U.S. capability to attack Chinese aircraft on the ground remains robust in the 2010 and the 2017 (projected) cases, but has declined somewhat since reaching a peak in 2003.

Owing to the smaller number of targets and paucity of sophisticated SAMs deployed in southern locations, PLA air bases relevant to the Spratly Islands scenario remain substantially more vulnerable to air attack throughout the time frame examined. With the exception of the single base with underground parking (present only in the 2010 and 2017 cases) and another possible exception, all parking areas at all bases can be effectively destroyed in all snapshot years. The one possible exception concerns bases with open-air parking in the 2017 time frame, when we posit that two HQ-9 batteries (or their equivalent) will be in place, complicating the offensive task of legacy bombers and necessitating the use of stealth. Depending on the RCS level assumed for these bombers, they may or may not be able to accomplish this task with moderate risk to themselves. (This uncertainty is reflected in the range of values given for the 2017 case.)

As noted in Chapter Five, the addition of additional HQ-9 batteries and, potentially, S-400s would make the offensive task more difficult. But the Spratly scenario's smaller, coastal target set would still make achieving results (in terms of the proportion of targets struck) easier than in the Taiwan case. However, a more comprehensive set of Chinese improvements could further complicate the U.S. task. The construction of additional air bases and hardened facilities could, for example, provide basing redundancy and a sink for U.S. standoff munitions. And a larger fleet of aerial tankers would expand the geographic scope (and number of air bases) from which Chinese fighter and attack aircraft could fly, forcing U.S. aircraft to fly farther and against more (and more difficult) targets to produce the same effect.

The percentages in Table 6.9 should not be taken as a direct indicator of the percentage of aircraft that could be destroyed on the ground by U.S. air and missile strikes. The results measure parking areas, hangars, and shelters destroyed, and there is not necessarily a direct correlation between those numbers and aircraft. The PLA could, to an extent, disperse aircraft to other hidden locations (including bases far inland). At the same time, in some of the cases in which U.S. attacks cannot do comprehensive damage, a more modest level of success might nevertheless be achieved. That said, these results indicate, in a very general sense, the degree of threat that U.S. air and missile attacks might pose to Chinese air bases and the trends in that capability over time.

If the United States attacks air bases in China, it is unlikely to focus on cratering runways to the exclusion of destroying parked aircraft, or vice versa. More likely, the U.S. Air Force would pursue both goals simultaneously, as well as prosecuting other missions, such as the bombardment of ports, troop mobilization points, air defense sites, and command centers. Our analysis did not simulate realistic bombing campaigns or address specific operational questions. Rather, it provided metrics and assessment tools that offer insight into the potential impact of U.S. strikes against Chinese air bases, some of which are outlined here.

Conclusions

This scorecard analysis shows that, in some areas, improvements to U.S. capabilities have been dramatic. Despite the very real challenges posed by Chinese IADS, U.S. forces can threaten Chinese targets on the mainland and complicate the task of Chinese air planners by closing or partly destroying a portion of their air bases. When considered in conjunction with the first three scorecards (which examine the Chinese missile threat, the air superiority battle, and the U.S. capability to penetrate Chinese air defenses), it is clear that any future air war would see thrust and counterthrust. Both sides would face the kind of losses, particularly in terms of air and naval systems, which neither has suffered in many years.

Although the two sides have applied new technologies differently—with the PLA devoting more effort to conventionally armed ballistic missiles and the United States emphasizing cruise missiles and direct attack—the continuing revolution in precision capabilities has given both sides greater capability to attack and deny key areas or neutralize critical assets, even those deep behind their adversary's lines. Given the geographic asymmetries involved in some East Asian scenarios, China will often enjoy a more robust basing structure and be relatively less affected by U.S. attacks than U.S. forces would be by similar types of Chinese strikes. Nevertheless, Chinese air bases and, especially, advanced aircraft are not unlimited in number, and with adequate ISR, the United States might seriously impede Chinese air efforts.

In this scorecard analysis, the differences between the two scenarios are notable. PLA air bases that support South China Sea operations are far more vulnerable than those that support Taiwan-area operations. For the most part, a dedicated U.S. campaign could effectively neutralize Spratly scenario air bases. The main questions, in that case, would be less operational than political. Would the U.S. military be permitted to strike these bases, and, if so, under what circumstances? U.S. strikes against air bases in the Taiwan scenario would be somewhat less effective, and U.S. capabilities in that scenario have declined somewhat since 2003. In a short war, U.S. air and missile strikes against Chinese bases would nevertheless significantly impinge on Chinese air operations, even in a Taiwan scenario. In a longer war, however, limitations on the number of standoff weapons would limit the U.S. ability to revisit key targets and sustain an air campaign over the mainland.

Scorecard Coding

Figure 6.3 provides our summary coding of the results of scorecard 4. Advantage in this scorecard is based on the modeling of attacks against runways, aircraft parking areas, and shelters, as well as the robustness of the results. In the Taiwan scenario, the coding for 1996 and 2003, *approximate parity* moving to *U.S. advantage*, reflects the improved results derived from increased precision. Having improved dramatically by 2003, the U.S. ability to damage runways and destroy aircraft on the ground remains relatively stable through 2017.[29] However, given the increasing reliance on a small number of stealthy aircraft and cruise missiles to achieve the same results, we coded the results for 2010 and 2017 as more modest U.S. advantage than was the case in 2003. In the Spratly scenario, the smaller number of targets and the lower density of high-end SAMs result in better prospects for U.S. ground attack. We therefore coded the scenario as major U.S. advantage throughout the period from 1996 to 2017, though this could change in the more distant future as China continues to improve its air defenses.

---

[29] In addition, the introduction of larger numbers of fifth-generation aircraft provides U.S. forces with a continuing lethal SEAD capability.

Figure 6.3
Scorecard 4 Summary Coding

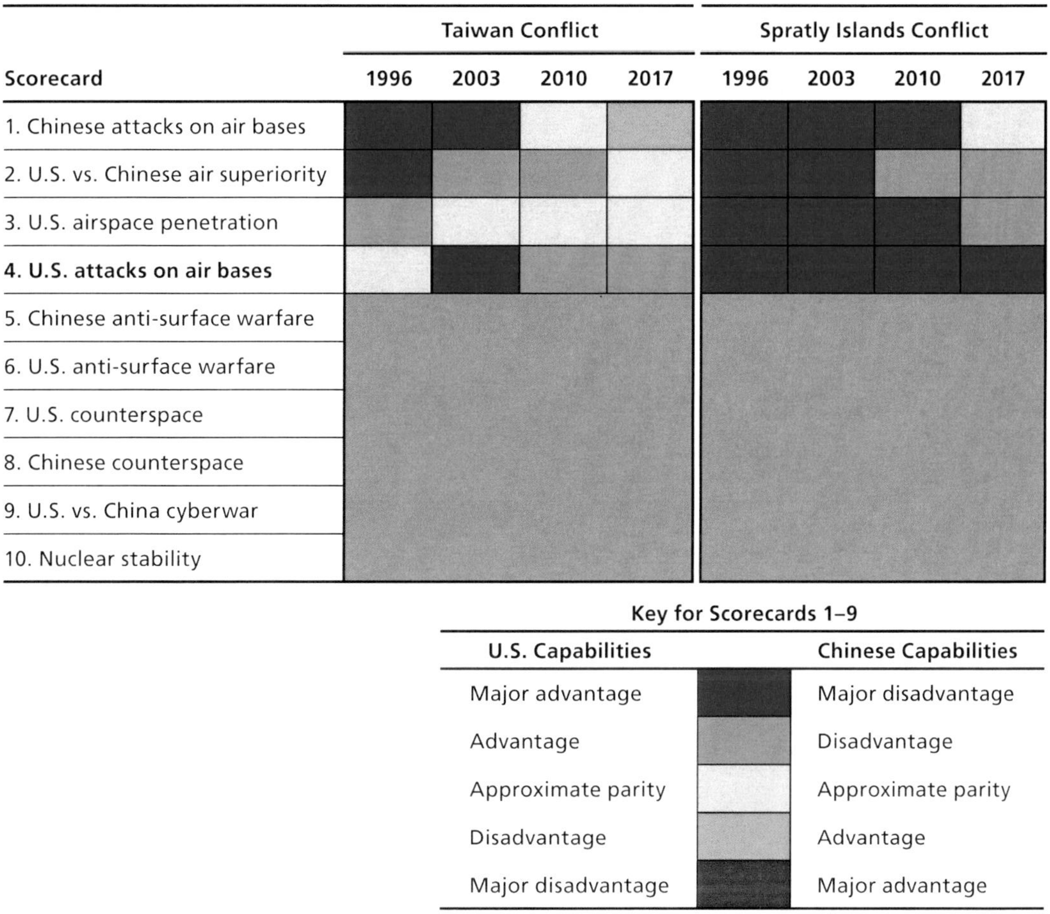

	Taiwan Conflict				Spratly Islands Conflict			
Scorecard	**1996**	**2003**	**2010**	**2017**	**1996**	**2003**	**2010**	**2017**
1. Chinese attacks on air bases								
2. U.S. vs. Chinese air superiority								
3. U.S. airspace penetration								
4. U.S. attacks on air bases								
5. Chinese anti-surface warfare								
6. U.S. anti-surface warfare								
7. U.S. counterspace								
8. Chinese counterspace								
9. U.S. vs. China cyberwar								
10. Nuclear stability								

Key for Scorecards 1–9

U.S. Capabilities	Chinese Capabilities
Major advantage	Major disadvantage
Advantage	Disadvantage
Approximate parity	Approximate parity
Disadvantage	Advantage
Major disadvantage	Major advantage

CHAPTER SEVEN

Scorecard 5: Chinese Anti-Surface Warfare

Recent Chinese military investments have focused, in part, on acquiring the capability to threaten the U.S. Navy's surface fleet and prevent it from operating in or near Chinese territorial waters in the event of a war. In this chapter, we examine China's ability to hold U.S. surface ships—particularly aircraft carriers and their associated CSGs—at risk at varying distances from the mainland. We also analyze U.S. defensive capabilities and their ability to protect U.S. surface assets. As in previous scorecard analyses, we assess each side's relative capabilities over time. This scorecard assesses Chinese developments in four areas:

- OTH targeting
- ASBM capability
- ASCM threats
- offensive submarine capabilities.

For each type of development, we also discuss current and future U.S. defensive systems and their ability to mitigate or neutralize the threat.

China's ability to threaten the U.S. surface fleet has improved dramatically over the periods considered here. In 1996, the United States could have deployed warships to just outside the range of China's land-based surveillance radars with little risk from Chinese attack. Today, U.S. surface forces could no longer operate freely without risk at close or even moderate ranges from the Chinese mainland.

While much recent press reporting has focused on the threat from China's development of an ASBM capability—in large measure owing to the novelty of this weapon—it may not prove to be the kind of one-shot, one-kill weapons often portrayed. The analysis in this chapter suggests that the incremental improvement of existing categories of systems, particularly submarines, may well prove a more serious challenge. Nevertheless, the challenge posed by China's anti-access forces is clearly multidimensional, greatly complicating the task of operating a fleet within several thousand kilometers of the Chinese coast, at least during the initial stages of a conflict.

We stress that U.S. commanders might, depending on the circumstances, accept a substantial degree of risk to close within effective range of the mainland. Nevertheless, the terms of engagement have changed, and U.S. surface ships are likely to face a variety of hurdles in making their presence felt as fully as they once could.

Chinese Over-the-Horizon Targeting Complex

Successfully engaging maritime targets well beyond the horizon is difficult even for U.S. forces and would pose real challenges for the PLA in the event of a conflict with the United States. The maritime region within 2,000 km of the Chinese mainland between Hainan Island and the Yellow Sea covers an area of more than 7 million sq km. This section examines the capability of Chinese strategic ISR—specifically, OTH radar and satellites—to detect, locate, and identify U.S. maritime surface forces and to transmit that information quickly to strike platforms.

The specific means employed in the entire kill chain could vary substantially, depending on the offensive systems involved. (A *kill chain* is defined as the process of finding, fixing, tracking, targeting, engaging, and assessing attacks on targets.) Some systems require external cueing (i.e., intelligence from another source) to be effective. For example, ASBMs, which will necessarily be located far from their targets, require near-real-time information to achieve targeting solutions. While such information could come from other operational assets, such as warships or fighter aircraft, specialized long-range ISR systems will often be better prepared to pass information quickly and accurately to the firing element. Hence, the long-range, dedicated systems discussed in this section will be particularly pertinent to ASBM attacks.

Other types of offensive systems (such as submarines, naval surface units, and land-based aircraft) may be less reliant on long-range, dedicated ISR systems. For example, in the case of Chinese submarines, which can loiter under the surface for days or weeks, the submarine itself could execute the entire kill chain. It could employ visual or electronic periscope searches to find and identify targets, launch onboard torpedoes or cruise missiles, and use visual or acoustic means to assess the effects of its attacks. Even in the case of submarines, naval surface vessels, and land-based aircraft, however, the intelligence derived from dedicated ISR systems could be employed. Indeed, the analysis of submarine attacks later in this chapter demonstrates that the regular external cueing of Chinese submarine forces would dramatically improve the effectiveness of a Chinese submarine campaign.

China's Over-the-Horizon Surveillance Systems

Prior to the Taiwan Strait confrontation in 1996, China had virtually no persistent surveillance capability beyond radar line of sight from its own shoreline. Furthermore, it had no credible airborne surveillance capability, and its submarine force was outdated and noisy. In short, China had virtually no OTH surveillance capability in 1996, and

U.S. Navy ships could operate close to the mainland with near impunity and little fear of Chinese forces gaining adequate situational awareness of their operations.

China's efforts to modernize its military following the Taiwan Strait confrontation in 1996 have greatly improved its sensing and surveillance capabilities. Many of these improvements have come as a secondary effect of combat systems with greater range and more capable onboard sensors. But the addition of dedicated ISR platforms has also contributed. The following is a timeline for selected Chinese acquisition milestones since 1995.

- *Modern diesel submarines (1995).* China commissioned the first of two Russian *Kilo 877* conventional submarines, the first submarines in Chinese service with teardrop-shaped hulls. It subsequently purchased ten of the improved *Kilo 636*, a significantly quieter boat.[1]
- *Modern surface combatants (1999).* The PLAN commissioned the first of four Russian *Sovremenny*-class (Type 956E and 956EM) destroyers equipped with modern OTH targeting systems and ASCMs.[2]
- *Modern indigenous submarines (1999).* The PLAN commissioned its first modern, indigenously developed submarine, the *Song* class. The first of the *Yuan* class, with features indicating *Kilo* influence and, possibly, with air-independent propulsion, was commissioned in 2004.[3]
- *Long-duration spy satellites (2000).* China launched the first of three *Ziyuan*-class military satellites.[4] These were China's first long-duration (life expectancy of two to four years) spy satellites that stored and returned digital images to the ground via data link.[5]
- *Modern indigenous surface combatants (2004).* The PLAN commissioned its first indigenous destroyer with modern air defense systems capable of detecting fighter aircraft at ranges of 300–450 km.[6]
- *Fourth-generation fighter ground attack aircraft (2004).* The PLAN took delivery of the first batch of Russian SU-30MK2 fighter ground attack aircraft, the first

---

[1] *Jane's Underwater Warfare Systems*, "Submarine Forces: China," June 16, 2011.

[2] SinoDefence.com, "Project 956/EM Sovremenny Class Missile Destroyer," web page, last updated February 28, 2009.

[3] *Jane's Fighting Ships*, "Song Class (Type 039/039G)," February 13, 2015; *Jane's Fighting Ships*, "Yuan Class (Type 41)," February 12, 2013.

[4] Also, referred to as the JiangBing 3A, 3B, and 3C satellites.

[5] *Jane's Space Systems and Industry*, "ZiYuan-2/JianBing-3 Series," November 24, 2014.

[6] As of January 2015, it operates five *Luyang II*–class (Type 052C), two *Luzhou*-class (Type 051C), and one *Luyang III*–class (Type 052D) vessels, with one additional *Luyang II*–class and five *Luyang III*–class (Type 052D) destroyers launched but not commissioned. See *Jane's Fighting Ships*, "Luyang II (Type 052C) Class," February 16, 2015; *Jane's Fighting Ships*, "Luyang II (Type 052C) Class," 2013; *Jane's Fighting Ships*, "Luzhou (Type 051C) Class," February 13, 2015; and *Jane's Fighting Ships*, "Luyang III (Type 052D) Class," February 16, 2015.

PLAN aircraft with beyond-visual-range air-to-air and precision-strike capabilities.[7]

- *SAR satellites (2006).* China launched its first SAR imaging satellite (Yaogan 1) and the first of a new class of military spy satellites.[8]
- *AEW platforms (2006).* The Chinese KJ-2000 AEW aircraft, which appears to be an effort to replicate the capabilities of the Israeli Phalcon system, conducted initial flight tests.[9]
- *OTH radar (2007).* China began constructing a long-range, OTH skywave radar system near Xiangfan, China.[10]
- *Electronic intelligence (ELINT) satellites (2010).* China launched its first ELINT satellite system (Yaogan 9), consisting of three elements flying in close formation capable of performing time-distance-of arrival (TDOA) against emitting maritime vessels.[11]
- *ISR unmanned aerial vehicles (2011).* China completed a trial program that used UAVs in Liaoning Province to take aerial imagery of 980 square miles of sea area.[12]

Using these acquisitions, all of which have been further developed since the milestones mentioned above, could China detect, identify, and target U.S. surface ships? If so, how quickly and reliably could it accomplish these tasks, and at what range from the mainland? Certainly, the U.S. Navy can no longer assume that its surface ships will remain unlocated and safe from earnest attack during a confrontation with China simply by positioning them beyond the radar horizon. Significant effort may now be required to track, engage, disrupt, or spoof Chinese surveillance assets. In the following sections, we analyze China's ability to detect, identify, and locate targets (and especially U.S. aircraft carriers) using its OTH radar and space-based surveillance assets.[13]

---

[7] *Jane's Defence Weekly*, "China Accepts Su-30MK2 Fighters," March 31, 2004.

[8] *Jane's Space Systems*, "Yaogan Series," January 20, 2015.

[9] SinoDefence.com, "KongJing-2000 Airborne Warning and Control System," web page, last updated January 4, 2009.

[10] Bussert and Elleman, *People's Liberation Army Navy*, 2011.

[11] These satellite formations are referred to as naval ocean surveillance systems (NOSS).

[12] Ian Easton, *China's Evolving Reconnaissance-Strike Capabilities: Implications for the U.S.-Japan Alliance*, Project 2049 Institute and The Japan Institute of International Affairs, February 2014.

[13] In the future, high-altitude, long-endurance, and stealthy UAVs could clearly support the expansion of China's reconnaissance as well as its strike capabilities, giving China the ability to locate, track, and strike U.S. Navy surface assets at greater distances in the Western Pacific. However, we assess it as highly unlikely that China will possess the necessary capabilities in the necessary numbers by 2017. Nonetheless, it is imperative that U.S. analysts continue to pay close attention to China's rapidly expanding and increasingly sophisticated UAV capabilities. On Chinese UAV programs and capabilities, see Office of Naval Intelligence, "The PLA Navy: New Capabilities and Missions for the 21st Century," Washington, D.C., April 2015, p. 22; and Ian Easton, *China's Evolving*

Cueing: OTH Radar and Naval Ocean Surveillance Systems

Both the Chinese ground-based skywave OTH radars and space-based NOSS provide China with highly capable broad-area surveillance systems that can detect and locate U.S. warships with some accuracy far from the Chinese coastline.[14] China could engage U.S. CSGs using data supplied exclusively by these assets, or it could use data from them to cue its space-based imaging assets to identify and more accurately locate CSG elements, such as the aircraft carrier, before engaging.

Skywave OTH radars are large, powerful systems with extremely long transmitting and receiving antennas. They operate in the high-frequency domain (3–30 MHz) and achieve extremely long detection ranges by reflecting radar energy off the ionosphere and onto a surface or airborne target. During the Cold War, the United States operated the AN/FPS-118 skywave OTH radar system near the Atlantic Coast. The system had minimum and maximum ranges of 925 km and 3,330 km (500–1,800 nm), respectively, and could scan an entire 60-degree sector (approximately 5.3 million sq km) in 24 seconds.[15] For analytical purposes, we assume that the Chinese systems have similar capabilities.[16] Figure 7.1 shows the potential coverage area for the skywave OTH radar near Xiangfan, China.

Skywave OTH radar performance is highly dependent on ionosphere conditions that vary from day to night, seasonally, and with the solar cycle. Ionosphere irregularities can degrade the radar's performance intermittently in both time and space.[17] However, large ships tend to have very large RCSs. Therefore, it is almost certain that Chinese OTH radars will detect U.S. surface ships at least occasionally at moderate ranges, assuming that U.S. forces do not take actions to mitigate their effectiveness by, for example, jamming or conducting kinetic strikes against them.

However, to launch attacks or cue space assets, Chinese OTH radar operators and analysts must also identify U.S. ships as potential targets of interest. OTH radars tend to have relatively large range resolutions and therefore cannot discriminate targets that are within several kilometers of one another. Therefore, the Chinese would not be able to identify a specific target from its radar return alone.[18] OTH radar can detect

---

Reconnaissance-Strike Capabilities: Implications for the U.S.-Japan Alliance, Project 2049 Institute and Japan Institute of International Affairs, February 2014.

[14] The skywave OTH radars can detect and locate ships out to 2,000 km from the Chinese coastline; NOSS coverage is virtually unlimited.

[15] George N. Lewis and Theodore A. Postol, "Long-Range Nuclear Cruise Missiles and Stability," *Science and Global Security*, Vol. 3, 1992.

[16] The Chinese system will likely be slightly less capable than the AN/FPS-118, according to Eric Hagt and Matthew Durnin. See Eric Hagt and Matthew Durnin, "China's Antiship Ballistic Missile," *Naval War College Review*, Vol. 62, No. 4, Fall 2009.

[17] Generally, skywave performance is more consistent during the day than at night due to increased solar activity and less environmental electromagnetic noise.

[18] It can measure the strength of the target's return and then estimate the target's size, however.

Figure 7.1
Potential Coverage Region for the Skywave Radar System Near Xiangfan, China

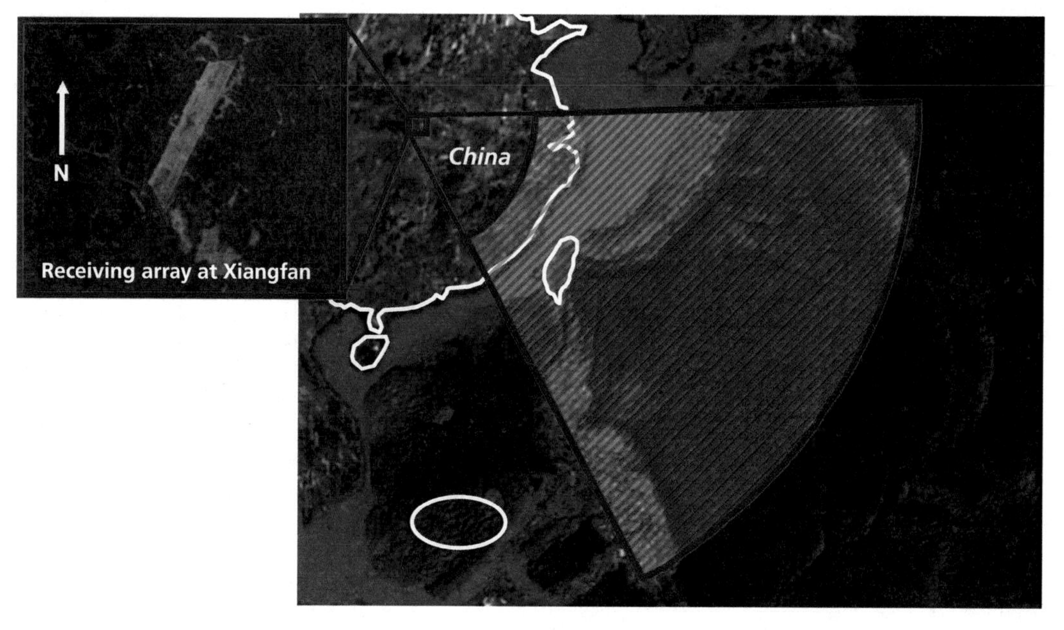

SOURCE: Google Earth with author overlay.
RAND *RR392-7.1*

aircraft in flight, however, thereby alerting OTH radar operators to carriers' radar returns during flight operations.

A second challenge for OTH operation is the target location error, or the difference between the actual target location and the expected location provided by the radar return.[19] According to one source, Australia's Jindalee skywave radar has expected targeting errors from 10 to 40 km in target range and one to five degrees in target bearing, depending on the ionospheric conditions.[20] Assuming that Chinese OTH radars have similar errors, these values yield a targeting location CEP of between 22 and 105 km at 2,000 km from the radar (or roughly 700 km southeast of Taiwan) and a

---

[19] Line-of-sight radars can ascertain a target's position with high precision. However, the ionosphere is a dynamic environment, and its characteristics (height and reflecting orientation) change quickly. OTH radars suffer as a consequence in terms of their ability to ascertain the precise location of potential targets.

[20] T. D. Keenan and S. J. Anderson, "Some Examples of Surface Wind Field Analysis Based on Jindalee Sky Wave Radar Data," *Australian Meteorological Magazine*, Vol. 35, December 1987. We assume that these accuracy values for the Jindalee are stated as 50-percent linear error values. In other words, we would expect the errors to be less than the stated value 50 percent of the time.

targeting location CEP of between 36 and 178 km at 3,300 km (2,000 km southeast of Taiwan).[21]

The impact of these targeting errors on China's ability to engage U.S. surface forces depends on the type of attack envisioned. If OTH radar data are used to vector persistent platforms with their own onboard maritime search capabilities (e.g., strike aircraft with maritime search radars), a general location with a relatively large error may be sufficient. On the other hand, if the maneuver and search capabilities of the strike asset are more limited, then it may be necessary to obtain a more precise target location. ASBMs, depending on the OTH radar accuracy and the parameters of their own autonomous search and maneuver capabilities, may fall into this latter category, a question addressed later in this chapter. The following section examines the likelihood that Chinese satellites could obtain a more precise target location, either with or without cueing from OTH radar.

Little information is available within the public domain concerning China's NOSS and their capabilities. NOSS generally consist of at least two ELINT satellites flying in close formation. These satellites receive signals from the maritime environment and perform TDOA analysis on the signals to geolocate the source of the emissions. The geolocation accuracy depends on both the system sensitivity and its ability to recognize and isolate specific signals of interest. It is likely that the Chinese systems have accuracies at least on the order of tens of kilometers and, hence, can potentially cue China's imaging satellites for additional targeting data. As of April 2015, China has launched a total of five NOSS, with an estimated three currently in operation.[22] Each system is said to have a surface coverage area with a 3,500-km radius. Assuming three NOSS constellations in operation, that would provide, on average, 18 overflights per day against a given target.[23]

Space-Based Imaging Assets

Although Chinese satellite imaging capabilities are becoming more robust, large maritime search areas would nevertheless make it difficult for PLA commanders to

---

[21] According to Hagt and Durnin, the Chinese skywave CEP ranges from 20 to 40 km. By definition, we expect the error distance between the actual target location and the radar-derived location to be less than the CEP 50 percent of the time. Note that it is possible to use natural or man-made objects, such as islands near the target, to improve (reduce) the skywave radar CEP. Hagt and Durnin, "China's Antiship Ballistic Missile," 2009.

[22] Launched NOSS include the Yaogan 9, 16, 17, 20, and 25. According to Chandrashekar and Perumal, the Yaogan 17 and Yaogan 25 likely replaced the Yaogan 9 and Yaogan 16. S. Chandrashekar and Soma Perumal, *China's Constellation of Yaogan Satellites and the Anti-Ship Ballistic Missile—An Update*, Bangalore, India: International Strategic and Security Studies Programme, National Institute of Advanced Studies, January 2015.

[23] Chandrashekar and Perumal, *China's Constellation of Yaogan Satellites and the Anti-Ship Ballistic Missile—An Update*, 2015.

locate targets using these satellites alone.[24] However, with cueing from OTH radars and NOSS about the general location of U.S. assets, Chinese imaging satellites could obtain more precise target identification and location data with some degree of regularity and reliability. These results highlight the importance of denying Chinese commanders the ability to cue satellites.

By 2003, China had launched two photo-reconnaissance satellites (the Jianbing 3A and 3B), operated by the PLA. These were China's first long-duration (life expectancy of three to five years) imaging satellites that returned digital images to the ground via data link. Both satellites were panchromatic, electro-optical (EO) imagers flying in low earth orbit (LEO) with three-meter imaging resolutions and 30-km imaging swaths, providing China with its first, albeit limited, OTH imaging capability.[25] Based on the orbital parameters, we estimate that, without external cueing, these satellites could have randomly imaged a given ship near China once every 35.1 days (median value).[26] Based on these calculations, then, the Jianbing 3 satellites represented an important technological advance for China, but they offered no real-time surveillance capability against U.S. Navy ships without cueing from broad-area surveillance assets.

China began launching its second-generation spy satellites in April 2006, including its first space-based SAR imaging satellites. By 2010, China had deployed three SAR (Yaogan 1, 3, and 6) and four high-resolution EO satellite imaging systems (Yaogan 2, 4, 5, and 7).[27] The first two SAR satellites were said to have resolutions from 5 to 20 meters and corresponding swath widths from 40 to 100 km,[28] while the last system was said to have a resolution of 1.5 meters.[29] Meanwhile, the EO satellite resolutions

---

[24] For another assessment of Chinese satellite capabilities, see Eric Hagt and Matthew Durnin, "Space, China's Tactical Frontier," *Journal of Strategic Studies*, Vol. 34, No. 5, October 2011. Hagt and Durnin assessed the coverage of a particular known location (e.g., a previously located and stationary target), rather than assessing the ability of space-based ISR to find unlocated (or imprecisely located) targets. Although the work initiated a very useful discussion of Chinese space-based ISR capabilities, it suffered from several methodological problems identified in David Wright, "Response to 'Space, China's Tactical Frontier' by Eric Hagt and Matthew Durnin," *Journal of Strategic Studies*, Vol. 34, No. 5, October 2011.

[25] *Dragon in Space*, "Ziyuan 2," web page, last updated April 3, 2012.

[26] The median value cited here assumes that the weather is clear and that the satellites are constantly imaging at nadir (perpendicular to the earth's surface, or at a 90-degree look angle) during the day as they pass over the region. The satellites' orbital parameters are estimates based on those of the Ziyuan 2 satellite (n2yo.com, Satellite Database, data from December 1, 2011).

[27] This assumes that all Jianbing 3 satellites were no longer operational by 2010. Also, the Yaogan 3 occupied the exact same orbit as the Yaogan 1, indicating that the Yaogan 1 may not have been operational as far back as late 2007. Incidentally, the Yaogan 1 suffered from an internal explosion in February 2010. See National Aeronautics and Space Administration, "Old and New Satellite Breakups Identified," *Orbital Debris Quarterly News*, Vol. 14, No. 2, April 2010.

[28] *Jane's Space Systems and Industry*, "Yaogan Series," January 20, 2015.

[29] The Yaogan 1 and 3 carried the Jianbing 5 SAR sensor whereas the Yaogoan 6 carried the Jianbing 7 SAR sensor.

ranged from 1.5 meters (Yaogan 2, 4, and 7) down to one meter (Yaogan 5).[30] Without external cueing and incorporating the same assumptions applied to the 2003 calculations, the four EO satellites in 2010 could randomly image a given ship once every 13.8 days (median value).[31] Adding the two SAR satellites (Yaogan 3 and 6), the median revisit time becomes 6.9 days for the entire six-satellite constellation.[32]

External cueing can shorten revisit rates significantly, depending on the satellites' abilities to slew sensors off nadir while maintaining sufficient resolution.[33] The more an EO imager looks off to the side (i.e., the shallower look angle), the more the line-of-sight path from the satellite to the target elongates, thereby reducing the image resolution.[34] This also produces "side-looking" images where the target-grazing angle (the angle between the image line of sight and the earth's surface at the target) is less than perpendicular.[35] For each EO satellite, we calculated the minimum look angle for which it can maintain a resolution of 5 meters or better (i.e., smaller). Unlike EO, SAR satellites cannot image directly beneath themselves (at nadir) and, instead, image off to the side using side-looking antennas. Image resolution and range depend on a number of factors, including image collection time (usually tens of seconds), radar power and bandwidth, and signal processing capabilities.[36]

---

[30] *Dragon in Space*, "Yaogan," web page, last updated May 29, 2012. The Yaogan 2, 4, and 7 carried the Jianbing 6 EO sensor, and the Yaogan 5 carried the Jianbing 8 EO sensor.

[31] These calculations assume that the Yaogan 2, 4, and 7 have imaging swaths of 40 km and that the Yaogan 5 has an imaging swath of 30 km, based on the flying altitudes of the satellites in question and the assumption that these satellites have the same field of view as the Jianbing 3C.

[32] Once again, these calculations assume clear weather conditions throughout the target area. Clouds and heavy fog can obscure targets from EO and IR sensors, reducing the number of imaging opportunities. However, the same is not true for SAR sensors, which can image both day and night and during most weather conditions. If clouds obstruct the target 25 percent of the time, then the 2010 median revisit rates become 18.4 days (EO only) and 7.8 days (EO and SAR). Increasing the obstruction to 50 percent changes these values to 27.7 days (EO only) and 9.5 days (EO and SAR).

[33] Although it is certainly possible to detect U.S. warships with coarser-resolution images, we require a resolution of at least five meters (EO, IR, and SAR) to identify them.

[34] Images with shallow grazing are usually harder to interpret than images with steeper grazing angles or at nadir; however, this may be somewhat mitigated by the fact that the U.S. warships are large and relatively isolated, leaving them unobstructed even at shallow grazing angles.

[35] According to David Wright, target grazing angles of five degrees (viewing the target almost entirely from the side) worsen the EO satellite image resolution by a factor of ten compared with an image at nadir. Increasing the grazing angle to 20 degrees improves the resolution, but it is still worse by a factor of three relative to its resolution at nadir. This decrease in resolution is due to longer image ranges, "stretching" of the image in the downrange direction, and increased atmospheric attenuation (loss in reflected target energy via atmospheric particles). Wright, "Response to 'Space, China's Tactical Frontier' by Eric Hagt and Matthew Durnin," 2011.

[36] According to Hagt and Durnin, SAR imaging is generally restricted to between 20- and 80-degree grazing angles. The size and range of SAR "sweet spots," where resolution is at an optimum, depend on the factors listed in the text. Eric Hagt and Matthew Durnin, "Space, China's Tactical Frontier," *Journal of Strategic Studies*, Vol. 34, No. 5, October 2011.

For 2010, assuming clear weather and an accurate, near-continuous target location cue that allows the PLA satellites to preemptively focus their sensors on the approaching target—as might be provided by OTH radar against targets within their arc of coverage—the median revisit time falls from 13.8 days to 8.0 *hours* (EO only) and from 6.9 days to 4.0 *hours* (EO and SAR).[37]

How would Chinese ISR fare in 2017? As of April 2015, China maintains three to four SAR and three to four high-resolution EO imaging satellites.[38] In addition, China began launching a new class of high-altitude (1,200 km), medium-resolution EO imaging satellites in 2010.[39] These satellites can image swaths 100 km wide at resolutions between three and ten meters making them well-suited for finding, locating, and possibly even identifying large distinctive ships such as U.S. aircraft carriers.[40]

Based on previous constellations and launch schedules, our best estimate is that, for the time being, China will maintain three to four SAR, three to four high-resolution EO, and an additional three to four high-altitude, medium-resolution EO satellites on station and focus on increasing the capability of those satellites.[41] If so, the 2017 median revisit times would be between 2.9 days (without cueing) and 2.6 hours (with near-continuous cueing) for the entire nine- to 12-satellite constellation.[42] The times above assume clear weather; cloud cover could make them somewhat longer.[43] Note that, for both 2010 and 2017, near-continuous satellite cueing reduces the median revisit times by as much as 95 percent. Such cueing could come from a variety of sources, including China's NOSS, skywave OTH radars, or ground-based ELINT.[44]

---

[37] A cloud cover of 25 percent increases the median revisit times from 8.0 hours to 10.7 hours (EO only) and from 4.0 hours to 4.6 hours (EO and SAR). Fifty-percent cloud cover further increases the median revisit times to 16.0 hours (EO only) and 5.2 hours (EO and SAR).

[38] These likely consist of the Yaogan 10, 13, 18, and 23 (SAR) and Yaogan 14, 21, 24, and 26 (EO); however, it is possible that the Yaogan 10 and 21 are no longer operational.

[39] These include the Yaogan 8, 15, 19, and 22.

[40] Due to their high altitudes, these satellites' fields of regard cover an area approximately 1,000 km in diameter. See Chandrashekar and Perumal, *China's Constellation of Yaogan Satellites and the Anti-Ship Ballistic Missile—an Update*, 2015.

[41] In 2009, Mark Stokes predicted that China would maintain a stable four-EO/four-SAR future satellite constellation. Stokes, *China's Evolving Conventional Strategic Strike Capability*, 2009.

[42] These revisit times are the averages between a nine- and 12-satellite constellation.

[43] A cloud cover of 25 percent increases the median revisit time from 2.9 days to 3.5 days (no cueing). Fifty-percent weather degradation increases it further to 4.2 days for the no-cueing case. For the near-continuous cueing case, the median revisit time increases from 2.6 hours to 2.9 hours (25-percent cloud cover) and 3.3 hours (50-percent cloud cover).

[44] China also has a variety of "nonmilitary" imaging space assets managed by different bureaucratic entities within the Chinese government. In a time of crisis, it is possible that control of these assets would be ceded to the PLA, though it is unclear what role these systems could play. See Hagt and Durnin, "China's Antiship Ballistic Missile," 2009.

The median satellite revisit times, with and without cueing, for three of the four snapshot years examined in this report are listed in Table 7.1.[45] Due to the many assumptions and unknowns in the calculations, these numbers should not be regarded as definitive. However, they do provide a good indication of China's evolving over the horizon ISR capability. At the same time, these figures do not account for U.S. countermeasures (discussed later) and should thus be considered optimistic from a Chinese perspective.

C4ISR Capabilities

China's C4ISR capabilities are not well documented, but multiple C4ISR delays would likely arise during the execution of the Chinese anti-surface warfare kill chain. Because U.S. ships will be constantly moving, uncertainty about their location will necessarily increase during any gap or delay in the kill chain. If China engages U.S. ships with, for example, ASBMs using only NOSS or OTH radar track data, then there will be some time delay from last target coordinate update to weapon arrival on target, further increasing the target's actual distance from the reported location.

If, on the other hand, Chinese commanders decide to use these sensors to cue its space imaging assets to more accurately identify and locate U.S. surface ships prior to engagement, additional delays will arise from the coordination of multiple and, sometimes, disparate assets. Because the targets are constantly moving, such delays decrease

Table 7.1
Bounding the Challenge: Median Time Between Satellite Images of a Given U.S. Target With and Without OTH Radar Cueing

Characteristic	1996	2003	2010	2017[a]
Median time without cueing	N/A	35.1 days	6.9 days	2.9 days
Median time with cueing[b]	N/A	N/A	4.0 hours	2.6 hours
Assets employed	N/A	2 EO (Jianbing 3A/B)	4 EO (Yaogan 2, 4, 5, 7) 2 SAR (Yaogan 3, 6)	3–4 EO (high-resolution) 3–4 EO (medium-resolution) 3–4 SAR

[a] It is possible that future Chinese EO satellites will incorporate infrared (IR) sensors, allowing them to image targets (unobstructed by weather) at night as well. If we assume that these satellites also carry IR imaging sensors with capabilities similar to those of EO sensors, then the median or average revisit times decrease by 10–40 percent, depending on the cueing assumptions, satellite mix, and weather conditions.

[b] Assumes clear weather over the target area and accurate, continuous target cueing with no satellite tasking or command-and-control delays.

[45] It is difficult to provide similar details for 2017 due to the lack of future satellite orbital data.

the likelihood that the next satellite pass will actually image the target. Hence, the revisit rates "with cueing" listed in Table 7.1 provide the most optimistic revisit rates from the Chinese perspective. However, these revisit rates do not account for cueing sensor error CEPs and command-and-control delays within the kill chain and are therefore not entirely realistic.[46]

Countering Chinese OTH ISR

Both China's OTH radar and space-based surveillance assets are vulnerable to active and, to a lesser extent, passive U.S. countermeasures. Such measures include active jamming (of electromagnetic sensors, such as radars and ELINT sensors) and dazzling (of EO/IR sensors), kinetic strikes, and passive avoidance. China's OTH radar, NOSS, and SAR satellites are potentially susceptible to active jamming techniques, including noise and coherent waveform (spoofing) jamming. Such jamming could partially mask U.S. Navy operations while increasing the level of location uncertainty, though the state of U.S. jamming and dazzling force development is uncertain at this time. Furthermore, the OTH radar receiving arrays, control stations, and power generators are well within the range of Tomahawk cruise missiles launched from the East and South China Seas. However, while militarily feasible, attacks on these targets located deep inside the Chinese mainland would be politically sensitive and would likely require high-level political approval.

Passive countermeasures, such as avoidance, are more readily available because they require limited (if any) additional systems and force structure. However, passive countermeasures may not be as effective, and any effect may be difficult to ascertain. China's current suite of imaging satellites flies in sun-synchronous orbits with predominantly north-south trajectories. If U.S. forces can accurately predict and track Chinese satellite orbits, then U.S. Navy surface ships may more effectively avoid impending satellite passes by maximizing speeds with east-west headings.

In the case of OTH radar, the radar illuminates large patches of the sea surface, which, in turn, creates a large sea clutter return. OTH radar uses Doppler effects to differentiate real targets from the sea clutter, and even when large ships, such as aircraft carriers, maintain a Doppler similar to that of the sea clutter, OTH radar operators will have difficulty deciphering them from the clutter background. (Ships may maintain a Doppler by moving laterally while remaining roughly at a constant distance from the radar.) However, staying within the sea clutter Doppler spread may be difficult and

---

[46] Under specific conditions, we can calculate the satellite imaging revisit rate against a particular ship. For example, if we assume that (1) the cueing sensor CEP equals *50 km*, (2) the cueing sensor detects the target an average of *four times a day* (once every six hours), (3) the weather is clear, and (4) there is an average one-hour tasking delay for satellites, then the median revisit rate for Chinese satellites in 2010 is once every 2.39 days. If the weather is 50-percent cloudy, then this value becomes once every 3.96 days. Note that both of these values fall directly between the corresponding extreme values (continuously cued and not cued), as expected.

restrictive operationally. Furthermore, flight operations may cue OTH radar operators to the carriers' locations.[47]

Ideally, U.S. naval forces can completely mask their operations and locations by utilizing both active and passive countermeasures. However, it is unclear whether this can be fully achieved. At the very least, such actions may significantly degrade Chinese OTH targeting sensors by decreasing the frequency of detection, as well as the accuracy of targeting.

Chinese OTH Targeting: Summary

Despite having no dedicated, long-range ISR assets in 1996, China now has a growing, multifaceted capability. However, each aspect of the PLA's long-range ISR remains relatively thin and, at present, underdeveloped, making it likely that prospects for U.S. countermeasures are relatively good in the event of a conflict over the next several years. China is clearly investing heavily in dedicated long-range ISR, and U.S. prospects in the longer term are less certain.

In the remainder of this chapter, we address three types of offensive systems that could potentially conduct attacks against U.S. surface vessels: ASBMs, air- and sea-launched anti-ship cruise missiles, and conventional or nuclear attack submarines.

The Chinese Anti-Ship Ballistic Missile Threat

For the past decade, China has been developing the capability to engage surface ships, such as U.S. aircraft carriers, at ranges well beyond the radar horizon from the mainland. In this section, we assess the PLA's ability to strike and disable a U.S. aircraft carrier using ASBMs, looking first at the relationship between key variables having to do with the capabilities of the missiles and the accuracy of intelligence, then at the Chinese ability to execute the entire kill chain under operational circumstances. We conclude by considering future developments through 2017 and beyond.

A 2006 Office of Naval Intelligence assessment noted, "China is equipping theater ballistic missiles with maneuvering reentry vehicles with radar or IR seekers to provide the accuracy necessary to attack a ship at sea."[48] As of early 2013, the system had been repeatedly tested over land, though it has not yet been tested against targets at sea.[49] Nevertheless, according to a 2011 statement by Vice Admiral Jack Dorsett, the

---

[47] It may be possible for aircraft to restrict their launch and landing headings long enough to mask the carrier's location. However, this, too, may not be viable operationally, especially for patrol aircraft, such as helicopters and AEW aircraft.

[48] Office of Naval Intelligence, *Sea Power Questions on the Chinese Submarine Force*, Washington, D.C., December 20, 2006.

[49] Andrew S. Erickson, "China Channels Billy Mitchell: Anti-Ship Ballistic Missile Alters Region's Military Geography," *China Brief*, Vol. 13, No. 5, March 2013.

head of naval intelligence, the system can be considered in "initial combat capability."[50] The DF-21D TBM is the initial platform for the weapon system and has a range of roughly 1,500–2,000 km. Some reports suggest that a brigade of DF-21Ds had been formed by the end of 2014 and that an estimated six missiles may have been deployed to it.[51] The system's range could be extended to a possible 3,000 km with an additional third "glide" stage.[52]

Much remains uncertain about the system and its capabilities. Nevertheless, by parameterizing several of the critical variables, we can get a rough sense of what ASBMs might or might not accomplish under a range of circumstances. The output examined here is the salvo size, or number of missiles in a single attack, needed to achieve an 80-percent probability of damaging or destroying a U.S. carrier. Key inputs include the following three factors:

- *Missile kill radius*, or the maximum distance from its prelaunch aim point within which the missile can still engage (find, locate, and maneuver to) the target (see Figure 7.2). The kill radius (or basket) depends not only on the missile's kinematics but also on its in-flight search, identification, tracking, and data-link capabilities. Published reports speculate that the DF-21D kill radius might be between 25 and 40 km.[53]
- *The targeting location CEP* is the expected difference between the actual target location and the targeted location at firing time.[54] As noted previously, OTH radar has a targeting location CEP of between 22 and 178 km, depending on the distance from the target and other variables, though delays in transmitting identification and location data to the firing elements may effectively add to the CEP.
- *The weapon Pk* is the probability that the weapon will hit the target given that the target is within the kill radius of the missile's aim point when the weapon arrives. It is a function of the weapon's capabilities (e.g., speed, reliability, maneuverability, sensors) and the target's defenses, both kinetic and nonkinetic.

To the extent that the kill radius of the missile is small in relation to the targeting location CEP, a larger salvo size (in terms of number of missiles) will be required to achieve the same probability of kill against the target ship. If, on the other hand, the missile's onboard search functions and maneuverability give it a greater kill radius rela-

---

[50] See comments by Deputy Chief of Naval Operations Vice Admiral David J. Dorsett in Tony Capaccio, "China's Ballistic Missile, Stealth-Fighter Advances Draw Attention of U.S.," *Bloomberg*, January 6, 2011.

[51] IISS, *The Military Balance*, 2015, p. 237.

[52] Hagt and Durnin, "China's Antiship Ballistic Missile," 2009.

[53] Hagt and Durnin, "China's Antiship Ballistic Missile," 2009.

[54] By definition, we expect the error distance between the actual target location and the ISR-derived targeted location to be less than the CEP 50 percent of the time.

Figure 7.2
DF-21D Flight Profile and Missile Kill Radius

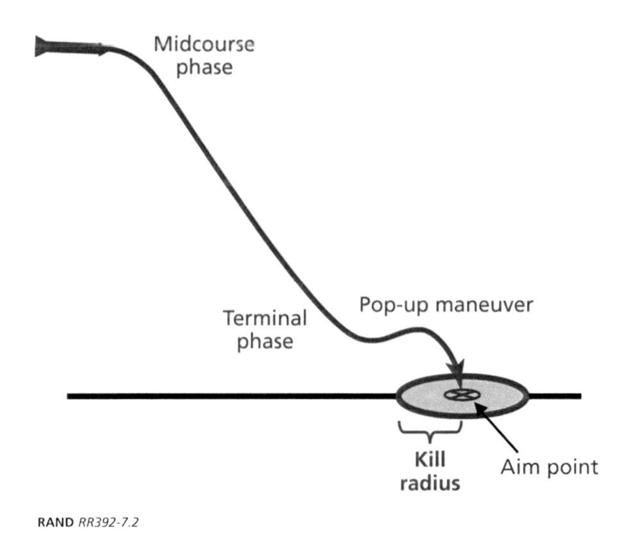

RAND *RR392-7.2*

tive to the targeting location CEP, fewer missiles will be required. Figure 7.3 provides a graphic representation of a case in which the missile kill radius is significantly smaller than the targeting system CEP.

Table 7.2 shows the DF-21D salvo sizes required to achieve an 80-percent probability of a mission kill against a given ship, using a variety of values for missile kill

Figure 7.3
Missile Kill Radius and Targeting Location CEP

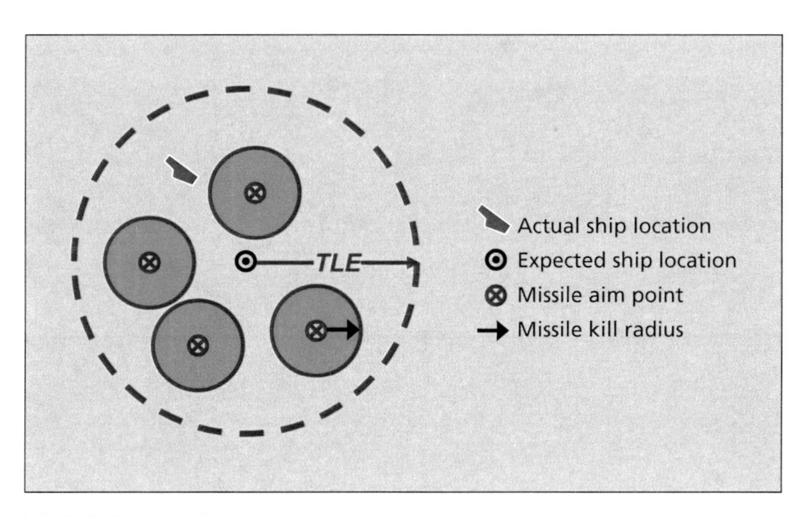

NOTE: TLE = target location error.
RAND *RR392-7.3*

Table 7.2
DF-21D Salvo Sizes Required to Achieve an 80-Percent Probability of Damaging or Destroying a U.S. Surface Ship

Ratio of Missile Kill Radius–to–Targeting Location CEP	ASBMs Required to Achieve Probability of Kill				
	0.2 Pk	0.4 Pk	0.6 Pk	0.8 Pk	1.0 Pk
0.25	100+	100+	100+	82	49
0.50	100+	55	33	21	13
1.00	31	14	8	5	4
2.00	9	4	3	2	1

radius, targeting system CEP, and weapon Pk.[55] (Here, we consider it a *mission kill* if at least one weapon in a salvo impacts the target.) To capture the three key variables without a complicated simulated three-dimensional display, the missile kill radius and the targeting system CEP are combined in the leftmost column and expressed as a ratio of the former to the latter (missile kill radius to targeting location CEP). The larger the number, the greater the ability of the warhead to maneuver is relative to the uncertainty about the location of the target at the time when the warhead arrives in the vicinity.

The upper-left cell in the body of Table 7.2 represents the best case considered from the U.S. perspective, while the lower right is the worst. To take a single cell, assuming the kill radius of the weapon equals 40 km and the targeting location CEP (including ship movement during missile flight) is 40 km (ratio = 1.00), then a weapon Pk of 0.6 yields a required salvo size of eight missiles. However, if the kill radius is 25 km and the targeting location CEP is 50 km (ratio = 0.50), then the same weapon Pk of 0.6 yields a required salvo size of 33 missiles to achieve the same 80-percent probability of mission kill.[56]

Executing the ASBM End-to-End Kill Chain

How well can China successfully execute the entire kill chain—detect, identify, track, target, engage, and assess—against a U.S. CSG? In part, the answer will be driven by the PLA's own rules of engagement and available DF-21D inventory. For example, PLA commanders may launch weapons using only NOSS or OTH track data. On the other hand, they may require a prestrike image of the target before missile launch. As discussed earlier, the latter is more challenging and, as such, will likely offer fewer engagement opportunities. Also, China's DF-21D inventory is not unlimited, and large salvo

---

[55] These calculations assume an optimal aim point distribution and that the weapons in each salvo behave independently and have the same weapon Pk independent of the salvo size.

[56] Note that decreasing the expected probability of mission kill decreases the missile salvo requirements.

requirements or target uncertainty could elevate the launch threshold, thereby reducing the number of ASBM engagements.

Once the decision to engage has been made, there is the question of the missile salvo size. Based on previous OTH radar and DF-21D performance estimates, we can calculate an estimate of the required salvo size. The *kill radius–to–targeting location CEP ratio* appears to vary from 0.23 to 1.20 when the PLA launches ASBMs based on the OTH radar track alone. As discussed earlier, U.S. forces may be able to diminish Chinese OTH radar targeting accuracy using active and passive means, decreasing the missile kill radius–to–targeting location CEP ratio.[57] Small ratios (less than 0.25) require large missile salvos (more than 49 missiles) even for the perfect weapon system (Pk = 1.0).[58]

Such large salvos may not be operationally feasible. Even if they were, inventory limits might deter the use of DF-21Ds under such uncertain conditions. Between 1991 and 2012, China deployed an estimated total of 122 DF-21 launchers of all variants (nuclear and non-nuclear), with an associated number of missiles of between roughly 250 and 500.[59] Given limitations on production capacity and, perhaps more importantly, force structure (i.e., missile brigades and their associated manpower and infrastructure), the total DF-21D build is unlikely to exceed 100 or, at most, 200 missiles.

On the other hand, large ratios (greater than 1.0) that favor the Chinese lead to much smaller salvo requirements, especially for high weapon Pk values (greater than 0.6). For example, if the missile kill radius–to–targeting location CEP ratio equals 1.20 and the weapon Pk equals 0.6, then a salvo size of six missiles is required to achieve an 80-percent probability of mission kill. In this case, U.S. forces may have to rely more on point defenses to reduce the weapon Pk. In the example here, if the missile defenses reduce the ASBM's Pk to 0.2, then the salvo size requirement becomes 21 missiles.

If the PLA can successfully image a U.S. carrier via satellite, then the kill radius–to–targeting location CEP ratio will depend largely on the time it takes for the image to be downloaded, processed, exploited, and disseminated to the proper action elements (e.g., command elements, missile brigades, OTH radar operators).[60] While these

---

[57] These calculations assume that the carrier is operating 700 km southeast of Taiwan. They also assume an additional 15-minute C4ISR delay from the last target coordinate to weapon launch and another 15-minute weapon time of flight. (In reality, the plausible range of C4ISR delay times could vary widely, depending on both technical and operational conditions.) During the additional 30-minute delay, we assume that the aircraft carrier targeting location CEP increased by 25 km based on an assumed average radial speed of 50 km per hour. Additional C4ISR delays may result from high command–level launch authority and multiple-launch-site coordination. These missile kill radius–to–targeting location CEP ratios decrease at longer ranges (favorable to the United States) because of the increasing OTH radar targeting error CEP.

[58] A similar analysis can be done for the Chinese NOSS and an estimated TLE.

[59] On the methodology for estimating DF-21 missile numbers, see Chapter Three (scorecard 1).

[60] Image exploitation includes finding and identifying the target of interest within the image and then deriving its location. Target coordinates derived from satellite imagery tend to be much more accurate, often on the order of meters.

activities occur, U.S. ships are on the move. The longer these processes take, the more dated the image becomes and the larger the effective target location becomes.

In recent years, U.S. forces have invested heavily in C4ISR architectures and data processing to expedite these very same processes. It is unclear how quickly China can currently perform each step in succession. If its forces can perform the entire process in 15 minutes, then the kill radius–to–targeting location CEP ratio would vary from 0.67 to 1.07 (under the same assumptions used earlier). If, instead, it takes two hours, then the missile kill radius–to–targeting location CEP ratio varies from 0.20 to 0.32, and the required salvo size would be large.

ASBM Summary and Future Developments

A few summary observations can be derived from the calculations. First, although ASBMs present a new challenge that has not been previously faced by U.S. surface combatants, they are likely not the one-shot, one-kill weapons sometimes portrayed in the popular literature. There may be limited opportunities for engagement, and when large salvo sizes are required to obtain a mission kill, such salvos may not be realistically possible. U.S. forces can mitigate Chinese ASBM capabilities by increasing the targeting location CEP (and thus reducing the kill radius–to–targeting location CEP ratio), decreasing the Chinese weapon Pk values through, for example, point defenses, or both. One approach may be more plausible than the other depending on the cost and technological maturity of the potential means employed. The examples cited here indicate that increasing the Chinese targeting errors may provide more "bang for the buck."

Furthermore, attacks based solely on China's NOSS or OTH radar track data without the confirmation of target identity from other sources may result in the engagement of false targets, a problem that is not included in our calculations but which would likely confront Chinese planners in real life. Finally, if the post-image processes take several hours, it is doubtful that satellite images of U.S. ships will have any real engagement value. However, if it is on the order of an hour or minutes, China's space imaging assets may be crucial in holding U.S. warships at risk.

China is continuing to modernize its maritime surveillance and strike assets. Possible near-term developments that might boost its ability to find and attack U.S. CSGs include

- improving SAR/EO/IR imaging and NOSS satellite capabilities
- building additional OTH radar facilities to create more overlapping coverage of key areas[61]

[61] China could build additional skywave radar systems that cover the South China and Yellow seas. In addition to expanding the area covered, overlapping coverage would create a more robust capability.

- fielding large, long-range UAVs with maritime search radar[62]
- incremental improvements to ASBM capability[63]
- improving C4ISR and reducing command delays.[64]

China's success in one or more of these areas could increase the number of engagement opportunities while improving the ASBM kill radius–to–targeting location CEP ratio, weapon probability of kill against the U.S. surface fleet, or both.

Nevertheless, it should be understood that the United States will also be working on a variety of potential counters to the ASBM threat.[65] These could include measures designed to degrade or disable China's OTH ISR capability, including jammers, dazzlers, and kinetic strikes against ground-based infrastructure. A similar range of point defenses could be used to counter the sensors carried onboard the ASBMs. Other counters may be passive in nature, including strict emission control on the part of surface combatants, the use of obscurants, or operational practices that maximize mobility. A variety of decoys could also be employed and could include integrated suites of decoy systems simulating an entire group of ships. Finally, anti-ballistic missile systems might be used selectively against incoming ASBMs, either midcourse or during terminal flight. Needless to say, China is likely to answer with its own counter-countermeasures. As in many other areas of conventional conflict, this is a competitive arena in which the balance of offensive and defensive capabilities is likely to see multiple changes over time.

The Chinese Anti-Ship Cruise Missile Threat

China has dramatically improved its ASCM strike capabilities through overseas acquisitions and indigenous development. ASCMs generally approach the target at low altitudes (5–10 meters above the surface) and either impact the target near the water line or dive into it during terminal flight. Modern ASCMs receive offboard targeting data via data link, locate the target using GPS and multimode sensor packages, and can

[62] A number of reports indicate that China places a high priority of developing UAVs with a variety of functions, to include maritime search. "China Flies its Largest Ever Drone: The Divine Eagle," *Popular Science*, February 6, 2015; and "Eyeing Exports, China Steps Up Research into Military Drones," Reuters, April 29, 2015.

[63] As China deploys its ASBM system, it will be looking toward improvements, such as the development of a quality multimode seeker capable of defeating U.S. passive and active point defenses.

[64] China will continue to improve its C4ISR systems and networks to effectively coordinate disparate assets, quickly process and disseminate crucial information, and synchronize multiple weapon systems (e.g., ASBMs, ASCMs) to potentially overwhelm CSG defenses.

[65] For more detail on potential measures to defend against ASBM attacks, see Marshall Hoyler, "China's 'Anti-access' Ballistic Missiles and U.S. Active Defenses," *Naval War College Review*, Vol. 63, No. 4, Fall 2010, and Jonathan F. Solomon, *Defending the Fleet from China's Anti-Ship Ballistic Missile: Naval Deception's Role in Sea-Based Missile Defense*, thesis, Washington, D.C.: Georgetown University, April 15, 2011.

strike maritime targets at long ranges (greater than 200 km). They can be launched by ground elements, aircraft, surface ships, and submarines. In the following sections, we discuss China's ability to hold U.S. warships at risk using a combination of modern launch platforms and ASCMs.

Air-Launched ASCMs

The PLAN has extensively modernized its aviation assets over the past 19 years. As of 1996, PLAN aviation was—much like the PLAAF—a large, second-generation air force comprising mostly 1950s- and 1960s-vintage aircraft. By 2015, however, a majority of PLAN fighters were fourth generation (though the PLAN still relied on the somewhat outmoded JH-7 for much of its strike capability). At the same time, PLAN aviation forces significantly upgraded their air-to-surface weapons with an emphasis on air-launched ASCM systems. Other recent developments have included AEW platforms, SEAD systems (such as anti-radiation missiles), ELINT platforms, and electronic attack systems. All told, PLAN aviation can now employ maritime strike packages made up of modern escort fighters, strike fighters, and support aircraft—all of which combine to pose a significant threat to U.S. warships at long range.

Table 7.3 shows PLAN combat aircraft in 1996, 2003, 2010, and 2015, along with an estimate for 2017. We break the fighters/tactical aircraft into two broad categories: (1) air supremacy fighters (FTR in the table), whose primary mission is air-to-air combat, and (2) strike fighters (FGA in the table), whose primary mission is air-to-surface strike.[66] In 1996, more than 75 percent of PLAN aviation tactical aircraft were second-generation; however, by 2010, more than 85 percent of its fighter force was third-generation or newer. By 2017, the PLAN fighter force will be overwhelmingly fourth-generation, though it will likely still operate obsolete JH-7s in a strike role. Two new aircraft, the J-15, based on the Russian Su-33, and the J-16, based on the Russian Su-30MK2, are joining other Chinese fourth-generation aircraft in the PLAN inventory.

The bomber force has also been modernized, though not as dramatically as the fighter and strike portion of PLA naval aviation. By 2013, the last of the H-5 light bombers, armed with torpedoes, had been retired. Although the PLAN's only remaining bomber, the H-6, is based on an old design (originally modeled on the Soviet Tu-16, which entered service in 1954), it has evolved over time. The PLAN's improved H-6G models are of recent manufacture (the aircraft made its first appearance in 2002), have greatly improved range, and can carry four anti-ship cruise missiles (as opposed to the two carried by the H-6D).[67] As in the PLAAF case, modernization has been accompanied by a reduction in the total number of PLAN aircraft, though the force structure

[66] Most strike fighters also have an air-to-air capability, especially after releasing their ordnance.

[67] *Jane's All the World's Aircraft*, "XAC H-6," October 20, 2014; GlobalSecurity.org, "H-6D Bomber," web page, last updated July 11, 2011. The H-6G may ultimately be replaced by the H-6K, which has still better performance in a number of areas and which is in service with the PLAAF.

Table 7.3
PLAN Aviation Combat Aircraft, 1996–2017

Aircraft Generation and Model	Type	Number				
		1996	2003	2010	2015 (current)	2017
Fighters and attack aircraft						
2nd generation						
J-6	FTR	311	200			—
Q-5	FGA	40	30	30	—	—
3rd generation						
J-7	FTR	70	26	36	—	—
J-8	FTR	30	48	48	24	24
JH-7	FGA		20	84	120	120
4th generation						
Su-30MK2	FGA			24	24	24
J-16 (Su-30MK2)[a]	FGA					24
J-11B/BS	FTR				60	72
J-10A/S	FTR				24	28
J-15 (Su-33)[a]	FTR				—	28
Total		451	324	222	252	320
Bomber aircraft						
H-5	Light	130	50	20	—	—
H-6	Medium	7	—	—	—	—
H-6D	Medium	9	18	30	—	—
H-6G	Medium				30	30
Total		146	68	50	30	30

SOURCES: IISS, *The Military Balance*, 1996, 2003, 2010, and 2015. Estimates for 2017 are based on IISS, *The Military Balance*, 2013, as well as the discussion of PLAN aviation programs in *Jane's Sentinel Security Assessment*, "China: Procurement," March 5, 2015.

NOTES: Most recent fighter and attack aircraft models have both a fighter and surface attack capability; designations represent the primary capability. The FTR's primary mission is air-to-air combat. The FGA's primary mission is air-to-surface strike.

[a] Indicates the Russian aircraft on which the Chinese model is primarily based.

appears to have stabilized and may begin to rise modestly as new-generation fighters and attack aircraft reach series production.

The quality of Chinese air-launched ASCMs has also significantly improved.[68] Since 1996, these weapons have transitioned from short-range, easy-to-intercept missiles with limited maneuverability and navigation capabilities to more lethal systems that are designed to challenge high-quality defenses from multiple approach axes and from longer ranges. Recent Chinese designs have smaller diameters and fly at altitudes as low as three to five meters above the surface, making them harder to track and destroy. Some Chinese air-launched ASCMs can fly at supersonic speeds. However, most supersonic ASCMs in the PLA inventory are shorter-range systems, like the YJ-91. The new longer-range missiles, such as the YJ-62, use turbofan engines to extend their range while flying at subsonic speeds. The most potent Chinese air-launched ASCM to emerge is the YJ-12, which combines the YJ-62's long range and YJ-91's supersonic terminal speed into one weapon, employing technologies that are similar to the Russian Kh-31. It has reportedly been test-fired but it is not yet in the inventory.[69] Table 7.4 provides a list of recent Chinese air-launched ASCMs, along with some of their capabilities and the approximate dates when they entered the fleet.

The net effect of these recent PLAN aviation advances is to increase the range and magnitude of threat posed to U.S. Navy elements by PLAN aviation strike packages. U.S. Navy assets face a higher likelihood of battle damage at longer ranges from the Chinese mainland. This increased threat may also force the U.S. fleet into a more defensive posture, reducing the fraction of U.S. Navy assets available for offensive tasks. A greater fraction of U.S. carrier aircraft might be held back for defensive tasks, and the location and weapon loadouts for surface combatants could be apportioned to favor defense against the low-flying cruise missile threat.[70]

Figure 7.4 plots PLAN ASCM effective range rings for 1996 and 2017 in both the Taiwan and Spratly Islands scenarios. The inner rings represent the maximum combat radius of the current strike fighter, and the outer rings add in the maximum weapon range. In 1996, the Q-5 strike fighter had a maximum combat radius of 600 km (inner ring), and the YJ-81 ASCM has a maximum range of 70 km, for a total combined range of 670 km (outer ring). In contrast, with the addition of the J-16 (China's Su-30MKK equivalent) and YJ-62 by 2017, maximum launch and launch-plus-missile ranges increased to at least 1,500 km and 1,780 km, respectively.

---

[68] *Jane's International Defence Review*, "Storm Force Warning: China's Anti-Ship Missile Range Spreads Its Wings," April 17, 2013.

[69] The Kh-31 flies subsonic for most of its mission. As the missile approaches its target, the warhead and an accompanying rocket system are jettisoned. The warhead then engages the target at supersonic speeds. *Jane's Weapons: Naval*, "YJ-12," March 11, 2015.

[70] U.S. Navy CGs (cruisers) and DDGs (guided missile destroyers) have a fixed number of vertical launch cells. Each weapon loadout consists of mixed numbers of SM-2s (for fleet defense), SM-3s (for other asset defense), and TLAMs.

Table 7.4
Recent and Current PLAN Aviation Air-Launched ASCM Systems and Capabilities

ASCM	Date	Range (km)	Speed (Mach) Cruise	Speed (Mach) Terminal	Altitude (m) Cruise	Altitude (m) Terminal	Guidance
YJ-81	1995	70	0.9	0.9	20	6	INS, active radar
YJ-82K	2002	130	0.9	0.9	20	6	INS, active radar
YJ-83	2004	250	0.9	1.4	20	5	Data link, active/passive radar
YJ-91[a]	~2008	50	2.5	2.5	20	7	INS, active radar
YJ-62	~2008	280	0.8	0.8	30	8	INS, GPS, active radar
YJ-12	2017 (?)	300–400	0.9	3.0			INS, active, passive, radar

SOURCES: *Jane's Strategic Weapons Systems*, "C-801 (CSS-N-4 'Sardine'/YJ-1/-8/-81), C-802 (CSSC-8 'Saccade'/YJ-2/-21/-22/-82/-85), and C-803 (YJ-3/-83/-88)," February 7, 2012; *Jane's Strategic Weapons Systems*, "YJ-91 (KR-1/Kh-31P/AS-17 'Krypton')," June 23, 2011; *Jane's Strategic Weapons Systems*, "C-602 (HN-1/-2/-3/YJ-62/X-600/DH-10/CJ-10/HN-2000)," March 24, 2014; *Jane's Strategic Weapons Systems*, "KD-88 (K/AKD-88)," October 24, 2012; *Jane's International Defence Review*, "Storm Force Warning: China's Anti-Ship Missile Range Spreads Its Wings," April 17, 2013; *Jane's Weapons: Naval*, "YJ-12," March 11, 2015.

NOTE: INS = inertial navigation system.

[a] The specifications are for the anti-ship version of the YJ-91; there is also an anti-radiation version with a maximum range of 120 km.

Longer-range strike systems provide PLAN aviation with more capability and options. However, a fuller appreciation of the implications will also acknowledge limitations. At longer ranges, PLAN aviation will have more difficulty projecting and maintaining air-to-air combat patrols that protect ASCM shooters from air attack. At the same time, longer ranges will likely place them closer to the center of the battle group's air defense umbrella while diminishing their ability to project forces in quantity.

Surface and Submarine-Launched ASCMs

Since 1996, the PLAN has also invested heavily in both its surface and submarine fleets, building or purchasing from abroad significant numbers of modern destroyers, frigates, and diesel or nuclear submarines. In the case of foreign purchases, the surface and submarine fleets obtained advanced ASCMs along with their launch platforms. The PLAN acquired its first two *Sovremenny*-class destroyers from Russia in 2000 and, beginning in 2005, acquired eight modern Russian *Kilo* (Type 636) diesel submarines and two additional advanced *Sovremenny*-class destroyers. Both platforms came equipped with advanced, long-range, supersonic, Russian-made ASCMs: the surface-launched SS-N-22 Sunburn and the submarine-launched SS-N-27 Sizzler, with maxi-

Figure 7.4
Maximum Chinese Air-Launched ASCM Engagement Ranges, 1996 and 2017

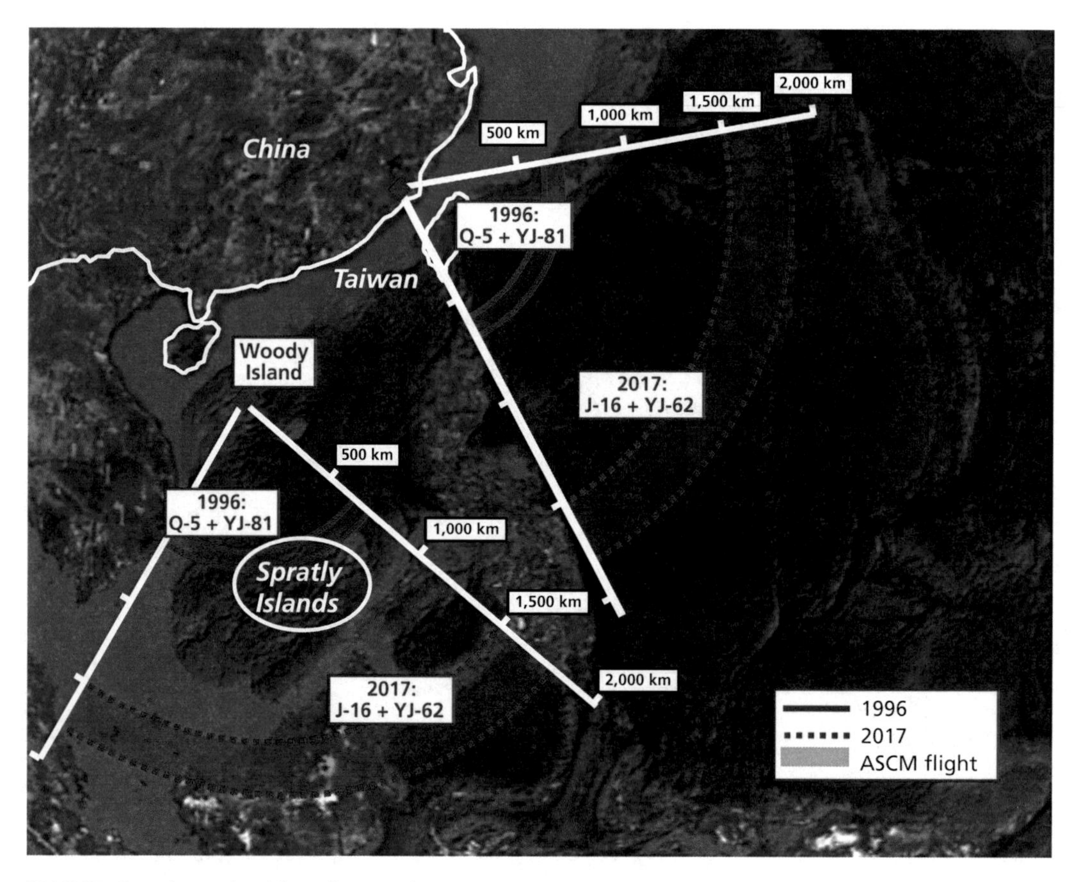

SOURCE: Google Earth with author overlay.
RAND *RR392-7.4*

mum ranges of 120 and 200 km, respectively.[71] As of 2015, there are 12 *Kilo*-class submarines in the PLAN, eight of which are equipped with the SS-N-27 ASCM.[72]

Since the early 1990s, China has also built several modern classes of destroyers, frigates, and attack submarines. Table 7.5 lists the PLAN's surface combatants (destroyers and frigates), their current ASCM weapons and capabilities, IOC, and the 1996, 2003, 2010, 2015, and 2017 (projected) force structures. The PLAN's building program has been somewhat uneven over time. By the late 1990s, new classes of frigates and submarines were reaching series production. However, in building destroyers, the PLA continued to experiment with different designs until roughly 2010, when it

---

[71] The SS-N-27 is also referred to as the *Klub-S*.

[72] Office of Naval Intelligence, *The PLA Navy*, 2015, p. 19.

Table 7.5
PLAN Destroyers and Frigates and Anti-Ship Cruise Missile Capabilities, 1996–2017

Ship Class[a]	IOC	Tons	1996	2003	2010	2015 (current)	2017	ASCM[b]	Range (km)	Speed (Mach)	ASCM Upgrade Year[c]
Destroyers											
Luda (Type 051)	1971	3,250	17	16	15	6	5	YJ-8/C-801 YJ-83/C-802	40 120	0.9 0.9	~2010
Luhu (Type 052A)	1994	4,600	1	2	2	2	2	YJ-8/C-801 YJ-83/C-802	40 120	0.9 0.9	~2002
Sovremenny (Type 956)	1999	7,940		2	2	2	2	SS-N-22	120	2.1	
Luhai (Type 051B)	1999	6,000		1	1	1	1	YJ-8/C-801 YJ-83/C-802	40 120	0.9 0.9	~2005
Luyang I (Type 052B)	2004	7,000			2	2	2	YJ-83/C-802	120	0.9	
Luyang II (Type 052C)	2004	7,000			2	5	6	YJ-62/C-602	280	0.8	
Sovremenny (Type 956EM)	2005	7,940			2	2	2	SS-N-22[d]	240	2.1	
Luzhou (Type 051C)	2006	7,000			2	2	2	YJ-83/C-802	120	0.9	
Luyang III (Type 052D)	2013	7,500				1	6	YJ-18	220	3.0	
Frigates											
Chengdu			2					None			
Jianghu	1968	1,702	33	37	30	16	10	HY-2/C-201	80	0.9	

Table 7.5—Continued

Ship Class[a]	IOC	Tons	1996	2003	2010	2015 (current)	2017	ASCM[b]	Range (km)	Speed (Mach)	ASCM Upgrade Year[c]
Frigates (cont.)											
Jiangwei (Type 053)	1991	2,250	4	12	14	14	14	YJ-8/C-801 YJ-83/C-802	40 120	0.9 0.9	~2007
Jiangkai I/II	2005	3,900			8	20	23	YJ-83/C-802	120	0.9	

SOURCES: Ship inventories and associated missiles are from IISS, *The Military Balance*, 1996, 2003, 2010, 2013, and 2015. Data on ship types and characteristics are from *Jane's Fighting Ships* articles: "Luda (Type 051DT/051G/051GII) Class," February 13, 2015; "Luhu (Type 052A) Class," February 13, 2015; "Jiangkai II (Type 054A) Class," February 16, 2015; "Luyang II (Type 052C) Class," February 16, 2015; "Luyang III (Type 052D) Class," February 16, 2015; "Jianghu I/II/V (Type 053H/053H1/053H1G) Class," February 13, 2015; "Jianghu III (Type 053 H2) Class," January 6, 2015; and Office of Naval Intelligence, *The PLA Navy*, 2015. Missile capabilities are from *Jane's Strategic Weapons Systems*, "SY-1 (CSS-N-1 'Scrubbrush'), HY-1 (CSS-N-2 'Safflower'/CSSC-2 'Silkworm'), HY-2 (CSSC-3 'Seersucker'), Fu-Feng-1/JL-9 (SS-N-22 'Sunburn')"; "CSS-N-4 'Sardine' (YJ-8/YJ-8A/C-801); CSS-N-8 'Saccade' (YJ-82/YJ-83/C-802/C-802A/Noor/Ghader)," May 15, 2015; Dennis M. Gormley, Andrew S. Erickson, and Jingdong Yuan, *A Low-Visibility Force Multiplier: Assessing China's Cruise Missile Ambitions*, National Defense University Press, 2014; Deagel.com, "YJ-18," web page, last updated May 5, 2015.

[a] Where different subclasses share common ASCMs, we have collapsed those subclasses together (e.g., *Jiangwei I* and *II*s), despite other differences (e.g., in size or air defense armament).

[b] The literature on ASCM types employs inconsistent nomenclature. We have not attempted to differentiate between different variants of individual missile types, though their ranges can differ significantly. Also, newer ASCMs are often retrofitted onto older warships (see discussion below).

[c] Newer ASCMs are often retrofitted onto older Chinese warships. We have indicated the older and newer missiles in separate rows, together with the approximate date of conversion (selecting the start date where the process took a period of years).

[d] The *Sovremenny*-class (Type 956EM) destroyers employ an improved version of the SS-N-22 known as the 3M80MBE with an extended range of 240 km.

began series production of the *Luyang II* (Type 052C). In 2012, it began series production of the *Luyang III* (Type 052D).[73] The opening of new lines and the introduction of modular construction are speeding production and the replacement of obsolete ships and boats.[74]

Most of the newly constructed frigates and destroyers are equipped with modern, indigenous ASCMs, such as the YJ-83 and YJ-62. These missiles have also been retrofitted on improved versions of a number of older classes. Significantly, the *Luyang III* guided-missile destroyer is equipped with the new vertically launched YJ-18, a long-range supersonic ASCM, reportedly capable of maneuvering at 10G acceleration to avoid enemy interception by air-to-air or surface-air missiles.[75] The *Sovremenny* (Type 956 and 956EM), *Luyang I* (Type 052B), *Luyang II* (Type 052C), *Luyang III* (Type 052D), and *Luzhou* (Type 051C) destroyers and the *Jiangkai II* (Type 054A) frigate are equipped with the Russian Mineral-ME OTH surveillance system. The Mineral-ME is a dual active and passive radar system with OTH target detection capability. The system also includes a missile data link that can relay target information to an in-flight ASCM at ranges limited by the horizon (~30 km). The passive radar has an optimal maximum range of 450 km, while the active radar has an optimal maximum range of 180 km, with a target location error of 50 meters in range and 0.25 degrees in bearing.[76] However, both systems depend heavily on environmental conditions and, at times, may have less capability than the specifications cited here. Figure 7.5 compares the maximum PLAN surface combatant detection and engagement ranges in 1996, 2003, 2010, and 2017.

Finally, Table 7.6 presents the PLAN's diesel and nuclear attack submarine force structures in 1996, 2003, 2010, 2015, and 2017 (projected). It also lists each vessel's primary ASCM and maximum range. Submarines could certainly target carrier group assets using external targeting information. However, this would come with an increased risk of detection by U.S. Navy anti-submarine warfare assets. On the other hand, PLAN submarines could find and target U.S. Navy assets autonomously, a topic we discuss in more detail in the next section.

---

[73] Between 2010 and the end of 2014, China launched four *Luyang II*s and six *Luyang III*s, two more destroyers than were produced in the seven previous years. (Note that the number launched do not equal the number commissioned during the same years. The latter are reflected in Table 7.5.) See *Jane's Fighting Ships*, "Luyang II (Type 052C) Class," February 16, 2015, and *Jane's Fighting Ships*, "Luyang III (Type 052D) Class," February 16, 2015.

[74] The Office of Naval Intelligence reports that in 2013 and 2014, "China launched more naval ships than any other country and is expected to continue this trend through 2015–16." It is unclear what standard was employed to support this conclusion, however. Office of Naval Intelligence, *The PLA Navy*, 2015, p. 15.

[75] Office of Naval Intelligence, *The PLA Navy*, 2015, p. 13; and Deagel.com, "YJ-18," web page, last updated May 5, 2015.

[76] Norman Friedman, *The Naval Institute Guide to World Naval Weapon Systems*, 5th ed., Annapolis, Md.: Naval Institute Press, 2006.

Figure 7.5
Maximum PLAN Destroyer Surface Detection Ranges and ASCM Ranges, 1996, 2003, 2010, 2017

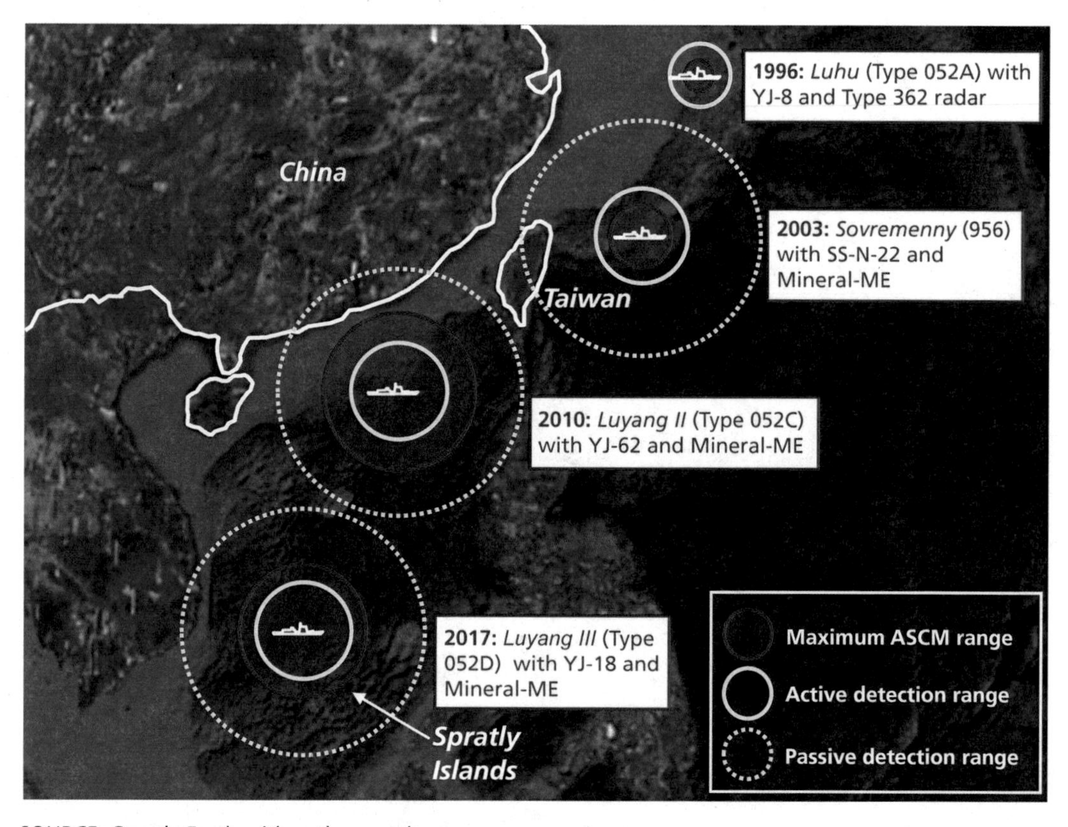

SOURCE: Google Earth with author overlay.
NOTE: Although the YJ-18 noted for 2017 has a shorter range than the YJ-62, it is a highly maneuverable, supersonic weapon.
RAND *RR392-7.5*

Developments in U.S. Naval Air and Missile Defenses

The Aegis Combat System is the centerpiece of the U.S. Navy's air and missile defense capability. Employed onboard both the *Ticonderoga*-class cruisers (CG-47) and *Arleigh Burke* destroyers (DDG-51), the system consists of the SPY-1 phased array radar, advanced computer software and hardware systems, and long-range SMs. It was originally designed to defeat air and ASCM threats. Recent software upgrades, installed on selected ships, provide the Navy with its own ballistic missile defense (BMD) capability as well. These BMD-capable ships can defend both fleet and land-based assets and installations from ballistic missile attack.

The Aegis fleet has expanded significantly since 1983, with 27 *Ticonderoga*-class (CG-47) ships entering service between 1983 and 1994 and 62 *Arleigh Burke*–class (DDG-51) ships entering service since 1991. The *Arleigh Burke* is the follow-on to

Table 7.6
PLAN Attack Submarine (Diesel and Nuclear) and ASCM Capabilities, 1996, 2003, 2010, 2015, and 2017

Ship Class	IOC	Tons	1996	2003	2010	2015 (current)	2017	ASCM	Range (km)	Speed (Mach)	Guidance
Attack submarines (diesel)											
Romeo (Type 033)	1962	1,830	63	35	8	—	—				
Ming (Type 035)	1971	2,113	10	19	19	19	19				
Kilo (877)	1995	2,350	2	2	2	2	2	N/A[a]			
Kilo (636)	1997	2,350		2	2	2	2	N/A[a]			
Song (Type 039)	1999	2,250		3	13	13	13	YJ-82[b]	40	0.9	INS, active radar
Kilo (636)[a]	2005	2,350			8	8	8	SS-N-27	200	2.5	INS, active radar
Yuan (Type 041)	2006	N/A			2	12	16	YJ-82[b]	40	0.9	INS, active radar
Attack submarines (nuclear)											
Han (Type 091)	1980	5,550	5	5	4	3		YJ-82	40	0.9	INS, active radar
Shang (Type 093)	2006	6,000			2	2	2	YJ-82[b]	40	0.9	INS, active radar
Shang, Improved (Type 093A)[c]	2016 (?)	6,000					2	YJ-82 or YJ-18	40 220	0.9 3.0	

SOURCES: Submarine inventories and associated missiles are from IISS, *The Military Balance*, 1996, 2003, 2010, and 2015. Information on characteristics from *Jane's Fighting Ships*: "Yuan Class (Type 041)," February 13, 2015; "Shang Class (Type 093/093A)," February 13, 2015; "Kilo Class (Project 877EDM/636)," February 13, 2015; "Song Class (Type 039/039G)," February 13, 2015.

[a] Only the last eight (of 12) Russian *Kilo*-class submarines are capable of employing the SS-N-27 "Sizzler" ASCM; however, several sources suggest that the remaining four will be retrofitted to employ the SS-N-27 as well.

[b] The Office of Naval Intelligence reports that these submarines could be retrofitted with the vertically launched, supersonic YJ-18 with a range of 220 km. As noted above, the *Luyang III*–class destroyer is equipped with the YJ-18 ASCM. Office of Naval Intelligence, *The PLA Navy*, 2015.

[c] Two nuclear attack submarines are currently under construction. They appear to be improved *Shang*-class boats, but could also be a new class, the Type 095. The Office of Naval Intelligence reports that the Type 095 SSN (nuclear attack submarine), when launched, may provide a generational improvement in quieting and weapon capacity." Office of Naval Intelligence, *The PLA Navy*, 2015.

the *Spruance*-class destroyer (DD-963), which exited the fleet in 2005. The *Spruance* class was capable of providing point defense for itself against air and missile attacks, whereas the Aegis system on the *Ticonderoga* and *Arleigh Burke* classes extends formidable air and missile defense across the entire battle group, completely transforming the Navy's surface combat fleet. The Navy continues to procure *Arleigh Burke* Flight II under its current shipbuilding schedule, and it will procure 24 *Arleigh Burke* Flight III ships between 2016 and 2031.[77] The Flight III ships will be equipped with the Air and Missile Defense Radar, providing substantially better detection, tracking, and engagement performance against high-flying ballistic missiles.

Table 7.7 shows the U.S. Navy's cruiser and destroyer force structures for 1996, 2003, 2010, 2015, and 2017 (projected). In 1996, the Navy had 34 Aegis-equipped ships and no BMD-capable ships. By 2010, the number of Aegis-equipped ships had doubled to 68, and the Navy had deployed 18 BMD-capable ships. And, by 2017, we estimate that those numbers will rise to 87 and 39, respectively. In addition, two of three planned *Zumwalt*-class destroyers may be commissioned by 2017, which the U.S. Navy claims will have triple the capability against cruise missiles.[78]

Table 7.7
U.S. Navy Cruisers and Destroyers, 1996–2017

Ship Class (Cruisers and Destroyers)	IOC	Tons	1996	2003	2010	2015 (current)	2017
Spruance (DD-963)	1975	8,040	36	29	—	—	—
Ticonderoga (DD-47)	1983	9,600	27	27	22	22	22
Arleigh Burke (DDG-51) Flight I/II	1991	8,184	7	21	21	28	28
Arleigh Burke (DDG-51) Flight IIA	1998	9,100	—	6	25	34	37
Zumwalt (DDG-1000)			—	—	—	—	2
Aegis-equipped (total)			34	54	68	84	87
BMD-capable (total)			0	3	18	33	39

SOURCES: IISS, *The Military Balance*, 1996, 2003, and 2010; *Jane's Fighting Ships*, "Spruance Class: Destroyers," March 7, 2006; Jane's Fighting Ships, "Ticonderoga Class," March 24, 2015; *Jane's Fighting Ships*, "Arleigh Burke (Flights I and II) Class," March 24, 2015; *Jane's Fighting Ships*, "Zumwalt (DDG 1000) Class," March 24, 2015; and Missile Defense Agency, "Aegis Ballistic Missile Defense," fact sheet, January 2015; Ronald O'Rourke, *Navy Aegis Ballistic Missile Defense (BMD) Program: Background and Issues for Congress*, June 12, 2015.

[77] U.S. Navy, *Report to Congress on Annual Long-Range Plan for Construction of Naval Vessels for FY 2011*, Washington, D.C., February 2010.

[78] Jane's estimates that the first *Zumwalt*, launched in October 2013, will be commissioned in February 2016 and that a second will be commissioned February 2017. *Jane's Fighting Ships*, "Zumwalt (DDG 1000) Class," March 24, 2015; see also U.S. Navy, "Destroyers—DDG," fact sheet, April 4, 2013.

Aegis ships employ several SM variants for air, cruise missile, and, most recently, BMD. These SM variants include the SM-2 class for air, cruise missile, and terminal ballistic missile intercept defense; the SM-3 class for midcourse ballistic missile intercept defense; and, beginning in 2011, the SM-6 for extended air and cruise missile defense. As in the case of China's offensive missile inventory, U.S. defenses have improved over time.

SM-2 Variants

The SM-2 Block III and IIIA, with a maximum range of 165 km and semi-active radar terminal guidance, entered service in 1981. SM-2 Block IV entered service in 2004 and is designed specifically to intercept ballistic missiles during their terminal phase. The SM-2 Block IIIB, which entered service in 2008, incorporates an IR seeker for improved performance against cruise missiles.[79]

SM-3 Variants

The SM-3 Block IA, the Navy's current Aegis BMD missile, entered service in 2007 and has a three-stage rocket with a maximum intercept range of 1,200 nm.[80] The SM-3 Block IB, which entered service in April 2014, incorporates an improved target seeker, advanced signal processing, and an improved divert/attitude-control system for course adjustment.[81] The SM-3 Block IIA missile, scheduled to enter service in 2018, increases the missile's terminal velocity by around 50 percent, enabling it to intercept medium- and intermediate-range ballistic missiles. The SM-3 Block IIB, which was to incorporate a lighter kill vehicle and intercept ICBMs, was effectively canceled in March 2013.

SM-6

The SM-6, which reached IOC in November 2013, is an entirely new missile class that receives guidance updates from offboard, non-Aegis systems, such as the Navy's E-2D AEW platform. This potentially extends the SM-6 engagement envelope against inbound ASCMs to 370 km, well beyond the Aegis radar horizon, allowing it to engage inbound ASCMs before entering the terminal phase.[82]

U.S. defenses will be layered. In addition to SM variants used to intercept missiles at long range, shorter-range point defenses will pick up "leakers." At the same time, jamming will be used to defeat cruise missiles by interfering with their receipt of positioning data (e.g., GPS data) or their use of radar for terminal guidance. Chaff and

---

[79] *Jane's Strategic Weapons Systems*, "RIM-66/-67/-156 Standard SM-1/-2, RIM-161 Standard SM-3, and RIM-174 Standard SM-6," July 30, 2013.

[80] A modified SM-3 Block IA missile successfully intercepted an inoperable U.S. surveillance satellite on February 20, 2008.

[81] *Jane's Weapons: Naval*, "Standard Missile 1/2/3/4/5/6 (RIM-66/67/156/161/174 and RGM-165)," September 19, 2014.

[82] This assumes that the E-2D operates at 25,000 feet.

decoys will also be employed to confuse the target picture. At the other end of the kill chain, U.S. Navy commanders will seek to destroy launch and ISR platforms.

As the Aegis system upgrades have entered the fleet, the U.S. Navy's ballistic and cruise missile defense capabilities have continued to improve. This improvement has not been equally distributed between ASBM and ASCM defense, however. Cruise missile defense is an enduring Aegis mission, for which the Aegis Combat System is quite capable, particularly against low numbers of incoming missiles. ASBMs are an emerging threat, and the Aegis system's utility against them is mostly unknown.[83] The capabilities required to support defense against ASBMs are resident in the Aegis system, and there is no reason to assume that they cannot provide some level of protection in the future.

Chinese Submarine Threat

China's submarine fleet is a centerpiece of its military modernization efforts and poses a significant and increasing threat to U.S. Navy surface assets operating in the region. The Chinese submarine fleet has been rapidly modernizing since 1996. U.S. anti-submarine warfare capabilities have improved to a lesser degree over the same period. To quantify recent trends, we developed an anti-submarine warfare model to measure how well and how often different Chinese submarines could find and engage U.S. aircraft carriers (with and without external cueing) operating at different ranges from the mainland. In the following sections, we discuss China's SSN and SSK (attack submarine) fleets and their evolution since 1996, as well as the U.S. anti-submarine assets charged with defending the U.S. surface fleet from underwater attack. We then briefly discuss the sonar acoustic characteristics of different submarine and anti-submarine warfare platforms. Finally, we discuss the mechanics of our anti-submarine warfare model, followed by its results for the Taiwan and the Spratly Islands scenarios during the four snapshot years: 1996, 2003, 2010, and 2017.

The Chinese Submarine Fleet

In 1996, the Chinese diesel submarine fleet consisted of poor-quality, 1960s-vintage, Soviet-designed *Romeo*-class and 1970s-vintage *Ming*-class submarines. Both are small and rarely, if ever, operated beyond China's littoral waters. China's nuclear submarine fleet consisted of the noisy, seldom deployed *Han*-class attack submarines and *Xia*-class ballistic-missile submarines. In the mid-1990s, China imported two modern *Kilo 877*–class diesel submarines from Russia and, by 2010, had also acquired ten high-end *Kilo 636*s. China has since supplemented the *Kilo*s with its own production of sophisticated indigenous diesel submarine designs, the *Song* and *Yuan* classes. In

[83] Ronald O'Rourke, *Navy AEGIS Ballistic Missile Defense (BMD) Program: Background and Issues for Congress*, Washington, D.C.: Congressional Research Service, June 12, 2015.

addition to its diesel submarine fleet, China continues to develop and field nuclear submarines. At the time of this report, China had built and delivered into service four *Jin*-class (Type-094) submarines, the PLA Navy's second-generation ballistic missile submarine.[84] It has taken delivery of two second-generation nuclear attack submarines, the *Shang* class (Type 093), and three improved models (Type 093A) are currently under construction. The *Shang* will be followed by a third-generation Chinese SSN, the Type 095, which is anticipated by 2020.[85]

While the Chinese fleet is modernizing, it has shrunk in size. In 1996, its diesel submarine fleet consisted of 75 submarines, of which only two were modern boats.[86] By 2010, the fleet had shrunk to 54 diesel submarines, but half (27) were modern designs. Since 2010, the rate of production appears to have risen, and by 2017, we anticipate that almost 70 percent (41) of the PLAN's 60 diesel boats will be modern. The nuclear attack submarine fleet has remained steady at four to six submarines, with the *Shang* class (Type 093 and Type 093A) replacing the older *Han* boats.

Acoustic performance is the single most important attribute in submarine performance. Acoustic performance includes both the level of noise generated by the submarine itself and its own sonar's sensitivity to sound from nearby submarines and surface vessels. Self-generated noise directly affects a submarine's survivability. Both self-generated noise and sonar sensitivity heavily influence its ability to detect and locate external sound sources. Submarines that are difficult to detect are often the most capable at searching for other submarines and surface targets. Finally, nuclear submarines have the size and electrical generation to support significantly larger and more capable sonar systems. In this analysis, we use the relative acoustic performance of Chinese and Russian submarines as described in an August 2009 report by the Office of Naval Intelligence (see Figure 7.6).

We excluded both the *Romeo* and the *Ming* classes from this analysis because they are unlikely to operate effectively against U.S. forces in our selected scenarios. It should also be noted that this analysis considers the capabilities of the Chinese submarine fleet but does not fully account for the state of training and tactics.[87] The PLAN submarine force may not be able to maintain the operational tempo assumed in this analysis. However, the Chinese attack submarine force has been far more active in recent years, with an average of fewer than three active patrols annually between 1996 and 2005

---

[84] A fifth is reportedly under construction. *Jane's Fighting Ships*, "Jin class (Type 094)," February 13, 2015.

[85] See Table 7.6 for the year-by-year PLAN submarine force structures.

[86] The initial *Kilo*-class submarines purchased by China were the export version (Type 877E). Subsequent purchases were the domestic Russian *Kilo* version (Type 636).

[87] For more on this topic, see Michael S. Chase, Tai Ming Cheung, Kristen A. Gunness, Scott Warren Harold, Susan Puska, and Samuel K. Berkowitz, *China's Incomplete Military Transformation: Assessing the Weaknesses of the People's Liberation Army (PLA)*, Santa Monica, Calif.: RAND Corporation, RR-893-USCC, 2015.

Figure 7.6
Relative Detectability of Chinese and Russian Submarines

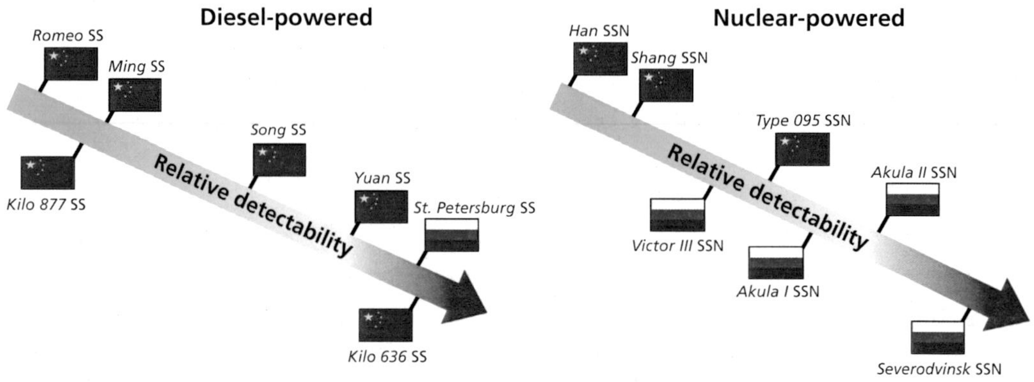

SOURCE: Office of Naval Intelligence, *The People's Liberation Army Navy*, 2009.
RAND *RR392-7.6*

and an average of more than 11 per year between 2006 and 2012.[88] The Chinese submarine community is growing in sophistication, and there is no reason to suppose that it cannot continue to narrow the gap in the competence of its operators.

U.S. Anti-Submarine Warfare Assets

We analyzed three types of U.S. Navy anti-submarine warfare assets in our model: U.S. SSNs (*Los Angeles*, *Seawolf*, and *Virginia* classes), maritime patrol aircraft (P-3 Orion and P-8 Poseidon), and tactical auxiliary general ocean surveillance (T-AGOS) ships. The SSN fleet operates independently, detecting Chinese submarines with passive (hull-mounted and towed-array) sonars, attacking with Mk-48 heavyweight torpedoes. Maritime patrol aircraft (MPA) operate independently and in conjunction with the T-AGOS ships. These aircraft detect submarines with sonar-buoys, radar, and magnetic detection and attack using Mk-54 lightweight torpedoes. T-AGOS ships detect submarines using large-volume passive towed sonar arrays and cue MPAs for the final search and attack.

Like its surface fleet, the U.S. Navy's SSN fleet continues to evolve, replacing older, earlier-class submarines with newer, more modern ones. In 1996, *Los Angeles*–class submarines made up the bulk of the U.S. submarine fleet. By 2003, three *Seawolf*-class SSNs had entered the fleet, and the latest U.S. attack submarine, the *Virginia* class, entered service by 2010.[89] The *Virginia* class, designed as a lower-cost alternative to the *Seawolf*, maintains similar acoustic performance but carries fewer weapons. The

---

[88] Hans M. Kristensen, "Chinese Nuclear Developments Described (and Omitted) by DoD Report," *FAS Security Blog*, May 14, 2013.

[89] The U.S. Navy operates two *Seawolf*-class submarines that are in SSN configuration. The *Seawolf*-class boats were not included in the anti-submarine warfare assets modeled for this analysis.

Virginia and a majority of the *Los Angeles* submarines are outfitted with 12 vertical launch tubes carrying TLAMs, in addition to torpedoes. As *Virginia*-class construction continues, the *Los Angeles* class is retiring. In 2017, the replacement of the *Los Angeles* class by the *Virginia* class will continue, though at a less than one-to-one ratio. The overall U.S. SSN fleet composition is summarized in Table 7.8. In response to Chinese submarine modernization, the U.S. Navy homeported three *Los Angeles*–class submarines in Guam to increase its Western Pacific presence. Further, the Navy has shifted the bulk of its SSN fleet to the Pacific.[90]

In addition to deploying larger and more capable platforms, the U.S. submarine community has replaced its entire submarine combat system. By 1996, the deployed submarine combat system had reached the performance limits of its military specification computing hardware. At the same time, Russian submarines had sufficiently advanced to challenge the acoustic supremacy of the U.S. SSN fleet. The U.S. Navy responded by developing an open-architecture combat system that combines commercial computer hardware with flexible combat system software. The vastly improved computing capacity allows the SSN fleet to utilize complex acoustic algorithms that were previously limited to large shore facilities in real time. The Acoustic Rapid Commercial, Off-the-Shelf Insertion program restored the U.S. acoustic advantage and has been backfitted to the entire SSN fleet.[91] As a result, the anti-submarine warfare capabilities of the U.S. submarine fleet have significantly improved.

Table 7.8
U.S. Navy Attack Submarines, 1996, 2003, 2010, 2015, and 2017

Attack Submarines (SSNs)	IOC	Tons	1996	2003	2010	2015 (current)	2017
Sturgeon (SSN-637)	1967	4,714	25	1	—	—	—
Los Angeles (SSN-688)	1976	6,927	37	28	22	19	12
Los Angeles, improved (SSN 688i)	1988	7,147	20	23	23	22	22
Seawolf (SSN-21)	1997	9,138	—	2	3	3	3
Virginia (SSN-774)	2004	7,800	—	—	5	11	15
Total			82	54	53	55	52

SOURCES: IISS, *The Military Balance*, 1996, 2003, 2010, and 2015; Federation of American Scientists, "SSN-637 Sturgeon Class," web page, undated; *Jane's Underwater Warfare Systems*, "Los Angeles Class," September 28, 2011; *Jane's Underwater Warfare Systems*, "Seawolf Class," March 24, 2015; Ronald O'Rourke, *Navy Virginia (SSN-774) Class Attack Submarine Procurement: Background and Issues for Congress*, Washington, D.C.: Congressional Research Service, June 1, 2015.

[90] U.S. Navy, "Attack Submarines—SSN," fact sheet, November 27, 2012.

[91] "USA Upgrades Submarine Fleet Acoustics Under A-RCI Program," *Defense Industry Daily*, April 30, 2012.

The U.S. MPA fleet for conducting anti-submarine warfare is currently composed primarily of P-3C Orion aircraft, which were first introduced in 1962.[92] The P-3C employs a variety of sensors in performing its anti-submarine warfare mission. It can deploy active and passive sonobouys and process the data using onboard systems. In addition, the P-3C can detect periscopes using its maritime surveillance radar (the APS-137). Finally, the aircraft employs a magnetic anomaly detection system to locate submerged submarines operating at relatively shallow depths.[93]

The Navy is in the process of replacing the P-3C Orion with the new P-8A Poseidon multimission maritime aircraft (MMA), which achieved IOC in December 2013. As of 2015, the Navy had 21 P-8As (four squadrons) in service and another 53 ordered.[94] Based on a modified Boeing 737-800ERX airframe, the P-8A is equipped with AN/APY-10 multifunction radar, which provides high-resolution radar images in maritime, littoral, and overland environments. It carries SLAM-ER cruise missiles, Harpoon anti-ship missiles, depth charges, MK-54 torpedoes from a dedicated bomb bay, and more than 100 deployable sonobuoys for submarine detection. While the P-8A lacks a magnetic anomaly detection system that is present on the P-3C, it is designed to operate in conjunction with the MQ-4C Triton UAV, the naval version of the RQ-4B Global Hawk UAV.[95] Expected to achieve IOC in 2018 (and therefore not factored into our modeling in this assessment), the MQ-4C will provide the Navy in the near future with a high-altitude (60,000 feet), persistent (30 hours) maritime ISR capability through a suite of advanced mission systems.[96] The Navy plans on procuring 68 MQ-4Cs, enabling the P-8A to focus on its core mission of anti-submarine and anti-surface warfare.[97]

T-AGOS ships (operated by the Navy's Military Sealift Command) carry the Surveillance Towed Array Sensor System (SURTASS). SURTASS, deployed at depths of 500–1,500 feet, is a large towed array that is capable of detecting acoustic signals at long ranges. In addition to passive detection, SURTASS has a low-frequency active sonar capability. The initial T-AGOS ships were the *Victorious* class, deployed in 1991, and the latest is the *Impeccable* class, deployed in 2001.[98] Sonar data from these

---

[92] The P-3C fleet has an average age of 30 years, and the airframes suffer from serious fatigue issues. After an assessment of airworthiness in 2005, the P-3C fleet was downsized from 227 to 177 airframes.

[93] *Jane's Aircraft Upgrades*, "Lockheed Martin (Lockheed) P-3 Orion," March 9, 2015.

[94] *Jane's World Navies*, "United States," March 6, 2015.

[95] Naval Air Systems Command, "MQ-4C Triton," undated.

[96] These will include inverse SAR, EO/IR full-motion video, maritime moving-target detection, electronic support measures, and Link-16.

[97] U.S. Government Accountability Office, *Defense Acquisitions: Assessments of Selected Weapon Programs*, Washington, D.C., GAO-13-294SP, March 28, 2013, p. 103; and Grant Turnbull, "The P-8 Poseidon Adventure: Delivering a New-Era of Maritime Aircraft," *Naval-Technology.com*, January 28, 2014.

[98] *Jane's Underwater Warfare Systems*, "AN/UQQ-2 SURTASS," September 16, 2014.

ships is preprocessed onboard and then provided to shore installations and nearby anti-submarine warfare ships. In our model, T-AGOS provides cueing to MPA aircraft.

Modeling Chinese and U.S. Sonar Performance

To provide a first-order assessment of the relative capabilities of Chinese submarine forces and U.S. anti-submarine warfare capabilities, we created a simple submarine engagement model. The model considers the relative noise levels of different classes of Chinese submarines and sensor capabilities on both sides to estimate detection distances, which we then used, in conjunction with weapon ranges and capabilities and platform movement speeds, to determine the number of potential engagements Chinese submarines might achieve against U.S. aircraft carriers. In this section, we describe the model in detail; the next section summarizes the results.

Modern submarines, both nuclear and diesel, detect surface targets using a variety of sensors, including sonar, radar, and passive radio-frequency (RF) detection. In the face of active anti-submarine warfare forces, submarines will rely heavily on sonar searches.[99] For the purposes of this analysis, we assumed that Chinese submarines would detect and target U.S. surface ships exclusively with passive sonar.[100] We also assumed that U.S. anti-submarine warfare forces rely on passive and active sonar to detect Chinese submarines.[101] We modeled detection ranges between two platforms using discrete acoustic convergence zones (CZs).[102] Sound in the ocean tends to bend and converge at a series of detection rings due to the effect of water pressure and temperature on sound propagation (see Figure 7.7). The sound intensity at these CZs is significantly stronger than at other nearby ranges. However, the sound intensity drops by approximately 20 dB (or 100-fold) at each successive CZ.[103]

In our sonar modeling, platform A's maximum detection range against platform B is described in terms of zones, or distance from the target: the direct source zone (less than 5 nm), first CZ (~25 nm), second CZ (~50 nm), and third CZ (~75 nm). (Platforms A and B can be either SSNs or anti-submarine warfare assets.) The maximum detectable distance depends on platform A's self-generated noise and sonar capabilities and platform B's noise level. The specific detection ranges that we used for individual platforms are presented in Table 7.9.

---

[99] Sensors requiring a mast to be raised above the water significantly increase the threat of detection.

[100] In some model runs, we allowed Chinese submarines to receive offboard cueing to the carriers' locations. However, they still must acquire the target with sonar before engagement.

[101] U.S. airborne anti-submarine warfare assets, such as the P-3C, are equipped with radars specifically designed to detect periscopes; however, we did not include periscope detection in our model.

[102] For a discussion of this and other acoustic issues, see Thaddeus G. Bell, *Probing the Oceans for Submarines: A History of the AN/SQS-26 Long-Range Echo-Ranging Sonar*, Newport, R.I.: Naval Undersea Warfare Center, 2010.

[103] Edward Tucholski, "Regions of the Sound Velocity Profile," course handout for SP411, Underwater Acoustics and Sonar, U.S. Naval Academy, Annapolis, Md., undated.

Figure 7.7
Sonar Convergence Zones

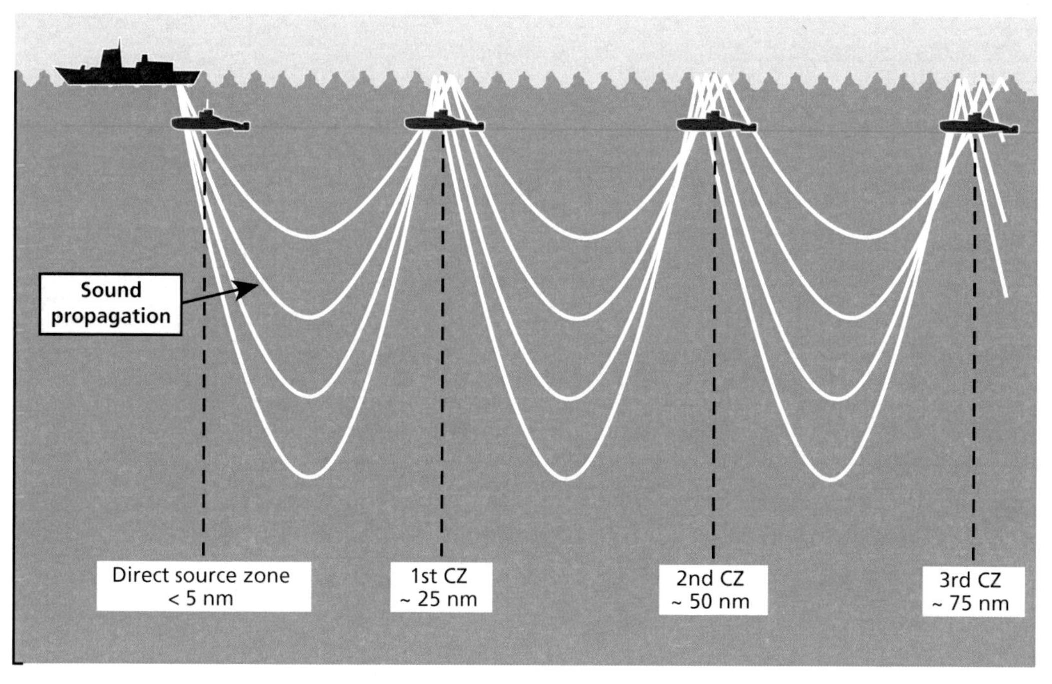

SOURCE: Tucholski, undated.
RAND *RR392-7.7*

The detection distances in Table 7.9 are intended to provide a basis for assessing the impact of the changing force structure over time and are based on the relative detectability cited in Figure 7.6.[104] They are not meant to represent the actual capabilities of individual systems. Nevertheless, the general range of distances is consistent with historical cases.[105] And the results may give us a first-order sense of the impact of Chinese submarine modernization and improvements in U.S. anti-submarine warfare systems.

---

[104] Office of Naval Intelligence, *The People's Liberation Army Navy*, 2009.

[105] Depending on acoustic conditions and the speed of the sensing platform and target, the best Cold War submarine hull–mounted sonar could detect targets out to 50 km, while less effective or poorly handled ones might have a range of 5–10 km. The towed arrays associated with *Los Angeles*-class submarines are estimated to have a range of up to 100 km. See James F. Dunnigan, *How to Make War: A Comprehensive Guide to Modern Warfare in the 21st Century,* 4th ed., New York: HarperCollins, 2003, pp. 260–262. In general, detection ranges by submarines' passive sonar against surface targets will be greater than those of similar ship-based systems against submarines. However, towed arrays can greatly improve the effectiveness of surface-based platforms. And T-AGOS ships are equipped with surveillance towed array sensor systems that have very significant detection distances against submarines using low-frequency active sonar. For other estimates of detection distances, see Cote, "Assessing the Undersea Balance Between the U.S. and China," 2011, and Federation of American Scientists, "Low Frequency Active (LFA)," web page, June 21, 1997.

Table 7.9
Model Inputs: Detection Distance by Chinese Submarines and U.S. Anti-Submarine Warfare Assets, by Convergence Zone and Approximate Distance

Detection of U.S. Carrier by PLAN Submarines		Detection of PLAN Submarines by U.S. Anti-Submarine Warfare Assets		
		By U.S. SSNs	By U.S. MPAs	By U.S. T-AGOS Ships
Kilo (877)	Direct source (< 5 nm)	2nd CZ (~50 nm)	1st CZ (~25 nm)	3rd CZ (~75 nm)
Song	1st CZ (~25 nm)	1st CZ (~25 nm)	Direct source (< 5 nm)	2nd CZ (~50 nm)
Yuan	1st CZ (~25 nm)	1st CZ (~25 nm)	Direct source (< 5 nm)	2nd CZ (~50 nm)
Kilo (636)	1st CZ (~25 nm)	DS (< 5 nm)	Direct source (< 5 nm)	1st CZ (~25 nm)
Han	Direct source (< 5 nm)	2nd CZ (~50 nm)	1st CZ (~25 nm)	3rd CZ (~75 nm)
Shang (093)	Direct source (< 5 nm)	2nd CZ (~50 nm)	1st CZ (~25 nm)	3rd CZ (~75 nm)
Shang (093A)[a]/ Type 095[a]	1st CZ (~25 nm)	1st CZ (~25 nm)	Direct source (< 5 nm)	2nd CZ (~50 nm)

SOURCES: Estimated detection distances of and for Chinese submarines are based on the relative scale of detectability provided for those submarines in Figure 7.6. For U.S. assets, the purpose of the system is also taken into account (e.g., the T-AGOS SURTASS is specifically designed for long-range detection of submarines).

[a] Based on the ten-year lag between the construction of the first two *Shang*-class boats and the improved *Shang*, we credit the latter with significant improvements, and anticipate that it may be as quiet as a 2009 Office of Naval Intelligence report judged that the forthcoming Type 095 would be. Office of Naval Intelligence, *The People's Liberation Army Navy: A Modern Navy with Chinese Characteristics*, 2009.

The left-most columns in the table present the Chinese submarines' detection distances against a U.S. aircraft carrier, with greater detection ranges providing Chinese submarines with more and better attack opportunities. To illustrate with an example, the *Kilo* (877s) submarines can detect U.S. carriers within the direct source zone (less than 5 nm), while the *Yuan* submarines can detect carriers within the first CZ (or to roughly 25 nm). The next three columns of the table present the distance at which U.S. assets (SSNs, MPAs, and T-AGOS ships) can detect Chinese submarines (by submarine type), with greater detection ranges allowing U.S. forces more opportunity to engage and either suppress or sink Chinese submarines before they can attack the U.S. carrier.

We assume that if platform A can detect platform B at a certain CZ, then the probability of detection is 100 percent at that distance and all nearer zones. Furthermore, we assume that the probability of detection is 0 at the ranges beyond the CZ given in Table 7.9 and in all areas between CZs. Again, this "cookie-cutter" approach is not meant to produce precisely accurate analysis of actual expected outcomes, but it is designed to provide a first-order assessment of the relative performance of Chinese submarines and U.S. anti-submarine warfare capabilities over time. We presume that using more accurate data on noise or detection capabilities would change the absolute results but would not likely affect broad trends.

Having generated a rough gauge of capabilities for individual assets, we then developed a model to simulate the interaction between packages of Chinese submarines and an individual U.S. CSG, supported by dedicated U.S. anti-submarine warfare assets.

The model assumes that U.S. SSNs provide a screen between the CSG and the Chinese mainland and that U.S. MPAs/MMAs and T-AGOS ships operate as a close-in anti-submarine warfare screen near the carriers. Chinese submarines must search the entire operating area (either 500 nm or 1,000 nm from the focus of conflict). We evaluated both cases in which Chinese submarines did not receive cueing from external sources as to the location of U.S. aircraft carriers and those in which they did. When cued, we allowed Chinese submarines to receive carrier position updates randomly once per day (24 hours). (More or less frequent cueing is, of course, possible in the real world.) Such cueing data could derive from OTH radar, satellites, or observation by manned or unmanned aircraft.

Detection of U.S. carriers by Chinese submarines occurs at discrete ranges based on the CZ data in Table 7.9. We did not model the actual engagement between Chinese submarines and U.S. carriers. Instead, the model's output metric is the number of potential torpedo or cruise missile "engagement opportunities" by Chinese submarines against U.S. carriers during a seven-day period. To engage, the submarine must not only detect the carrier but also be within its weapon range (given weapons' flight times, system movement speeds, and relative directions).

Meanwhile, as Chinese submarines search for the carriers, U.S. anti-submarine warfare assets attempt to locate and destroy Chinese submarines. When these assets detect a Chinese submarine, they engage it with a Pk that depends on the location uncertainty and the submarine's speed. The Pk values assume that an air-dropped torpedo must land within 3–4 nm of a Chinese nuclear and diesel submarine, respectively, to achieve a kill.[106] U.S. SSN Pk values depend on the detection range (second CZ, first CZ, or direct source) and the enemy submarine's maximum speed. The model does not account for the loss of any U.S. anti-submarine warfare assets, either from Chinese submarines or from other platforms.

Submarine Model Results

We first modeled the results of a single Chinese submarine attempting to find and engage a single U.S. carrier. Results for each type of modern Chinese submarine were modeled separately. (Given the lack of quieting and other limitations of the legacy *Romeo*- and *Ming*-class submarines, they would have little realistic offensive role and

---

[106] The difference is driven by the submarines' maximum or "escape" speed, which we assume is 30 and 20 nm per hour (knots) for the nukes and diesels, respectively.

we therefore did not include them in our modeling.)[107] We then modeled a larger submarine campaign in which China either surges the bulk of its modern submarine force into the operational area or conducts a sustained campaign, keeping a relatively even number of submarines on station. In all cases, we assumed that U.S. anti-submarine warfare assets (SSNs, MPAs, and T-AGOS ships) would attempt to protect the carrier by trying to locate and either destroy or drive away attacking Chinese submarines.[108]

Table 7.10 presents the results for contests between one carrier and its anti-submarine warfare complement against one enemy submarine (one-on-one engagement) at different operating ranges (500 nm and 1,000 nm from the Chinese coast) during a seven-day period. One can think of the values in Table 7.10 as the number of potential engagements *per submarine per carrier* every seven days. Naturally, higher numbers, reflecting more engagement opportunities, are better for the PLA. It should be emphasized that the output metric, "engagement opportunities," addresses only part of the overall problem. Chinese submarine commanders would not necessarily take all of the engagement opportunities presented, and not all of the engagements that occurred would result in a hit, much less critical damage to or the sinking of a U.S. aircraft carrier.

The individual submarine results in Table 7.10 show that different noise levels and armaments can produce dramatically different results. In the table, results for different submarines under the same operational circumstances differ by as much as a factor of 30. Although the input parameters (especially detection ranges) represent very rough estimates, and specific results must be treated with caution, China's newer submarines are becoming quieter and better armed, and there is every reason to believe that their capability to find and attack U.S. surface ships has vastly improved over the period in question. The table also suggests that cueing could dramatically improve the number of engagements achieved (a topic we explore later) and that the U.S. Navy can mitigate its vulnerability somewhat by operating at greater distances from the Chinese coast (though vulnerability and risk must be weighed against the carriers' ability to project power forward).

We next turn to an assessment of how many engagement opportunities a larger force of Chinese submarines might achieve against U.S. carriers operating near Taiwan or the Spratly Islands. To do this, we used the data on individual submarines from Table 7.10, combined with data on the Chinese force structure in Table 7.6. Here,

---

[107] Noise levels are discussed in the text. See also Ronald O'Rourke, *China Naval Modernization: Implications for U.S. Navy Capabilities*, Washington, D.C.: Congressional Research Service, August 8, 2013, p. 16.

[108] For the 1996, 2003, and 2010 cases, we assume that two *Los Angeles*–class submarines, two P-3C MPAs, and two T-AGOS ships provide anti-submarine warfare protection for each carrier. For 2017, we assume two *Virginia*-class submarines, two P-8 MMAs, and two T-AGOS ships. The generations of U.S. equipment are differentiated primarily by their patrol speeds. We assume that the *Los Angeles*– and *Virginia*-class SSNs patrol at 15 and 25 knots, respectively, and that the P-3C and P-8 effectively patrol at 75 and 100 knots, respectively. Meanwhile, the P-3C and P-8 maximum speeds are 320 and 440 knots, respectively.

Table 7.10
One-on-One Results: Engagement Opportunities by a Single PLA Submarine Against a Single U.S. Carrier, Seven-Day Campaign

		1996		2003		2010		2017	
Boat Class	**Cueing**	**500 nm**	**1,000 nm**	**500 nm**	**1,000 nm**	**500 nm**	**1,000 nm**	**500 nm**	**1,000 nm**
Kilo (877) SSK	None	0.003	0.003	0.003	0.003	0.003	0.003	0.002	0.003
	24 hours	0.021	0.029	0.021	0.029	0.021	0.029	0.016	0.026
Han SSN/ *Shang* (093) SSN[a]	None	0.018	0.014	0.018	0.014	0.018	0.014	0.013	0.012
	24 hours	0.048	0.067	0.048	0.067	0.048	0.067	0.035	0.058
Song SSK/ *Yuan* SSK	None			0.073	0.040	0.073	0.040	0.068	0.038
	24 hours			0.326	0.323	0.326	0.323	0.303	0.309
Kilo (636) SSK	None					0.111	0.045	0.108	0.044
	24 hours		N/A			0.487	0.365	0.475	0.360
Shang (093A) SSN/Type 095 SSN	None							0.217	0.102
	24 hours							0.538	0.645

NOTE: The table displays the number of submarine/carrier engagements over the course of the campaign, assuming a single carrier within the search zone. Additional carriers will result in more engagements.

[a] The *Shang*-class SSN entered the inventory in 2006.

we focus primarily on two cases: one with no cueing and one with cueing once every 24 hours. In both cases, we assume that a single U.S. carrier operated within 1,000 nm of the focal point of conflict. Shrinking the operational area would produce more engagements, while increasing the number of carriers would increase the total number of engagements against carriers (while also increasing the number of Chinese submarines destroyed by U.S. anti-submarine warfare assets).

In addition to data on the overall size of the force and the effectiveness of individual submarines, it is also necessary to estimate how many submarines would or could deploy to the operational area for operations. We assume that Chinese submarines can deploy for a maximum of 30 days and that their effective transit speeds are 8 knots for diesel submarines and 25 knots for nuclear submarines. Figure 7.8 shows the location of China's three primary submarine bases and approximate distances from potential operating locations near Taiwan and the Spratly Islands. On balance, Chinese submarines are based closer to operational areas that are relevant to the Taiwan scenario than

Figure 7.8
Chinese North, East, and South Sea Fleet Locations and Ranges to the Taiwan and Spratly Islands Operating Areas

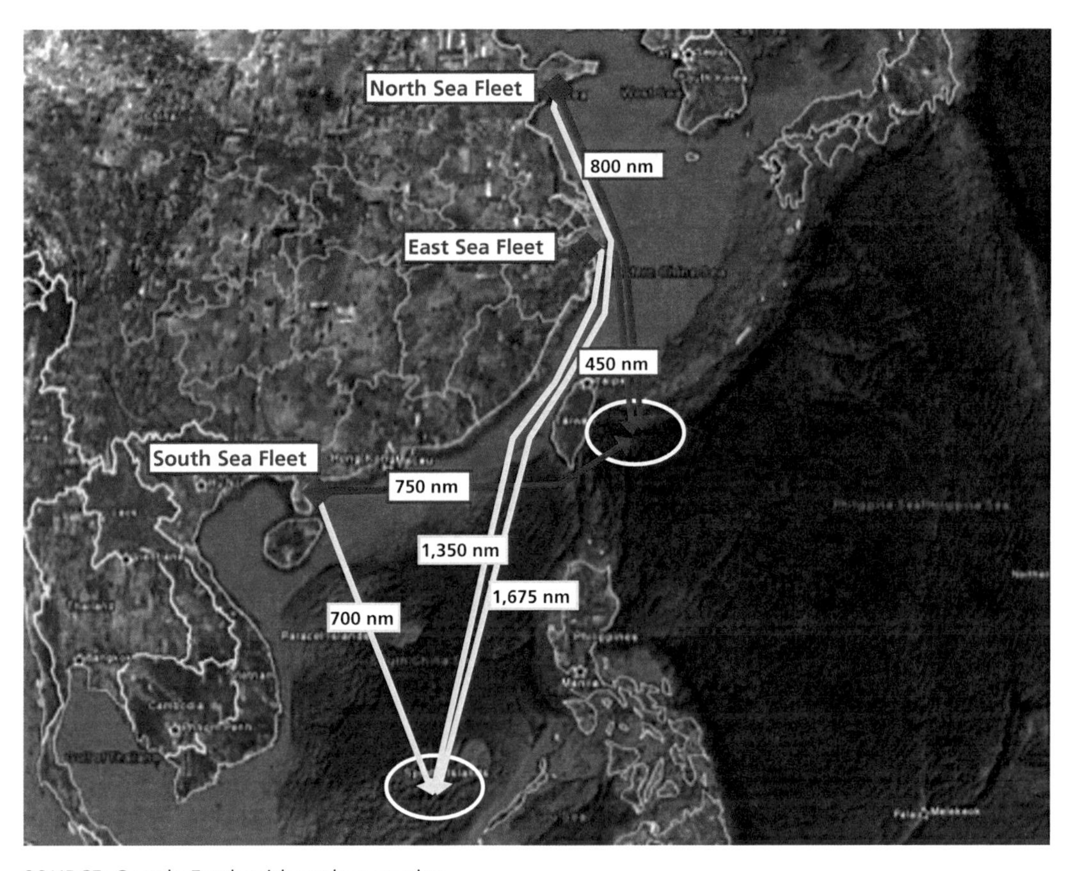

SOURCE: Google Earth with author overlay.
RAND RR392-7.8

they are to areas that are relevant to the Spratly scenario (though relevant distances vary by submarine type).[109]

Based on the speeds given here and the starting port locations of diesel and nuclear submarines, we calculated the number of diesel and nuclear submarines that would be required in the inventory to keep a single submarine on station (see Table 7.11). For example, we found that approximately 2.59 diesel submarines would be needed to maintain one submarine on continuous patrol within 1,000 nm and to the southeast of Taiwan, while it would take 3.35 submarines to keep a single submarine on station southeast of the Spratly Islands.

---

[109] China maintains its nuclear submarine force in its North Sea Fleet based at Qingdao and in its South Sea Fleet, headquartered at Zhanjiang, and most of its modern diesel fleet is maintained by its East Sea Fleet (based at Ningbo) and South Sea Fleet.

Table 7.11
Number of Submarines Required to Maintain One Continuous Patrol Within 1,000 nm of Taiwan and the Spratly Islands

Chinese Submarine Type	Taiwan	Spratly Islands
Diesel	2.59	3.35
Nuclear	1.95	2.17

Using the ratios in Table 7.11, we calculated the number of submarines that could participate in an extended submarine campaign. While the number varies, depending on the scenario and the losses experienced by the submarine force over time, the on-station number totals roughly 35–50 percent of all modern submarines. It is possible that the Chinese could attempt to surge all available submarines in the event of a conflict. However, although this is theoretically possible, we find such an eventuality unlikely, given the natural military inclination to maintain a reserve and hedge against future uncertainty, as well as the historical record of poor Chinese maintenance.

Based on deployed submarine numbers, together with the total force structure in 1996, 2003, 2010, and 2017 (projected) and single-submarine engagement data, we derived a total number of engagement opportunities that all deployed Chinese submarines might achieve in a Taiwan scenario and in a Spratly Islands scenario over a seven-day period (see Table 7.12).

We illustrate the findings with two examples. In 2017, Chinese submarines operating within 1,000 nm of Taiwan might be expected to achieve an average of 0.6 engagement opportunities (total) over a seven-day period without cueing; with cueing to submarines of the carrier location, PLAN submarines might achieve 4.7 engagements over the same period. We emphasize again that the specific numbers

Table 7.12
Total Expected Chinese Submarine Engagement Opportunities Against a Single U.S. Aircraft Carrier, Seven-Day Campaign

Cueing	Scenario	1996	2003	2010	2017
No cueing	Taiwan	0.04	0.09	0.42	0.58
	Spratly Islands	0.03	0.07	0.33	0.45
With cueing	Taiwan	0.19	0.59	3.25	4.68
	Spratly Islands	0.17	0.48	2.54	3.63

NOTE: Green shading indicates cases 0–0.2 engagements, yellow indicates 0.21–0.5, and red indicates more than 0.5.

derived from the model are not intended to represent fully developed or exact predictions. The inputs are based on publicly available information that is only crudely represented in those sources, and the model is a rough representation of interactions between forces. Nevertheless, by introducing basic dynamics of submarine and anti-submarine warfare, the analysis goes beyond static analysis and, combined with qualitative assessments, allows us to draw the following tentative conclusions.

First, the Chinese submarine fleet has made major gains relative to U.S. defensive capabilities. Under any single set of assumptions assessed within the model, the number of expected potential engagements by Chinese submarines against U.S. carriers increases by more than an order of magnitude (and, in some cases, by more than 20 times) between 1996 and 2017.

Second, cueing could also substantially improve the Chinese ability to engage U.S. targets. Daily cueing increases the average number of engagement opportunities by a factor of five to eight, depending on other assumptions. (Interestingly, the impact of cueing is greater than the impact of surging all available submarines into the operational area.)[110] It should be acknowledged that, depending on the nature of the communication systems employed, attempting to coordinate submarines at sea can also increase the submarines' vulnerability, as the Germans learned during World War II.[111] Nevertheless, PLA commanders may be happy to accept such a trade-off. Given the potential impact of cueing, the ability of U.S. forces to either disrupt the Chinese ISR system or jam Chinese submarines' reception will be particularly critical to the security of U.S. surface forces.

Third, the modeling results suggest that not only is the threat increasing rapidly, but it has also become significant in absolute terms, a fact that may have implications for how the United States employs its carriers. Even without cueing, Chinese submarines might have close to an even chance of engaging a single U.S. carrier over a seven-day period. With cueing, submarines might expect to gain several offensive opportunities over the same period. Moreover, if more than one U.S. carrier were in the operational area, the number of Chinese attack opportunities would rise almost proportionately. Given the cost, number of personnel, and symbolic importance of U.S. aircraft carriers, this level of risk could prompt U.S. commanders to hold carriers back until areas closer to China could be sanitized by U.S. anti-submarine assets.

---

[110] Although we did not include the results in Table 7.12, we considered a surge case in which roughly 80 percent of the submarine fleet moved to the operational area prior to the outbreak of conflict (as opposed to roughly 40–50 percent of the fleet in the sustained case). In the surge case, the average number of engagement opportunities roughly doubles. Whatever the number of boats deployed, getting them into position prior to the start of hostilities is important.

[111] U.S. submarines operating against Japan during World War II operated relatively autonomously and faced fewer vulnerabilities as a result of their communication systems (though they suffered from other technical flaws, especially at the beginning of the conflict).

Conclusions

In contrast to the situation in 1996, China can now hold the U.S. Navy's surface fleet at risk at significant ranges from the mainland. The extent of the threat to the U.S. surface fleet continues to grow. China's anti-surface capability is founded on four developments: (1) the establishment of an increasingly capable long-range surveillance system, which improves the PLA's ability to detect and track surface ships at long ranges; (2) the deployment of sophisticated anti-ship cruise missiles and the development of an ASBM with a range of 2,000 km; (3) the acquisition of strike aircraft and surface ships with greater range and power; and (4) the deployment of new classes of larger and quieter submarines armed with both torpedoes and cruise missiles.

The U.S. military has a variety of means to mitigate specific threats, and it will improve them over time. New counterspace and cyber capabilities may enable U.S. forces to degrade Chinese space-based ISR and OTH radar. The U.S. Navy is almost certainly hard at work on technical counters to China's budding ASBM threat, including both anti-missile systems and, perhaps more importantly, ways to defeat Chinese ISR. Aircraft carriers can provide their own defensive combat air patrols to defeat the threat from enemy aircraft, and the United States is acquiring more and better anti-submarine warfare assets. As important, U.S. surface forces can also adjust their operational practices. U.S. surface forces may, for example, stand off farther from the Chinese coast, thereby reducing China's ability to find and target them.

However, several of these defensive measures distract from or diminish the ability of U.S. forces to project power. Holding carriers farther from the scene of the main battle area would entail longer transit times for combat aircraft, fewer aircraft on station, and an increased demand for U.S. Air Force tanker support. Particularly in light of the Chinese missile threat to forward U.S. air bases (see Chapter Three, scorecard 1), finding basing for more tankers to support U.S. Navy air operations would be difficult. And the reduction in time on station for U.S. naval combat aircraft as they are forced to fly greater distances would further complicate an increasingly challenged air superiority battle (see Chapter Four, scorecard 2), as would withholding aircraft to protect the carriers.

The impact of Chinese threats to carriers will likely be greatest during the first stages of a conflict. In a protracted fight, U.S. forces would probably be able to progressively mitigate the threat, allowing U.S. aircraft carriers to approach closer to the main battle areas with less risk to themselves. During the critical first days of a conflict, however, this would leave U.S. and partner forces less well protected from air attack. Moreover, as Chinese capabilities grow in both sophistication and numbers, it will take longer to achieve the same level of mitigation. Together with the Chinese missile threat to U.S. air bases (Chapter Three, scorecard 1), the growing threat to U.S. surface ships outlined in this chapter is arguably the most serious challenge facing U.S. forces in any potential China scenario.

Scorecard Coding

Figure 7.9 provides our summary coding of the results of scorecard 5. Advantage in this scorecard is determined by the ability of Chinese naval and air forces to hold U.S. aircraft carriers at significant risk within operational ranges of the scenario in question (*Chinese advantage*) and the U.S. ability to operate with relative impunity (*U.S. advantage*). In making this judgment, we consider Chinese ISR capabilities and the offensive potential of submarine, air, surface, and (in 2017) ASBM attacks against U.S. aircraft carriers protected by their complements of destroyers, submarines, and fighter aircraft.

In 1996, antiquated Chinese submarine, air, and surface assets—together with an almost complete lack of long-range ISR—limited the PLA's capability to hold U.S. aircraft carriers at risk. We coded both the 1996 Taiwan and Spratly Islands scenarios as *U.S. advantage*. By 2003, Chinese capabilities had improved marginally but not enough to shift coding decisively in either scenario. However, our modeling of Chi-

Figure 7.9
Scorecard 5 Summary Coding

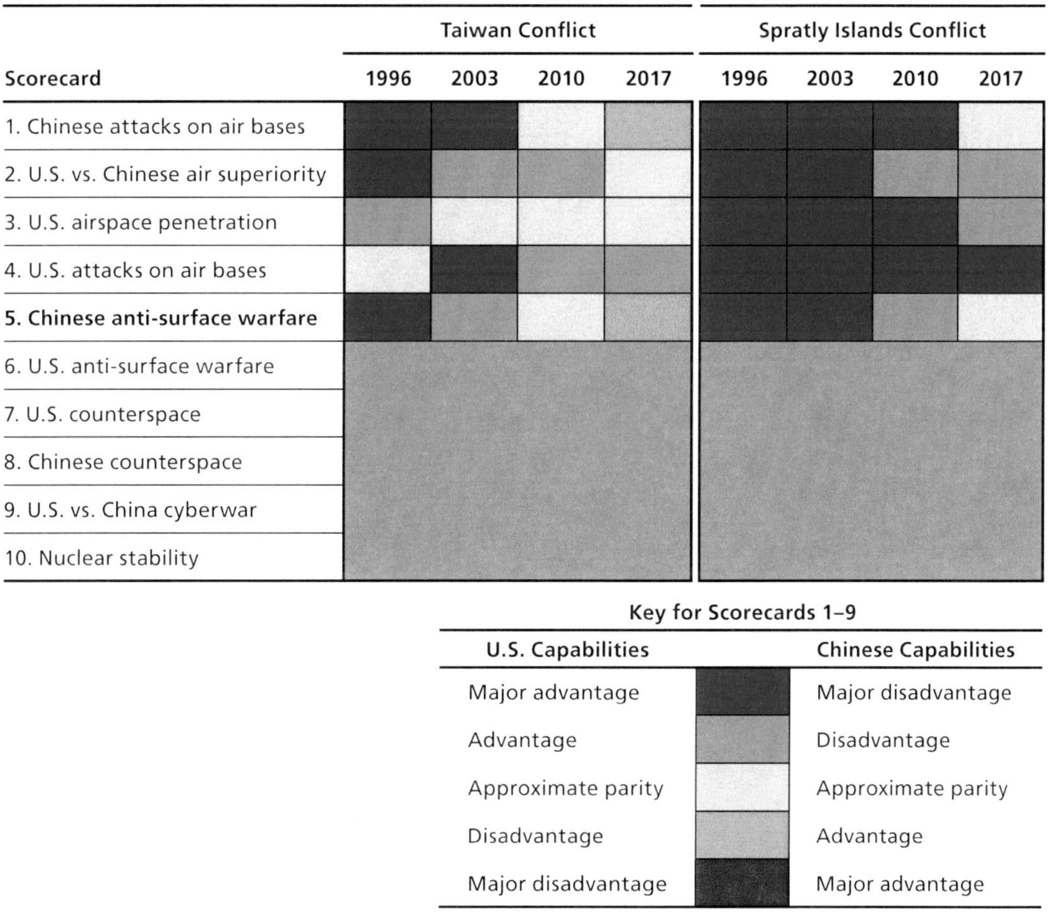

	Taiwan Conflict				Spratly Islands Conflict			
Scorecard	1996	2003	2010	2017	1996	2003	2010	2017
1. Chinese attacks on air bases								
2. U.S. vs. Chinese air superiority								
3. U.S. airspace penetration								
4. U.S. attacks on air bases								
5. Chinese anti-surface warfare								
6. U.S. anti-surface warfare								
7. U.S. counterspace								
8. Chinese counterspace								
9. U.S. vs. China cyberwar								
10. Nuclear stability								

Key for Scorecards 1–9

U.S. Capabilities		Chinese Capabilities
Major advantage		Major disadvantage
Advantage		Disadvantage
Approximate parity		Approximate parity
Disadvantage		Advantage
Major disadvantage		Major advantage

nese ISR and offensive submarine attacks showed more marked PLA improvements by 2010. In the Taiwan scenario, we coded the 2010 case as one of *approximate parity.* In the Spratly scenario, the ability of U.S. carriers to remain farther from the Chinese coast and still contribute to the air battle results in lower levels of vulnerability to air attack, so we coded the 2010 Spratly case as *U.S. advantage.*

By 2017, further improvements across all areas of Chinese anti-surface warfare, especially submarine capabilities, lead to *Chinese advantage* in the Taiwan case and *approximate parity* in the Spratly Islands scenario. It should be remembered that Chinese advantage refers only to the situation at the first few weeks of conflict. While this period could prove critically important to, for example, a ground campaign in Taiwan, it does not necessarily suggest that China would be able to hold U.S. warships at a similar degree of risk during a more protracted conflict. Also, even at the outset of conflict, U.S. commanders could reduce the risk to carriers by holding them farther from the coast, though this would reduce their contribution to the air battle.

CHAPTER EIGHT

Scorecard 6: U.S. Anti-Surface Warfare Capabilities Versus Chinese Naval Ships

Having examined the Chinese ability to threaten the U.S. surface fleet, we now turn to an assessment of the U.S. capability to do the same to the Chinese fleet. We focus here on the U.S. military's capability to thwart a Chinese seaborne invasion by sinking or disabling elements of the amphibious fleet as it transits toward Taiwan or the Spratly Islands. As in all of the scorecard analyses, this chapter provides only a partial picture of the larger area addressed in the scorecard, and other types of analysis could also be profitably pursued (such as U.S. capabilities against Chinese surface action groups). Nevertheless, the U.S. ability to attack an amphibious invasion fleet defended by escorting aircraft and warships is an important problem in its own right, and it also provides a view of the larger problem of U.S. anti-surface warfare.

We examine the evolution of four component aspects of the problem in both the Taiwan and Spratly Islands scenarios. First, we outline the evolution of China's amphibious fleet and its capacity to transport PLA ground forces to their assault areas or beachheads. Second, we assess the ability of U.S. submarines to sink ships of the amphibious fleet in the face of PLAN anti-submarine warfare defenses, as well as the losses that might be incurred by U.S. submarines. Third, we evaluate the ability of U.S. aircraft to strike and destroy a PLA invasion fleet. Finally, we consider attacks by U.S. surface ships. In all areas, we consider the evolution of both sides' forces and their impact on net results.

We find that although the U.S. ability to strike and destroy Chinese surface units has declined somewhat relative to the Chinese fleet's ability to defend itself from attack, U.S. anti-surface warfare capabilities nevertheless remain formidable. U.S. nuclear attack submarines remain an effective and efficient anti-surface warfare platform. The heavyweight torpedoes carried by submarines are the most reliable and punishing anti-surface weapons available. To be sure, recent advances in Chinese anti-submarine warfare capabilities have somewhat degraded the ability of U.S. submarines to sink PLA amphibious invasion ships. Nevertheless, U.S. submarines, with support from air and surface elements, would likely inflict terrible punishment on a Chinese invasion force.

For almost two decades after the Cold War, little attention was paid to the anti-surface warfare capabilities of U.S. naval surface warships and Navy and Air Force strike aircraft. The U.S. inventory of ASCMs shrank and aged, and new systems were not optimized well for the current threat environment near China. The Chinese ability to hold U.S. air bases and aircraft carriers at risk and contest air superiority (discussed in Chapters Three, Four, and Seven) would make it difficult for U.S. forces to launch attacks. Nevertheless, U.S. surface and air attacks could complement U.S. submarine anti-surface warfare, and new ASCMs are currently in development or reaching maturity. Although the trend lines run counter to the United States in this scorecard, the results remain more positive for the United States than those of several other scorecards and suggest that, even in 2017, any large-scale amphibious invasion would be extraordinarily risky for China.[1]

Taiwan Scenario

In this section, we examine the U.S. military's capability to thwart a Chinese seaborne invasion of Taiwan by sinking or disabling elements of the amphibious fleet as it transits the Taiwan Strait.[2] In the following sections, we first quantify the Chinese amphibious fleet's capabilities and their evolution since 1996. We then discuss each of the primary anti-surface warfare options. We begin with an assessment of submarine attacks against a PLA invasion fleet, for which we have developed a model that simulates U.S. attack submarines engaging PLAN amphibious ships as they transit back and forth across the Strait (as well as losses to U.S. submarines). We then discuss trends in the U.S. offensive air and surface capability against China's amphibious fleet.

PLA Amphibious Fleet

To conquer Taiwan, the PLA must first establish a secure lodgment on the Taiwanese shore and then repeatedly resupply and reinforce its forces until a breakout can be achieved. The PLA would then have to defeat Taiwanese forces and establish control over the island. In such a scenario, the bulk of Chinese troops and equipment would come from the sea.[3] To build up sufficient forces to conquer Taiwan, PLAN amphibious ships would have to ferry ground forces multiple times, making repeated transits back and forth across the Strait. Table 8.1 presents China's amphibious landing fleet force structure in 1996, 2003, 2010, 2015, and 2017 (projected).

---

[1] As in the other scorecard analyses, we do not explicitly include training in this assessment. While this allows us to focus on an objective, quantitative assessment of the impact of changes to each side's equipment inventories, it also leads us to understate the degree of U.S. advantage, since amphibious operations are highly complex operations.

[2] Other RAND work has touched on the possibility of a PLA amphibious invasion of Taiwan. See, for example, Roger Cliff, Phillip C. Saunders, and Scott Harold, *New Opportunities and Challenges for Taiwan's Security*, Santa Monica, Calif.: RAND Corporation, CF-279-OSD, 2011.

[3] Other PLA elements, such as airborne units, would also play a role, but the majority of the invasion forces—troops, equipment, supplies—would come from the sea.

In 1996, the PLAN operated a total of 54 amphibious ships, many of which were antiquated craft. Some 13 (*Shan* class) were U.S. ships built between 1942 and 1945, and the overall average age of the amphibious fleet was more than 22 years. By 2015, the PLAN's inventory of amphibious ships had grown to 89 craft. Over time, the average age of the fleet has declined as older ships have gradually been retired and new classes of larger and more seaworthy craft are being introduced. In 2007, the first of *Yuzhao*-class (Type 071) landing platform dock ships entered service. At an estimated 17,000 tons, it dwarfs any previous Chinese amphibious ship (the largest up to that point was less than 5,000 tons). The *Yuzhao*, which has been deployed on counter-

Table 8.1
Growing Number and Capacity of PLAN Amphibious Landing Ships

					Number of Ships in Service				
Ship Class	**Type**[a]	**IOC**	**Troops**	**Armored Fighting Vehicles**	**1996**	**2003**	**2010**	**2015 (current)**	**2017**
Shan	LST	1950s	159	16	13	3	—	—	—
Yudao	LSM	1980s	500	10	1	1	1	—	—
Yuliang	LSM	1980	250	5	30	22	31	28	26
Yukan	LST	1980	200	10	5	7	7	7	6
Yuting	LST	1992	250	10	1	9	10	9	9
Yudeng	LSM	1994	180	6	4	1	1	1	1
Yuhai	LSM	1996	250	2	—	13	13	10	10
Yuting II	LST	2004	250	10	—	—	10	10	10
Yubei[b]	LCU	2004	200	10	—	—	10	10	10
Yunshu	LSM	2004	500	6	—	—	10	10	10
Yuzhao	LPD	2007	600	15	—	—	1	4	5
Type 081[c]	LHD	2017 (?)	900		—	—	—	—	2
Total					54	56	94	89	89

SOURCES: IISS, *The Military Balance*, 1996, 2003, 2010, and 2015; Office of Naval Intelligence, *The PLA Navy*, 2015, p. 185; *Jane's Amphibious and Special Forces*, "Sea Lift (China)," April 22, 2014. See O'Rourke, China Naval Modernization, 2013; and Cole, *The Great Wall at Sea*, 2012, pp. 106–107.

[a] LSM = landing ship, medium. LCU = landing craft, utility. LPD = landing platform dock. LHD = landing helicopter dock.

[b] Although the *Yubei* is an LCU, and we do not include other LCUs in this table because of their generally smaller size and limited function, we include *Yubei* here because it is a modern design and, at 1,200 tons (loaded), is as large as some older classes of LSMs, such as the *Yuliang*.

[c] Reports differ on the progress of China's first amphibious assault ships, but PLA leaders have expressed clear interest in larger and more capable amphibious craft.

piracy operations in the Gulf of Aden, accommodates four air-cushion vehicles, which exit through a stern gate, as well as four helicopters on its rear deck.

A number of sources suggest that China is currently working on, and may be building, a significantly larger (perhaps 40,000-ton) landing helicopter dock (commonly referred to as an *amphibious assault ship*), designated the Type 081.[4] This ship might deploy eight helicopters and up to 1,100 soldiers, as well as armored vehicles. In November 2013, Yin Zhuo, the director of the PLA Navy's Expert Consultation Committee, told China Central Television that China was developing an amphibious assault ship that would be some one and a half times larger than Japan's *Izumo*-class helicopter destroyer (which is 27,000 tons).[5] A Hong Kong military analyst wrote in early 2015 that the first such ship could be built by the end of the year.[6] Given this and other efforts to upgrade amphibious capability, we credited China with two large "next generation" amphibious ships in 2017, though we note that the timing of delivery, and IOC, is uncertain.[7]

Figure 8.1 shows the fleet's total transport capacity in infantry division equivalents.[8] The one-way total transport capacity rose from 1.2 divisions in 1996 to 2.6 divisions in 2010. Our projections indicate that it will increase to 2.7 divisions by 2017, though the more important change is the further replacement of rudimentary landing ships with far more modern, capable craft. The totals assume that all PLA amphibious ships participate and that the space onboard can be devoted to lifting divisional personnel and equipment (as opposed to supplies and nondivisional personnel). Given the age of some ships, their dispersion among different fleets, and the other material requirements of an amphibious invasion, the number of divisions should be treated as a theoretical maximum derived primarily to facilitate the current evaluation of U.S. anti-surface warfare capabilities.[9]

---

[4] Office of the Secretary of Defense, *Military and Security Developments Involving the People's Republic of China*, 2015 p. 10; Office of Naval Intelligence, *The PLA Navy*, 2015, p. 15.

[5] "Nation Starts Research on Naval Jet," *China Daily*, May 13, 2015.

[6] "Nation Starts Research on Naval Jet," *China Daily*, May 13, 2015; and "PLA to Build Amphibious Assault Ships: Report," Focus Taiwan News Channel, January 25, 2015. In 2013, a separate report suggested that the ship was under construction. See "China Building 1st Amphibious Assault Ship in Shanghai," *Global Post*, August 26, 2013.

[7] In addition to the amphibious assault ship discussed in the text, China appears to be building mobile landing platforms of roughly 5,000 tons. See "Mobile Landing Platform Being Constructed for PLA Navy," *Want China Times*, June 6, 2015.

[8] Divisions vary widely in strength. For analytical purposes, we counted one infantry division as equal to 10,000 troops, in addition to vehicles.

[9] Note, however, that the PLA could land more forces if it seized a port. China is building a variety of potentially dual-use civilian ships (including roll-on, roll-off ferries), but these would likely require secure ports for disembarkation and thus are probably not relevant to a beach landing.

Figure 8.1
PLAN Amphibious Transport Capacity

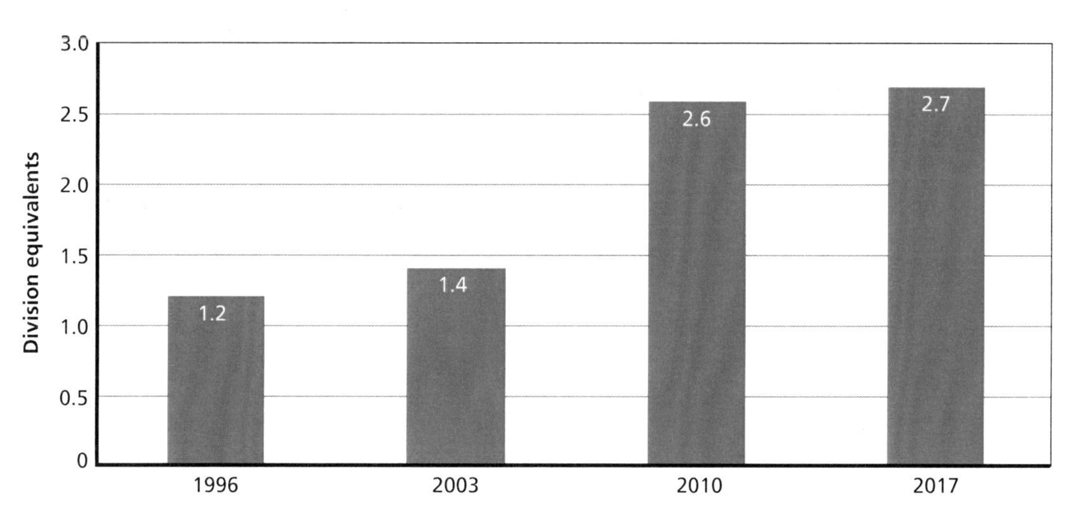

SOURCE: Data on the capacity of ship classes are from *Jane's Amphibious and Special Forces*, "Sea Lift (China)," April 22, 2014.
NOTE: The figure shows the number of division equivalents that could be lifted in a single trip using every amphibious ship in the inventory.
RAND *RR392-8.1*

We then calculated the maximum number of infantry division equivalents China could land on Taiwan over a seven-day period. In addition to single-sortie amphibious lift capacity, this result depends on how quickly the fleet can transit the strait, unload, transit back to the mainland, reload, and line up in formation for the next sortie. We incorporated the following assumptions:[10]

- The average transit distance across the strait is 120 nm; if the invasion fleet averages 15 knots, then each crossing (to and from Taiwan) takes approximately eight hours.
- Given the generally small size of these craft, each wave landing on the beach takes one hour to unload.
- Reloading, refueling, and maintenance upkeep require 12 hours for each amphibious ship before it is ready for the next transit.
- The amphibious column consists of ten-ship waves separated by approximately 24 minutes.[11]

---

[10] These assumptions are geared primarily toward the 2010 and 2017 cases and apply less to 1996 and 2003, when the fleet was composed primarily of legacy vessels. As such, we somewhat overestimate the PLA's lift capabilities in 1996 and 2003.

[11] These parameters are driven by the assumption that, at most, 20 ships can land and unload at any given time.

The amphibious fleet includes other vessels, such as escorts/surface combatants, auxiliary vessels, and decoy ships.[12] In each case, the fleet is able to make five transits to Taiwan. Assessing total lift capacity in a standard seven-day period, the PLA could land a total of 6, 7, 13, and 13.5 divisions in 1996, 2003, 2010, and 2017, respectively, without opposition from U.S. and Taiwanese forces.

Large-scale amphibious invasions are extremely difficult and complex operations. Success requires much more than just a fleet of ships to ferry troops and equipment. Logistics, training, coordination, command and control, and operational experience all play vital roles. Once ashore, Taiwan's military awaits with an active-duty army of 13 brigades of armor, mechanized infantry, and light infantry, backed up by a reserve force of 21 infantry brigades (down from roughly 39 active brigades and seven reserve divisions in 2010).[13] Although China had a single trip transport capacity of roughly one and a half divisions in 1996 and 2003, it is highly unlikely that the Chinese could have successfully invaded Taiwan, even without U.S. intervention.

Even today, China does not regularly conduct large-scale amphibious exercises—though it has done so on occasion and its tactical skills are improving.[14] In addition to the PLA Navy's three marine brigades, China formed two amphibious mechanized infantry divisions between 2007 and 2012, and Taiwanese sources reported in early 2015 that the number of such divisions would be doubled to four.[15] These formations will presumably conduct tactical amphibious assault training on a more routine basis. Nevertheless, gaining the full repertoire of skills necessary to coordinate a large landing will take substantial time.

U.S. Submarine Campaign Against PLA Amphibious Fleet

We begin with the assessment of U.S. submarine capabilities against Chinese amphibious forces. We built a model that assesses the capacity of U.S. submarines to attack amphibious ships in the Taiwan Strait while Chinese forces attempt to employ their anti-submarine warfare capabilities to locate and destroy U.S. submarines. The modeling dynamics described here are highly simplified and stylized, but they are nevertheless intended to at least capture some of the key aspects of submarine and anti-submarine warfare.

---

[12] We assume that five to 20 escorts or surface combatants transit with the amphibious fleet.

[13] IISS, *The Military Balance*, 2010 and 2015. The reduction in the size of the force has been driven by the progressive shortening of conscription (presently 12 months) and efforts to create an all-volunteer force (a program that has not been fully implemented).

[14] The PLA's Mission Action–2013 exercise employed a regiment of an amphibious mechanized infantry division moving from the Guangzhou Military Region to the Nanjing Military Region in a mock attack amphibious assault. Jonathan D. Pollack and Dennis J. Blasko, "Is China Preparing for a 'Short, Sharp War' Against Japan?" Brookings Institution, February 25, 2014.

[15] IISS, *The Military Balance*, 2015, p. 238; and "China Just Doubled the Size of Its Amphibious Mechanized Infantry Divisions," *The Diplomat*, January 9, 2015.

U.S. Submarine Patrols

We modeled two near-continuous U.S. submarine combat patrols inside the strait (see Figure 8.2) and their efforts to sink as many amphibious ships as possible. Attack submarines rotate into the strait from other missions and then rotate out on a predetermined rotation schedule. The total number of rotations depends on the available force structure in theater. For both 1996 and 2003, we assume that *Los Angeles*–class submarines occupy both patrols. In 2010 and 2017, after *Virginia*-class submarines have become operational, that class conducts one of the two patrols.[16] The *Virginia* class has a larger torpedo magazine than the *Los Angeles* class (38 versus 26). With its new design, we assume that its periscope also has a smaller RCS and is therefore harder to detect. Figure 8.2 shows U.S. Navy submarine patrols in relation to the PLA amphibious fleet transit lanes.

For modeling purposes, we made several assumptions about the conditions and conduct of a submarine campaign. We assume that the impending invasion would provide U.S. forces with indications and warnings and enough time for U.S. submarines to arrive on station. U.S. submarines identify and target PLAN ships while in transit by periscope and engage two targets at a time. U.S. submarines must be within 12 nm of the amphibious column to visually identify and differentiate the amphibious ships

Figure 8.2
Notional Diagram of U.S. Submarine Patrols and PLA Amphibious Fleet Transit Lane Locations

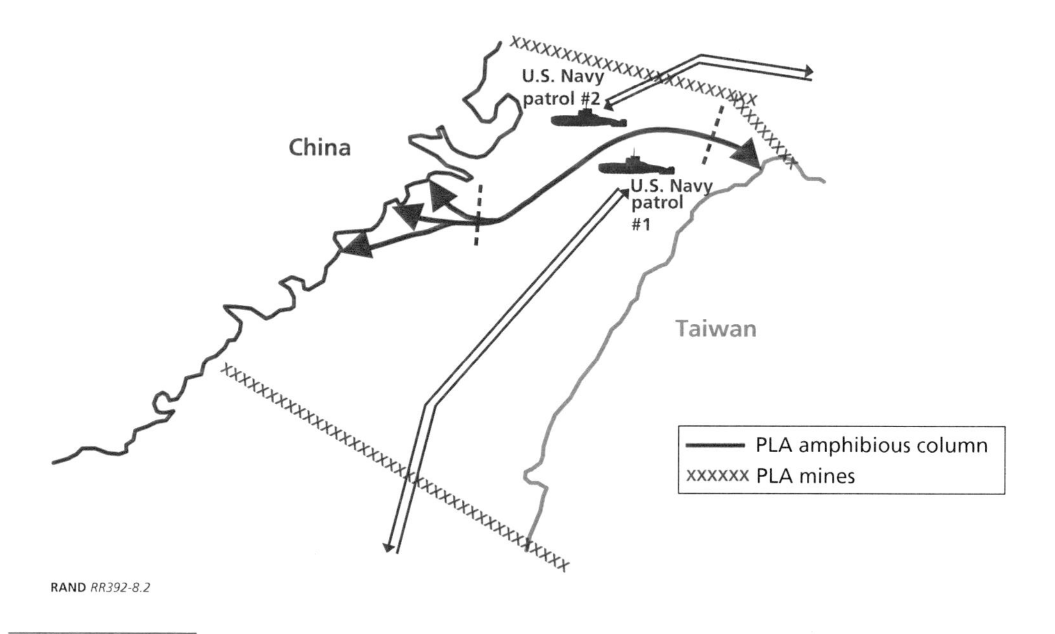

RAND *RR392-8.2*

[16] The *Seawolf* class would also be effective in this role, but given that its production stopped with only three in service (of a total of 58 tactical submarines in mid-2013), it is quite possible that the ships will be either unavailable or employed in other missions.

from other ships. The maximum torpedo range is 24 nm. Because they can visually discriminate the targets, we assume that U.S. submarines have a Pk equal to 0.8 per torpedo, including false target engagements. After each attack, it takes U.S. submarines two hours to change locations before their next engagement attempt. Finally, we assume that each submarine rotation requires eight hours to complete, during which time the combat patrol remains empty.[17]

PLAN Anti-Submarine Warfare Forces

PLAN anti-submarine warfare capabilities have not improved as much as capabilities associated with other maritime missions, though the pace of change appears to have accelerated considerably over the past several years.[18] Initially, investments largely focused on airborne platforms, such as helicopters and MPA equipped with maritime radars. Table 8.2 shows China's airborne anti-submarine warfare force structure for helicopters and MPAs. Only limited open-source information is available on the maritime radars employed by the PLAN.[19] The information that is available suggests that although the total number of aircraft assigned to the mission has declined as the H-5 and Z-5s have been retired, capabilities have improved. The total number of PLA anti-submarine warfare aircraft equipped with specialized maritime search radar has increased from only a handful (five aircraft) in 1996 to 68 by 2010 and 77 by 2015. The PLA Navy is currently testing a new and heavier anti-submarine warfare helicopter, the Z-18F, which will be equipped with dipping sonar, 32 sonobuoys, and surface search radar.[20] In 2015, China commissioned its first four-engine fixed-wing ASW aircraft, the Gaoxin-6, equipped with many of the features found on the P-3C Orion, including a magnetic anomaly detector, sea-search radar, and anti-submarine weapons. In addition to airborne platforms, China has also deployed a new purpose-built anti-submarine warfare corvette, the *Jiangdao*-class (Type 056), equipped with towed array sonar and platform for a Ka-28 helicopter. Perhaps as a measure of recent commitment to the anti-submarine warfare mission, 22 of these ships were launched between 2012 and mid-2014, though some are for export.[21]

For modeling purposes, we assume that PLAN helicopters fly parallel paths 12 nm offset from both sides of the amphibious column at 100 feet above the sur-

---

[17] The long rotation time is, in part, because of the assumption that the PLA has mined both entrances to the Taiwan Strait.

[18] See, for example, Office of Naval Intelligence, The *PLA Navy*, 2015, pp. 17 and 18. See also Bussert and Elleman, *People's Liberation Army Navy*, 2011, pp. 127–139.

[19] Available information suggests that the Chinese have employed a Doppler radar and the AN/APS-504(V) onboard their SH-5 and Y-8X aircraft and the Agrion-15 onboard their Z-8 and Z-9C helicopters. See *Jane's All the World's Aircraft*, "Airborne ASW Platforms (China)," June 15, 2011; and *Jane's All the World's Aircraft*, "SAC Y-8 and Y-9 (Special Mission Versions)," February 3, 2015.

[20] *Jane's Navy International*, "China Unveils ASW Version of Z-18 Helicopter," August 20, 2014.

[21] *Jane's Fighting Ships*, "Jiangdao (Type 056/056A) Class," March 11, 2015.

Table 8.2
PLAN Anti-Submarine Warfare Helicopters and Maritime Patrol Aircraft

Aircraft Type		1996	2003	2010	2015 (current)	2017
Helicopters	Z-5	40	—	—	—	—
	Z-8	3	8	25	27	27
	Ka-28	—	8	10	19	19
	Z-9C	—	6	25	25	25
	Z-18F[a]					2 (?)
MPAs	H-5	130	50	20	—	—
	SH-5	5	4	4	3	3
	Y-8X	—	4	4	3	3
Total		178	80	88	77	79
Total with maritime search radars		5	22	68	77	79

SOURCE: IISS, *The Military Balance*, 1996, 2003, 2010, and 2015.

NOTES: Shading indicates that the aircraft had no specialized maritime search radar. The Z-8 was upgraded with radar by 2010.

[a] The Z-18F is a redesigned and heavier version of the Z-8 that will carry surface search radar, dipping sonar, and sonobuoys.

face. Meanwhile, the MPAs fly directly over the amphibious column at an altitude of 300 feet. The helicopters and MPAs have maximum radar ranges of 12 and 21 nm, respectively, as shown in Figure 8.3. Given the importance of protecting amphibious forces, we assume that Chinese commanders will allocate a large portion of their anti-submarine warfare assets to this task and that they will maintain continuous, 24/7 air patrols using roughly 15–20 percent of the helicopters and 25 percent of the MPAs in the total PLAN anti-submarine warfare force structure. (The total inventory is shown in Table 8.2.)

Other tactics that might be employed to counter U.S. submarines include "floating mines" and active sonar. Floating mines are stationary submarines that sit (or float) quietly and wait to attack enemy submarines that pass within range. We did not model floating mines in this scenario. Because of the strait's shallow waters, we also judge that active sonar will be largely ineffective against U.S. submarines. As a consequence, we did not model the employment of active sonar.

Although the development of Chinese anti-submarine warfare capabilities has lagged behind that of other mission areas, they still pose a serious threat to U.S. attack submarines in this scenario. By operating in tightly constrained waters against assets that PLA planners understand will constitute a key target set, U.S. submarines will repeatedly expose their general locations to relatively large numbers of enemy assets.

Figure 8.3
Notional Diagram of PLA Airborne Anti-Submarine Patrol Routes and Radar Coverage Areas Over the Taiwan Strait

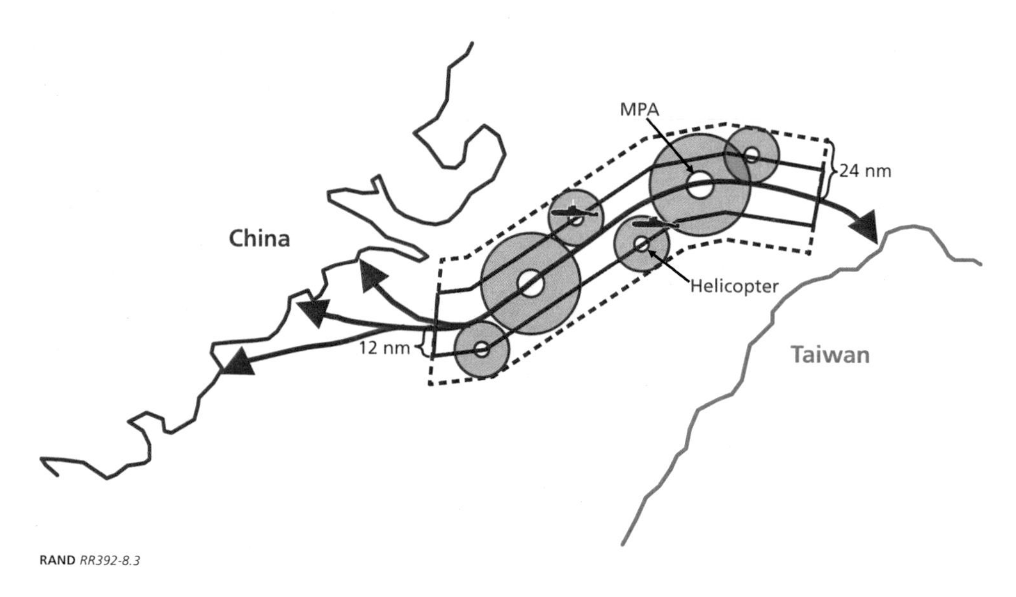

RAND *RR392-8.3*

In this setting, even the least capable platforms (such as the H-5 MPA, which relies exclusively on visual search) can find and engage U.S. submarines by random chance if deployed in large enough numbers.

Model Parameters and Characteristics

U.S. submarines engage Chinese amphibious ships as they transit the Taiwan Strait by visually acquiring the targets using their periscopes. To do so, a submarine must get within 12 nm of the amphibious column and raise its periscope. We assume that during each engagement, the U.S. submarine acquires two distinct targets. This requires the submarine to keep its periscope up for approximately two minutes, during which time the submarine is vulnerable to detection (by visual means or radar).[22] Once the submarine has identified the target vessels and correlated them with the proper sonar contacts, U.S. submarines can then engage targets within a maximum range of 24 nm (including separation after firing).

PLAN anti-submarine warfare assets can engage U.S. submarines after detecting them in one of three ways: radar, visual, or submarine attack (i.e., being alerted by seeing torpedo wakes or ships being struck). In each case, PLAN assets drop torpedoes on and around the submarine's suspected location, where their effectiveness depends on the degree of location uncertainty and the weapon employed. To achieve a kill,

[22] We also modeled an alternative concept of operations in which U.S. submarines engage on sonar contacts alone. This eliminates the risk of periscope detection; however, it also increases the chance of mistakenly engaging false targets, thereby decreasing their Pk value.

PLAN torpedoes must land within 0.5 nm of the U.S. submarine, and, within that distance, we assume that the Pk per weapon is 50 percent.[23]

It is difficult to detect periscopes with radar because their small RCSs compete with the background sea clutter. In addition, U.S. periscopes are equipped with ELINT sensors that detect radar energy and alert the submarine to quickly lower its periscope.[24] Such evasive actions do avert attacks against the amphibious column, and we assume an additional hour delay before the submarine's next engagement attempt. If an anti-submarine warfare asset does detect (via radar or visual means) and engage a submarine, then we postulate an additional two-hour delay before the next submarine engagement (assuming the submarine survives the engagement).

If a submarine is lost, its patrol remains empty until the next scheduled rotation. The model assumes three rotations or at least six U.S. submarines in theater in 1996 and 2003 and four rotations/eight submarines in 2010 and 2017.

Results: U.S. Submarine Attacks

Figure 8.4 presents the results of the modeling, depicted as the percentage of total PLAN amphibious ships destroyed in a seven-day campaign and the percentage reduction in the number of Chinese infantry division equivalents delivered to Taiwan. (U.S. submarine losses are discussed later.) The reduction in infantry division equivalents is measured against the total number of division equivalents that could be transported to Taiwan in the absence of U.S. opposition. Even in the case of 100-percent ship losses, only a portion will be sunk on each crossing, and some of the invasion ground force will therefore succeed in making the crossing.

In 1996, U.S. submarines sink 54 of 54 Chinese amphibious ships (100 percent) and decrease the PLA force buildup by 70 percent. In 2003, they sink 56 of 56 Chinese ships (100 percent) and reduce the PLA force buildup by 60 percent. In 2010, U.S. boats sink 69 of 94 (73 percent) ships and decrease the buildup by 38 percent. And projected figures for 2017 show the U.S. side sinking only 36 of 89 Chinese amphibious ships (41 percent) and decreasing the PLA force buildup by 22 percent.

Figure 8.5 shows the PLA force buildup on Taiwan in the face of opposition by U.S. submarines. In both 1996 and 2003, U.S. submarines sink the entire Chinese amphibious fleet within the first five days of the invasion. However, by 2010, China's amphibious fleet has nearly doubled in size. At the same time, the PLAN's anti-

[23] The difference between the top submerged speed of nuclear submarines and anti-submarine torpedoes is relatively small, and the range of most torpedoes is limited. U.S. submarines also have effective countermeasures against pursuing torpedoes. Hence, if the torpedo does not land close to the submarine and detonate before the submarine can accelerate and begin evasive action, the submarine has an excellent chance of evasion. For representative figures on submarine and torpedo speeds and other characteristics, see *Jane's Fighting Ships*, "Los Angeles Class," March 24, 2015; *Jane's Naval Weapons Systems*, "Mk 48 (YU-6)," July 29, 2011; and U.S. Navy, "Torpedoes: Mark 46, Mark 48, Mark 50," fact sheet, undated.

[24] See the discussion of electronic warfare (AN/WLR-10) in *Jane's Underwater Warfare Systems*, "Los Angeles Class," March 24, 2015.

Figure 8.4
Model Results: PLA Amphibious Ships Destroyed by U.S. Submarines and Decrease in PLA Buildup, 1996, 2003, 2010, and 2017

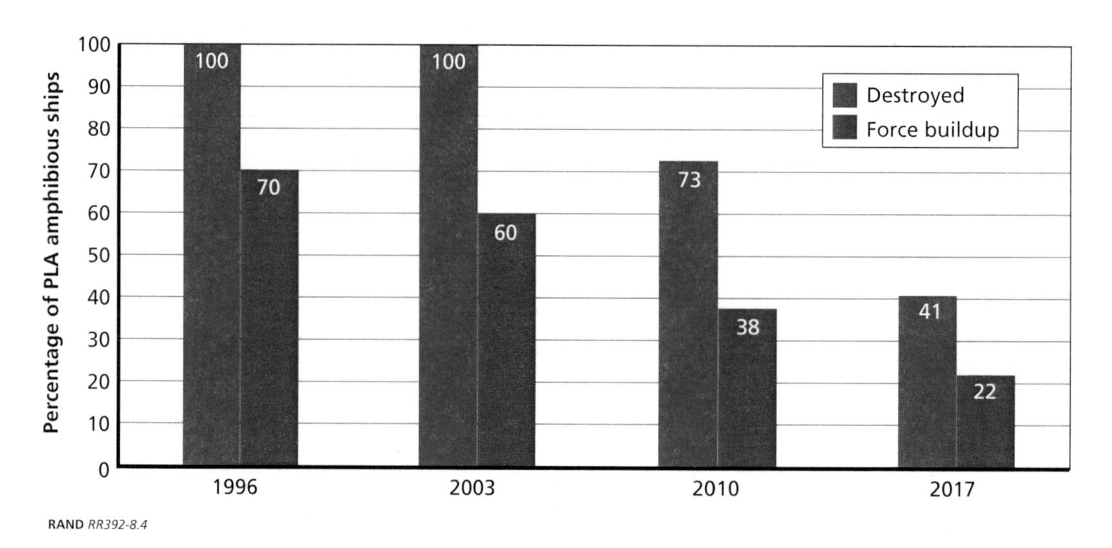

RAND *RR392-8.4*

Figure 8.5
Model Results: PLA Amphibious Force Buildup Against U.S. Submarine Opposition, 1996, 2003, 2010, and 2017

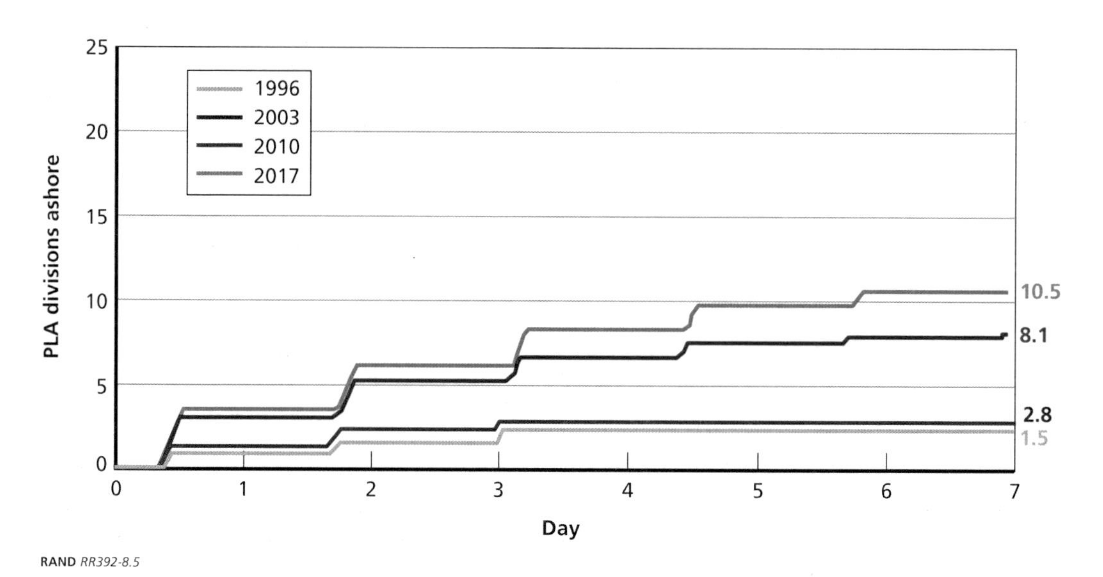

RAND *RR392-8.5*

submarine warfare capabilities become more robust, improving its ability to thwart U.S. attack submarines and, to a lesser extent, kill the attackers (see Figure 8.6). The net effect is diminishing U.S. submarine effectiveness against the Chinese invasion fleet, and that trend holds in 2017.

Figure 8.6
Model Results: U.S. Submarine Losses, 1996, 2003, 2010, and 2017, Seven-Day Campaign

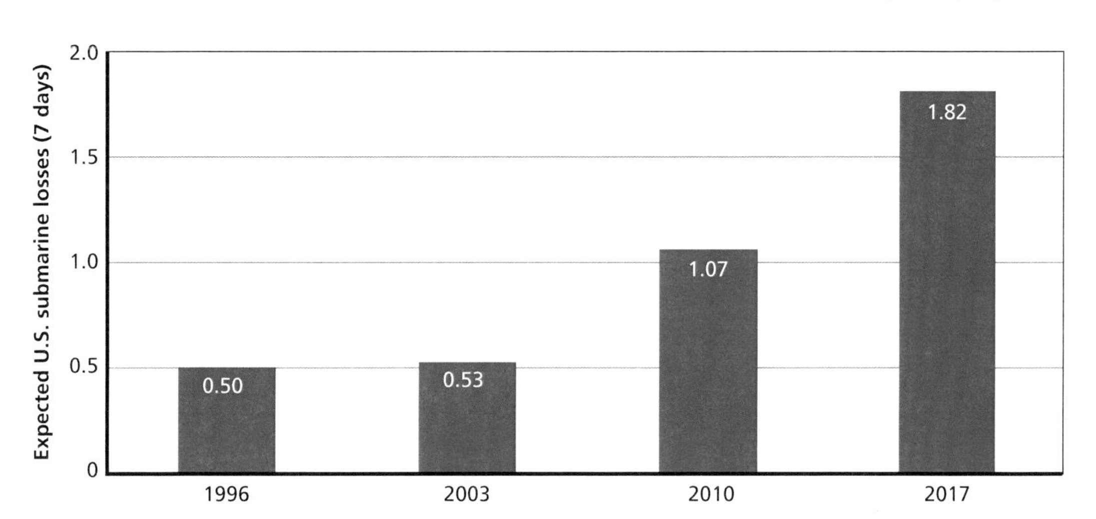

NOTE: The figure assumes three submarine rotations (a total of six U.S. submarines) in 1996 and 2003 and four rotations (a total of eight U.S. submarines) in 2010 and 2017.
RAND *RR392-8.6*

The impact of submarine attacks, especially in the 2010 and 2017 cases, would depend greatly on the course of the ground battle on shore. In the 2017 case, U.S. submarines reduce the PLA invasion force delivered to Taiwan by 22 percent, significantly less than the 70-percent reduction achieved in 1996. Nevertheless the losses would be terrible. At the end of the first week, 40 percent of the Chinese amphibious fleet would have been sunk (and perhaps as many as 5,000 personnel lost at sea if even half of these transports were loaded when sunk). If Chinese ground forces had failed to capture a port intact by that time, the PLA's ability to continue to supply its forces on shore would become increasingly tenuous, as it would be more difficult to deliver supplies to an expanding force. U.S. submarines, for their part, would be in a position to prosecute the battle with the same intensity as they did during the first week of combat. If, however, a port had been captured intact by the end of the first week, then the PLA could begin to ship supplies using regular merchant vessels.

Figure 8.6 shows expected U.S. submarine losses in campaigns conducted during each of the four snapshot years. For 1996 and 2003, U.S. expected average losses are approximately 0.5 submarines out of a total of six engaged. In 2010 and 2017, the United States loses 1.12 and 1.82 subs (out of a total of eight engaged), respectively. These losses are by no means operationally crippling, given that the United States had 57 tactical submarines as of 2010, with roughly 60 percent home-ported in the Pacific. From a moral and symbolic perspective, however, the stakes are high. Each submarine carries a crew in excess of 100, and the loss of a submarine—perhaps second only to aircraft carriers as symbols of national power—would be a severe blow to the United

States. Presumably, however, the United States would not enter a war with China over Taiwan unless it had decided that the stakes were worth the significant cost.

Alternative tactics to those outlined here might reduce the likelihood or magnitude of U.S. losses, but that could carry other disadvantages. Engaging PLA amphibious ships using only sonar, for example, would eliminate the risk of periscope detection. However, the Chinese would likely deploy a large number of vessels with the amphibious column to serve as decoy ships.[25] Engaging ships without confirming their type or identity would increase the probability of engaging false targets, decreasing the effective Pk against the amphibious column. Moreover, while the submarines' overall chance of survival would increase, submarines would still face some threat from air-dropped torpedoes after attacking the amphibious column.

Nevertheless, our analysis of this alternative concept of operation—engaging without raising the periscope—suggests that if at least half of the U.S. torpedoes launched, struck, and sank an amphibious ship (weapon Pk = 0.5), it would improve the results for the 2010 and 2017 cases (though not for 1996 or 2003). In the 2017 case, the total number of PLA divisions that would reach Taiwan via amphibious ship would fall from 10.5 (using periscope identification) to 8.2 (using sonar only)—a reduction of 22 percent.[26] Moreover, U.S. forces would lose fewer submarines: roughly 0.6 lost out of eight submarines engaged in both 2010 and 2017. If, however, the effective weapon Pk equals 0.25 (with only one in four torpedoes sinking an amphibious ship), then this tactic would be less effective in sinking Chinese amphibious ships than the baseline method (periscope engagement), though U.S. forces would still incur fewer losses.

The submarine analysis presented in this section is, for the most part, independent of the other scorecards discussed in this report. Submarines operate on their own, with little assistance from other assets.[27] They provide their own access and onboard ISR/targeting capabilities and do not necessarily require nearby, potentially vulnerable basing support. Their only limitations in this particularly stressing scenario are weapon throughput and, to a lesser extent, platform survivability.[28] However, Chinese anti-submarine warfare effectiveness would be largely dependent on the degree of local air superiority achieved in the area, as that effort relies largely on slow-moving airborne assets.

---

[25] These decoy ships would also serve as weapon soaks for U.S. air attacks.

[26] In the 2010 case, the number of PLA divisions successfully transported fell modestly, from 7.9 to 7.6.

[27] It is true, however, that if U.S. airpower could deny or inconvenience PLA helicopters over the amphibious fleet, U.S. submarines would face fewer risks in prosecuting attacks.

[28] Submarines are naturally limited in both their weapon loadout capacity and the rate at which they can engage surface vessels. Survivability is an issue for all platforms. However, the loss of even one nuclear submarine could have a substantial impact on U.S. military and public mindsets.

U.S. Air Strikes Versus PLA Amphibious Invasion Fleet

Air strikes can complement submarine operations against amphibious forces by delivering a large number of weapons in a relatively short period. However, in the face of a credible air defense threat, airborne strike assets may require support from other assets, including air-to-air fighters, AWACS, and SEAD platforms. Moreover, if forced back beyond the strait, U.S. aircraft may need to rely on external targeting from theater and national ISR assets. And, finally, many of the air assets (strike and support) will have to be based within theater to generate the sorties necessary to deliver large quantities of weapons when needed. Such basing may not be available or secure from attack.[29]

The Chinese amphibious fleet has nearly doubled in size since 1996 and the air defenses of its escorts have improved dramatically. The anti-surface capabilities of U.S. carrier- and ground-based aircraft have not improved as rapidly. Nevertheless, U.S. aircraft could contribute significantly to anti-surface operations, and new weapons may reinforce that capability in the future.

Over the past 15 years, China has built a system of multilayered air defenses that are capable of reaching out to the Taiwanese shore and beyond, providing a protective shield for its amphibious fleet that may be difficult to penetrate. Table 8.3 documents the evolution of Chinese destroyers and frigates and their air defenses. It shows the increasing size of the ships in each successive class: PLAN frigates today are larger than the Chinese destroyers of the 1990s. It also highlights the increasing range of PLAN air defense systems. Over the past 15 years, the range of the SAM systems deployed on the newest classes of destroyers has increased from 15 km (for the HQ-7) to 150 km (for the SA-N-20). *Jiangwei I* or *Jiangkai I/II* frigates, deploying shorter-range but nevertheless modern SAM systems, could act as escorts, operating within and around the amphibious column to defend against aircraft and anti-ship cruise missiles.

Many of the more recent classes of PLA amphibious ships are equipped with close-in weapon systems, providing fleet point defense. For longer-range defenses, the PLAAF and PLAN have a growing inventory of fourth-generation fighters (discussed in Chapters Four and Seven, scorecards 2 and 5) and advanced land-based, double-digit SAMs (discussed in Chapters Five and Six, scorecards 3 and 4). Advanced fighters could wreak havoc on U.S. strike packages unless the latter are defended by their own contingent of air-to-air fighters. China has also acquired a small number of AEW platforms that could, from secure "racetracks" over the mainland, coordinate and direct fighters against inbound strike packages.

In addition to putting U.S. strike missions at risk, China's more robust air defense capability will make it difficult to operate some types of maritime ISR assets in support of the anti-surface warfare mission. This is especially true in the case of dedicated MPAs, like the P-3C and P-8A, but could also apply to standoff platforms like Global

---

[29] See Chapter Four (scorecard 2) for a discussion of the air-to-air competition and Chapter Three (scorecard 1) for a discussion of U.S. theater basing.

Table 8.3
Improvements to PLAN Destroyer and Frigate SAM Capabilities

Ship Class	IOC	Tons	1996	2003	2010	2015 (current)	2017	SAM	Range (km)	Missile Launch/Loadout
Destroyers										
Luda (Type 051)	1971	3,250	16	14	12	2	1	N/A		
Luda II/III (Type 051DT/G)	1987	3,250	1	2	3	4	4	HQ-7	15	8 CSA-N-4s
Luhu (Type 052A)	1994	4,600	1	2	2	2	2	HQ-7	15	8 CSA-N-4s; 32 missiles
Sovremenny (Type 956/956EM)	1999	7,940		2	4	4	4	SA-N-7	30	2 9M38M2s; 44 missiles
Luhai (Type 051B)	1999	6,000		1	1	1	1	HQ-7	15	8 CSA-N-4s
Luyang I (Type 052B)	2004	7,000			2	2	2	SA-N-12	60	2 9M38M2s; 48 missiles
Luyang II (Type 052C)	2004	7,000			2	5	6	HHQ-9	120	8 HQ-9s; 48 missiles
Luzhou (Type 051C)	2006	7,000			2	2	2	SA-N-20	150	6 48N6s; 48 missiles
Luyang III (Type 052D)	2013	N/A				1	6	HHQ-9	120	8 HQ-9B; 64 missiles
Frigates										
Chengdu (Type 053)			2					N/A		
Jianghu I (Type 053H)	1968	1,702	33	37	30	16	10	N/A		
Jiangwei I (Type 053H2G)	1991	2,250	4	4	4	4	4	HQ-61	10	6 CSA-N-4s
Jiangwei II (Type 053H3)	1998	2,250		8	10	10	10	HQ-7	15	8 CSA-N-4s
Jiangkai I (Type 054)	2005	3,900			2	2	2	HQ-7	15	8 CSA-N-4s
Jiangkai II (Type 054A)	2007	3,900			6	18	21	HHQ-16	38	32 HQ-16s; 32 missiles

SOURCES: Data on SAMs associated with each ship model are from the respective ship entries in *Jane's Fighting Ships*. Data on SAM capabilities are from *Jane's Weapons: Naval*, "HHQ-7/FM-80N(CSA-N-4) and HHQ-7A/FM-90N," April 27, 2015; *Jane's Strategic Weapons Systems*, "Urugan (SA-N-7 'Gadfly')," July 23, 2013; *Jane's Strategic Weapons Systems*, "Smerch/Shtil-1/-2 (SA-N-7B/C or SA-N-12 'Grizzly')," July 23, 2013; *Jane's Strategic Weapons Systems*, "HQ-9/-15 and HHQ-9 (RF-9/-15, FD-2000 and FT-2000)," August 19, 2013; *Jane's Weapons: Naval*, "SA-N-6 'Grumble' (V601 Fort/Rif)/SA-N-20 'Gargoyle' (Fort-M/Rif-M)," March 20, 2012; *Jane's Strategic Weapons Systems*, "HQ-6/HHQ-6 (RF6, SD-1 and CSA-N-2)," August 19, 2013; *Jane's Strategic Weapons Systems*, "HQ-16/-17 (HHQ-16/-17 and MD-2000)," August 19, 2013.

Hawk and U-2. One option for U.S. air forces is to strike the Chinese amphibious ships with short-range, line-of-sight weapon systems—an approach that mitigates the need for cueing from external ISR assets.[30] However, fighting through China's multilayered air defenses could be difficult and costly. U.S. forces would likely have to suppress the most capable of these defensive systems before such attacks would be viable. While this might be possible, it would slow the commencement of strikes against the amphibious fleet during the early crucial phases of an invasion.

A second option is to strike with long-range ASCMs, which would allow launch platforms to remain outside the Chinese naval and land-based SAM envelopes. The development of U.S. ASCMs lagged for a number of years after the end of the Cold War, leaving the United States with an aging and (probably) smaller force than it once enjoyed. Over the last several years, however, the development of such missiles has been reinvigorated. In 1990, the United States had roughly 4,000 Harpoon (AGM-84) missiles in its inventory.[31] The current number in service is unclear, but it may be significantly smaller.[32] Portions of the Harpoon inventory have been upgraded over the years, and Boeing's most recent proposed upgrade (dubbed "Harpoon Next Generation") would increase the Harpoon Block II's range from 130 km to 240 km.[33]

With the cancellation of other proposed replacements, the SLAM (AGM-84E) and, later, the extended-range version (the SLAM-ER, or AGM84-H) became the Harpoon's primary successor. Unlike the Harpoon, SLAMs were originally manufactured as "man-in-the-loop" systems, useful for avoiding collateral damage but less well suited to a high-threat environment where the imperative is to launch large numbers of systems and get out of harm's way.[34] This limitation has since been remedied with the addition of an automatic target recognition unit, and the SLAM-ER provides greater

---

[30] Long-range, offboard ISR may still be required to decipher feints from the actual amphibious columns. However, its requirement to identify and precisely locate targets is reduced in this context.

[31] Norman Friedman, *The Naval Institute Guide to World Naval Weapons Systems, 1991/92*, Annapolis, Md.: Naval Institute Press, 1991, p. 188. More than 7,500 have been produced to date, including export models.

[32] Portions of the inventory have been upgraded over the years. For example, 2,000 missiles were upgraded to AGM-84D Block IG, which integrated new radar modes, after 1995. It also appears that some older models have been retired. A proposed "Block III" upgrade to 850 missiles was designed to return 850 systems to the inventory. The upgrade was subsequently canceled. See *Jane's Air-Launched Weapons*, "AGM-84 Harpoon," March 25, 2015, and Boeing, "Boeing Awarded Contract for Next-Generation Harpoon Block III Missile," press release, January 31, 2008.

[33] *Jane's 360*, "Navy League 2015: Boeing Developing Kit to Upgrade Harpoon Missiles for Extended Range," April 15, 2015.

[34] On the missile's original lack of "fire-and-forget" capability, see U.S. General Accounting Office, *Precision-Guided Munitions: Acquisition Plans for the Joint Air-to-Surface Standoff Missile*, Washington, D.C., GAO/NSIAD-96-144, June 28, 1996.

range (270 km) and more targeting options than the Harpoon.[35] Roughly 1,000–1,500 SLAM-ERs have been procured to date.[36]

The U.S. Navy is also pursuing a long-range anti-ship missile (LRASM) program that is scheduled to reach early operational capability in 2018. Although not much data are publicly available about specific parameters, the missile will have significant common elements with the JASSM-ER and will have a significantly longer range than the SLAM-ER. It will have autonomous targeting capabilities, be optimized for anti-surface warfare, and be resistant to GPS and other forms of jamming.[37]

As Figure 8.7 shows, we estimate the number of ASCMs needed to inflict damage to 30 or 50 percent of the Chinese amphibious fleet in each of our snapshot years. The figure uses the number of amphibious ships for each year listed in Table 8.1. In addition, we assume an equal number of "decoy ships" that absorb a portion of the missiles fired. We further assume that the fleet is protected, in each case, by ten warships.

Figure 8.7
Model Results: ASCM Requirements to Damage 30 or 50 Percent of the Chinese Amphibious Fleet

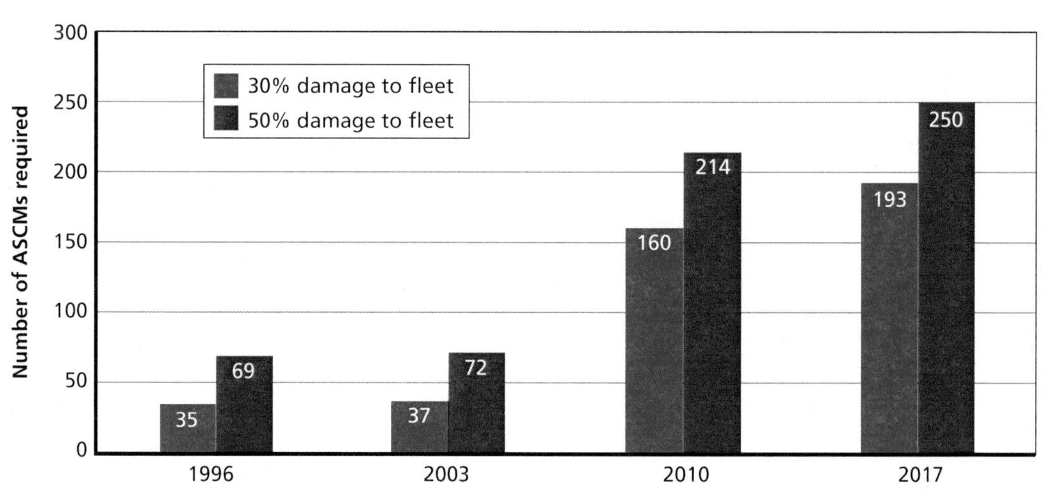

RAND *RR392-8.7*

---

[35] *Jane's Air-Launched Weapons*, "AGM-84 Harpoon," March 25, 2015; *Jane's Air-Launched Weapons*, "AGM-84E SLAM, AGM-84H/K SLAM-ER," October 22, 2014.

[36] An October 2014 Jane's report puts the number of SLAM missiles delivered at 600 and the number of SLAM-ERs at 700. All SLAMs will be converted to SLAM-ERs. See *Jane's Air-Launched Weapons*, "AGM-84E SLAM, AGM-84H/K SLAM-ER," October 22, 2014.

[37] *Jane's International Defense Review*, "Back Into the Blue: LRASM Honed for Extended Reach, Precision Punch," September 10, 2014; *Jane's Air-Launched Weapons*, "AGM-158A JASSM, AGM-158B JASSM-ER and LRASM," May 26, 2015; and "LRASM Missiles: Reaching for a Long-Range Punch," *Defense Industrial Daily*, June 3, 2015.

(These warships include the best-available frigates in each year, as well as somewhat older destroyers.)[38] The number of weapons needed to inflict either a 30- or 50-percent damage expectancy against the Chinese fleet has increased dramatically with both the increase in the number of potential ships to engage increase and the ships' improved protection from attack.

The air attack results can be viewed from several perspectives. On the one hand, the required number of missiles remains well below the total U.S. inventory of ASCMs (which probably totals several thousand). On the other hand, delivering these missiles would be problematic in the context of improved Chinese air defenses and tactical combat aircraft, though it may become somewhat easier after the LRASM enters the inventory. Despite the challenges, it is clear that U.S. aircraft could deliver a powerful punch in conjunction with other forms of attack on an amphibious fleet.

U.S. Surface Strikes Versus PLA Amphibious Invasion Fleet

As in the case of air-launched ASCMs, the development of surface-launched ASCMs lagged for a number of years after the end of the Cold War. U.S. surface shooters have been carrying Harpoon missiles (RGM-84s) since 1977. However, the U.S. Navy received its last Harpoon in 1992, and it has since undertaken only relatively modest upgrades to the system.[39] The Harpoon lacks the range that would enable surface ships to attack an amphibious fleet without significant risk to itself. In large measure, the failure to undertake more substantial upgrades or to replace the Harpoon was a function of the U.S. surface navy's post–Cold War shift in focus toward land attack. For this purpose, it primarily employs the TLAM, a long-range (> 1,000-nm), GPS-guided, subsonic cruise missile that flies to a predesignated coordinate.[40]

One option that would obviate the need to bring U.S. surface ships into or close to the strait (as a Harpoon attack would require) is to engage Chinese amphibious ships with TLAMs once they have anchored off the Taiwanese coast. However, unlike Harpoon missiles, TLAMs require precise targeting from external sources prior to arrival in the target area. Depending on the specific location of the anchorage and the time spent unloading, this support might be provided by individuals on the ground.

---

[38] Until recently, China had only a handful of destroyers with long-range SAMs. We assume that these systems would have formed the backbone of two surface action groups providing air defense to the north and south of the Taiwan Strait. In 1996, we assume that four *Luda-*, two *Jianghu-*, and four *Jiangwei-*class ships would have been employed as escorts. In 2010, we assume two *Luhu-*, two *Luyang I-*, two *Jiangwei II–*, and four *Jiangkai II–*class ships. In 2017, we assume two *Luhu-*, two *Luyang I–*, and six *Jiangkai II–*class ships.

[39] The Block II introduced land attack capability and some improvements to guidance, but the Block III, which would have increased range, better target discrimination, and anti-spoofing capability was canceled in 2009. *Jane's Strategic Weapon Systems*, "AGM/RGM/UGM-84 Harpoon/SLAM/SLAM-ER," July 14, 2014.

[40] The TLAM has a range in excess of 1,000 nm and can also guide to the target using terrain contour matching and digital scene-matching area correlations.

Alternatively, U.S. military assets might provide the required ISR. Both the U.S. Navy's MQ-4C Triton (expected IOC 2017) and the U.S. Air Force's RQ-4 Global Hawk are high-flying, long-endurance UAVs capable of providing high-resolution SAR images at ranges in excess of 100 nm. The MQ-4C also has an inverse SAR imaging capability, enabling it to identify ships and other objects. These assets would have to operate in an increasingly hostile environment, within range of Chinese fighters and, potentially, long-range naval SAMs. Nevertheless, using land- or carrier-based air defense to protect two or three high-flying UAVs would be less demanding than, for example, escorting multiple strike packages over the Chinese mainland. Figure 8.8 depicts this particular concept of operations, including potential ISR orbit locations and TLAM launch positions at 100 nm and 250 nm, respectively, from a PLA landing area on the northern side of Taiwan.[41]

Figure 8.8
Schematic of ISR Orbit and Launch Range for a TLAM Strike on a Hypothetical Landing Area

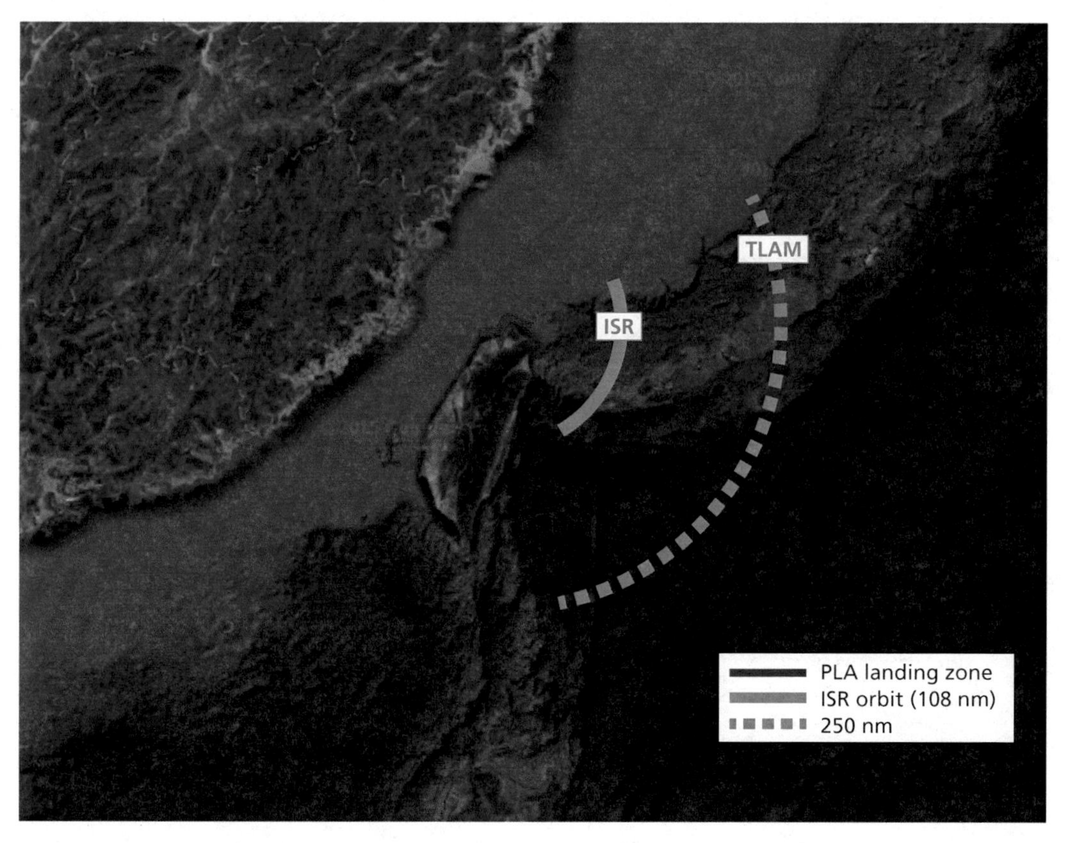

SOURCE: Google Earth with author overlay.
RAND *RR392-8.8*

---

[41] Flying at 500 knots, the Tomahawk missile can travel 250 nm in approximately 30 minutes.

Multiple current and future U.S. Navy ship classes can fire Tomahawk missiles, including *Ticonderoga*-class cruisers (CG-47); *Arleigh Burke*– (DDG-51) and *Zumwalt*-class (DDG-1000) destroyers; *Los Angeles*– (SSN-688), *Virginia*- (SSN-774), and *Seawolf*-class (SSN-21) attack submarines; and the converted *Ohio*-class (SSGN-726) cruise missile submarines. The SSGN-726 and SSN-21 submarines and the DDG-1000 destroyer have the largest TLAM capacities, with maximum loadouts of 154, 50, and 80 cruise missiles, respectively.[42] All three have varying degrees of stealth, allowing them to operate at closer ranges than traditional cruisers and destroyers.[43] Also, unlike early variants, current TLAMs can fly at higher altitudes, greatly simplifying and expediting the mission planning process. In fact, missions can be mostly planned beforehand and quickly completed and launched once a target coordinate has been derived.

Assuming persistent ISR and two or three large TLAM shooters within relatively close range, U.S. forces may be able to rain Tomahawk missiles down on the Chinese amphibious fleet as it anchors off the Taiwanese shore. The Navy has recently tested an even more demanding variant of this concept, firing a TLAM Block IV from a guided missile destroyer and employing an F/A-18 to guide the missile to a *moving* maritime target.[44] With more rudimentary ISR, these ships could salvo missiles into a landing area or lodgment, greatly complicating the attacker's task at a critical point in the battle. In the longer term (perhaps as early as 2018), the LRASM, which will have a surface-launched variant as well as an air-launched one, will provide the capability to attack Chinese amphibious forces at sea or at the beachhead from roughly 1,000 km and with less precise ISR.

Having discussed U.S. submarine, air-launched, and naval surface warfare capabilities in the context of a Taiwan invasion scenario we turn to a discussion of these capabilities in the Spratly Islands scenario.

---

[42] TLAMs compete for space with other weapons, so these boats and ships—especially the SSN-21 and DDG-1000—may not carry the maximum TLAM loadout. The SSN-21 can carry a mix of 50 TLAMs and Mk 48 advanced-capability torpedoes, or up to 100 mines. The DDG-1000 will have 80 vertical missile launchers, which can be loaded with TLAMs, Evolved SeaSparrow Missiles, or Standard-type air defense missiles. See *Jane's Underwater Warfare Systems*, "Seawolf Class," March 24, 2015, and Donna Lyons, "Construction Contract Awarded for USS Michael Monsoor (DDG 1001)," *Defense Media Network*, July 28, 2011. On the SSGN-726, see *Jane's Navy International*, "Striker Beneath the Sea," March 18, 2003.

[43] It should be noted, however, that firing off large salvos of Tomahawk missiles may jeopardize the launchers' positions, requiring surface or airborne platforms to defend them from anti-submarine or anti-surface attack. In addition, the U.S. Navy has or has planned for only limited numbers of these ship classes (four SSGNs, three SSN-21s, and three DDG-1000s).

[44] U.S. Naval Institute News, "Video: Tomahawk Strike Missile Punches Hole Through Moving Maritime Target," February 9, 2015.

Spratly Islands Scenario

Distance from the Chinese mainland and local geography sharply differentiate the Spratly Islands scenario from the Taiwan scenario. Even by 2017, the United States should have little difficulty denying a successful amphibious landing in the Spratly Islands (assuming that hostilities have commenced prior to the landing) or preventing resupply to forces already located there. In this section, we briefly discuss the impact of those differences, focusing primarily on the ability of U.S. submarines and air forces to locate, target, and sink PLAN amphibious ships and combat ships in the South China Sea.

The most versatile and, arguably, most effective anti-surface warfare platform in the U.S. inventory is the attack submarine, with its stealth, large-aperture sonar, and heavyweight torpedoes. However, geography and the size of the amphibious fleet limit the effectiveness of this platform in the Taiwan Strait. The shallow and constrained waters of the strait restrict the number of submarines that can operate against an amphibious force. The large number of ships that would be involved in an invasion would overwhelm the magazine size of U.S. submarines. Finally, many of China's most capable anti-submarine platforms are land-based aircraft that would be within easy range of the patrol areas, though they would potentially be at high risk in contested airspace.

In the Spratly Islands scenario, these challenges are either nonexistent or dramatically reduced. Whereas the danger of blue-on-blue engagements would severely limit the number of U.S. boats that could operate in the Taiwan Strait—we postulated two submarines at a time in our modeling—there is no such limitation in the Spratly Islands scenario. All U.S. submarines assigned to the operation can participate simultaneously, with each assigned to patrol a distinct geographic area. It is also possible that U.S. boats could detect and trail any Chinese ships that might sortie before the commencement of hostilities. Finally, the hydrography of the South China Sea is also conducive to submarine anti-surface warfare operations. The water around the Spratly Islands is deep, permitting submarine acoustic detection and tracking of surface ships at long ranges. The converse is also true, with Chinese escort ships better able to detect U.S. submarines.

The two scenarios also differ in terms of the number of amphibious ships that would likely be mobilized. A Taiwan invasion would involve the entire amphibious fleet, plus decoys, escort ships, and possibly elements of the merchant marine. While this would make for a target-rich environment, it would also limit the extent of damage that submarines could inflict relative to the force as a whole. With a limited number of submarines operating in the strait, each limited to 38 torpedoes for *Virginia*-class submarines or 26 torpedoes for the *Los Angeles* class, the large number of Chinese ships in the fleet could overwhelm the submarines' ability to destroy them. And although the scenario is target-rich, congested waters might also raise the risk of visual detection, slowing target prosecution. Given the small size of the South China Sea islands and their distance from the mainland, the Chinese fleet operating there would be much smaller, removing the U.S. submarine magazine size from consideration. While it may take longer to locate Chinese surface ships, unless U.S. submarines are in trail at the beginning of the conflict, the combined weapon inventory of the U.S. submarines is more than sufficient to threaten the entire Chinese fleet.

In the Taiwan Strait scenario modeling presented earlier, U.S. submarine losses increased from an expected average value of 0.5 submarines in the 1996 case to nearly two in 2017. These submarine losses are primarily the result of Chinese air-based anti-submarine forces operating in the strait. Operations in the South China Sea will restrict China's ability to provide these air-based forces. Not only are Chinese air bases farther from the conflict area, but the air balance would be more challenging for the PLA, making operations by land-based MPAs hazardous at any distance from the coast. Moreover, unlike the tightly packed amphibious operations in the Taiwan scenario, in which ship-based anti-submarine helicopters can provide overlapping coverage, South China Sea operations will involve fewer numbers of discrete Chinese surface groups. The air-based anti-submarine warfare threat to the U.S. submarine fleet, while not overwhelming in the Taiwan Strait, is dramatically reduced in the South China Sea.

Like the submarine campaign, U.S. air-based anti-surface warfare operations also face fewer challenges in the Spratly Islands scenario, and the greater distance from the mainland is again a key driver. In contrast to the Taiwan scenario, in which SAM coverage extends from the mainland coast over much of the battle area, Chinese long-range SAM coverage over the Spratly Islands is limited to that carried by its surface ships. As late as 2010, naval versions of the so-called double-digit SAMs were carried on only four of China's destroyers. However, that is changing rapidly with the introduction of new ships. By 2017, it is likely that China will have a dozen ships with such armament (including the *Luzhou*, *Luyang II*, and *Luyang III* classes). Nevertheless, for the immediate future, only a handful of these ships would likely be present in China's South China Sea task forces. China may deploy tactical air defenses to its newly created islands in the South China Sea, but the lack of concealment and space to maneuver makes it unlikely that they would put high-value air defense assets on them.

China's other anti-access capabilities also have less effect at these longer ranges. Outside the ASBM threat ring, U.S. airpower operating off carriers can maintain a larger presence over the Spratly Islands. The same is true for U.S. Air Force aircraft operating outside the Chinese TBM range. A similar pattern holds in the case of ISR. China's ability to deny U.S. ISR assets visibility into the Taiwan Strait vastly complicates the targeting and destruction of the Chinese amphibious fleet. The Chinese are far less capable of denying U.S. air-breathing ISR assets from operating in the Spratly Islands area. Taken together, the distance of the Spratly Islands from the Chinese mainland changes the nature and magnitude of the threat to U.S. military operations.

U.S. anti-surface warfare efforts in the Spratly Islands scenario probably would have met with great success throughout the period considered. In the longer future, further developments in China's destroyer forces, aerial refueling capabilities, AWACs, anti-submarine warfare forces, and submarine forces (among others) may increase the degree of difficulty confronting U.S. forces in the South China Sea, but the U.S. ability to prevail will likely remain stronger here than in the Taiwan case.

Conclusions

Since 1996, China's amphibious fleet and supporting elements have grown in size and sophistication. Both the number of ships in the fleet and their collective transport capacity have increased significantly since 1996 (with carrying capacity more than doubled), and the fleet is adding modern ships capable of delivering troops more quickly and safely onto enemy occupied beaches. China has also introduced a range of modern fighters, support aircraft, SAMs, and, to a lesser extent, anti-submarine warfare assets—all of which serve to complicate the U.S. anti-surface warfare task. Nevertheless, despite these challenges, U.S. anti-surface warfare capabilities against Chinese amphibious forces remain relatively robust.

Although the analysis highlights the particular difficulties of the Spratly case for China, with that country's fleet vulnerable to air, surface, and submarine attack during a much longer transit, it also underlines the uncertainties and dangers for China in a Taiwan conflict. Even in the 2017 Taiwan case, U.S. anti-surface warfare could sink Chinese ships carrying thousands of Chinese soldiers and sailors, as well as tons of vital supplies.[45] Amphibious invasions of this scope are extremely complex and difficult, especially for a military with limited experience, and the loss of cohesion and order in the attacking force could pose as many challenges to Chinese commanders as the material damage itself. Overall, the results on this scorecard are better for the United States than is the case for other key scorecards.

Scorecard Coding

Figure 8.9 provides our summary coding of the results of scorecard 6. Advantage in this scorecard is based on the ability of U.S. forces to destroy or disrupt a Chinese amphibious landing force sufficiently to jeopardize its success (U.S. advantage) and the ability of a Chinese amphibious fleet to land an invasion fleet with modest losses

[45] Even if only half of the amphibious ships were loaded when sunk, roughly 5,000–6,000 could be lost at sea. Conceivably, many could be rescued, but torpedo attacks often result in rapid sinking with very high loss of life, and the PLA would be unlikely to stage extensive rescue operations in the midst of a high-intensity, fast-paced operation.

Figure 8.9
Scorecard 6 Summary Coding

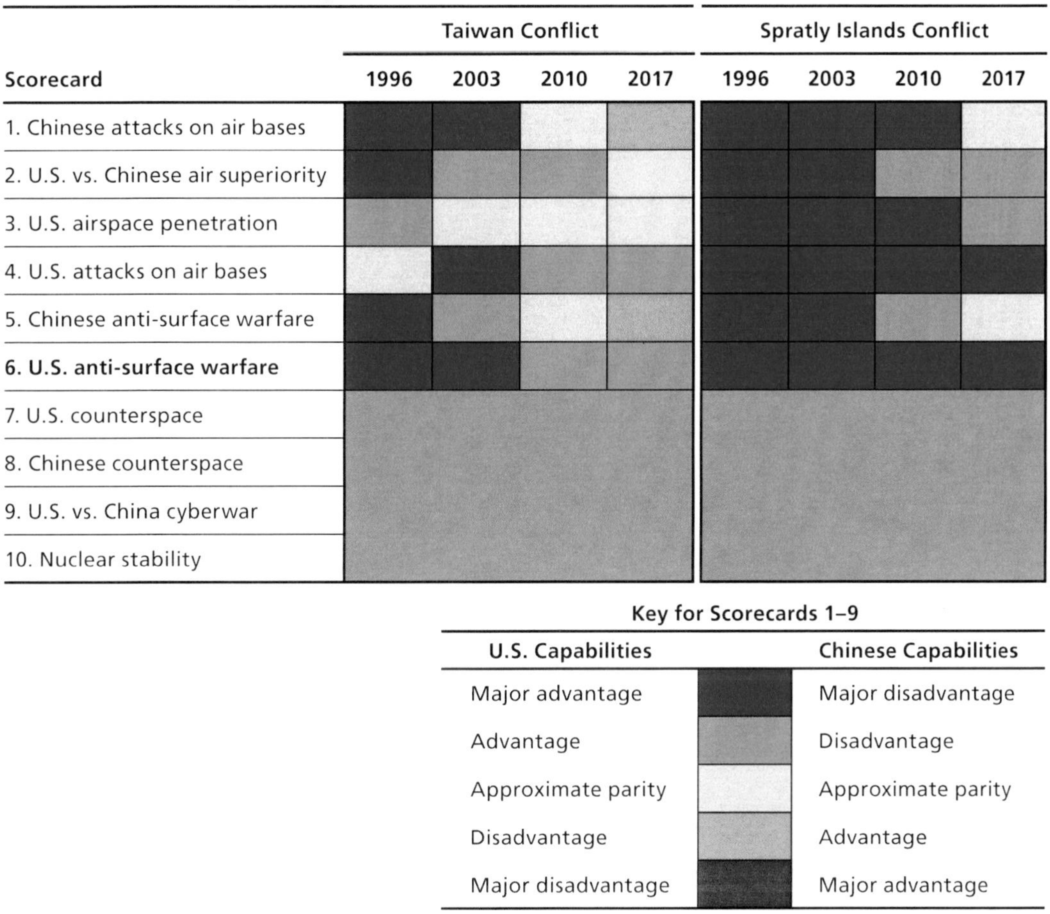

	Taiwan Conflict				Spratly Islands Conflict			
Scorecard	**1996**	**2003**	**2010**	**2017**	**1996**	**2003**	**2010**	**2017**
1. Chinese attacks on air bases								
2. U.S. vs. Chinese air superiority								
3. U.S. airspace penetration								
4. U.S. attacks on air bases								
5. Chinese anti-surface warfare								
6. U.S. anti-surface warfare								
7. U.S. counterspace								
8. Chinese counterspace								
9. U.S. vs. China cyberwar								
10. Nuclear stability								

Key for Scorecards 1–9

U.S. Capabilities		Chinese Capabilities
Major advantage		Major disadvantage
Advantage		Disadvantage
Approximate parity		Approximate parity
Disadvantage		Advantage
Major disadvantage		Major advantage

at sea and to sustain that force during the first weeks of ground operations (Chinese advantage).

Relative Chinese capabilities improved in virtually all component parts of the scorecard, especially after 2003. Nevertheless, U.S. air, surface, and, especially, submarine attacks remained capable across the snapshot years of doing enough damage to the Chinese fleet to jeopardize an amphibious operation, leading us to code this scorecard as *U.S. advantage* in both scenarios across all periods. To be sure, the degree of *U.S. advantage* by 2017, especially in the Taiwan case, would be significantly lower than it would have been in the 1996 case. Nevertheless, even in 2017, a combination of air, submarine, and, possibly, surface attack against the fleet in transit and at the landing beaches would likely take a severe toll on Chinese landing elements. U.S. advantage in the Spratly Islands case would be more robust.[46]

---

[46] Chinese forces operating around the Spratly Islands would be more vulnerable to attacks by U.S. air, submarine, and surface assets. Coordination among the three could produce synergies, with, for example, submarines destroying PLAN surface action groups providing air cover for the amphibious fleet. Air assets could facilitate submarine operations by attacking China's long-range MPAs, such as the Y-8X, should those assets venture deep into the South China Sea.

CHAPTER NINE

Scorecard 7: U.S. Counterspace Capabilities Versus Chinese Space Systems

Space and counterspace operations would be important elements in any armed confrontation between the United States and China. The transformational warfighting capabilities that U.S. military forces have developed since the end of the Cold War are largely enabled by satellite support, and space-based ISR and communication connectivity would be especially important in the broad expanses of the Western Pacific theater. U.S. military leaders are aware that other countries also use satellites to enhance their military capabilities. Therefore, the United States is developing systems to interfere with some aspects of future opponents' space support operations.

This chapter assesses the risks that U.S. counterspace capabilities pose to Chinese space assets and functions. To provide context for that assessment, it first reviews how many military and nonmilitary satellites China and the United States have put into orbit each year from 1996 to 2010. It identifies trends in those efforts, compares the size and composition of each country's current orbital infrastructure, and briefly discusses the contribution that each side's satellite capabilities might play in the two scenarios addressed elsewhere in this report. The chapter then examines the U.S. capability to conduct counterspace operations and estimates the risks that capability presents to the Chinese space-force enhancement systems in six key areas: imagery; SIGINT; ocean surveillance; communication; position, navigation, and timing (PNT); and weather.

Given the limitations of open-source inputs and the broad scope of our study, the assessment of risk to space functions is not based on quantitative modeling. (The same is also true, for the same reasons, of the next two chapters, on Chinese counterspace capabilities and cyber warfare.) Rather, the evaluation is based on the nature of the satellite constellations in which the six space-based functions mentioned here are embedded. (We also considered orbital characteristics, the number of satellites deployed, and the ownership of the satellites in question.) Finally, we examined the nature and magnitude of the U.S. offensive capabilities that are most relevant to operations against various satellite constellations. In both this chapter and the next, we consider dual-use satellite systems and counterspace capabilities (e.g., experimental systems that could be employed in offensive operations), along with efforts by both sides to make their satellites or satellite constellations more robust.

Ultimately, the evaluations in this chapter are somewhat more subjective than the preceding scorecard analyses. However, we hope that by disaggregating each side's space capabilities and evaluating the risks to different components, the assessment will contribute to the current understanding of space and counterspace dynamics.

U.S. and Chinese Orbital Infrastructures

While the United States and China both use satellites to enhance the capabilities of their terrestrial military forces, they depend on space support to different degrees. The United States gets more support from space systems than does China. China, fighting near its own territory, could rely more on terrestrial-based assets for these functions, particularly communication. This asymmetry in dependence would be especially pronounced in a conflict over Taiwan, just 100 miles from China's shore, but it would be somewhat less significant in a conflict in the South China Sea, where PLA forces would have to operate farther from the Asian mainland. There, the PLAAF and PLAN would likely rely more heavily on satellite communication to provide command and control for their forces.[1]

As a consequence of these differences, as well as the developmental history of the two countries, orbital infrastructure of the United States is far more developed than that of China. According to the Union of Concerned Scientists, as of January 31, 2015, the United States had 526 operational satellites in orbit while China had 132—and such figures may not fully capture the U.S. quantitative advantage, given that U.S. entities are, on balance, more transparent about which satellites may no longer be functioning.[2] China is, however, developing its space assets at a pace and scope that suggests a desire on the part of Beijing to become a major space power. Table 9.1 provides the total number of satellites put into orbit by year, broken down over three time intervals between 1997 and 2014.

As Table 9.1 illustrates, the United States has consistently put more satellites into orbit than has China since 1996. Nevertheless, China is making a concerted effort to increase its space-based ISR, navigation, and communication capabilities, as evidenced

---

[1] The range of effective radio communication in the very-high- and ultra-high-frequency (VHF and UHF) bands is the distance to the radio horizon, which is just beyond the visual horizon. Depending on atmospheric conditions and sea state, some refraction, ducting, or tropospheric scattering may occur when radio waves are transmitted over water, increasing signal ranges somewhat. However, these phenomena are too intermittent for reliable communication. The range in the radio horizon varies with the height of the transmitting antenna. With an antenna 100 feet tall, the radio horizon is 14 miles away. On the other hand, if one transmits from an aircraft flying at 30,000 feet or from the ground to an aircraft at that altitude, the radio horizon is 245 miles away. See Chow Yen Desmond Sim, *The Propagation of VHF and UHF Radio Waves Over Sea Paths*, thesis, Leicester, UK: Leicester University, November 2002.

[2] Union of Concerned Scientists, UCS Satellite Database (official names only), data current through January 31, 2015.

Table 9.1
U.S. and Chinese Satellites Placed into Orbit, 1997–2014

Country	1997–2002	2003–2008	2009–2014
United States	349	142	253
China	33	54	111
Ratio	10.6:1	2.6:1	2.3:1

SOURCES: Satellitedebris.net, "Satellites by Country, China"; data from Space-Track.org. Other sources that track satellite launches include Jonathan McDowall, Jonathan's Space Report, and Gunter Kirk Krebs, "Spacecraft by Country," *Gunter's Space Page*.

by a more than threefold increase in launches between the periods 1997–2002 and 2009–2014. While the pace of U.S. military and nonmilitary launches has fluctuated since the late 1990s, China has significantly increased the tempo of its satellite launches. Table 9.2 describes in general terms what missions these Chinese satellites are believed to perform.

As Table 9.2 indicates, most Chinese satellites are used for military or other governmental purposes. ISR and remote sensing, communication, and earth observation represent the three largest contributors to China's satellite development program. There is a great deal of uncertainty and imprecision in these categories. The Union of Concerned Scientists categorizes satellites primarily by their stated purpose and indicates that seven of the total 34 Yaogan satellites are used by the government, primarily for land survey and resource management purposes. Yet all Yaogan satellites are listed as "PLA-operated" and are most likely surveillance and reconnaissance satellites used for

Table 9.2
China's Operational Satellites, by Mission and Owner

Mission	Government	PLA	Commercial	Civil	Total
ISR and remote sensing	9	28			37
Navigation		15			15
Communication	8	4	11	1	24
Earth observation	28			1	29
Space sciences	8			2	10
Tech development	14	1	1	1	17
Total	67	48	12	5	132

SOURCE: Union of Concerned Scientists, UCS Satellite Database (official names only), data as of January 31, 2015. The UCS data are compiled primarily from the United Nations Registry of Space Objects and the U.S. Space Objects Registry, augmented by news reports and blog posts.

military purposes, though some may have limited capabilities.[3] Although China denies employing these assets to collect intelligence on other countries, numerous analysts have concluded that they are primarily military reconnaissance platforms.[4] Table 9.2 employs definitions provided in the Union of Concerned Scientist database, except in the case of the Yaogan series and three other satellites specifically listed in the database as having likely military functions.

Available launch and orbital data indicate that there are several different spacecraft designs in the Yaogan series. Yaogan 1 (no longer operational), 3, 10, and 13 are thought to be radar imaging satellites. In contrast, the lower orbital parameters and physical conformations of Yaogan 5, 12, 14, and 21 suggest that these satellites collect optical imagery. All Yaogans, except for ELINT satellites (Yaogan 9, 16, 17, 20, and 20 series), operate from near-polar, sun-synchronous orbits, and they most likely provide multi-wavelength, overlapping, global imagery of military targets.[5] Many of the launches have been organized in pairs, with a radar satellite and an optical satellite flying only a few weeks apart. Other clusters of satellites have been launched on single rockets. The Yaogan 9, 16, 17, 20, and 25 series are each comprised of three satellites and each set was lifted into close formation on a single rocket. Analysts believe that they perform ocean surveillance using ELINT and SIGINT sensors.[6] Although China currently possesses no missile launch warning satellites, some Chinese operational writings suggest that there is an aspiration to acquire such systems.[7]

In addition to the PLA's dedicated surveillance and reconnaissance platforms, the Chinese military and other governmental agencies operate a number of other remote sensing satellites. One is jointly owned and operated by the China Space Agency and Brazil's National Institute for Space Research.[8] With optical, IR, and SAR sensors that are less capable than those on the Yaogan series, these satellites are primarily used for such functions as environmental monitoring, mapping, and disaster management. However, they could—and some of them probably do—also support some military functions, such as mission planning and general situational awareness. Similarly, the eight communication satellites, 28 earth observation, and eight space sciences satellites

---

[3] Although the Union of Concerned Scientists satellite data include operational satellites only, it is unclear how accurate that determination is, and some of the 19 listed Yaogan satellites may no longer be operational.

[4] See, for instance, Rui C. Barbosa, "China Launch YaoGan Weixing-9, Announce Increase in Vehicle Production," *NASASpaceFlight.com*, March 5, 2010; *Jane's Space Systems and Industry*, "Yaogan Series," January 20, 2015.

[5] *Jane's Space Systems and Industry*, "Yaogan Series," January 20, 2015.

[6] *Jane's Space Systems and Industry*, "Yaogan Series," January 20, 2015.

[7] Fravel and Medeiros, "China's Search for Assured Retaliation," 2010.

[8] This satellite is one of the China-Brazil Earth Resources Satellites (CBERS) series, which are used for monitoring land use, agricultural development, and environmental degradation in the two countries. A total of five satellites had been launched (including one launch failure) as of March 2015, but only one remains in operation. See *Jane's Space Systems and Industry*, "China-Brazil Earth Resources Satellite (CBERS)/Ziyaun Series," December 17, 2014; Union of Concerned Scientists, UCS Satellite Database, data as of January 31, 2015.

that support other government users could also provide services and data to military users. Finally, the PLA and other government users lease channels on China's 11 commercial communication satellites.[9]

The U.S. orbital infrastructure is proportioned differently in terms of numbers of satellites performing each mission. As previously mentioned, as of January 31, 2015, the United States had 526 operational satellites in orbit. Table 9.3 profiles in general terms the missions that these satellites are believed to perform.

As Table 9.3 indicates, the overwhelming majority of U.S. satellites support communication functions, and the vast majority of those (196) are owned and operated by commercial concerns. Most of the satellites owned and operated by the U.S. military are divided into three categories: 45 that support surveillance and reconnaissance functions (optical and radar imagery, electronic surveillance, ocean surveillance, and early warning), 36 that support navigation (GPS satellites that provide PNT data to military and civilian users), and 42 that support military satellite communication (MILSATCOM).[10] The United States gathers earth observation data from 55 satellites. Twenty-five earth observation satellites are government assets, most of which directly support national and international weather and environmental research. Civil and scientific users also receive data from defense meteorological satellite program (DMSP) satellites, which the National Oceanic and Atmospheric Administration has operated, along with its civilian weather satellites, since 1998.

Table 9.3
U.S. Operational Satellites, by Mission and Owner

Mission	Government	Military	Commercial	Civil	Total
ISR		45			45
Navigation		36			36
Communication	80	42	196	2	320
Earth observation	25	7	22	1	55
Space sciences	15			3	18
Technology development	7	30	10	5	52
Total	127	160	228	11	526

SOURCE: Union of Concerned Scientists, UCS Satellite Database (official names only), data as of January 31, 2015. The UCS data are compiled primarily from the United Nations Registry of Space Objects and the U.S. Space Objects Registry, augmented by news reports and blog posts.

[9] Union of Concerned Scientists, UCS Satellite Database, data as of January 31, 2015.

[10] Union of Concerned Scientists, UCS Satellite Database, data as of January 31, 2015.

As is the case of China, the U.S. military obtains images from commercial remote sensing satellites to supplement those from its dedicated reconnaissance platforms and leases channels on commercial communication satellites to supplement the bandwidth provided by dedicated military communication satellites. For comparison, 36 percent of Chinese satellites and 30 percent of U.S. satellites are used for military purposes. But the dual-use capacity of many of the satellite platforms on both sides suggests that these percentages probably undervalue the military use of space assets.

Satellites operate in a variety of orbits around the earth. Figure 9.1 illustrates those most frequently used by China and the United States.

As the figure illustrates, there are four main satellite earth orbits. Most satellites operate in LEO, between roughly 300 and 2,000 km in altitude.[11] This puts the systems whose missions require detailed views of the earth's surface or atmosphere—such as reconnaissance, meteorology, and other types of remote sensing and earth science—close to their targets of interest. Often, these satellites are placed in polar orbits, which pass over or near the north and south poles. That allows them full coverage of the earth's surface over time as the globe rotates beneath them.

Figure 9.1
Primary Satellite Earth Orbits

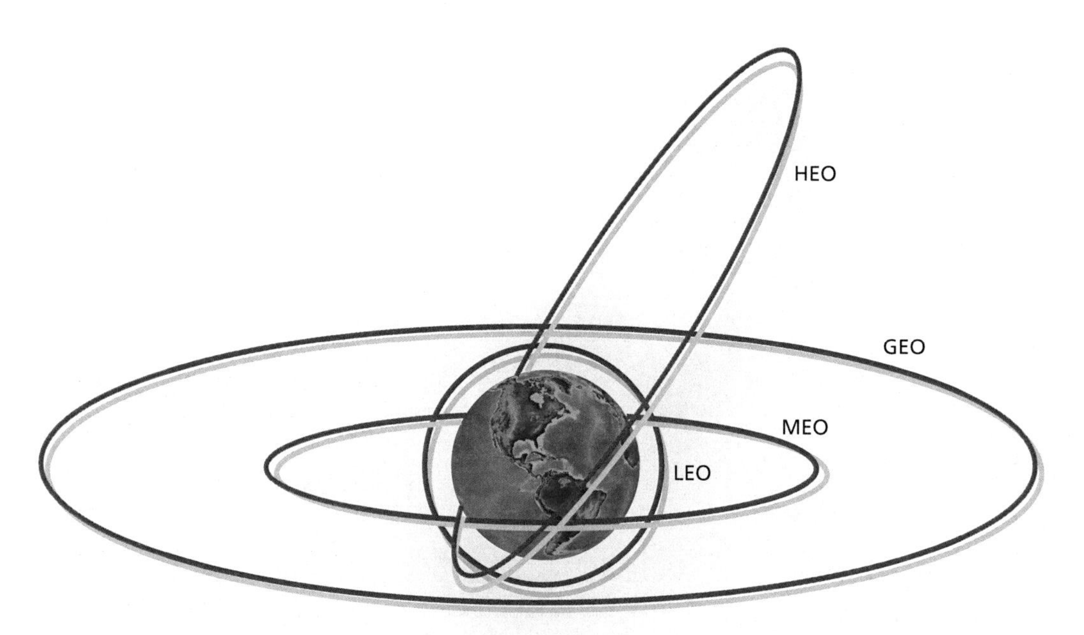

SOURCE: Adapted from U.S. Joint Chiefs of Staff, Space Operations, Joint Publication 3-14, Washington, D.C., May 29, 2013, p. G-4, Figure G-1.
RAND *RR392-9.1*

[11] There is no standard definition of LEO, and some place the lower bound at 100 km while others designate it as 300 km.

Another important region is geostationary earth orbit (GEO). This circular orbit, about 36,000 km in altitude, gives each satellite there a period of 24 hours. With zero degrees of inclination, this period allows the satellite to seemingly hover over a designated point on the equator, giving it a view (and putting it in view) of almost an entire hemisphere of the globe.[12] GEO is heavily populated with communication satellites.[13] Most U.S. early warning satellites, some of China's Beidou navigation satellites, and some of each country's weather satellites also operate in GEO.

Medium earth orbit (MEO) resides between LEO and GEO. Satellites there have circular orbits about 20,000 km in altitude, giving them 12-hour observation periods. The United States' 32 GPS satellites and some of China's Beidou navigation satellites operate in MEO.

The inclined ellipse in the figure illustrates a highly elliptical orbit (HEO). Satellites in these orbits pass the earth very quickly at perigee and have long dwell times around apogee.[14] The high inclinations and long dwell times give them good coverage of the high northern latitudes, making them ideal for specialized surveillance and communication missions. For instance, some of the new U.S. Space-Based Infrared System (SBIRS) early-warning satellites are in HEO.

U.S. Counterspace Versus Chinese Space Capabilities

Space control is one of four military mission areas cited in U.S. national space policy, and it is addressed extensively in joint and Air Force space doctrine.[15] According to the U.S. *Joint Doctrine for Space Operations*,

> Space control supports freedom of action in space for friendly forces and, when necessary, defeats adversary efforts that interfere with or attack US or allied space systems and negates adversary space capabilities.[16]

The U.S. Air Force considered the denial and negation elements of the space control mission area important enough to publish a doctrine manual dedicated to coun-

---

[12] *Inclination* is the tilt of a satellite's orbital plane, measured in degrees relative to the earth's equatorial plane. A satellite's *period* is the time it takes it to complete one full orbit around the earth.

[13] A substantial number of communication satellites also operate in LEO.

[14] *Perigee* is the point in an elliptical orbit at which the satellite passes closest to the earth. *Apogee* is the point at which it is farthest from the earth.

[15] See *National Space Policy of the United States of America*, The White House, June 28, 2010, p. 14; U.S. Joint Chiefs of Staff, *Space Operations*, 2013; and Air Force Doctrine Document 2-2, *Space Operations*, Washington, D.C., November 27, 2006, pp. 5, 23, 28.

[16] U.S. Joint Chiefs of Staff, *Space Operations*, 2013, p. II-8.

terspace operations.[17] Yet, as we explain later, current U.S. counterspace capabilities are relatively limited.

Dedicated Counterspace Systems

The U.S. Air Force, Army, and Navy have all attempted to develop counterspace capabilities at various times over the course of the space age, but they have frequently faced stiff opposition from the White House, DoD, and Congress. The principal source of opposition in the late 1950s and early 1960s was the Eisenhower and Kennedy administrations' desire to establish and preserve "freedom of space" for unhindered overflight of U.S. reconnaissance satellites. Given a 1958 United Nations Committee on the Peaceful Uses of Outer Space opinion that such freedom legally extended only to satellites used for "peaceful purposes," DoD put constraints on military programs that might suggest other-than-peaceful U.S. activities in space. Later in the Cold War, concerns arose that developing space weapons might be destabilizing, and, in the 1980s, Congress restricted testing and ultimately cut funding for the Air Force's air-launched, direct-ascent, kinetic satellite interceptor out of concern that further development might start an arms race in space. Post–Cold War administrations have also resisted space weapon development, largely due to concerns about the political ramifications of "weaponizing" that domain.[18]

Nevertheless, in 2002, the Air Force received funding approval for two new counterspace systems, which DoD and Congress allowed because they were ground-based and designed to create "reversible effects"—that is, their use would temporarily interfere with the operation of targeted systems without causing permanent damage. One was the Counter Surveillance Reconnaissance System (CSRS), a mobile platform that could temporarily deny enemy satellites the ability to collect information on U.S. forces. The second was the Counter Communications System (CCS), a mobile satellite communication (SATCOM) jammer designed to interrupt enemy command-and-control communications.[19]

Both of these efforts were funded in 2003; however, Congress cut all funding for CSRS in 2004, effectively terminating that program. CCS, on the other hand, achieved IOC that year, and, by 2007, the Air Force had three first-generation CCS platforms in the field, with four more on order and a second-generation system in

---

[17] Air Force Doctrine Document 2-2.1, *Counterspace Operations*, Washington, D.C., August 2, 2004.

[18] On the history of U.S. counterspace development efforts, see Curtis Peebles, *High Frontier: The United States Air Force and the Military Space Program*, Washington, D.C.: Air Force History and Museums Program, 1997, pp. 59–67; Paul B. Stares, *The Militarization of Space: U.S. Policy, 1945–1984*, Ithaca, N.Y.: Cornell University Press, 1985; and David W. Ziegler, "Safe Heavens: Military Strategy and Space Sanctuary," in Bruce M. DeBlois, ed., *Beyond the Paths of Heaven: The Emergence of Space Power Thought*, Maxwell AFB, Ala.: Air University Press, September 1999.

[19] U.S. Air Force, Exhibit R-2, RDT&E Budget Item Justification, PE No. 0604421F, "Counterspace Systems," February 2004.

development.[20] At least seven of these systems are now fully operational and available for deployment with two Air Force space control squadrons.[21]

Dual-Use Capabilities

Although the United States currently enjoys only modest counterspace capabilities from dedicated systems, those capabilities could, theoretically, be augmented in a time of war by systems designed for other missions. For instance, like China, the United States could use ground- or sea-based radio transmitters to beam high-power radio noise in selected frequency bands at Chinese satellites, jamming their receivers. Similarly, the United States could use lasers designed for other purposes to track or attack Chinese satellites. Like China, the United States operates a network of laser ranging stations. Both countries' tracking sites are part of the International Laser Ranging Service and feed information to international data centers, but the tracking sites themselves function under national control.[22] These were not designed for use as weapons and would not be useful even as low-power dazzlers because they are in fixed locations and not deployable to the Western Pacific or South China Sea areas of operation. They could, however, be used to generate precise satellite tracking data to supplement data provided by the Air Force's dedicated Space Surveillance Network. Such data would be needed to identify, track, and target Chinese satellites for attack by other U.S. counterspace systems.[23]

There are more significant capabilities in the high-power laser research programs under way in the U.S. Navy, Army, and Air Force. The Navy's Mid-Infrared Advanced Chemical Laser (MIRACL), developed in the 1970s, is a megawatt-class, continuous-wave deuterium fluoride laser.[24] It has been used in directed-energy weapon experiments to shoot down airborne drones, missiles, and artillery rockets.[25] In 1997, MIRACL

---

[20] Col Donald E. Wussler, Vice Commander, Space and Missile Systems Center, "Space Superiority Systems Wing," briefing, Space and Missile Systems Center Industry Days, El Segundo, Calif., April 18, 2007.

[21] CCS platforms are operated by the 4th Space Control Squadron at Holloman AFB, New Mexico, and the 76th Space Control Squadron at Peterson AFB, Colorado. See Air Force Space Command, "The 4th Space Control Squadron," fact sheet, Peterson AFB, Colo., undated, and Air Force Space Command, "The 76th Space Control Squadron," fact sheet, Peterson AFB, Colo., undated.

[22] On the International Laser Ranging Service, see Michael Pearlman, Carey Noll, Jan McGarry, Werner Gurtner, and Erricos Pavlis, "International Laser Ranging Service," undated.

[23] Five of the 40+ International Laser Ranging Service sites are located in the United States, and several others are in countries that are allied or friendly with the United States. See International Laser Ranging Service, "Stations Site Listing," undated.

[24] According to one source, the MIRACL program has been terminated. See Office of Naval Research, "Navy Solid State Laser Program Overview," February 22, 2013.

[25] *Jane's Electro-Optic Systems*, "Northrop Grumman Mid-InfraRed Advanced Chemical Laser (MIRACL)," September 21, 2010; *Jane's Strategic Weapon Systems*, "Ship-based laser," July 25, 2014.

illuminated a U.S. satellite at a range of 420 km in a spacecraft vulnerability test.[26] The Army's Tactical High-Energy Laser (THEL) system offers a similar capability. This deuterium fluoride laser, also located at White Sands, began testing in 1999 and has been used to successfully shoot down Katyusha rockets, 152-mm artillery shells, and mortar bombs.[27] A mobile version of the THEL, called Skyguard, is currently being developed, and the U.S. Department of Homeland Security is studying concepts for using it as a defense against SAM threats to U.S. civil aviation.[28] In September 2014, the Army tested a High Energy Laser Mobile Demonstrator (HEL MD), which it sees as a step toward ultimately developing a 100 kW–class laser for base defense against artillery, rockets, and mortars.[29] Finally, the Air Force has several laser systems at its Starfire Optical Range at Kirtland AFB, New Mexico, where it conducts research on laser guide star adaptive optics, beam control, and space object identification.[30] According to a 2006 *New York Times* article, federal officials have indicated that this research also has counterspace applications.[31]

A final category of systems that could offer dual-use capabilities for counterspace operations consists of interceptors in the U.S. BMD program. In February 2008, the U.S. Navy destroyed a failed U.S. reconnaissance satellite with a specially modified SM-3 block 1A missile launched from a *Ticonderoga*-class cruiser in the Pacific Ocean. This intercept demonstrated that the SM-3, a sea-based component of the U.S. BMD system, could have some dual-use capability as a kinetic anti-satellite (ASAT) weapon.[32] The SM-3 IIA will have significantly greater range and altitude, with correspondingly greater potential as an ASAT weapon, and is expected to reach IOC by 2018. Work on the SM-3 Block IIB, which was intended to intercept ICBMs and would have had greater range, was halted in March 2013.[33]

Other BMD systems that might contribute to space control capabilities include the Terminal High-Altitude Area Defense (THAAD) system and the Ground-Based Midcourse Defense (GMD) segment of the program. THAAD is a mobile, ground-

---

[26] Bryan Bender, "Army Successfully Fires MIRACL Laser at Satellite," *Defense Daily*, October 21, 1997.

[27] *Jane's Land Warfare Platforms*, "Tactical High-Energy Laser (THEL)," August 14, 2012.

[28] *Jane's Electro-Optic Systems*, "Northrop Grumman Skyguard," September 3, 2010; *Jane's Land Warfare Platforms*, "Skyguard (Laser Air Defence)," March 10, 2015.

[29] *Jane's Defence Weekly*, "HEL MD Laser Continues Testing, Moves Towards 60 kW system," September 10, 2014.

[30] U.S. Air Force, "Starfire Optical Range at Kirtland Air Force Base," fact sheet, March 9, 2009.

[31] William J. Broad, "Administration Researches Laser Weapon," *New York Times*, May 3, 2006.

[32] Amy Butler, Michael Bruno, David A. Fulghum, and John M. Doyle, "Ambiguous Intercept: Impact of the Satellite Shootdown Both Lauded and Damned," *Aviation Week and Space Technology*, Vol. 168, No. 8, February 25, 2008, p. 30; *Jane's Strategic Weapon Systems*, "RIM-66/-67/-156 Standard SM-1/-2, RIM-161 Standard SM-3, and RIM-174 Standard SM-6," July 30, 2013.

[33] O'Rourke, *Navy Aegis Ballistic Missile Defense (BMD) Program: Background and Issues for Congress*, 2015.

based, medium-range BMD system operated by the U.S. Army. The current version of THAAD is designed to intercept ballistic targets at altitudes of between 20 and 150 km at a range beyond 200 km.[34] By the end of 2015, five batteries, each with six launchers and 48 missiles, are expected to be in service, with equipment for a sixth to be delivered in 2016.[35] Lockheed is currently working to develop an improved, two-stage variant of THAAD, which could potentially more than double the range and ceiling of the missile.[36] GMD is the long-range, silo-based, component of the BMD system. GMD interceptors have a range of approximately 5,000 km and have been test-fired to an altitude of 1,875 km. The system is considered to be in IOC, with silos at Fort Greely, Alaska, and Vandenberg AFB, California. Current plans call for the full complement of 44 interceptors to be in service by FY 2017.[37]

Resultant Risks for Chinese Space Capabilities

U.S. dedicated and dual-use counterspace capabilities would pose risks to China's space capabilities in the event of a conflict. The precise levels of risk are difficult to measure and predict without a substantial modeling effort that was beyond the scope of this study. However, we can roughly approximate the threat by considering trends in U.S. capabilities, evaluating what threats those capabilities represent to satellites of each type and orbit, and considering what kinds of attacks the United States might choose to carry out against those systems in a limited war.

Figure 9.2 illustrates the approximate levels of risk that U.S. counterspace capabilities present to Chinese space systems. Red and orange indicate U.S. counterspace systems have little ability to hold Chinese systems at risk; yellow indicates that that the United States has the ability to hold Chinese systems at moderate risk; while light and dark green (not present in the figure) indicate U.S. systems could hold Chinese space systems at high risk.[38] This color scheme, while somewhat counterintuitive, is intended to be consistent with coding relative capabilities from the U.S. perspective throughout the document.

---

[34] "THAAD," *Jane's Land Warfare Platforms: Artillery and Air Defence*, March 3, 2015.

[35] Joakim Kasper Oestergaard Balle, "About the THAAD System," *Aerospace and Defense Intelligence Report*, October 22, 2014; Missile Defense Agency, "Terminal High Altitude Area Defense," fact sheet, May 2014.

[36] "Lockheed Working to Extend Range of U.S. Missile Interceptors," *Defense One* (online), January 7, 2015.

[37] *Jane's Defence Weekly*, "US MDA to Take Delivery of First CE II Block 1 Interceptor by End of December," December 16, 2014; and *Jane's Strategic Weapons Systems*, "Ground-Based Mid-course Defense (GMD) Segment," September 12, 2014.

[38] We caution that these estimates do not necessarily reflect the operational vulnerabilities of Chinese space systems. That is, yellow and green do not indicate that U.S. counterspace efforts would succeed in degrading or destroying Chinese space capabilities. Determining each Chinese space system's actual vulnerability to U.S. attack would require an assessment of factors beyond the scope of this study, such as an in-depth technical analysis of the passive and active defensives on Chinese space systems and an operational analysis of the defensive tactics employed by Chinese operators.

Figure 9.2
Estimated Risk Posed to Chinese Space Systems by U.S. Counterspace Systems

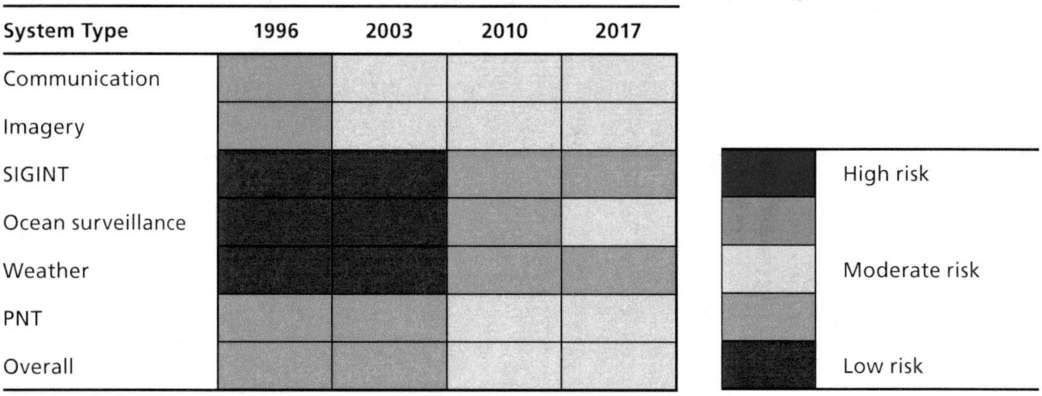

System Type	1996	2003	2010	2017
Communication				
Imagery				
SIGINT				
Ocean surveillance				
Weather				
PNT				
Overall				

High risk

Moderate risk

Low risk

Arguably, the category of Chinese space systems that is most in danger from U.S. counterspace capabilities is space-based communication. Chinese communication satellites were already at an elevated level of risk between 1996 and 2003, stemming from the U.S. ability to use high-powered radio transmitters as improvised jammers, and that level increased in the next period as deployable CCS jammers were developed and delivered to Air Force space control squadrons. Risks to Chinese communication capabilities are likely to remain at least as high in the 2010–2017 period and may continue to rise as CCS platforms increase in number and sophistication.[39]

CCS and dual-use jammers also pose risks to other Chinese space systems that rely on RF technology, such as PNT systems. China obtains PNT data from its Beidou ("Northern Dipper") system, also known as "Compass." First-generation Beidou satellites are not equipped with atomic clocks and, therefore, operate in a fashion that is different from GPS or Russia's navigation satellite system, GLONASS. Instead of generating precise timing signals at each satellite, Beidou uses two GEO satellites to relay timing signals between a master control station and user terminals.[40] This operating concept makes first-generation Beidou satellites susceptible to uplink jamming, and the improved U.S. capability against PNT between 2003 and 2010 (shown in Figure 9.2) reflects the effect of the U.S. Air Force acquiring CCS during that period. Between 2010 and 2015, China launched 14 of its planned 35 second-generation Beidou satellites, which carry their own atomic clocks and operate more like GPS.[41] This will make the system more resilient to uplink jamming and will likely reduce China's PNT risk.

[39] The United States contracted for upgrades to five CCS systems in November 2012 and two more in April 2013. See Strategic Defense Intelligence, "U.S. Awards Communications Systems Upgrade Contract to Harris," November 12, 2012; "Contracts" U.S. Department of Defense, press release No: 276-13, April 26, 2013.

[40] *Jane's Space Systems and Industry*, "Beidou/Compass Series," April 13, 2015.

[41] Union of Concerned Scientists, UCS Satellite Database, data as of January 31, 2015; and Tai Ming Cheung, *China's Emergence as a Defense Technological Power*, New York: Routledge, 2014, p. 115.

However, the United States could develop downlink Beidou jammers (or buy them from a commercial vendor, such as Russia's Aviaconversia), which could push the risk to China's PNT capabilities higher.[42]

China's SIGINT and ocean surveillance satellites also rely on RF technology, so the degree of risk to these systems is also affected by U.S. access to CCS and dual-use jammers. However, the risks to these systems are not as pronounced, because the United States would face the same challenges in jamming them that China would in efforts to jam comparable U.S. systems. Jammers would probably not know which frequencies are monitored by passive SIGINT collectors at any given time, and jamming China's ocean surveillance satellites, which operate in LEO and have small footprints, would require a jammer to be located close to the vessels it is trying to conceal, thereby defeating the purpose of jamming.

Figure 9.2 also illustrates the risks that U.S. laser developments pose to China's space systems. Because the United States has been experimenting with MIRACL and other dual-use lasers since before 1996, the figure depicts some risk at the outset for imagery satellites, which rely on optical sensors.[43] Those levels rose between 1996 and 2003 due to new U.S. dual-use laser program developments, such as THEL and the 1997 MIRACL satellite vulnerability test, combined with CSRS program initiation in 2002.[44] Although the CSRS program was canceled in 2004, dual-use laser development has continued to advance.[45]

The potential impact of U.S. laser development on Chinese imagery satellites is limited by the fact that MIRACL, THEL, and the Starfire Optical Range are located at fixed sites in the United States and, therefore, could not be used to dazzle Chinese satellites in the Western Pacific or South China Sea. Some of these lasers are certainly powerful enough to damage optical sensors if they are active when satellites overfly them, which would make satellite sensors unavailable for later use in the area of operations. But in wartime, Chinese operators would likely shutter their sensors or point their telescopes away when known laser sites are in view. It is unknown whether U.S. dual-use lasers are powerful enough to damage Chinese satellite components other than optical sensors. In any event, risks to Chinese satellites with optical sensors could increase if the United States deploys operational high-energy lasers. In 2014, the Navy installed a 30-kW Laser Weapon System aboard the amphibious transport dock

---

[42] On the advantages of downlink jamming, see Jeff Harley, *Space Control and Information Operations*, Huntsville, Ala.: U.S. Army Space and Missile Defense Command/Army Forces Strategic Command, 2002; and *AU-18 Space Primer*, Air University Press, September 2009.

[43] Although weather satellites also rely on optical sensors, other factors mitigate the risks to those systems, as explained later.

[44] "U.S. Test-Fires 'MIRACL' at Satellite Reigniting ASAT Weapons Debate," *Arms Control Today*, October 1997.

[45] Adolfo J. Hernandez, *Military Role in Space Control: A Primer*, Washington, D.C.: Congressional Research Service, September 23, 2004.

USS *Ponce* for testing. The U.S. Navy declared the laser an operational asset by the end of 2014 and reportedly has plans to test a 100- to 150-kW version in 2016 or 2017.[46]

The final source of risk reflected in Figure 9.2 is the dual-use kinetic capability provided by U.S. BMD systems. The modified SM-3 missile used to destroy the failed satellite achieved the intercept at 247 km in altitude. An unmodified SM-3 Block 1A missile has a maximum theoretical ASAT intercept altitude of about 500 km, and that of the Block 1B, which reached IOC in April 2014, is marginally higher.[47] The SM-3 Block IIA, which is expected to deploy in 2016 and reach IOC in 2018, is a substantially larger missile and will have a theoretical ceiling several times that of the Block 1A.[48] Current THAAD systems are reported to have a maximum engagement altitude of 200 km, and although an extended-range version could more than double that figure, it is currently in research and is not a program of record. Of the confirmed systems, only the improved versions of the SM-3, and especially the SM-3 Block IIA, will reach the orbits of even China's lowest-flying military satellites. Alternatively, U.S. GMD interceptors, with their maximum engagement altitude of 1,875 km, could easily reach any of China's LEO satellites, but it is unclear whether the United States would use a key strategic defense asset sited in the homeland for an ASAT mission or, for that matter, use any kinetic weapon that would litter the highly trafficked LEO belts with debris.

Should the United States be sufficiently provoked to commence kinetic attacks on Chinese satellites, China's imagery and ocean surveillance assets would be at greatest risk, being high-value, low-density resources in LEO.[49] China's weather satellites, on the other hand, would be least likely to suffer a U.S. attack of any kind for some of the same reasons that U.S. weather satellites are at low risk of attack. Although these assets operate in LEO and are vulnerable to both kinetic and directed-energy weapon attack, China, like the United States, contributes data to the UN World Meteorological Organization.[50] This gives it access to weather data from other countries in the event of the loss of any of its own satellites. Moreover, given that prevailing weather patterns in the Northern Hemisphere move from west to east, China could use terrestrial instruments in its own territory to monitor atmospheric developments and predict the weather

---

[46] "U.S. Navy Allowed to Use Persian Gulf Laser for Defense," *U.S. Naval Institute News*, December 10, 2014; and Ronald O'Rourke, "Naval Shipboard Lasers for Surface, Air, and Missile Defense: Background and Issues for Congress," Washington, D.C.: Congressional Research Service, June 27, 2013.

[47] *Jane's Strategic Weapons Systems*, "RIM-66/-67/-156 Standard SM-1/-2, RIM-161 Standard SM-3, and RIM-174 Standard SM-6," July 30, 2013.

[48] Brian Weeden, "The Space Security Implications of Missile Defense," *The Space Review*, September 28, 2009; and *Jane's Navy International*, "Surface Navy 2015: SM-3 Block IIA Program Set of CTV Test," January 14, 2015.

[49] Some of China's Yaogan satellites are under 500 km in altitude. Orbital data on Chinese satellites from Union of Concerned Scientists, UCS Satellite Database, data as of January 31, 2015.

[50] Ministry of Foreign Affairs of the People's Republic of China, "China and the World Meteorological Organization (WMO)," April 3, 2012.

in conflicts off its eastern shores. Consequently, Chinese weather satellites would not be lucrative targets for U.S. attack, as reflected in the low levels of risk shown in Figure 9.2.

U.S. counterspace capabilities would vary little across the two scenarios considered in this study. To some extent, the United States might be better positioned to jam communication uplink signals from its Japanese bases in a Taiwan scenario. But assuming U.S. forces would be intervening in support of one or more friendly states in Southeast Asia in a Spratly Islands scenario, it is reasonable to surmise that those states would allow the United States to operate jammers from their soil.

The major difference between the two scenarios would be less one of capability than the importance of space support to China and, therefore, of counterspace. In a Taiwan scenario, China would be far less dependent on space support than it would be in a Spratly Islands conflict (and it would be less dependent on such support than would U.S. forces). It could use terrestrially based systems and methods for most of its communication and air-breathing systems to meet many of its ISR requirements. This pattern of difference is changing with time, as Chinese strike assets gain greater range and as space systems become a better option for many support requirements, even in the Taiwan case. Nevertheless, the importance of space and counterspace in a Spratly Islands scenario, in which much of the action would be beyond the range of Chinese land-based support systems, will remain much greater than in a Taiwan contingency.

Conclusions

While dedicated U.S. counterspace capabilities are limited, the United States could considerably augment its existing capabilities with dual-use systems. Whether U.S. leaders would choose to do so in a war with China is questionable, given China's ability to respond with attacks on U.S. space capabilities and the greater degree to which the U.S. military depends on those systems for force enhancement. However, should the PLA attack U.S. space systems first, U.S. leaders might have little to lose in striking back.

Scorecard Coding

Figure 9.3 provides our summary coding of the results of scorecard 7. The assessment of advantage in this scorecard is based on the ability of U.S. counterspace systems to hold China's military or dual-use satellite capabilities at significant risk (*U.S. advantage*) versus the ability of China to utilize its space assets unimpeded (*Chinese advantage*).

In 1996, China had only a single satellite deployed, and none of its previous satellites were long-duration. Hence, the coding for that period, indicating U.S. disadvan-

Figure 9.3
Scorecard 7 Summary Coding

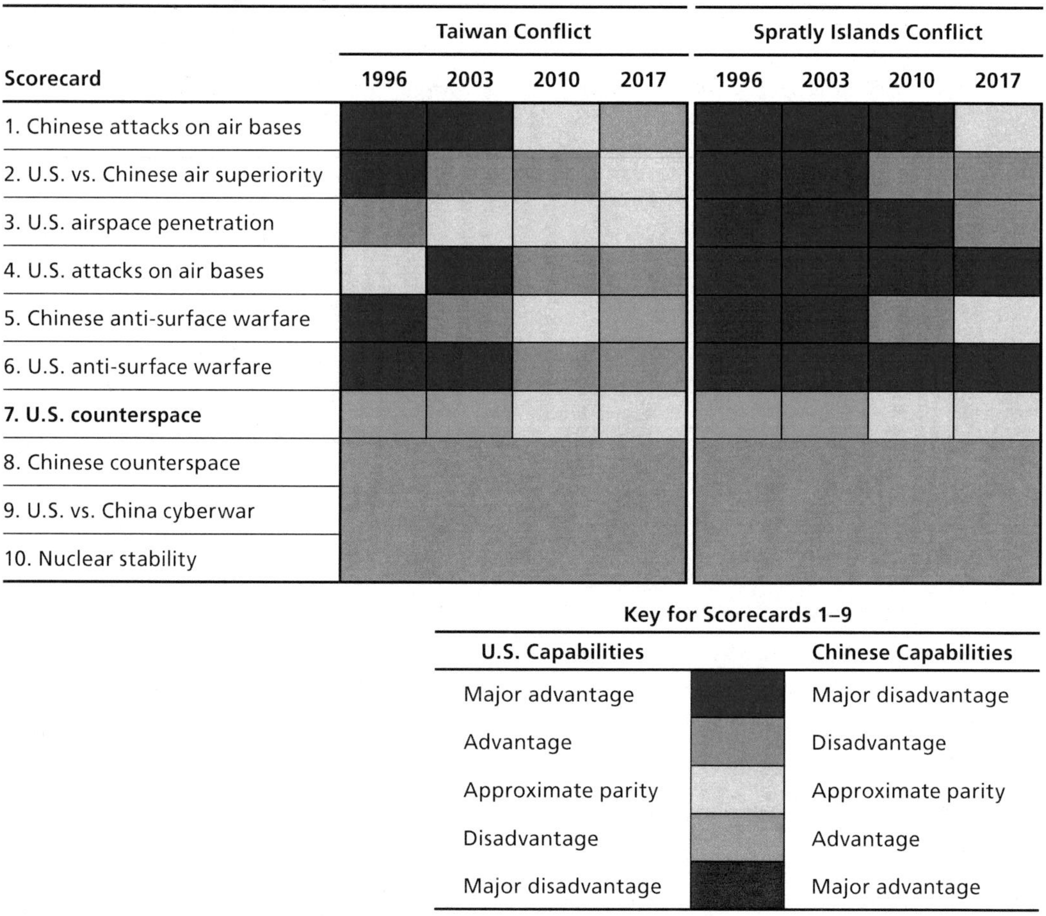

	Taiwan Conflict				Spratly Islands Conflict			
Scorecard	**1996**	**2003**	**2010**	**2017**	**1996**	**2003**	**2010**	**2017**
1. Chinese attacks on air bases								
2. U.S. vs. Chinese air superiority								
3. U.S. airspace penetration								
4. U.S. attacks on air bases								
5. Chinese anti-surface warfare								
6. U.S. anti-surface warfare								
7. U.S. counterspace								
8. Chinese counterspace								
9. U.S. vs. China cyberwar								
10. Nuclear stability								

Key for Scorecards 1–9

U.S. Capabilities		**Chinese Capabilities**
Major advantage		Major disadvantage
Advantage		Disadvantage
Approximate parity		Approximate parity
Disadvantage		Advantage
Major disadvantage		Major advantage

tage, simply serves as a reminder of the relatively weak U.S. counterspace capabilities during that period rather than as a measure of capability against a fully developed set of Chinese satellite capabilities. By 2003, China had deployed six military satellites (with three in orbit at that time), while U.S. ASAT capabilities remained at the nascent research stage. Although the United States might have been able to employ some experimental systems during a conflict, the scale and effectiveness of these efforts would have been limited in practice. Therefore, we coded 2003 as *Chinese advantage*. By 2010, the United States had deployed operational systems designed to jam satellite communications and was continuing work on other types of experimental counterspace systems. Given these developments, we coded 2010 as parity, indicating that U.S. counterspace capabilities could begin to hold some Chinese space functions at risk more reliably. While this coding is borderline for 2010, it is somewhat more robust by 2017, with further improvements to U.S. jamming capabilities and further development of high-energy laser systems. Although the impact of U.S. counterspace activities on Chinese operations would be greater in a Spratly Islands scenario than in the Taiwan case, given the greater dependence on satellite-based ISR and communication systems in the former, the U.S. ability to conduct counterspace operations would differ only marginally in the two cases. Therefore, we scored the two scenarios equally.

CHAPTER TEN

Scorecard 8: Chinese Counterspace Capabilities Versus U.S. Space Systems

Chinese leaders understand the importance of space to their country's potential adversaries. Space assets provide critical enabling functions that have made recent U.S. conventional military successes possible and greatly reduced the cost of those operations. The PLA has absorbed lessons from those conflicts and is developing weapons for kinetic and nonkinetic attacks on enemy space systems to interdict the support they provide to terrestrial forces. This chapter reviews the development of Chinese counterspace capabilities and assesses the risks that those capabilities pose to U.S. space force enhancement systems in several key areas: imagery, SIGINT, missile warning, ocean surveillance, communication, PNT, and weather.

As in our analysis of U.S. counterspace capabilities (Chapter Nine, scorecard 7), the limitations of open-source inputs and the scope of our study prevented us from developing a quantitative model of the threat. Hence, our evaluation is based on the characteristics of U.S. satellite constellations, including orbital characteristics, the number of satellites deployed, and the ownership of the satellites in question, in relation to the nature and magnitude of Chinese offensive capabilities.

Chinese Efforts to Develop Counterspace Capabilities

The PLA understands how much U.S. forces rely on space support to enhance their warfighting capabilities. Consequently, "China is developing a multidimensional program to improve its capabilities to limit or prevent the use of space-based assets by potential adversaries during times of crisis or conflict."[1] China's effort to develop counterspace capabilities first came to U.S. public attention in the late 1990s, when an increasing number of open-source publications reported on Chinese efforts to purchase or develop low- and high-powered laser technology, RF jammers, and other capabilities that could be used to attack satellites.[2]

---

[1] Office of the Secretary of Defense, *Military Power of the People's Republic of China 2010,* August 2010, p. 7.

[2] See, for instance, Matthew Campbell, "Chinese 'Death Ray' Threatens U.S. Satellites," *Sunday Times* (London), December 6, 1998; Shawn L. Twing, "U.S. Defense Intelligence Agency Report Accuses Israel of

Chinese Laser Developments

In September 2006, *Defense News* reported that China fired a "high power laser at a U.S. spy satellite." Later reports suggested the satellite was not permanently damaged and that China may have simply been conducting laser range finding.[3]

According to an often-cited 1999 U.S. Army War College study, China's effort to develop laser weapons actually began as early as the 1960s, with most of it being administered by the Chinese Academy of Sciences and the Commission of Science, Technology, and Industry for National Defense.[4] Western analysts surveying Chinese academic journals have since concluded that China has been undertaking a major research effort to develop the state-of-the-art technology that would be needed for such weapons.[5] The extent to which current laser research is sponsored by the PLA is difficult to ascertain, but nearly all of the Chinese journal articles published on the topic in the late 1990s were written by authors affiliated with institutions subordinate to the Chinese Academy of Sciences, the Commission of Science, Technology, and Industry for National Defense, or the China Academy of Engineering Physics—all of which are known to be involved in some amount of defense-related high-technology research.[6]

A related area of technology development on which more information is available is China's network of satellite laser ranging stations. This network consists of five fixed stations located at space observatories in Shanghai, Changchun, Beijing, Wuhan, and Kunming. At least two mobile systems are also available.[7] The Chinese satellite laser ranging program began in 1972 and has gone through several generations of development, bringing single-shot precision from the one- to two-meter accuracy of the first-generation system to 5-cm accuracy by 1986 and 12- to 30-mm accuracy by 2000.[8] Today, the Chinese are able to calculate satellite orbital parameters with sub-centimeter precision by combining measurements from these stations with those from other time-synchronized stations located around the world. China has access to such

---

Laser Technology Transfer to China," *Washington Report on Middle East Affairs*, April–May 1999, p. 44; "China Develops New Light Weapons—Hong Kong Press," *BBC Summary of World Broadcasts*, January 13, 2000; and Robert Wall, "Directed-Energy Threat Inches Forward," *Aviation Week and Space Technology*, Vol. 153, No. 18, October 30, 2000.

[3] David Axe, "Chinese Laser vs. U.S. Sats?" Military.Com, September 25, 2006; "Bachmann's Claim that China 'Blinded' U.S. Satellites," *Washington Post*, October 4, 2011.

[4] Mark A. Stokes, *China's Strategic Modernization: Implications for the United States*, Carlisle, Pa.: Strategic Studies Institute, U.S. Army War College, September 1999, pp. 195–196.

[5] See, for instance, Carlo Kopp, *High Energy Laser Directed Energy Weapons*, Air Power Australia, Technical Report APA-TR-2008-0501, updated April 2012.

[6] Stokes, *China's Strategic Modernization*, 1999, pp. 195–196.

[7] Yousaf Butt, *Satellite Laser Ranging in China*, technical working paper, Cambridge, Mass.: Union of Concerned Scientists, January 8, 2007, p. 2.

[8] Yang Fumin, "Current Status and Future Plans for the Chinese Satellite Laser Ranging Network," *Surveys in Geophysics*, Vol. 22, No. 5–6, 2001.

data through its participation in the International Laser Ranging Service, a worldwide network of more than 40 stations that collects, merges, analyzes, and shares data with its 23 member countries.[9]

The Union of Concerned Scientists argues that China's satellite laser ranging network poses little direct threat to U.S. satellites, maintaining that satellite laser ranging "cannot be considered an anti-satellite (ASAT) weapon and, in fact, would be ineffective in this role."[10] The network's stated mission is scientific research, and scientists involved with the project have published papers on such topics as the earth's crustal movements and gravity irregularities, changes in sea level, and other geodetic and geophysical phenomena. In fact, the International Laser Ranging Service's main operations center is NASA's Goddard Spaceflight Center in Greenbelt, Maryland. Most of China's satellite laser ranging stations operate at an average power level of only one watt, the minimum needed to range on a constellation of about 30 satellites described as "cooperative," in that they have attached reflective mirrors. Union of Concerned Scientists analyst Yousaf Butt maintains that even the 40-watt power level that the Shanghai station uses to track "uncooperative" space debris poses little risk to optical sensors, given that the laser's 20-hertz pulse rate and ten-nanosecond pulse duration create an exposure risk so brief that the probability of damaging sensor cells or filters on a 1-meter-resolution imaging satellite in a chance illumination is only about one in 1,000, even if the laser is active and pointed at the sensor as it passes overhead.[11]

Nonetheless, China's satellite laser ranging network represents a greater risk to U.S. satellites than the Union of Concerned Scientists suggests. Although the energy that its lasers emit is currently very low, scaling the power up would not present a difficult technical challenge, and each station might have that capability even now. Moreover, Butt's risk analysis is based on the assumption of a chance illumination. Should China position one of its mobile ranging lasers next to a point of U.S. interest and deliberately illuminate an imaging satellite as it is collecting data, the exposure time would likely be longer than what Butt used in his calculations. Butt concedes that longer exposure would increase the probability of damage to satellite sensors, even at the power levels the Chinese satellite laser ranging network currently uses.[12] Finally, even if China does not intend to use its satellite laser ranging network as a weapon in itself, it could still be an important element in the counterspace "kill chain," provid-

---

[9] Butt, *Satellite Laser Ranging in China,* 2007, pp. 1–2. China's satellite laser ranging network also participates in several other international data-sharing projects, such as the Asia-Pacific Space Geodynamics program and the Western Pacific Laser Tracking Network. See Yang, "Current Status and Future Plans for the Chinese Satellite Laser Ranging Network," 2001, p. 465.

[10] Butt, *Satellite Laser Ranging in China,* 2007, p. 1.

[11] Indeed, a continuous-wave laser would be a more efficient weapon than the pulse lasers used in China's satellite laser ranging network, further supporting the argument that these stations were not designed for use as weapons. See Butt, *Satellite Laser Ranging in China*, 2007 p. 5.

[12] Butt, *Satellite Laser Ranging in China,* 2007, pp. 10–11.

ing data of sufficient precision to target U.S. satellites with other weapons. This range of considerations is likely what alarmed U.S. leaders when a Chinese laser, probably at one of its satellite laser ranging stations, illuminated a U.S. reconnaissance satellite in September 2007, adding one more to several reported incidents in which Chinese lasers have illuminated U.S. reconnaissance satellites in recent years.[13]

Kinetic Anti-Satellite and Ballistic Missile Defense Capabilities

Developments that pose more obvious counterspace risks to U.S. satellites are emerging from China's direct-ascent kinetic ASAT and ground-based BMD programs. On January 11, 2007, China destroyed an expended Fengyun-1C weather satellite at about 850 km in altitude using a kinetic-kill vehicle on a modified two-stage, solid-fuel MRBM. U.S. officials have since designated this ASAT weapon the SC-19.[14] This event followed three previous tests, conducted in October 2005 and April and November 2006, which either failed or were not intended to destroy the target satellite.[15] The previous tests were noted primarily by the scientific and security communities and received no public attention, but the January 2007 event created a substantial cloud of orbital debris, endangering satellites from all spacefaring nations and provoking a storm of international criticism.

Subsequently, in January 2010, January 2013, and July 2014, China announced the successful test of ground-based, kinetic BMD interceptors.[16] Because this capability was nominally defensive and the test created no orbital debris, these events generated relatively little international criticism. But considering the intercepts took place at altitudes similar to that of the ASAT intercept, and that the interceptor in each case was probably an SC-19, some analysts question the distinction and refer to both as "hit to kill" systems.[17] However these systems are viewed, China's kinetic BMD interceptors use the same tracking, targeting, and guidance infrastructure as its ASAT system. This infrastructure includes four large phased array radars (LPARs) that provide coverage into Russia, Central Asia, South Asia, and Southeast Asia.[18] Taken together, ABM

---

[13] The previous alleged illuminations were reported by unnamed U.S. government sources. See Francis Harris, "Beijing Secretly Fires Lasers to Disable U.S. Satellites," *Daily Telegraph* (London), September 26, 2006; "Alleged Laser Test Sparks Debate on U.S./China Space Cooperation," *Aerospace Daily and Defense Report*, Vol. 219, No. 3, October 2, 2006; and "China Targets U.S. Satellite," *Courier Mail* (Australia), October 7, 2006.

[14] Shirley Kan, *China's Anti-Satellite Weapon Test,* Washington, D.C.: Congressional Research Service, April 23, 2007, p. 1.

[15] Stephanie Lieggi and Erik Quam, "China's ASAT Test and the Strategic Implications of Beijing's Military Space Policy," *Korean Journal of Defense Analysis,* Vol. 19, No. 1, Spring 2007, p. 9.

[16] "China Says Third Missile-Defense Test in Four Years Successful," *Bloomberg Business*, July 24, 2014.

[17] Catherine Dill, "Korla Missile Test Complex Revisited," Arms Control Wonk (Blog), March 26, 2015; Brian Weeden, "Through a Glass, Darkly: Chinese, American, and Russian Anti-Satellite Testing in Space," Secure World Foundation, March 17, 2014.

[18] *Jane's Intelligence Review*, "Space Invaders—China's Space Warfare Capabilities," July 3, 2014.

systems and their associated infrastructure represent a dual-use technology that poses essentially the same risk as China's ASAT capability to U.S. satellites.

Finally, China's May 13, 2013, launch of a ground-based missile into space was widely regarded as a test of a significant new Chinese ASAT capability. The rocket reportedly reached at least 10,000 km in altitude and may have reached 35,000 km.[19] U.S. defense officials expressed concern that the capability could be used to destroy a satellite in orbit. One experienced independent analyst concluded, "The system appears to be designed to place a kinetic kill vehicle on a trajectory to deep space that could reach medium earth orbit (MEO), highly elliptical orbit (HEO), and geostationary earth orbit (GEO)." "If true," the analyst continued, "this would represent a significant development in China's ASAT capabilities."[20]

Radio-Frequency Jammers and Other Capabilities

China also appears to be developing RF jammers and other directed-energy weapons to attack SATCOM, GPS, and other space-based capabilities. In 2007, DoD's *Annual Report to Congress: Military Power of the People's Republic of China*, stated that

> in recent years Beijing has pursued a robust, multidimensional counterspace program. UHF-band satellite communications jammers acquired from Ukraine in the late 1990s and probable indigenous systems give China today the capacity to jam common satellite communications bands and GPS receivers.[21]

Subsequent reports have carried the same statement, adding, "In addition to the direct-ascent anti-satellite weapon tested in 2007, these counterspace capabilities also include jamming, laser, microwave, and cyber weapons."[22] The 2008 report included the following paragraph on China's counterspace capabilities:

> The PLA has developed a variety of kinetic and non-kinetic weapons and jammers to degrade or deny an adversary's ability to use space-based platforms. China also is researching and deploying capabilities intended to disrupt satellite operations or functionality without inflicting physical damage. The PLA is also exploring satellite jammers, kinetic energy weapons, high-powered lasers, high-powered

[19] Chinese press sources claimed at least 10,000 km. In July 2013, *Air Force Magazine* quoted Lieutenant Colonel Monica Matoush as stating that the rocket nearly reached geosynchronous earth orbit. See *Jane's Intelligence Review*, "Space Invaders—China's Space Warfare Capabilities," 2014; and Andrea Shalal-Esa, "U.S. Sees China Launch as Test of Anti-Satellite Muscle: Source," Reuters, May 16, 2013.

[20] Weeden, "Through a Glass, Darkly: Chinese, American, and Russian Anti-Satellite Testing in Space," 2014.

[21] Office of the Secretary of Defense, *Military Power of the People's Republic of China* 2007, May 2007, p. 21.

[22] Office of the Secretary of Defense, *Military and Security Developments Involving the People's Republic of China 2012*, May 2012, p. 9.

microwave weapons, particle beam weapons, and electromagnetic pulse weapons for counterspace application.[23]

Moreover, China could supplement these dedicated capabilities with dual-use technologies developed for other purposes. For instance, China could use high-powered radio transmitters to beam radio noise in selected frequency bands at U.S. satellites to jam their receivers. As Cuba demonstrated in 2003 when it jammed satellite uplinks supporting Voice of America broadcasts into Iran, and as Libya has demonstrated on several occasions, such crude, "brute-force" approaches to SATCOM jamming can effectively block the use of multiple unprotected channels simultaneously.[24]

Resultant Risks for U.S. Space Capabilities

China's emerging counterspace capabilities would pose certain risks to U.S. space capabilities in the event of a conflict. The precise levels of risk are difficult to predict without detailed modeling of different types of attacks on different parts of the U.S. space infrastructure. However, as we did in Chapter Nine for U.S. counterspace capabilities, we can roughly approximate the threat to U.S. space functions by reviewing trends in Chinese counterspace development, evaluating the nature of the potential threat to satellites of each type and orbit and considering what kinds of attacks the PLA might attempt to carry out against those systems in a limited war.

Figure 10.1 illustrates the approximate levels of risk that Chinese counterspace capabilities pose to U.S. space systems. Green indicates Chinese counterspace capabilities that represent a low risk to hold U.S. space-based functions, yellow indicates areas in which the PLA has the ability to hold U.S. systems at moderate risk, and orange represents Chinese counterspace capabilities that could hold U.S. space functions at high risk.[25] Based on the factors mentioned earlier, the risk to most U.S. space functions appears to be growing faster than the U.S. ability or effort to mitigate them. We discuss these areas first before addressing PNT and missile warning, two areas in which recent U.S. efforts to make the "system of systems" more robust may significantly mitigate the

---

[23] See, for instance, Office of the Secretary of Defense, *Military Power of the People's Republic of China 2008,* Washington, D.C., March 2008, p. 21.

[24] See Broadcasting Board of Governors, "BBG Condemns Cuba's Jamming of Satellite TV Broadcasts to Iran," press release, July 15, 2003; "Libya Jamming 'Exposed Vulnerability,'" BBC News, January 13, 2006; and "Thuraya Satellite Telecom Says Jammed by Libya," Reuters, February 24, 2011.

[25] We caution that these estimates do not necessarily reflect the operational vulnerabilities of U.S. space systems. That is, yellow and green do not indicate that Chinese counterspace efforts would succeed in degrading or destroying U.S. space capabilities. Determining each U.S. space system's actual vulnerability to Chinese attack would require an assessment of factors beyond the scope of this study, such as an in-depth technical analysis of the passive and active defensives on U.S. space systems and an operational analysis of the defensive tactics employed by U.S. operators.

risk through 2017—or at least offset improvements to Chinese offensive counterspace capabilities.

In 1996, Chinese counterspace capability developments posed little risk to U.S. space systems. However, China's pursuit of technology for RF jammers, low- and high-powered laser weapons, and other directed-energy weapon capabilities created increasing levels of risk in the latter half of the decade. The category of systems facing the highest level of risk in the 2003 snapshot year appears to have been space-based communication. At that time, MILSATCOM was concentrated on a limited number of GEO platforms, and most of the channels it provided were in frequency bands susceptible to jamming. Moreover, military operational demands for bandwidth were growing at a rate that far outstripped the capacity provided by dedicated MILSATCOM systems, and excess demand was met by leasing channels from commercial SATCOM carriers whose transceivers were also concentrated in GEO and even more susceptible to jamming. During Operation Desert Storm in 1991, about 80 percent of U.S. military satellite communications were carried on this MILSATCOM network. However, during peak periods of combat operations in Iraq in 2003, more than 80 percent were carried on commercial SATCOM systems.[26]

Operational demands for communication bandwidth remained high between 2003 and 2010, and continued Chinese jammer development drove the risks for U.S. space-based communication to higher levels by the latter part of that period. However, a couple of developments occurred that have mitigated those risks somewhat since 2003. First, some U.S. military clients began shifting their tactical communication to commercial SATCOM systems operating in LEO, such as Iridium and Globalstar. The distributed nature of those systems—multiple platforms with small footprints and,

Figure 10.1
Estimated Risk Posed to U.S. Space Systems by Chinese Counterspace Capabilities

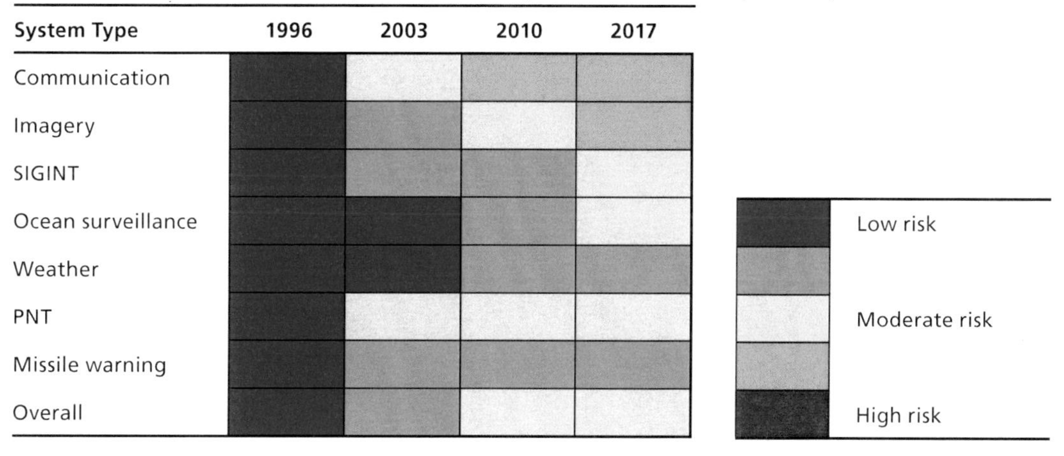

[26] Brian Eberhardt, Kenneth Kemmerly, and Paul Konyha III, "Satellite Communications," *Space Primer*, Maxwell AFB, Ala.: Air University, 2009, p. 183.

in the case of Iridium, satellite cross-linking—makes them inherently more robust against jamming threats than systems that concentrate communications into major nodes in GEO.[27] Second, MILSATCOM capacity increased considerably during this period. Between 2007 and 2015, six Wideband Global SATCOM (WGS) satellites were launched, with four more planned for future deployment.[28] WGS is the follow-on to the Defense Satellite Communications System, which was previously the U.S. military's main source of dedicated wideband connectivity. Using new technologies to improve the efficiency of bandwidth usage, a single WGS satellite can handle more communication throughput than the entire Defense Satellite Communications System constellation.[29] The U.S. military has also launched three Advanced Extremely High Frequency System satellites, which will provide robust communication capabilities (including in nuclear environments). When the entire constellation of five satellites is in place, it will have ten times the throughput of the Milstar system, which it will replace.[30] Finally, the U.S. military has increased emphasis on equipping commercial SATCOM platforms with components that met military specifications in an effort to increase their security and resilience to jamming threats.[31] The sum of these efforts will not eliminate all risk from Chinese RF jammer developments, but greater capacity and resilience can reduce the benefits of attacking these systems, thereby contributing to deterrence.[32]

China's RF jamming capabilities also pose risks to other U.S. space systems that use the electromagnetic spectrum, such as SIGINT collectors and ocean surveillance satellites. In each of these cases, however, system configuration and employment approaches mitigate the risks substantially. Unlike communication satellites,

> the United States deploys SIGINT spacecraft in all orbits—geosynchronous orbits to pick up ultra-high frequency (UHF) and very high frequency (VHF) communications, and low to medium Earth orbits to collect signals from air defense and early warning radars. Highly elliptical orbits give satellites both long dwell times at

---

[27] The Iridium constellation consists of 66 satellites, and Globalstar operates a constellation of 40 satellites. The size of a satellite's footprint—that is, the area encompassed by the satellite's 360-degree line of sight to the horizon—is important because a jammer would have to be within that footprint to attack the satellite.

[28] U.S. Air Force, "Sixth Wideband Global SATCOM Satellite," Air Force Space Command Fact Sheet, March 25, 2015. Launched," press release, August 7, 2013.

[29] Naval Network Warfare Command Public Affairs, "Navy Transitions to Wideband Global System," *CHIPS Magazine*, Vol. 26, No. 3, September 2008; *Jane's Space Systems and Industry*, "Wideband Global SATCOM," August 27, 2009.

[30] "Advanced Extremely High Frequency System," Air Force Space Command Facts Sheet, March 25, 2015.

[31] See Terry Costlow, "Meeting NATO's Satcom Needs Is No Simple Task," *Defense Systems*, February 25, 2011.

[32] For a fuller discussion on how improving system capacity and resilience contributes to space deterrence by denial, see Forrest E. Morgan, *Deterrence and First-Strike Stability in Space: A Preliminary Assessment*, Santa Monica, Calif.: RAND Corporation, MG-916-AF, 2010.

high altitudes and short dwell times at low altitudes, maximizing signals collection over multiple regions for specific and repeating durations or frequencies.[33]

Also unlike communication satellites, SIGINT sensors are "passive" in that they quietly monitor signals without transmitting RF energy that would reveal their presence. The highly distributed nature of these sensors, combined with uncertainty about what frequencies any one of them is monitoring at any given time, would complicate efforts to locate and jam them.

Some ocean surveillance satellites also rely on SIGINT, while others employ other RF technologies, such as SAR.[34] However, these assets operate in LEO, minimizing the size of their footprint, and the operational portions of their orbits are over the ocean. Jamming them would require putting the jammer on a ship and positioning it close to the vessel whose location the jammer was attempting to conceal. This would tip off U.S. ISR operators to the likely location of PLA vessels and largely defeat the purpose of the jammer. Operating in LEO, ocean surveillance satellites are also threatened by China's kinetic ASAT and BMD weapons developments. Given this vulnerability—and China's 2007 ASAT test and its ABM tests in 2010, 2013, and 2014—we designated elevated risk levels for both 2010 and 2017.[35]

The relatively higher risk assigned to U.S. satellite-based imaging capabilities in Figure 10.1 reflects the fact that imagery satellites are low-density, high-demand assets, potentially making them attractive targets in war—and a prominent driver of the PLA's development of kinetic ASAT and BMD weapons and laser illumination capabilities. Producers and consumers of U.S. space-based imagery supplement the data they receive from dedicated government platforms with images purchased from civil and commercial earth observation systems, such as GeoEye, Quickbird, and Worldview.[36] This distributes the load, mitigating risks somewhat. However, these assets also operate in LEO and are relatively few in number.

Interestingly, the category of space support that is probably at lowest risk of Chinese attack is weather satellites. These assets provide important support to military operations, and some of them operate in LEO, exposing them to China's ASAT

---

[33] Brian Crothers, Jeff Lanphear, Brian Garino, Paul P. Konyha III, and Edward P. Byrne, "U.S. Space-Based Intelligence, Surveillance, and Reconnaissance," *Space Primer*, Maxwell AFB, Ala.: Air University, 2009, p. 175.

[34] Crothers et al., "U.S. Space-Based Intelligence, Surveillance, and Reconnaissance," 2009, p. 175.

[35] On China's ASAT tests, see Catherine Dill, "Korla Missile Text Complex Revisited," Arms Control Wonk (blog), March 26, 2015. On the 2013 test, see Shalal-Esa, "U.S. Sees China Launch as Test of Anti-Satellite Muscle," 2013.

[36] Tamar A. Mehuron, "2009 Space Almanac: The US Military Space Operation in Facts and Figures," *Air Force Magazine*, August 2009, p. 64. DigitalGlobe is a prominent digital imagery provider that partners with a variety of industries and government entities. At the time of this writing, it owned and operated the IKONOS, QuickBird, WorldView-1, GeoEye-1, WorldView-2, and WorldView-32 satellites. "Satellite Information," DigitalGlobe, accessed April 9, 2015.

and BMD weapons, as reflected in the increased risk assessment assigned to weather satellite functions between 2003 and 2010. However, these satellites' distribution—multiple military and civilian platforms in LEO, with additional civilian satellites in GEO—and their participation in an international weather service, a global function coordinated by the UN World Meteorological Organization, make them relatively low-payoff targets with a significant degree of political risk for the attacker. Losses in weather satellite coverage from attacks on the DMSP system could largely be made up by the National Oceanic and Atmospheric Administration's Polar Orbiting Environmental System (POES) and Geostationary Orbiting Environmental System (GOES), as well as data from weather satellites belonging to other countries in the UN World Meteorological Organization. Attacking those satellites, on the other hand, would escalate the conflict with the United States and risk horizontal escalation with non-U.S. satellite owners.[37]

Although Chinese counterspace developments have created risks for all U.S. space systems to varying degrees, U.S. improvements have had an effect in mitigating those risks in two categories of space support: PNT and missile warning.

Risks increased for PNT and missile warning systems in the 1996–2003 time frame, but they were largely mitigated between 2003 and 2010. If further U.S. system enhancements arrive on schedule, these risks will likely continue to decline, though unforeseen Chinese developments could preclude that. The increased risk depicted in 2003 largely reflects Chinese RF jammer and laser development efforts, as discussed earlier. However, PNT (specifically, GPS) systems were subjected to additional threats. Although the risks they faced at the beginning of the period were low, they grew significantly when Russian manufacturer Aviaconversia began marketing portable equipment designed to jam downlink signals from GPS satellites in 1997 and China began working on indigenous versions of these systems.[38]

Between 2003 and 2010, however, improvements to GPS mitigated risks, largely offsetting the impact of further developments in Chinese offensive systems during that period. The Air Force had begun replacing the older generation of GPS Block IIA satellites with Block IIR (Replenishment) birds in 1997, and, by 2003, enough entered service to provide some benefit. These satellites were equipped with reprogrammable processors, enabling problem fixes and upgrades in flight, and intersatellite ranging for greater accuracy. By September 2005, 12 Block IIR satellites were operational when the first of eight GPS Block IIR-M satellites was launched. This satellite featured more powerful downlink transmitters and two new M-Code signals for improved accuracy,

---

[37] Morgan, *Deterrence and First-Strike Stability in Space*, 2010, p. 20.

[38] Daniel Kimmage, "Up in Arms Over Iraqi Arms," *Russia Weekly*, No. 251, April 3, 2003; Manuel Cereijo, "China and Cuba and Information Warfare (IW), Signals Intelligence (SIGINT), Electronic Warfare (EW), and Cyber-Warfare," Coral Gables, Fla.: Cuban-American Military Council, 2003.

encryption, and anti-jamming capabilities.[39] The seven remaining Block IIR-M satellites were launched over the next four years, with the last in the series put into orbit in August 2009.[40] Meanwhile, work also proceeded in developing technologies for making terrestrial GPS receivers more resistant to attack, such as controlled-reception-pattern antennas to null out jamming signals and Selective Availability Anti-Spoofing Module (SAASM) software upgrades at GPS ground stations to better protect the system from enemy intrusion and misdirection.[41]

The GPS network continued to become more resilient between 2010 and 2015, and more improvements are planned. In May 2010 the first Block IIF satellite was launched. Block IIF series enhancements include a new onboard-encrypted military code, crosslink improvements, signal power increases, faster processors, and more memory. As of March 2015, nine Block IIF satellites had been launched, with the last three scheduled for launch by January 2016. Block IIIA will be the first in a new constellation of GPS satellites and will provide higher data rate cross-links, a 10-dB signal strength increase, and a high-powered spot beam to meet more stringent anti-jam requirements.[42] Originally scheduled for deployment in 2015, the schedule has slipped to March 2017.[43] Given the limitations of five-color coding, and the lack of reliable information on the specific trajectory of Chinese jamming capabilities, Figure 10.1 does not show a reduction in risk to U.S. PNT capabilities for 2017, though at least limited diminution is likely.

A similar if, perhaps, less pronounced pattern can be seen in the levels of risk that missile warning satellites have faced. China's development of lasers, discussed earlier, resulted in heightened risk by 2003. The laser illumination reported in 2006 (with unnamed sources alleging that several have occurred earlier) represented an increased challenge to missile warning satellites between 2003 and 2010, but that period also saw the launch of the first two SBIRS HEO payloads in June 2006 and March 2008.

SBIRS is the follow-on to the Defense Support System satellite constellation, the central pillar of U.S. strategic missile warning since the early 1970s. Whereas each Defense Support System satellite hosts an IR sensor that scans in a single frequency band (short-wave), each SBIRS payload includes scanning and staring sensors that operate in multiple frequency bands.[44] Missile warning satellites also operate in sev-

---

[39] *Jane's Space Systems and Industry*, "GPS (NAVSTAR) Constellation," March 6, 2015.

[40] U.S. Air Force, "Global Positioning System," fact sheet, August 2010.

[41] Craig Covault, "Navigation Warfare," *The Year in Defense: Aerospace Edition*, Summer 2010.

[42] *Jane's Space Systems and Industry*, "GPS (NAVSTAR) Constellation," March 6, 2015.

[43] "First GPS III Launch Slips to FY17," *Inside GNSS*, November 14, 2014.

[44] *Jane's C4ISR and Mission Systems*, "Space-Based Infrared System," November 20, 2014.

eral orbits—with SBIRS, operating in HEO and GEO.[45] These characteristics would complicate Chinese efforts to dazzle, or blind, the system with lasers, thereby reducing the risk of attack.[46] The U.S. Air Force launched the first two SBIRS GEO satellite in 2011 and 2013, with the third and fourth expected to launch in 2015 and 2016.[47] If additional SBIRS GEO and HEO satellites are launched as expected, they will reduce risks to the U.S. missile warning system during the period to 2017.[48]

For this scorecard, Chinese counterspace capabilities would vary less by scenario (Taiwan and Spratly Islands) than they would for several of the others. Nevertheless, scenario characteristics would have some impact on outcomes. Dazzling of imaging systems, for example, requires that the dazzler be located close to the area or target being protected from observation. A Spratly Islands scenario would involve more distant Chinese naval operations than would a Taiwan scenario. China might want to counter U.S. space-based capabilities for locating its ships, but it would need to dazzle U.S. imagery satellites or jam U.S. SAR satellites from (or very close to) the vessels it is trying to obscure, thus revealing their locations and defeating the purpose of the counterspace attacks.

As a result, China would either have to abstain from attempting to counter U.S. space-based ISR or escalate to nonreversible effects attacks that it could execute from the mainland without revealing the locations of its naval forces. Given that a Spratly scenario would be less critical to China's regime survival than would a Taiwan conflict, Beijing would be less inclined to escalate to such levels there. And even if it were equally willing to conduct such attacks, China's practical counterspace capability would be somewhat weaker in a Spratly scenario than in the Taiwan case.

Conclusions

Chinese counterspace capabilities are increasing across the board, though not necessarily at a uniform pace. In a number of areas, the U.S. military is taking steps to mitigate the threat. Whether these efforts succeed in making U.S. systems safe or, at least, unat-

[45] A constellation of Precision Tracking Space System satellites operating in LEO was originally slated to complement SBIRS, but the program was canceled in 2013 after the launch of two demonstration satellites. Amy Butler, "PTSS Kill Leaves Hole in Missile Defense Sensor Plan," *Aviation Week and Space Technology*, Vol. 175, No. 14, April 29, 2013.

[46] The term *dazzle* refers to an attack on a satellite's optical sensor with a low-power laser that temporarily blinds the satellite without causing permanent damage. When space operators refer to "blinding" attacks, they are talking about laser illuminations of sufficient power to permanently damage the satellite's optical sensors.

[47] "SBIRS Tech Update Would Be Costly," *Space News*, August 4, 2014. William Graham, "ULA Atlas V Launch with SBIRS GEO-2 Successful," *NASASpaceflight*, March 19, 2013.

[48] The Air Force has also contracted for development of the long-lead-time components for SBIRS HEO-4 and GEO-4, though it is unclear if those will be launched within the time frame of this study.

tractive targets should a U.S.-China conflict occur will depend on what investments the United States makes in space defense in the coming years and whether it can find ways to reduce its systems' vulnerabilities.

Scorecard Coding

Figure 10.2 provides our summary coding of the results of the scorecard 8. The assessment of advantage in this scorecard is based on the ability of Chinese counterspace systems to hold U.S. military or dual-use satellite capabilities at risk (*Chinese advantage*) versus the ability of U.S. forces to utilize space assets unimpeded (*U.S. advantage*).

Chinese counterspace capabilities in 1996 were primitive, and we coded that period as one of *U.S. advantage*. By 2003, reports had indicated that the PLA had purchased jammers and was experimenting with lasers. Nevertheless, the Chinese showed

Figure 10.2
Scorecard 8 Summary Coding

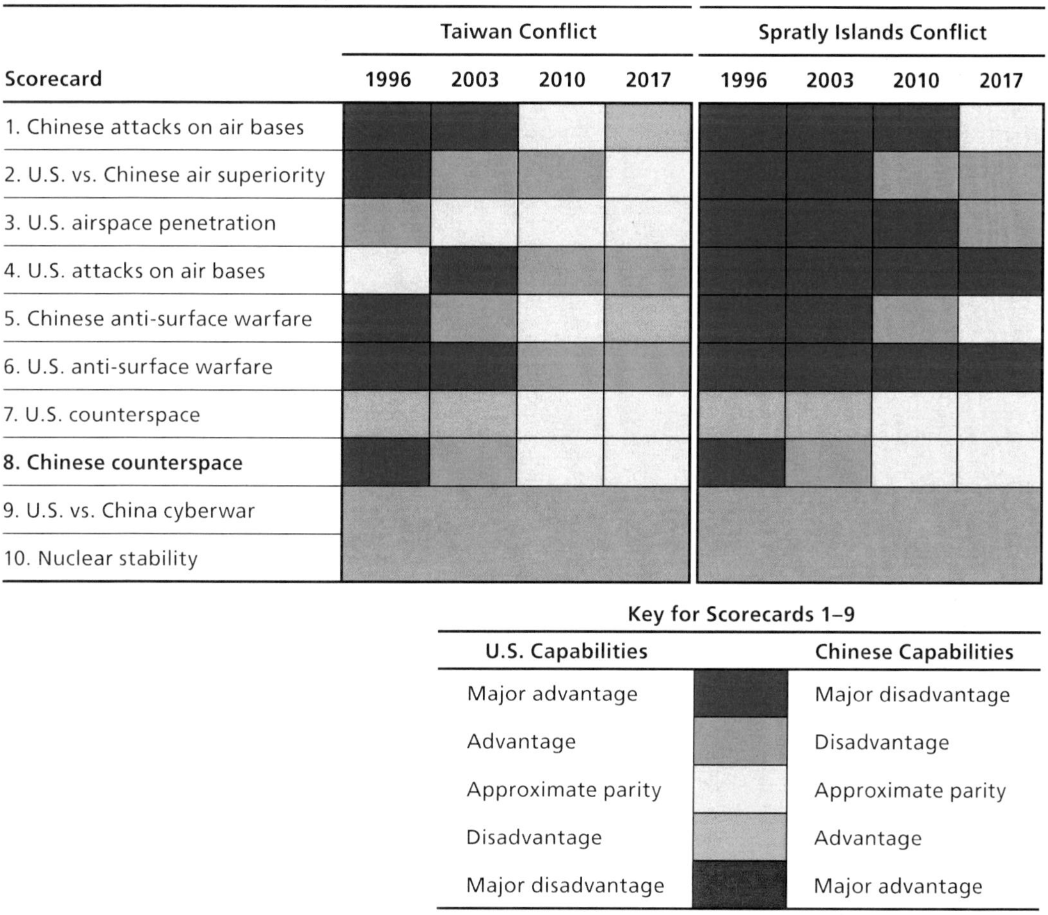

	Taiwan Conflict				Spratly Islands Conflict			
Scorecard	**1996**	**2003**	**2010**	**2017**	**1996**	**2003**	**2010**	**2017**
1. Chinese attacks on air bases								
2. U.S. vs. Chinese air superiority								
3. U.S. airspace penetration								
4. U.S. attacks on air bases								
5. Chinese anti-surface warfare								
6. U.S. anti-surface warfare								
7. U.S. counterspace								
8. Chinese counterspace								
9. U.S. vs. China cyberwar								
10. Nuclear stability								

Key for Scorecards 1–9

U.S. Capabilities	**Chinese Capabilities**
Major advantage	Major disadvantage
Advantage	Disadvantage
Approximate parity	Approximate parity
Disadvantage	Advantage
Major disadvantage	Major advantage

only limited, practical counterspace capability, leading us to code that period as one of *U.S. advantage*, though of reduced degree. Between 2003 and 2010, the PLA demonstrated a wider array of counterspace capabilities, testing kinetic ASAT approaches, illuminating U.S. satellites, and improving jamming. With a significant portion of U.S. satellite capabilities exposed to rising risks, we coded 2010 as one of *approximate parity*. China has continued to develop its counterspace capabilities, and the threat has grown. At the same time, however, the U.S. military has recently redoubled its commitment to protecting space assets, mitigating the threat in some areas. Overall, we consider 2017 to be characterized by continuing parity, though that may be edging relatively closer to Chinese advantage than it was in 2010. Because U.S. forces would be heavily dependent on space in all Asian contingencies, coding for the Taiwan and Spratly Islands scenarios is identical, though China may be less likely to employ kinetic attacks in a Spratly scenario than in a Taiwan one, in which the political stakes would be higher.

CHAPTER ELEVEN

Scorecard 9: U.S. and Chinese Cyberwarfare Capabilities

In the event of a military conflict, how would the U.S. military fare against China in the cyberspace domain? The short answer is "not as badly as some assume." The more precise answer is highly dependent on context, each side's goals, and features of each side's cyber-defense and cyber-offense capability that are, unfortunately, not well understood.

Following some general observations offered earlier in this report, this chapter addresses three component parts of the cyberwar problem. In the first two sections, we address what each side would like to achieve in the operational and strategic cyberwarfare domains, respectively, and how realistic such hopes might be given the nature of military and other information systems. Operational cyberwarfare entails attacks on military systems for the purpose of degrading the adversary's *means* of fighting. Strategic cyberwarfare largely comprises attacks on other government and nongovernment systems, largely for the purpose of degrading the adversary's *will* to fight. In the third section, we evaluate those factors that will affect relative U.S. and Chinese cyberwarfare capabilities, including doctrine, organization, materiel, leadership, network management, and approaches to zero-day vulnerabilities.

We conclude that, as a general proposition, the United States brings a much better foundation to the battle than China does. This is likely true in the offensive domain, and it is almost certainly true defensively. The bad news, not surprisingly, is that China's cyberwarfare capabilities are improving faster, and U.S. efforts cannot slacken.

Before moving on to the analysis of relative capabilities, some additional context is in order. First, the U.S. military divides computer network operations into three component parts: computer network exploitation, computer network attack, and computer network defense.[1] We focus on network attack and defense (and divide them into operational- and strategic-level subcategories), but computer network exploitation, or the use of information derived from penetrating the other sides' networks for intelligence purposes, is equally important. Indeed, since the exploitation of intelligence derived from penetrations is, on balance, less likely than network attack to tip off the

[1] For a more detailed introduction to these terms and concepts, see Jason Andress and Steve Winterfield, *Cyber Warfare: Techniques, Tactics and Tools for Security Practitioners*, Waltham, Mass.: Syngress, 2011.

adversary that its networks have been compromised, there may be a general preference for exploiting the information gained from being inside an enemy's network over measures to attack that network or the systems and capabilities it supports. The ultimate choice, though, will depend on a wide range of circumstances.

A more general point about cyberwar is that capabilities in this domain cannot be assessed as easily from quantitative indicators as they might be in the conventional and nuclear realms. Cyber operations are primarily support operations, and most *direct* effects of offensive cyber operations can be reversed relatively quickly.[2] Cyber operations also almost never involve force-on-force confrontations. One side's offensive cyber operators try to penetrate the other side's information systems, and the other side's *defensive* cyber operators and general systems engineers take action to oppose these efforts. As relatively new and rapidly evolving forms of warfare that are not readily amenable to quantification, cyberwarfare and counterspace operations share common characteristics.

Operational Cyberwarfare

Operational cyber attacks are directed against military systems and are designed to influence other, largely kinetic types of military operations. We first assess the prospects for China's use of cyberwarfare against U.S. logistics systems, as well as operationally relevant civilian infrastructure control systems—so-called supervisory control and data acquisition (SCADA) systems.[3] Because both U.S. military logistical and supporting civilian SCADA systems operate on unclassified networks, we devote particular attention to this topic. We then consider Chinese attacks against U.S. command-and-control systems, which operate on classified networks, as well as attacks designed to degrade U.S. weapon systems more directly. Finally, we address U.S. operational cyberwarfare against Chinese forces. It should be noted that although we make the distinction between operational and strategic cyberwarfare, some operational attacks, especially those against SCADA targets, may have collateral effects on national infrastructure that make it difficult for the victim to determine that a strategic attack has not been launched—a point with profound implications for escalation control.

China's Operational Cyberwarfare Against Logistics Targets

Given the U.S. dependence on a large and lengthy logistical tail, as well as the fact that U.S. logistics networks are unclassified, China's cyberwarfare efforts are likely to

---

[2] There are some exceptions to this general rule. One example is the Stuxnet virus, which did substantial physical damage to Iran's nuclear reprocessing facilities.

[3] While *SCADA systems* can refer to computerized systems controlling any civilian infrastructure (and we address attacks designed to undermine civilian morale and functioning in the section "Strategic Cyberwar," later in this chapter) here, we restrict our comments to systems that support military bases and operations.

devote considerable effort to attacking this type of target. To exploit this U.S. vulnerability, China's cyber operators would look to disrupt access to U.S. servers or to corrupt the data files that are necessary for such movements. By doing so, they would hope to force the U.S. military to revert to earlier methods of logistical operation, such as hand-carried data and manpower-intensive double-checking of inventory levels. Even better from the Chinese perspective would be a scenario in which U.S. forces ignored the potential for corruption and sent material to the wrong places often enough to slow down movement.

How much damage *could* the Chinese do to U.S. logistics operations? This is a complicated question with no simple answer, but we can derive some key observations from the information that is available. It is not impossible to penetrate DoD's unclassified network, the Non-Secure Internet Protocol Router Network (NIPRNet). Prior incidents include Moonlight Maze, a series of intrusions in the late 1990s that were thought to be the work of Russians; Titan Rain, a series of intrusions into the networks of the U.S. government and defense contractors in the mid-2000s thought to be of Chinese origin; and the more recent Byzantine series of NIPRNet intrusions, thought to be primarily of Chinese origin. In 2007, computers used by the Office of the Secretary of Defense were penetrated. Since then, reported penetrations have receded in volume and scope, though DoD's cyber defenders are far from convinced that the problem has disappeared.[4]

Nevertheless, from the perspective of Chinese cyber operators, the leap from penetration (and stealing information) to interference is nontrivial. As noted, Chinese hackers could have two possible goals: disruption and corruption.

Disruption could occur without even penetrating the NIPRNet if the Chinese can engineer a distributed denial-of-service attack that targets the NIPRNet's gateways to the outside world.[5] This could cut off DoD from its suppliers or isolate some parts of DoD from the rest of the NIPRNet. Such an attack can be nullified through architectural fixes, such as dedicated circuits between DoD and its key suppliers, gateway filters that drop mail going to isolated sites, or the use of commercial services that specialize in network load-balancing.[6]

Disruption could occur if the Chinese could infect enough clients to flood others in the network, thereby congesting it.[7] The effectiveness of a flooding attack is proportional to the number of infected clients. Clients themselves are not hard to infect.

---

[4] Sean Bodmer, Max Kilger, Gregory Carpenter, and Jade Jones, *Reverse Deception Organized Cyber Threat Counter-Exploitation*, New York: McGraw-Hill, 2012.

[5] This was the type of attack that knocked many Estonian government websites offline in 2007.

[6] The Air Force uses a company called Akamai to manage its web sites and has been spared many such incidents.

[7] Computer networks have clients (user-operated machines), servers (that hold data or provide services), routers (network connection machines), and devices (e.g., printers). These components are often but not always separate machines.

However, as more are infected, the odds are greater that anomalies in a system's performance or the random appearance of unexpected processes may clue system administrators to the attack. Furthermore, if flooding is detected, the offending systems can be disconnected.

Corruption can be carried out by infected clients or servers that make bogus changes to one or more databases using the authorization of its unwitting legal operator. The extent to which bogus information could interfere with operations depends in large part on the facilities in place to detect it. Credit card companies, for instance, have sophisticated algorithms to detect possible fraudulent activity. Similarly, the military services may have algorithms that flag suspicious activity in logistics systems and subject such transactions to human review. If so, the corrupter's tack may be to adopt a strategy that is subtle enough that the trail of corruption is hidden in noise.

To what extent could errors in the logistics system affect the ability to fight? The answer would depend on which assets with what levels of supply are in or near the theater at the time of combat. Although certain supplies (e.g., critical electronic spare parts) are brought into theater by air, most will come by sea. Crossing the Pacific takes well over a week, which means that forward-deployed units must have at least enough inventory to cope with high-intensity combat during that period. Thus, the burden of logistical errors is likely to fall lightly during the first week of combat and more heavily thereafter.

Traditionally, the U.S. military used an "iron mountain" approach, which erred on the side of overstocking at the front. Under such circumstances, logistical errors have modest effects. However, the U.S. military, like many commercial organizations, has moved from "just-in-case" to "just-in-time" inventory management. Therefore, induced logistical errors, unless caught quickly, are now more likely to interfere with operations.

U.S. logistics systems are more vulnerable to cyber attack than are many other parts of the military establishment. Nevertheless, to affect U.S. operations, Chinese cyber operators have to execute a three-bank shot: penetrating systems, translating penetrated systems into undetected (and, hence, uncorrected) logistical system errors, and inducing a significant, negative effect on operations. In preventing a Chinese attacker from accomplishing each of these steps, the role of alert and agile users cannot be overemphasized. If they are skilled or lucky, they may detect and reverse instances of penetration, catch anomalous changes to logistics databases, or preserve the ability of warfighters to carry out missions in the face of unexpected perturbations in logistics supplies.

China's Operational Cyberwarfare Against SCADA Systems

We address strategic cyber attacks on infrastructure targets in the United States, designed to undermine the U.S. will to fight, in the section "Strategic Cyberwarfare," later in this chapter. Here, we simply observe that U.S. military bases are largely dependent on external civilian infrastructure for full functionality. Although many critical base functions have backup systems that may be relatively impervious to attack (e.g., backup power systems that employ gasoline- or diesel-powered generators), the loss of external power, water services, and telecommunication support would degrade base capabilities. Hence, cyber attacks against a limited set of civilian targets could have significant operational effects, especially if an attack (or its effects) could be sustained.

SCADA targets are on unclassified systems and may be even less secure from penetration and attack than logistics systems. Moreover, U.S. bases in Japan and other overseas locations rely on external infrastructure support and may be particularly vulnerable. Not only will the infrastructure systems of host countries be characterized by varying standards of security, but U.S. military authorities overseas may also have limited options in negotiating with local utilities or governments on the sequence and terms of the restoration of services. While many of the relevant infrastructure targets could also be struck with kinetic attacks, using cyber attacks may, in addition to conserving lethal systems for other uses, be viewed as less escalatory, particularly if the country hosting U.S. forces has not entered active hostilities.

China's Operational Cyberwarfare Against Command-and-Control Systems

Attempts to interfere with command-and-control systems would require Chinese hackers to penetrate DoD's Secure Internet Protocol Router Network (SIPRNet). Were they able to do so, the results would be more immediately harmful to mission assurance than in the case of attacks on logistics systems because the impact on operations would be direct rather than indirect. But can Chinese hackers get into the SIPRNet? Publicly available evidence suggests that Chinese analysts believe that it would be difficult, though perhaps not impossible, to do so directly. China's primary cyberwarfare goals are to interfere with unclassified NIPRNet systems, foster propaganda, and, perhaps, leverage the ability to penetrate NIPRNet systems to interfere with the SIPRNet.[8] Nevertheless, consider some alternative methods by which Chinese hackers *could* attack the SIPRNet.

First, they could penetrate the unclassified NIPRNet and corrupt enough clients to flood the dual-use routers that carry both (unencrypted) unclassified traffic and (encrypted) classified traffic. The immediate effects of such an attack would depend on the configuration of the SIPRNet (the details of which are not publicly available). Some examples include which routers are susceptible to flooding attacks, what components of the SIPRNet are their own enclaves (and can therefore communicate with each other if the wider SIPRNet bogs down), and the extent to which the SIPRNet's services are cached or otherwise accessible if the primary servers are unavailable. The

---

[8] A Chinese attack on the NIPRNet may impede encrypted traffic flow if it moves across an unclassified backbone, which may be easier than penetrating the SIPRNet itself. See Bryan Krekel, *Capability of the People's Republic of China to Conduct Cyber Warfare and Computer Network Exploitation*, McLean, Va.: Northrop Grumman Corporation, October 9, 2009, p. 28.

effects also depend on how network defenders react.[9] It is also possible that while large portions of the SIPRNet would be inaccessible, those sections capable of supporting command and control may be usable.

Second, the Chinese could try to attack the SIPRNet directly. In theory, the SIPRNet is air-gapped, which means that it is physically and electromagnetically isolated from other, unsecure networks. In principle, this means that onsite access is required for an attack. However, as an intrusion reported in Afghanistan in late 2008 demonstrated—and Stuxnet reconfirmed—supposedly air-gapped systems can be infected by malware loaded onto removable media.[10] Notably, USB sticks could infect an air-gapped machine if plugged into such a device. Such an infection would move at an unpredictable and often slow pace, however.[11]

Using such an attack to transfer information from the SIPRNet to the attacker would require two such transfers: one, inbound, to infect the machine and the other, outbound, to carry the information sought. Timing an attack on an air-gapped system to launch when a war starts would require triggering the malware when needed. Success would depend on a very uncertain chain of events under conditions in which information security protocols are likely to be even tighter than normal. Stuxnet, by contrast, was a one-way weapon designed to make centrifuges self-destruct as soon as possible.

Third, China could attempt physical penetration. The level of physical protection across the SIPRNet, with its hundreds of thousands of clients, varies greatly. China's confidence in its ability to carry out such an attack requires a technologically proficient commando force. More to the point, once the war started, conducting such operations would be difficult and risky. If it were unable to successfully execute such an attack after the war started, the PLA would have to rely on the iffy triggering mechanisms discussed earlier.

---

[9] If the source of the flooding is outside the NIPRNet, local system administrators might be able to detect the sources of traffic and disconnect them. They do not have to catch every such machine if their goal is simply to reduce the load on servers to a level that ensures functionality. If all else fails, disconnecting all NIPRNet clients and servers from routers that also carry SIPRNet traffic might permit the SIPRNet to reclaim its functionality. But substitutes for NIPRNet functions, such as situation and logistics reports, would have to be found.

[10] In this incident, malware infected a computer on the NIPRNet. An infected thumb drive was plugged into a machine on the SIPRNet, leading to multiple infections of other SIPRNet-connected machines. Although infections within the SIPRNet are quite uncommon they are not unprecedented. See C. C. Mann, "The Mole in the Machine," *New York Times Sunday Magazine,* July 25, 1999.

[11] Computers in an air-gapped system can also be configured to avoid being infected. Prior to Stuxnet, it was widely believed that preventing systems from booting up from a removable drive would suffice, but Stuxnet revealed a flaw in Microsoft Windows that permitted an infection even if the system was so configured. The hole was found and patched, but there is no guarantee that someone will not discover another vulnerability. Since the Stuxnet attack, the military has clamped down on the use of removable devices on computers accessing the SIPRNet.

Fourth, the Chinese could manufacture a corrupted device and find some way of getting it purchased and inserted into the SIPRNet. When triggered, the device would carry out instructions designed to corrupt or disrupt SIPRNet operations. Concern about this possibility within the cyber defense community has resulted in increased scrutiny of DoD purchases. Thus far, no such corruption has been brought to light.[12]

Classified networks are clearly becoming more important to warfighting. Aircraft targeting packages, Navy ship radar operations, and Army force trackers are all growing increasingly dependent on these networks. Yet even in the unlikely event that the SIPRNet collapsed, the United States would have powerful forces that would be relatively unaffected. The nuclear establishment is hardened against such failures. Submarines can operate largely autonomously. RF networks can supply a great deal of information that might otherwise be carried by the network. If a network is inoperable, other means of command and control can be employed. Voice-to-voice communication is an alternative conduit for command and control.

The military services and U.S. Cyber Command (USCYBERCOM), working together, are beginning to study the potential impact of cyber attacks on mission effectiveness. The early emphasis was on understanding communication vulnerabilities, but officials are also considering the corruption of databases and message traffic. The U.S. military is devoting increasing levels of attention and funds to cyber defense. Despite increased austerity in military programs, the 2016 defense budget proposal includes an increase in "cyber spending" from $4.9 billion to $5.5 billion.[13] The Department of Defense has been given authority to "fast track" the hiring of roughly 3,000 civilian cyber operators to fill out the half-full Cyber Command.[14] And the National Security Agency (NSA) is working to provide hands-on experience in cyber defense to young military officers through an annual cyber defense competition that, since 2001, has pitted teams from U.S. military academies and graduate institutions against one another.[15]

China's Operational Cyberwarfare Against Weapon Systems

Even harder to assess is the possibility that China could attempt to degrade the performance of U.S. weapon systems through network attack. U.S. military forces are highly networked, and individual air and naval platforms are now large data pro-

---

[12] In early July 2011, Greg Shaffer, acting Deputy Under Secretary of Homeland Security, was asked whether any software or hardware components embedded with security risks had been installed. He replied, "I am aware that there have been instances where that has happened." See Aliya Sternstein, "Threat of Destructive Coding on Foreign-Manufactured Technology Is Real," Nextgov, July 7, 2010. However, without specifics, it is hard to evaluate the accuracy of that claim.

[13] "The Military's Cybersecurity Budget in 4 Charts," *Defense One*, March 16, 2015.

[14] "Pentagon Moves to Hire 3K Cyber Workers," *The Hill*, March 8, 2015.

[15] National Security Agency, "U.S. Military Academy Wins NSAs 14th Annual Cyber Defense Exercise," press release, April 11, 2014.

cessing centers. Aircraft now have computerized rather than mechanical controls, and missiles and aircraft are networked to one another as well as to external data centers. For example, F-35 operations depend on roughly 24 million lines of code, with some 9.5 million onboard the aircraft (six times the number onboard the F-16).[16] While this networking has greatly improved the performance of U.S. military forces, it also potentially makes them vulnerable to cyber attack. Corrupting software to disable systems, inducing false targets, making small changes to GPS feeds, or even hijacking drones or other subsystems are all possibilities. Depending on the nature of the attack, it may or may not be possible to quickly restore functionality by, for example, rebooting systems using backed-up or older software.

U.S. Operational Cyberwarfare Capabilities

U.S. offensive cyber operations against the PLA would be more challenging to execute than the PLA's operations against the United States. China sits much closer to the relevant operational areas and depends less on computer networks for warfighting. Proximity means that China's command and control is not as dependent on long-distance communication, and China has more options for getting messages to its units. Nevertheless, the cost of carrying out cyber attacks is relatively low, once the requisite intelligence has been gathered on the targets. Thus, U.S. forces will seek to penetrate Chinese connected and air-gapped networks to gather intelligence, if not to disrupt or corrupt operations.

Two other factors are also likely to govern the penetrability of China's connected networks. One is that most of China's personal computers use bootleg versions of Microsoft® Windows®,[17] and those pirated systems are harder to keep patched than their legal counterparts. According to a 2015 report by the Spanish security vendor Panda Labs, 49 percent of Chinese computers are infected with malware, the highest proportion of infected computers in the world.[18] A 2012 spot check by the Microsoft Corporation found that not only were pirated versions of its software preloaded onto new computers sold in China, but malware was also embedded in the software.[19] The other factor is that China places less emphasis on providing its warfighters with broad

---

[16] Michael J. Sullivan, Acquisition and Sourcing Management, U.S. Government Accountability Office, *Joint Strike Fighter: Restructuring Added Resources and Reduced Risk, but Concurrency Is Still a Major Concern*, statement before the Subcommittee on Tactical Air and Land Forces, Committee on Armed Services, U.S. House of Representatives, Washington, D.C., GAO-12-525T, March 20, 2012.

[17] Jon Brodkin, "Ballmer to Hu: 90% of Microsoft Customers in China Using Pirated Software," *Network World*, January 21, 2011; "Windows 10 Will Be a Free Upgrade for All Users Worldwide," *Ars Technica*, March 18, 2015.

[18] Panda Security, *PandaLabs Annual Report 2014*, Maitland, Fla.: Panda Security, 2015. See also Phil Muncaster, "China Is the World's Most Malware-Ridden Nation," *The Register*, February 7, 2013.

[19] Jeremy Kirk, "Microsoft Finds New Computers in China Preinstalled with Malware," *IDG News Service*, September 13, 2012.

access to information, though the recent PLA emphasis on "informationized" warfare may be changing this condition.[20]

Obvious targets for U.S. cyber operators would be the Chinese IADS and maritime ISR systems. Attacks on IADS could disconnect those systems from one another or create false radar images. Russian-made IADS systems in Syria and Iraq have purportedly been successfully attacked in cyberspace.[21] China's IADS, many of which have similar origins, may therefore be vulnerable. Alternatively, now that such vulnerabilities have been widely reported, the Chinese may have patched whatever problems previously existed. Cyber attacks against China's maritime ISR capabilities may prevent the PLA from targeting U.S. ships or incoming aircraft. Given the temporary nature of cyber effects and the difficulty of fooling the same system twice in the same way, U.S. cyber attacks against Chinese IADS and ISR targets are most likely to be successful early in a conflict.

Cyber attacks may achieve more permanent effects if the targets are similar in nature to industrial controllers—that is, computers or other devices that can be reprogrammed *in situ* and that can command equipment to carry out self-destructive operations (as happened to Iran's centrifuges in Natanz) or make the equipment easier to kill (e.g., by emitting telltale RF energy).[22]

Because China would be conducting operations in or close to its own territory, dual-use infrastructure is likely to be more important to China's ability to wage war than it would be for the United States. From the U.S. perspective, this creates opportunities and challenges. The opportunities arise from the prospect that owners of dual-use infrastructure will be less attuned to security requirements and may be concerned about the high financial cost of security.[23] Once a war starts, security would likely assume greater importance. Yet unless the Chinese employ dramatic short-term responses (e.g., pulling all network connections), they will have a difficult time ensuring total security in the initial weeks of a conflict. However, such measures would likely affect Chinese operations in other ways.

Weighed against their likely impact, cyber attacks against dual-use targets raise escalatory concerns. To be sure, this is true of many other forms of warfare (e.g., interdiction), but there is particular danger in cyberwar, with its inchoate rules of engage-

---

[20] Bryan Krekel, *Capability of the People's Republic of China to Conduct Cyber Warfare and Computer Network Exploitation*, 2009.

[21] See Richard Clarke and Robert Knake, *Cyber War: The Next Threat to National Security and What to Do About It*, New York: HarperCollins, 2010.

[22] This can also include reprogramming on a specialized networked device that is separate from the equipment itself.

[23] However, as Timothy Thomas, an expert on Russian and Chinese cyberwar capabilities, argues, the tendency of Chinese leaders to be engineers rather than lawyers and financiers (as is more typical in the United States) may sensitize them to the importance of engineering issues, such as good security. See Timothy L. Thomas, "China's Electronic Long-Range Reconnaissance," *Military Review*, November–December 2008.

ment, lack of international consensus on the legitimacy of different types of targets, and lack of meaningful experience to underpin the understanding of collateral effects.

Strategic Cyberwarfare

Strategic cyberwarfare involves attacks on other government and nongovernment systems for the purpose of affecting the adversary's *will* and *capacity* to sustain combat.

Chinese Strategic Cyberwarfare Activities

There is considerable disagreement over whether the Chinese would try to use cyber attacks to erode the *will* of the U.S. public to wage war over Taiwan. Authors such as Richard Clarke assume they would, admittedly in a scenario in which the United States appears to have attacked China's dual-use infrastructure first.[24] The media has widely trumpeted reports that various state actors have embedded implants in the U.S. electric power infrastructure to activate attacks that would lead to mass blackouts.[25] Published research by Chinese nationals on the vulnerability of the U.S. power grid has also been cited as evidence of Chinese interest in the topic.[26]

However, there is no broader evidence of Chinese interest in developing a strategic cyber capability, and good reasons why it may not pursue such capabilities.[27] The growing interconnectedness of the U.S. and Chinese economies would make strategic warfare very costly to both sides. In 2013 and 2014, U.S. firms invested about three and a half times as much in China as Chinese firms did in the United States.[28] Foreign investment, including U.S. investment, was critical to China's ability to sustain export growth, especially in high-tech sectors.[29] Even the threat of strategic cyberwar could persuade U.S. and other Western companies to limit their exposure in China. If so,

---

[24] See Clarke and Knake, *Cyber War,* 2010, pp. 179–218.

[25] Siobhan Gorman, "Electricity Grid in U.S. Penetrated by Spies," *Wall Street Journal*, April 8, 2009.

[26] In early 2010, U.S. observers discovered a study titled "Cascade-Based Attack Vulnerability on the U.S. Power Grid," by Jian-Wei Wang and Li-Li Rong of the Institute of System Engineering (at Dalian University of Technology) and announced it as proof of China's malign intentions. See John Markoff and David Barboza, "Academic Paper in China Sets Off Alarms in U.S.," *New York Times*, March 20, 2010.

[27] A 2009 report on Chinese cyber warfare capabilites prepared for the U.S.-China Economic and Security Review Commission conspicuously fails to mention Chinese interest in developing such a capability. Krekel, *Capability of the People's Republic of China to Conduct Cyber Warfare and Computer Network Exploitation*, 2009.

[28] David Dollar, "Why So Little Investment Between the United States and China?" Brookings Institution, February 26, 2015; Thilo Hanemann and Cassie Gao, "Chinese FDI in the United States: Q4 and Full Year 2014 Update," *Rhodium Group*, January 15, 2015.

[29] A 2012 report found that fully 85 percent of China's high-technology exports were generated by foreign-invested firms, a figure that had changed only slightly since the early 2000s. Yuqing Xing, "The People's Republic of China's High-Tech Exports: Myth and Reality," ABDI Working Paper, April 2012.

China would suffer a long-term loss in competitiveness (although that loss will decrease as Chinese firms become more capable of producing high-value-added goods).

The potential growth in cloud computing—the tendency of organizations to outsource their data to third parties—could also militate against strategic cyberwarfare tactics. In the future, many organizations may not know which country stores their data, but national authorities may still be able to tell owners of server farms to disable such services for wartime foes. Service could be lost even for organizations that otherwise practice faultless cyber security. Thus, a state that starts a strategic cyberwar may be subject to uncertain but potentially crippling consequences, even if the other side does not retaliate in kind. Unpredictability may be reminiscent of the fallout from the 2008 banking crisis or the unforeseen effects of the March 2011 tsunami on Japanese industrial production.

China's strategists may also contemplate the strong possibility that a strategic cyber attack may *raise* rather than *lower* the likelihood that the United States will persist in its defense of Taiwan. The Chinese narrative on Taiwan emphasizes that the conflict is an internal struggle between two parts of the same country and therefore of no concern to others. Once the United States has been subject to a large strategic cyber attack, the struggle over Taiwan would be seen as part of a broader strategic conflict in which U.S. credibility is at stake, making it impossible for Washington to remain on the sidelines.

U.S. Strategic Cyberwarfare Capabilities

Many of the same constraints on China's decision to engage in a strategic cyberwar with the United States will affect U.S. decisionmaking, but there are some key differences.

First, attacking China's dual-use infrastructure, as noted, would likely have a more immediate impact on the course of a conventional conflict than comparable attacks by China on the U.S. civilian infrastructure. However, if China interprets such attacks as an attempt by the United States to shift its goals from preserving Taiwan's status to destabilizing China, then the conflict could escalate from the operational level (warfighting) to the strategic level in cyberspace.

Second, in any limited engagement, the United States is likely to be sensitive to the fact that many aspects of strategic war contravene the laws of armed conflict—notably, those that enjoin states not to attack civilian targets. Therefore, the United States, which is sometimes willing to subordinate the search for advantage to the promotion of international norms, might be inclined to forfeit the possible advantages of strategic cyberwar, especially if China has not attacked first. That said, the United States has a long history of using strategic bombing, despite debates about its status under the law.

Comparative Military Cyberwarfare Capabilities

We now turn to factors that are likely to affect relative U.S. and Chinese warfare capabilities, including doctrine, organization, materiel, leadership, network management, and approaches to zero-day vulnerabilities. Before turning to those factors, however, several broad issues related to each country's cyberwarfare capabilities merit note.

First is the current U.S. dominance in software. China has made great progress in hardware (e.g., products from Huawei and Lenovo) and has largely nationalized its portion of the web (e.g., with what is effectively a near-monopoly of the search market by Baidu). However, U.S. firms (such as Microsoft, Intel, Google, Adobe, Facebook, Symantec, RSA, Apple, and Cisco) still dominate international software and the de facto software-related standardization process. No Chinese firm even comes close to challenging this corporate influence, and none is likely to be able to do so within the next five years. Of the top 100 companies ranked by software revenue in 2014, U.S. companies occupied 67 spots. Chinese companies held only two, and the largest of those, Neusoft, was ranked 71st and had software revenue 0.8 percent as large as Microsoft's.[30]

Software leadership tends to be self-sustaining. Many Chinese leaders and analysts believe that U.S. companies answer to the U.S. government.[31] They also believe that the United States has the ability to corrupt or disrupt the functioning of any device with U.S.-made software. Their appetite for a full-fledged cyberwar may be proportionately limited, and their receptivity to U.S. hints that it could unleash such cyberwar itself may be heightened.[32] Such considerations would nevertheless be unlikely to affect significantly their approach to operational cyberwar.

Second, China is narrowing the gap in science and technology overall. According to statistics provided by the SCImago Journal and Country Rank, which tracks worldwide scientific publications, the number of Chinese scientific papers was 19 percent that of the United States in 2003. By 2013, it was 76 percent that of the United States in the sciences overall, and China published 12 percent *more* papers than the United States in the area of computer science.[33] The gap in the importance of those papers, as measured by the average number of times each paper was cited, also narrowed over the past decade.[34] Although there are important qualifications in these data—such as

---

[30] PwC Technology Institute, "PwC Global 100 Software Leaders," March 2014.

[31] This information comes from our interviews with Chinese strategic thinkers. The belief persists even though the Chinese have requested and received (most of the) source code for Microsoft Windows and may well be stealing source code from the other firms, such as Google, as in the case of Operation Aurora.

[32] Making overt threats would be chancy because U.S. firms would quickly address vulnerabilities once they have been brought to their attention.

[33] SCImago Journal and Country Rank Database, accessed April 11, 2015.

[34] Chinese computer science papers published in 2003 were cited at a rate 37 percent that of the United States. For papers published in 2013, the rate was 66 percent. SCImago Journal and Country Rank Database, accessed April 11, 2015.

the high incidence of plagiarism in Chinese scientific publications—the broader trends are nevertheless clear.[35]

Third, Chinese cyber operations have gained enormous critical attention for their wholesale theft of industrial data, as well as their relentless efforts to access military secrets. There has been speculation in public sources for many years that Chinese military hackers were involved in corporate cyber espionage. Those suspicions were effectively confirmed by a 2013 report by the cybersecurity company Mandiant, which employed a variety of forensic methods to identify the sources, scale, and yields of specific PLA cyber espionage efforts against commercial targets. The report concluded that Unit 61398, of the Second Bureau, belonging to the 3rd Department of the PLA General Staff Department, had stolen "hundreds of terabytes of data from at least 141 organizations across a diverse set of industries beginning as early as 2006."[36] In May of 2014, the U.S. Justice Department handed down indictments for five members of the same unit, listing the alleged crimes and victims of each.[37]

But penetrating networks, while an important component of cyberwarfare, is far from cyberwarfare itself. Creating effects requires being able to understand the relationship between the target system and the overall warfighting effort, as well as the ability to create and monitor effects and counter the target's ability to mitigate them. In other words, nonkinetic and kinetic effects must be integrated. Whether Chinese forces—or U.S. forces, for that matter—have mastered these tasks is unknown. Among cybersecurity experts, some see the United States and Russia (with its cadres of top-flight mathematicians and high-prestige security bureaucracies) in a category by themselves, with China ranked one step below them.[38]

Doctrine

Doctrine is an important component of a nation's cyberpower. It not only indicates the relative importance of cyber operations, but it may also provide clues about their

[35] According to an investigation by the editor of the *Journal of Zhejiang University–Science*, fully 40 percent of papers on computer science topics submitted over two years were not publishable due to plagiarism issues. The investigation employed plagiarism detection software. See Louisa Lim, "Plagiarism Plague Hinders China's Scientific Ambition," *National Public Radio*, August 3, 2011.

[36] Mandiant Corporation, "APT1," 2013, p. 20.

[37] U.S. Department of Justice, "U.S. Charges Five Chinese Military Hackers for Cyber Espionage Against U.S. Corporations and a Labor Organization for Commercial Advantage," Press Release, Washington, D.C., May 19, 2014.

[38] In testimony to the U.S. House of Representatives, Frank J. Cilluffo, director of the Homeland Security Policy Institute and co-director of the Cyber Center for National and Economic Security at George Washington University, groups China with Russia but stipulates that "Russia's cyber capabilities are, arguably, even more sophisticated than those of China." Frank J. Cilluffo, "Cyber Threats from China, Russia and Iran," testimony before the Committee on Homeland Security, U.S. House of Representatives, Washington, D.C., March 20, 2013. On Russian organization for cyberwarfare, see Jeffrey Carr, *Inside Cyber Warfare: Mapping the Cyber Underworld*, 2nd ed., Sebastopol, Calif.: O'Reilly Media, 2011.

effectiveness.[39] Twenty years ago, China introduced the concept of "local wars under informationized conditions," calling for "a fighting force capable of winning high-technology wars under modern conditions, and [providing] the asymmetric means by which the weak could defeat the strong."[40]

Chinese leaders have since worked to modernize a broad spectrum of conventional warfighting capabilities. They have also contemplated how to defeat the United States by attacking its center of gravity—particularly its dependence on networks. Authoritative Chinese sources describe information warfare (which includes both electronic warfare and cyber operations) as the most important form of warfare. The 2013 *Science of Military Strategy*, a seminal PLA-wide document published roughly once a decade, stipulates that "The side holding network warfare superiority can adopt network warfare to cause dysfunction in the adversary's command system, loss of control over his operational forces and activities, and incapacitation or failure of weapons and equipment—and thus seize the initiative within military confrontation, and create the conditions for . . . gaining ultimate victory in war."[41]

It is perhaps not surprising that the Chinese put more *relative* emphasis on information warfare in general, and cyberwarfare in particular, than the United States does. U.S. military strength is sufficiently broad and deep that the country can talk plausibly about "full spectrum dominance" (a term from *Joint Vision 2020*).[42] China, by contrast, is more inclined to look for niche capabilities that can thwart plans for the United States to operate in East Asia. The *Science of Military Strategy* argues that, in contrast to conventional forces, "computer network operations require only small numbers of personnel and relatively low investment of funds to achieve operational goals; computer network operations thus have the characteristics of low cost, high benefit, and low risk."[43]

---

[39] Several factors complicate judgments about the quality of a state's cyberwarfare capabilities derived from its cyberwar doctrine. It is not always clear which doctrinal statements or authors are influential, and distinguishing between useful and harmful doctrine before either is used in combat is a fraught exercise. It is possible to determine whether some doctrine is more or less sophisticated, but sophistication is not the same as efficacy. Furthermore, the gap between good doctrine and its implementation can be large. The Soviet Union invented the concept of "reconnaissance-strike complex" before concluding that the United States was building such a capability while the Soviets could not. See Paul S. Giarra, *A Chinese Anti-Ship Ballistic Missile: Implications for the USN*, statement before the U.S.-China Economic and Security Review Commission, Washington, D.C., June 11, 2009.

[40] Ashley J. Tellis, "China's Military Space Strategy," *Survival*, Vol. 49, No. 3, September 2007, p. 51.

[41] 战略学, [*The Science of Military Strategy*], 3rd ed., Beijing: Military Science Press, 2013, p. 189.

[42] *Joint Vision 2020* was released in 2000, and the language of "full spectrum dominance" may have since fallen from favor. Nevertheless, the point remains that the United States maintains a wide array of fully mature capabilities.

[43] 战略学 [*The Science of Military Strategy*], 3rd ed., 2013, pp. 190–191.

Chinese strategists have been thinking and writing about cyberwarfare for more than two decades.[44] Authoritative Chinese writings categorize network operations in ways that partly overlap with those of the United States. The *Science of Military Strategy* describes three types network operations: network reconnaissance, network attack and defense operations, and network deterrence.

The same source describes network reconnaissance as the "most common" form of military cyber operation today. The text also suggests a strong identity between what the U.S. military would regard as computer exploitation activities on the one hand and reconnaissance on the other. "Reconnaissance" includes, for example, "exploiting loopholes in the adversary's computers to sneak into the adversary's network systems, and via spyware collect and steal information stored and processed in those computers."[45] Other Chinese sources would seem to agree with the identity between reconnaissance and exploitation. Zhu Wanguan and Chen Taiyi argue in their 1999 book *Information War*, "we can . . . enter networks as different users to do the surveillance . . . and borrow hackers to finish computer surveillance tasks."[46] Chinese texts generally do not highlight the tension between reconnaissance and exploitation or the risk that exploitation could alert an adversary to vulnerabilities and lead him to patch them.

The *Science of Military Strategy* suggests that network attack and defense operations are "the highest form of military struggle in the network domain."[47] Although Chinese authors emphasize the importance of defense, they also see cyber warfare as inherently offense-dominant: "Networks in integrated-whole terms have the features of susceptibility to attack and difficulty of defense, and this asymmetric quality of network attack and network defense is prominent."[48] One function of network attack is to retard the movement of U.S. logistics across the Pacific by targeting "harbors, airports, means of transportation, battlefield installations, and the communications, command and control and information systems."[49] A second function is to frustrate enemy C4ISR more broadly, including communications, radar, space-based systems, and military command and control.[50]

---

[44] George Patterson Manson III, "Cyberwar: the United States and China Prepare for the Next Generation of Conflict," *Comparative Strategy*, Vol. 30, No. 2, 2011; and James Mulvenon, "Chinese Information Operations Strategies in a Taiwan Contingency," testimony before the U.S.-China Economic and Security Review Commission Hearing, Washington, D.C., September 15, 2005.

[45] 战略学 [*The Science of Military Strategy*], 3rd ed., 2013, p. 192.

[46] Cited in Thomas, "China's Electronic Long-Range Reconnaissance," 2008.

[47] 战略学 [*The Science of Military Strategy*], 3rd ed., 2013, pp. 192–193.

[48] 战略学 [*The Science of Military Strategy*], 3rd ed., 2013, pp. 192–193.

[49] Lu Linzhi, "Preemptive Strikes Crucial in Limited High-Tech Wars," *Jiefangjun Bao* [*PLA Daily*], February 14, 1996.

[50] Kevin Pollpeter, "Chinese Writings on Cyberwarfare and Coercion," in John R. Lindsay, Tai Ming Cheung, and Derek S. Reveron, eds., *China and Cybersecurity: Espionage, Strategy, and Politics in the Digital Domain*, London: Oxford University Press, 2015, p. 152.

In this view, cyberwarfare is part of integrated network electronic warfare, the aim of which is to maintain information superiority on the traditional battlefield by controlling the flow of information available to the enemy.[51] China's cyber strategists appear to put greater emphasis on stopping the flow of 0s and 1s (i.e., computer data) than they do on corrupting data and making them harmful or untrustworthy.[52] This physical approach to the virtual world suggests that cyberwarfare is understood in terms that would be comfortable to the longer-established electronic warfare community.

The *Science of Military Strategy* defines network deterrence as "actions which display network attack and defense operational capability, and the firm resolve for retaliation, to prevent the adversary from daring to carry out large-scale network attacks." The text then goes on to make clear that deterrence operations are designed primarily to prevent "large-scale network attacks," which are defined as attacks with a "strategic" quality, or those that could have an impact on "the security and development interests" of the state.[53] Notably, the means of deterrence are not limited to network attack and defense capabilities "but also include the traditional military strike forces and means."[54]

The most frequently employed Chinese term for "deterrence," 威慑, can also mean "coercion." The authors of the *Science of Military Strategy* are clear that they use the term in the former sense, but other Chinese writers have employed it in the latter. For example, the 2000 version of the *Science of Military Campaigns*, a book published by the Chinese National Defense University Press, suggests, "We must send a message to the enemy through computer network attack, forcing the enemy to give up without fighting."[55] A similar theme can be found in the 1999 book, *Unrestricted Warfare*, which argued that, if circumstances warrant, China should not refrain from a no-holds-barred attack on China's adversaries that transcends the traditional battlefield.[56] More broadly, Chinese strategists highlight the psychological impact of cyber warfare and the potential for exploiting such effects.[57]

Significant tensions exist between different Chinese cyber warfare priorities. Chinese strategists emphasize the importance of surprise, yet they understand that revealing capabilities would be critical to network deterrence. More broadly, Chinese strategists see cyber warfare as inherently offense-dominant and may understand that that

---

[51] Jean-Loup Samaan, "Beyond the Rift in Cyber Strategy," *Strategic Insights*, Vol. 10, No. 1, Spring 2011, p. 10.

[52] Krekel, *Capability of the People's Republic of China to Conduct Cyber Warfare and Computer Network Exploitation*, 2009, p. 15.

[53] 战略学 [*The Science of Military Strategy*], 3rd ed., 2013, p. 193.

[54] 战略学 [*The Science of Military Strategy*], 3rd ed., 2013, p. 194.

[55] Cited in Mulvenon, "Chairman Hu and the PLA's 'New Historic Missions,'" 2009.

[56] Qiao Liang and Wang Xiangsui, *Unrestricted Warfare: China's Master Plan to Destroy America*, Panama: Pan American Publishing Company, 2002.

[57] Kevin Pollpeter, "Chinese Writings on Cyberwarfare and Coercion," in Lindsay, Cheung, and Reveron, eds., *China and Cybersecurity*, pp. 150–151.

the boundaries between military and civilian targets can be vague. Yet they appear to place considerable faith in, and certainly emphasize, network deterrence operations. The tensions between different aspects of Chinese doctrine are perhaps not surprising, given that the cyber domain is relatively new and that it is rapidly evolving. To an extent, contradictory writing may reflect debates within the Chinese cyber community or the evolution of its thinking.

U.S. doctrine on cyberwarfare is in its infancy, a fact that has slowed the evolution of military occupational specialties and the development cyber hardware. At the broadest operational level, cyberspace is considered a fifth warfighting domain, and the doctrine for that domain has yet to depart dramatically from the doctrine associated with the physical domains.[58]

Organization

China has a large cadre of cyber warriors, but the broad array of organizations and interests involved almost certainly represents a daunting challenge to Chinese leaders. The West's understanding of different parts of this structure varies greatly, depending on the Chinese mission and organizations in questions. The most important actors involved in cyber warfare are thought to be the PLA, the Ministry of State Security (MSS), and the Ministry of Public Security (MPS), each with different but overlapping functions.

Although Chinese officials continue to deny cyber espionage, the 2013 *Science of Military Strategy* acknowledged, for the first time, the existence and basic structure of Chinese organization for cyber warfare. The text provides a window into the thinking of professional military officers on an appropriate hierarchy of organization for cyber warfare:

> The forces employable in network operations can be divided into three types: armed forces professional network warfare forces, authorized forces, and civilian forces. Professional network warfare forces are armed forces operational units specially employed for carrying out network attack and defense; authorized forces are organized local forces authorized by the armed forces to engage in network warfare, mainly built within the associated government departments, including the Ministry of State Security and the Ministry of Public Security; and the civilian forces are nongovernmental forces which spontaneously carry out network attack and defense, and which can be employed for network operations after mobilization.[59]

In general, the roles of the MSS and MPS in cyber operations are less well understood than those of the PLA. The MPS is effectively the national police force and is

---

[58] On the five objectives established for U.S. military cyber security forces, see U.S. Department of Defense, *The Department of Defense Cyber Strategy*, April 2015.

[59] 战略学 [*The Science of Military Strategy*], 3rd ed., 2013, p. 196.

primarily charged with the maintenance of law and order. It monitors, and may attack, foreign hostile forces, and its 11th Bureau has a role in enforcing regulations related to the defense of critical infrastructure, cryptography, and information security systems. The MSS is charged with foreign intelligence, as well as counterespionage and counterintelligence.[60] In October 2014, a Western cyber security firm identified the MSS as behind the Axiom attacks, which targeted computer firms, government agencies, and overseas dissidents.[61]

Relatively more is known about the PLA's organization for network operations, especially for "reconnaissance," an activity that includes exploitation (and theft). The PLA General Staff Department (GSD) 3rd Department (originally signals intelligence and technical capabilities) is tasked with reconnaissance and is roughly analogous to the U.S. NSA.[62] The 3rd Department is divided into number of elements: 12 operational bureaus, each with either geographic or functional responsibilities; three research institutes; and 16 technical reconnaissance bureaus that support the seven military region headquarters and PLA services and branches.[63] Many of these elements are almost certainly involved in ongoing cyber espionage.[64]

Less is known about the PLA's organization for computer network attack. Potential peak organizations for network attack include the GSD's 3rd Department, 4th Department (originally ECM and radar), or Second Artillery.[65] A 2007 DoD report to Congress stated that the PLA's information warfare units "develop viruses to attack enemy computer systems and networks." After 2005, the report continues, "the PLA began to incorporate offensive CNO [computer network operations] into its exercises, primarily in first strikes against enemy networks."[66]

Much has been made of the PLA's so-called cyber militias and the employment of civilian hackers. As early as 2002, the PLA "recruited civilians into its 'net militia units' (Militia Information Technology Battalions)." These units are staffed by aca-

---

[60] Nigel Inkster, "The Chinese Intelligence Agencies: Evolution and Empowerment in Cyberspace," in Lindsay, Cheung, and Reveron, eds., *China and Cybersecurity*, p. 32.

[61] "China-Linked Hacking Foiled by Private-Sector Sleuthing," *Bloomberg Business*, October 28, 2014.

[62] Mark A. Stokes, "The Chinese People's Liberation Army Comptuter Network Operations Infrastructure," in Lindsay, Cheung, and Reveron, eds., *China and Cybersecurity*, p. 164.

[63] Mark A. Stokes, "The Chinese People's Liberation Army Comptuter Network Operations Infrastructure," Lindsay, Cheung, and Reveron, eds., *China and Cybersecurity*, pp. 168–174; Mandiant Corporation, "APT1," 2013, p. 8.

[64] Mandiant Corporation, "APT1," 2013, p. 20.

[65] The former director of the GSD 3rd Department, Lt. Gen. Wu Guohua, was promoted to deputy commander of the Second Artillery, and the two organizations appear to have a close working relationship—much like the U.S. NSA and Cyber Command. Mark A. Stokes, "The Chinese People's Liberation Army Comptuter Network Operations Infrastructure," in Lindsay, Cheung, and Reveron, eds., *China and Cybersecurity*, pp. 164–175.

[66] Office of the Secretary of Defense, *Military Power of the People's Republic of China 2007*, Washington, D.C., May 2007, p. 22.

demics and civilians with information technology backgrounds.[67] The PLA's collaboration with civilian organizations in recruiting talent and conducting research is critical to its success. But a careful study of information warfare militia—militia units organized for computer network operations within civilian universities and corporations—indicates that the primary wartime task of reserve organizations would be conducting network maintenance and network defense. The study finds no positive evidence of their having been used in peacetime reconnaissance, though it is difficult to prove a negative finding of this kind.[68] The *Science of Military Strategy* strongly supports the idea that the PLA favors professionalism in network operations: "Military confrontation in the network domain is a trial strength in terms of the knowledge, intelligence, and professional capability of the cream of network talent, and is a field with extremely strong professionalism. . . . Network warfare, although having a certain mass foundation, nevertheless cannot generate 'a nation in arms,' and network attack and defense strength lies in streamlined forces."[69] It should be noted that although this statement is found in an authoritative source, it may also be biased in ways that emphasize the centrality of the PLA role in overseeing network operations.

One prominent issue for China appears to be the coordination of its cyber effort. In a governing system characterized by a high degree of stove piping, the involvement of multiple organizations and ministries is likely to make coordination difficult. The divergent interests of the parties involved compound that problem. To be sure, during wartime, operational choices between exploitation and attack would loom large for any country. In the Chinese case, other non-military interests, including the control of information regarded as potentially destabilizing and surveillance of dissident activists, also loom large. Implicitly acknowledging the problem, China formed a Leading Small Group on Cybersecurity in February 2014. The group is chaired by Xi Jinping and includes 19 Politburo or minister-level members.[70] But while the group may establish leadership priorities, it will not have the staff to generate detailed policy or enforce existing regulations.

The United States, for its part, reorganized for cyberwar with the creation of USCYBERCOM in late 2009.[71] USCYBERCOM is a sub–unified command under

---

[67] *Jane's Intelligence Review*, "Breaching Protocol: The Threat of Cyberespionage," February 17, 2010; "Telecom Experts in Guangzhou Doubling as Militia Information Warfare Elements," *Guofang*, PLA Academy of Military Science, September 15, 2003 (cited in Krekel, *Capability of the People's Republic of China to Conduct Cyber Warfare and Computer Network Exploitation*, 2009, p. 33).

[68] Robert Sheldon and Joe McReynolds, "Civil-Military Integration and Cybersecurity: A Study of Chinese Information Warfare Militias," in Lindsay, Cheung, and Reveron, eds., *China and Cybersecurity*, pp. 193–200.

[69] 战略学 [*The Science of Military Strategy*], 3rd ed., 2013, pp. 191–192.

[70] Jon R. Lindsay, "China and Cybersecurity: Controversy and Context," in Lindsay, Cheung, and Reveron, eds., *China and Cybersecurity*, pp. 13–14.

[71] USCYBERCOM reached operational status in May 2010.

U.S. Strategic Command and is led by a four-star general. It comprises elements from each of the military services, including the Air Force's 24th Air Force, the Navy's 10th Fleet, the 2nd Army, and the Marine Corps Cyberspace Command. In March 2013, U.S. Army General Keith Alexander, who at the time commanded both the USCYBERCOM and the NSA, said that USCYBERCOM was developing 40 specialized teams. Significantly, he said, "This is an offensive team that DoD would use to defend the nation if it were attacked in cyberspace. Thirteen of the teams that we're creating are for that mission alone."[72] The 2015 DoD Cyber Strategy stipulated that the military's Cyber Mission Force will include nearly 6,200 military, civilian, and contractor support personnel and be organized into 133 teams. Teams will be grouped into three categories: Cyber Protection Forces to augment traditional defensive measures and defend DoD networks; National Mission Forces to defend the United States and its interests against cyberattacks of significant consequences; and Combat Mission Forces to support combatant commands by integrating cyber operations into operational plans.[73]

With cyberwarfare still in its infancy, organizational debates are legion. There are, for example, disagreements about the division of responsibilities among USCYBERCOM, the U.S. Department of Homeland Security, and the NSA; between USCYBERCOM and the regional combat commands; and between USCYBERCOM and the various services. Because USCYBERCOM handles top-level computer network defense and all computer network attack operations—and because it is very closely tied to the NSA, which carries out computer network exploitation—U.S. cyberwarfare elements appear much more coordinated with one another than is the case in China. Despite rumors that the direction of USCYBERCOM and NSA would be split after the allegations made by Edward Snowden, they remain linked at the top. Admiral Michael S. Rogers succeeded Keith Alexander in leading both organizations in April 2014.[74]

Materiel

All parties engaged in cyberwarfare use largely the same *materiel* but with some potentially interesting exceptions. China, wary of Microsoft, is developing its own operating system, Kylin, which may also find its way into military equipment. This would, in theory, offer two advantages: It may have been created without legacy design flaws, and its architecture may be obscure to outside inspection. However, closer examination of the software suggests that most of the kernel code of early versions was identical

---

[72] "U.S. Military Creating Cyberwarfare Teams," United Press International, March 13, 2013.

[73] U.S. Department of Defense, *The Department of Defense Cyber Strategy*, April 2015, p. 6.

[74] See Nedra Pickler, "US May Split Command of Spy and Cyber Agencies," *Associated Press*, November 7, 2013.

to that of FreeBSD5.3 (a Unix-like system).[75] The most recent version is based on the Linux operating system.[76] To the extent that it has new code, cyber security experts argue that security systems untested by the outside world are likely to have serious flaws that more widespread scrutiny would have uncovered.[77] Worse, even a secure operating system can be penetrated if its associated applications have weaknesses. Cryptographic algorithms and code-breaking machinery should also be considered. The United States is generally ahead in algorithms (in part because it can call on the service of NATO partners), but it no longer holds the lead in hardware speed as measured by supercomputing power.

Leadership

The *leadership* component of cyberwarfare is difficult to evaluate without assessing the relative competency of specific individuals, which naturally changes every few years. A broad U.S. advantage is that its flag officers are more likely than their Chinese counterparts to have an intuitive grasp of things cyber. Widespread U.S. exposure to computers 30 years ago (during the flag officers' formative years) greatly exceeds that of their Chinese counterparts. Even today, the awareness of cyber threats is low in China, with one 2014 article describing China as "a fish barrel for cyber criminals."[78] Needless to say, this advantage is eroding, though computer ownership remains far more widespread in the United States than China.

Network Management

Keeping systems operating correctly in the face of accidents, buggy software, user errors, and poor administration may be the best indicator of a country's ability to mitigate the effects of a concerted attack. In this area, the United States holds important advantages. As a general rule, U.S. network management is quite sophisticated, on par with global best practices, despite heterogeneous hardware and software (a legacy of long experience). Companies such as AT&T and Verizon maintain well-instrumented data centers and can reach back across decades of experience to keep networks running over a wide variety of circumstances.

By contrast, China's network administration in the face of today's level of cyber mischief can be spotty. As in other areas, China's network management is typical of

---

[75] Dancho Danchev, "China's 'Secure' OS Kylin—A Threat to U.S. Cyber Capabilities?" *ZDNet*, May 13, 2009; see also Neal Krawetz, "Kylin Time," *Hacker Factor Blog*, May 23, 2009.

[76] "China's 'Home-Made' Operating System Isn't Home Made at All, But Maybe That's OK," *TechinAsia*, December 23, 2014.

[77] This is a variant on a principle first put forward by Auguste Kerckhoffs in 1883: In a well-designed cryptographic system, only the key needs to be secret. There should be no secrecy in the algorithm. See Bruce Schneier, *Crypto-Gram Newsletter*, May 15, 2002.

[78] "China, A Fish Barrel for Cyber Criminals," *New York Times*, December 2, 2014.

a country partway between underdevelopment and modernity. The following passage from the book *Pirates of the ISPs* merits inclusion here:

> According to one local report, "nearly all internal networks used by Chinese firms have been attacked at least once during the past year, and hackers managed to take control of at least 85 percent of them." In 2010, more than 45 percent of Chinese Internet users complained of viruses or Trojans on their computers, according to a recent study from the China Internet Network Information Center, which reports to the Ministry of Industry and Information Technology. Nearly 22 percent reported that their accounts or password had been stolen. And that is actually an improvement over recent years. In 2009, the rate of infection was nearly 60 percent, and the rate of account theft was almost one in three.[79]

Hackers accidentally disabled Internet service in six provinces for several hours when a Chinese online game provider sought to cripple the servers of its rivals.[80] Other incidents are more ambiguous in nature. Several times in the past few years, erroneous instructions from Chinese ISPs have caused traffic that would normally be routed inside the United States to flow through China.[81] This was possible because the Internet allows ISPs to announce that they are the shortest distance between two points. In these cases the instructions were wrong. If it is true what the China Telecom and other Chinese officials have claimed—that these instructions were accidentally garbled—then these companies have serious problems managing their networks. If the Chinese prevaricated, then this incident speaks to their testing of Western networks, rather than their network management.[82]

China's efforts to maintain the functionality of its "Great Firewall" provide other indictors of its level of cyber competence. Here, the record is mixed. Internet users in China use a variety of methods, including virtual private networks (VPNs), mirror sites of blocked pages hosted on U.S. cloud computing services, and simple proxy servers, to circumvent censorship.[83] Recently, however, China has become more aggressive in thwarting such workarounds. It has made the blocking of VPNs more automated

[79] Noah Schactman, *Pirates of the ISPs: Tactics for Turning Online Crooks into International Pariahs*, Washington, D.C.: Brookings Institution Press, July 2011.

[80] Owen Fletcher, "China Game Boss Sniped Rivals, Took Down Internet," *IDG News Service*, August 28, 2009.

[81] See Elinor Mills, "Web Traffic Redirected to China in Mystery Mix-Up," *CNet*, March 25, 2010, and Elinor Mills, "Facebook Detour Through China: Accident or Not?" *CNet*, March 24, 2011.

[82] John Dunn at *Techworld* has argued, "The origin of this [early 2010] manipulation is believed to have been a third-party ISP, IDC China telecommunications, which makes it a certainty that this was a deliberate act." John Dunn, "Internet Hijack Claims Denied by Techcom," *Techworld*, November 18, 2010. For a different view, see Kit Eaton, "China Behind Yesterday's YouTube, Facebook, Twitter Outage," *Fast Company*, March 26, 2010.

[83] "Activists Are Finding New Ways Around China's Great Firewall," *Time*, November 21, 2013.

and dynamic.[84] And it has developed a new offensive tool, labeled the Great Cannon by Western observers, designed to divert traffic to denial-of-service attacks against sites hosting mirror sites of blocked web pages.[85]

Zero-Day Approach

As a general rule, once a software vulnerability has been found, it gets fixed by the vendor, sometimes within days but rarely in more than a matter of weeks. Shortly thereafter, the fix is installed, in many cases automatically. Microsoft and Mozilla, for instance, push patches to users and make opting out more difficult than accepting the patch. As a general rule, DoD is quite diligent about patch management, and, presumably, the Chinese are aware of its importance.

Indeed, against well-maintained core corporate systems, attacks almost always require using zero-day exploits (vulnerabilities that even the writers of the software were unaware of at the time they were exploited).[86] Chinese hackers are believed to have used zero-day attacks to penetrate Google in early 2010; however, some analyses of this attack suggest that it was flawed.[87] By contrast, almost nothing is known about U.S. zero-day exploits.

Conclusions

A military confrontation between the United States and China would almost certainly include operational cyberwarfare operations. Attacks on logistics and base-related civilian infrastructure, which are on the unclassified network, may yield particularly notable results. Cyberwarfare matters only to the extent that it affects the outcome of other military operations. Although the relative competencies of the United States and China *do* make a difference, one must take into account each side's dependence on networks and each side's ability to operate if networks function badly or not at all.

In view of the potential for escalation, it is uncertain whether either side will resort to strategic cyberwarfare. If they do, results may be highly unpredictable. The outcome will depend not only on each side's competence but also on chance factors (e.g., cascading affects), the defensive posture and resiliency of each side's organizations and infrastructure, how each side's public reacts, how political leaders factor public

---

[84] "China's Great Firewall Gets Taller, *Wall Street Journal*, January 30, 2015.

[85] "Massive Denial-of-Service Attack on GitHub Tied to Chinese Government," *ArsTechnica*, March 31, 2015.

[86] By contrast, cyber criminals, who tend to go after the more feckless users, do not need and therefore rarely use zero-day exploits.

[87] Kelly Jackson Higgins, "Flaws in the 'Aurora' Attacks," *DarkReading*, January 25, 2010. This was not the only example; see the story of the malware known as Poison Ivy and other instances in Brian Grow, Keith Epstein, and Chi-Chu Tschang, "The New E-spionage Threat," *BusinessWeek,* April 21, 2008.

opinion into their strategic calculus, and whether and how one side or the other escalates as a result.

Scorecard Coding

Figure 11.1 provides our summary coding of the results of scorecard 9. The assessment of advantage in this scorecard is based on the ability of each side to find flaws and openings in the cyber defenses of the other, exploit or attack those weaknesses to degrade adversary military capabilities, and maintain discipline and operational security in both the offensive and defensive.

Both U.S. and Chinese military forces are working to improve their offensive and defensive cyber capabilities. These efforts began later in China than in the United States, and we coded 1996 as *U.S. advantage*. By 2003, China had established its "Net

Figure 11.1
Scorecard 9 Summary Coding

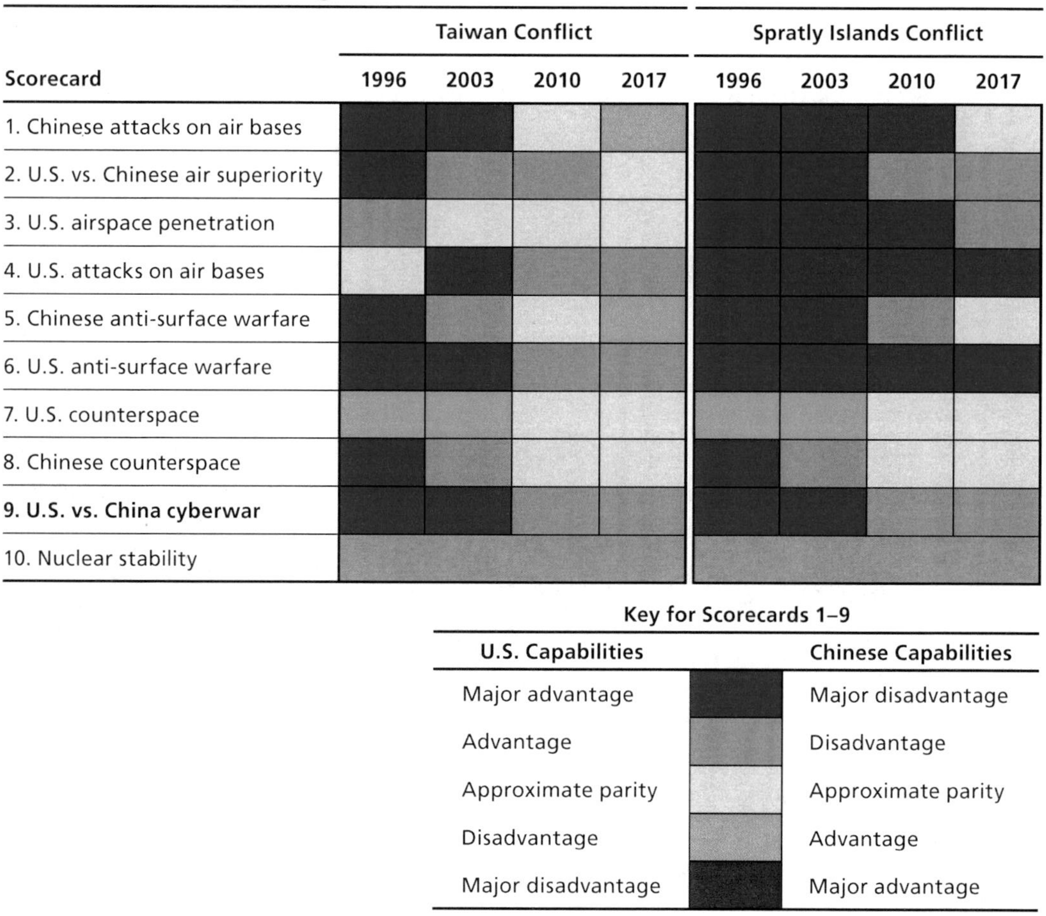

	Taiwan Conflict				Spratly Islands Conflict			
Scorecard	1996	2003	2010	2017	1996	2003	2010	2017
1. Chinese attacks on air bases								
2. U.S. vs. Chinese air superiority								
3. U.S. airspace penetration								
4. U.S. attacks on air bases								
5. Chinese anti-surface warfare								
6. U.S. anti-surface warfare								
7. U.S. counterspace								
8. Chinese counterspace								
9. U.S. vs. China cyberwar								
10. Nuclear stability								

Key for Scorecards 1–9

U.S. Capabilities	Chinese Capabilities
Major advantage	Major disadvantage
Advantage	Disadvantage
Approximate parity	Approximate parity
Disadvantage	Advantage
Major disadvantage	Major advantage

Force," but U.S. cyber-related elements at the service level (e.g., the U.S. Air Force's 67th Information Operations Wing) remained more extensive and almost certainly more experienced than their Chinese counterparts. We therefore coded 2003 as *U.S. advantage*. Both sides' organization for cyberwarfare has continued to evolve since 2003, with the establishment of large, well-staffed organizations.

The PLA has been heavily involved in large-scale cyber espionage since the mid-2000s, which has made it the subject of much media attention.[88] Because the most common targets of these attacks have been lightly defended corporate and unclassified government systems, this activity may have created an exaggerated sense of the capabilities that China might bring to bear in an operational military context. Nevertheless, growing challenges do also exist on the operational military side, and the 2013 DoD annual report on Chinese military power revealed increasing concern on the part of U.S. strategists over PLA capabilities to map and exploit U.S. networks.[89]

The United States, for its part, has not been idle. It stood up USCYBERCOM in 2009, acknowledged the formation of offensive (as well as defensive) teams in 2013, granted USCYBERCOM the authority to fast track the hiring of cyber specialists to fill its many vacancies in March 2015, and published a cyber strategy document the following month. While USCYBERCOM has responsibility for offensive cyber operations, the foundation for that capability will reside on the very significant intelligence and surveillance capabilities of the NSA.

We rated the 2010 and 2017 periods as ones of continuing, if diminishing, *U.S. advantage*, given U.S. relative strengths in the areas of network defense and resiliency, command understanding of cyber issues, and software development. At the same time, however, we emphasize that these judgments are based on necessarily limited information.

---

[88] See, for example, David E. Sanger, "U.S. Blames China's Military Directly for Cyberattacks," *New York Times*, May 6, 2013; and "Chinese Hack U.S. Weather Systems, Satellite Network, *Washington Post*, November 12, 2014.

[89] The DoD report warned, for example, that PLA information-gathering could easily be used for "building a picture of U.S. network defense networks, logistics, and related military capabilities that could be exploited during a crisis." Office of the Secretary of Defense, *Military and Security Developments Involving the People's Republic of China 2013*, May 2013, p. 36.

CHAPTER TWELVE

Scorecard 10: U.S. and Chinese Strategic Nuclear Stability

China conducted its first nuclear weapon test in October 1964.[1] It has maintained a force of fewer than 250 nuclear warheads since that time, even in the shadow of vastly superior U.S. and Soviet arsenals, numbering (at one time) in the tens of thousands.[2] China has been implicitly or explicitly threatened with nuclear weapons on several occasions, yet it has long maintained a no-first-use policy and a minimum deterrent posture.[3] China espouses a substantially different and more constrained view of the efficacy of nuclear weapons than many states, limiting their role to dissuading other states from using nuclear weapons against it or from using the threat of their use to coerce China.

While China's nuclear policy has, in many ways, been restrained, larger international changes have changed the context and implications of its policy direction. At the end of the Cold War, the United States and Russia sharply reduced their inventories of strategic nuclear weapons. With the New Strategic Arms Reduction Treaty (START), further modest reductions are under way in Russia and the United States. China, however, has not changed course and has continued to modernize its nuclear forces and add to its number of delivery systems (and possibly warheads).

This chapter assesses changes to U.S. and Chinese nuclear inventories and models counterforce nuclear attacks against one another using their respective 1996, 2003, 2010, and 2017 inventories. It also comments on other issues (such as missile defense) relevant to each side's second-strike capabilities. This approach to the assessment of stability is not meant to reflect the doctrine of either side. China's no-first-use policy ostensibly precludes a first strike. And neither China's policy nor its nuclear force struc-

---

[1] For more background on the early development of China's nuclear arsenal, see Lewis and Xue, *China Builds the Bomb*, 1988.

[2] See Fravel and Medeiros, "China's Search for Assured Retaliation," 2010.

[3] Chinese officials refer to the requirement for a "lean and effective" nuclear force. They reject the term "minimum deterrent," because China has not adopted specific criteria for the required level of retaliatory capability (like the French and British did during the Cold War). However, with the understanding that China's requirement has not been as rigorously defined (nor historically as large) as those of others who have employed the phrase, *minimum deterrent* nevertheless remains an apt descriptor for China's nuclear policy.

ture suggests a counterforce strategy launched primarily against nuclear weapons.[4] Similarly, DoD's latest *Nuclear Posture Review Report* commits the United States to reducing the role of nuclear weapons in its national security strategy, with the ultimate "objective of making deterrence of nuclear attack on the United States or our allies and partners the sole purpose of U.S. nuclear weapons."[5]

Rather, we evaluate first-strike counterforce capabilities to better highlight the level of structural stability in the strategic nuclear relationship. *First-strike stability* refers to the ability of both sides to absorb a first strike by the other and still have sufficient capability remaining to launch a second, retaliatory strike. States that lack such a capability may be more prone to build more weapons to correct the deficiency during peacetime, possibly driving others to build up their inventories as well. And during crises, these states may be tempted to use one or more weapons in a demonstration or preemptive attack if they think that the other side may deprive them of the ability to retaliate later. Mutual secure second-strike capability is not the only ingredient in a stable strategic relationship, but it is an important component.

Although we did not model the declaratory policies of the respective sides, our analysis may reflect each side's concerns about the other. While strategists on both sides will take the declaratory policies of the other into account when considering policy (including decisions about force structure and force posture adjustments), they will also pay great attention to capabilities. Because of the nature of nuclear weapons, which makes "winning" a meaningless concept in almost all cases of actual employment, we do not assess scorecard 10 in terms of relative "advantage" but, rather, in the degree of confidence that each side might reasonably have in the survivability of its own second-strike capability.

By departing from the emphasis on relative advantage, this chapter differs fundamentally from the other scorecard assessments in this report. Indeed, in this case, diminished advantage—even diminished U.S. advantage—can produce outcomes that are "improved," at least from the perspective of first-strike stability.

---

[4] *The Science of Second Artillery Campaigns*, an internal-use document published by the PLA Press, suggests that such campaigns will generally occur "under conditions in which the enemy is strong and we are weak." The priority in targeting is "to cause huge losses for the enemy and to cause the enemy to be very shaken psychologically in order to weaken their will to wage war." These targets would include "enemy command centers, communications hubs, transportation hubs, military bases, political centers, economic centers, important industrial bases, and other strategic and campaign targets." See 中国人民解放军第二炮兵部队 [PLA Second Artillery], 第二炮兵战役学 [*Science of Second Artillery Campaigns*], 2004, pp. 298, 304.

[5] U.S. Department of Defense, *Nuclear Posture Review Report*, Washington, D.C., April 2010.

Methodology

To evaluate first-strike stability, we employed a modeling methodology first developed at RAND to evaluate the nuclear dynamics between the United States and Soviet Union during the Cold War, but we modified it for use in the U.S.-China context.[6] The model was designed to determine whether a competition is characterized by first-strike stability or instability. Stable situations are defined as those in which both sides possess a survivable second-strike capability. Under stable circumstances, striking first cannot guarantee that the attacker will escape devastating retaliation. Striking first to gain advantage is, therefore, not an appealing option. Perhaps more importantly, when first-strike stability exists, each side is secure in the knowledge that it can retaliate against the other even if the crisis should escalate and it is attacked first. Each side understands that the other has little incentive to attack, and the weaker party is not driven by "use-them-or-lose-them" incentives to attack before its own nuclear arsenal is destroyed.

We are not currently in a Cold War, and present-day China is not the former Soviet Union. The prospect of a conventional war with China is distant, and the possibility of a nuclear war is even more remote. Nevertheless, both sides have concerns about the other's strategic direction. U.S. analysts and policymakers question whether a significant increase in Chinese warhead numbers might accompany the Chinese introduction of multiple independently targetable reentry vehicles (MIRVs) and make future reductions in the U.S. arsenal risky. Some are also concerned that as Chinese second-strike capability becomes more reliable, Beijing, more confident that the United States will not escalate to the nuclear level, might accept greater risks with regard to conventional conflict. China, for its part, is concerned that U.S. missile defense, advanced conventional precision weapons, and more accurate and "usable" U.S. nuclear forces could threaten the viability of a Chinese second-strike capability.

Beyond this assessment of stability, this chapter sheds light on the evolving Chinese-U.S. nuclear balance and provides a more complete historical understanding of the two evolving nuclear force structures and their capabilities over time. That the modeling shows severe structural imbalances during every snapshot year is not surprising, given that China possesses substantially fewer strategic weapons than the United States, which built up its forces in a decades-long superpower competition with the Soviet Union. Even by 2017, when the inventory of U.S. strategic nuclear warheads will have fallen to less than a third the 1996 level, U.S. warheads will outnumber China's by at least 13 to one (2,144 warheads to somewhere between 106 and 160 warheads).

---

[6] For a description of the model as it was originally conceived and used, see Kent and Thaler, *First-Strike Stability*, 1989. Another source that uses drawdown curves to model a counterforce first-strike interactions is Michael M. May, George F. Bing, and John D. Steinbruner, "Strategic Arsenals After START: The Implications of Deep Cuts," *International Security*, Vol. 13, No. 1, Summer 1988.

Our modeling of nuclear exchanges intentionally excludes elements of China's nuclear arsenal that cannot range U.S. counterforce targets.[7] While these weapons and platforms have and will continue to play an important role in China's nuclear deterrent strategy, especially as it relates to neighboring states, such as India and Russia, they are less relevant to U.S. and Chinese first-strike capabilities against one another. Similarly, we did not look at elements of the U.S. nuclear arsenal that cannot range Chinese counterforce targets.[8]

Alert Levels

To valuate the effectiveness of a nuclear first strike, the model postulates two possible alert levels for the nuclear forces being attacked (see Table 12.1): a low-alert level and a medium to high-alert level. Within the model, these alert levels stipulate the percentage of particular types of nuclear systems subject to attack. Alert levels reflect, for example, the proportion of submarines in port, as opposed to on patrol, and the number of bombers parked on the tarmac versus those that are either on standby or in the air. Because ballistic missiles launched from underground silos are fixed in a particular geographic location, they are always susceptible to an attack under both alert levels and thus have to rely on other means, such as silo hardening and dispersion, to increase their survivability. As a result of these alert levels, the model derives an expected value of surviving nuclear weapons.

We have only partial knowledge of U.S. and, especially, Chinese alert procedures and levels. Available Chinese writings discuss alert principles and procedures in some detail, but not enough is known about how these principles might be applied in times of crisis to make specific predictions in support of this modeling effort.[9] Nevertheless, two points are clear. First, China's normal peacetime alert level is lower than that of the United States, and, second, both countries maintain a range of options for increasing alert levels for different parts or elements of their nuclear force structure during crises.

We posit different parameters for each country at each level, as indicated in Table 12.1. China is not currently believed to maintain many or possibly any of its

---

[7] Thus, we omit MRBMs and IRBMs, such as the CSS-2 (DF-3), CSS-3 (DF-4), and CSS-5 (DF-21). We also omit nuclear bombs and nuclear ALCMs on China's past and present nuclear-capable airframes, including the Q-5 attack aircraft and the H-6 medium bomber because they cannot reach U.S. counterforce nuclear targets.

[8] Most notably, we exclude the MX Peacekeeper ICBM deployed during the 1996 and 2003 periods in study. As a weapon optimized to hit targets in the former Soviet Union, the MX Peacekeeper has a relatively limited range of 9,600 km. Combined with a deployment location in southeastern Wyoming, it is unable to strike Chinese nuclear counterforce targets. *Jane's Strategic Weapons Systems*, "LGM-118 Peacekeeper," October 13, 2011.

[9] For the most part, Chinese writing focuses on the degree of preparation for launch at bases ordered to different alert levels, but these sources do not provide any indication of how much of the force might be alerted under different types of contingencies. See, for example, 中国人民解放军第二炮兵部队 [PLA Second Artillery],《第二炮兵战役学》 [*Science of Second Artillery Campaigns*], 2004, pp. 284–286. For an English-language discussion that draws on a wider range of Chinese source material, see John W. Lewis and Xue Litai, "Making China's Nuclear War Plan," *Bulletin of the Atomic Scientist*, September 21, 2012, pp. 57–59.

nuclear forces on alert and has not yet achieved proficiency in SSBN operations. Indeed, China appears to keep all of its warheads stored in depots, separate from launchers and launch facilities, during peacetime.[10] Hence, Chinese nuclear forces are not only smaller, but they are also more susceptible to a "bolt from the blue" nuclear attack than U.S. strategic forces. U.S. alert levels, though lower than they were during the peak of the Cold War, remain high.

We focus on the low-alert level in our analysis, and, except where otherwise noted, we refer to this alert level when discussing outcomes. In a crisis, both sides might raise their alert levels, but in considering vulnerability to a potential first strike by the other, both might be inclined to consider cases in which they were caught relatively unprepared. In the Chinese case, a low-alert level already indicates a level that is above the alert level that it practices under normal circumstances. It is, however, a level that Beijing could maintain during peacetime if it chose to do so. And it is a level to which Beijing could move relatively quickly in the event of a crisis. Washington, for its part, maintains higher peacetime alert levels. This is partly out of concern that raising the alert level during a crisis (which might become necessary if the normal level were reduced) might be destabilizing. It would, however, consider raising the alert level further under a variety of conditions.[11]

Table 12.1
U.S. and Chinese Forces in Fixed Locations and Subject to Attack, by Alert Posture

	% of Assets	
Country	Posture A (low alert level)	Posture B (high alert level)
United States		
SSBNs	40	25
Bombers	95	50
ICBMs	100	100
China		
SSBNs	70	25
ICBMs (silo)	100	100
ICBMs (mobile)	75	25

[10] On China's highly centralized system of warhead management, see Mark A. Stokes, *China's Nuclear Warhead Storage and Handling System*, Arlington, Va.: Project 2049 Institute, March 12, 2010.

[11] The logic of the U.S. peacetime alert posture is laid out in U.S. Department of Defense, *Nuclear Posture Review Report*, 2010, pp. 26–27.

Warhead Salvo Success

Successfully destroying an entire set of counterforce targets is a function of both the previously mentioned alert rate of the counterforce targets and the characteristics of the attacking force. We modeled warhead salvo success based on the following criteria:

1. *Weapon reliability* is the probability of a single warhead delivered successfully to its target without malfunction. A reliable missile will be launched successfully, attain proper trajectory, successfully separate from its booster(s), and detonate when it reaches the target area. Because the reliability of U.S. and Chinese nuclear weapons is unknown, we made a simplifying assumption and set weapon reliability at an uncorrelated 80 percent for both countries.
2. *Single-shot Pk* for point targets (such as silos) is a function of three things: the hardness of the target, the yield of the nuclear warhead, and its accuracy. Using these three factors, we can determine the likelihood that a nuclear warhead will create a sufficient overpressure near enough to a particular target to destroy it.[12] Multiple designated ground zeros can be used when the lethal radius and CEP of a single warhead are insufficient to achieve necessary overpressure over a particular area. The single-shot Pk of different U.S. and Chinese nuclear weapons will, therefore, depend on the specific nature of the targets, in addition to their own characteristics.
3. *The multiple-shot Pk* is the cumulative probability of destroying a target when more than one warhead is directed against it (designated ground zero). Multiple warheads may be used when the target hardness or size results in a low single-shot Pk or when high overall Pks are required.
4. *Confidence level* against the entire target set is the probability that every target in the target set will be destroyed in an attack. Even when offensive warhead salvos have high multiple-shot Pks, minor possibilities of failure become magnified as the number of targets increases. In the model, we set the confidence level that both the United States and China seek against the entire target set at 80 percent.

Because neither China nor the United States can plausibly be argued to have any intention or desire to initiate nuclear war with the other, these calculations pertain more to the consideration of their own vulnerability than they do to the probability of destroying the other. If a state feels that a potential opponent cannot achieve nuclear force suppression with high confidence of complete success, then it will be more sat-

---

[12] Formulas for calculating probabilities of kill can be found in Samuel Glasstone and Philip J. Dolan, eds., *The Effects of Nuclear Weapons*, 3rd ed., Washington, D.C.: U.S. Department of Defense and U.S. Energy Research and Development Administration, 1977, pp. 111–113, and Lynn Etheridge Davis and Warner R. Schilling, "All You Ever Wanted to Know About MIRV and ICBM Calculations but Were Not Cleared to Ask," *Journal of Conflict Resolution*, Vol. 17, No. 2, June 1973, p. 212.

isfied with the existing level of its nuclear forces, less likely to pursue an arms race or buildup, and less likely to feel pressure during a conflict or crisis to "use or lose" its nuclear weapons. Targeting is not meant to reflect how either side might fight a nuclear war, even in the extraordinarily unlikely event that either commenced on such a course. The methodology employed here would not, for example, always yield the greatest expected damage to the adversary's nuclear forces, especially in the case of Chinese attacks in 2010 and 2017. Rather, the methodology is designed to provide a comparable and consistent lens for the assessment of confidence in first-strike potential.

Targeting

Targeting in this model is not focused on U.S. or Chinese nuclear warheads, but, rather, the platform that launches or drops those missiles or bombs. As a result, we consider a counterforce first strike by an attacking country successful if it can destroy the silos, ballistic missile submarines, TELs, or aircraft that carry warheads.[13] (We did not analyze other operational possibilities, such as attacking warhead depots or transportation networks and infrastructure designed to deliver warheads to launch units.) We consider warheads to have been functionally defeated, if not actually destroyed, if the delivery means are destroyed. While other nuclear warheads, missiles, or bombs might be in storage or maintenance, they would lack an obvious and immediate way of being used in a retaliatory second strike.[14]

The United States and China do not publicly share certain data about their nuclear forces that are relevant to our model. As a result, we needed to make assumptions when this information was not available. First, as a simplifying assumption, we gave both sides perfect targeting knowledge. That is, both China and the United States know exactly where to strike to destroy the other's counterforce capability (provided such forces are in port or garrison) in every snapshot year.[15] Second, we assumed that both the United States and China know how to successfully strike each other's counterforce targets.[16] That is, each side knows the hardness and dimensions (and, thus, the

---

[13] With the exception of silos, all of these platforms are mobile and therefore require area rather than point targeting.

[14] One can think of novel alternative warhead delivery methods that a country could use to exact a level of revenge, though this is not generally referred to as a second-strike capability and was outside the scope of our research.

[15] For Chinese strikes, this is an easy assumption to make. The locations of all U.S. counterforce targets are publicly known. In the Chinese case, substantial open-source research has identified firing positions for many Second Artillery units through Google Earth and other sources. See, for example, Sean O'Connor, *PLA Second Artillery Corps*, Air Power Australia, Technical Report APA-TR-2009-1204, April 2012. We assume that the U.S. government possesses additional information.

[16] Open-source imagery confirms that up to three *Jin*-class (Type 094) submarines have docked at one time or another at a third port, Xiaopingdao Submarine Base, since 2006. Instead of being a permanent basing option for Chinese SSBNs, Xiaopingdao is the likely location for sea trials after the subs' launch from the shipyards at Huludao. As a result, we do not target Xiaopingdao during any of the snapshot years, and we only

yield and accuracy) it would take to destroy the other's counterforce capabilities, again provided that the resident nuclear launch platforms (if capable) are not deployed and sufficient weapons are available for the counterforce first strike.[17]

Exploring the U.S.-China Nuclear Balance, 1996–2017

In this section, we evaluate the survivability of Chinese and U.S. nuclear forces in the snapshot years chosen for this study: 1996, 2003, 2010, and 2017. We do so by modeling a counterforce first strike by each side, using the force structures appropriate for the period in question. Table 12.2 lists the missiles and warheads relevant during these periods.

China and the United States in 1996

The U.S. nuclear force structure is much larger and more robust than that of China in all the periods considered. The U.S. advantage is greatest in the 1996 case (see Table 12.3). In 1996, China possessed up to 19 strategic weapons that could target U.S. nuclear assets. These weapons included seven silo-launched CSS-4 Mod 2 (DF-5A) ICBMs and 12 CSS-N-3 (JL-1) SLBMs. China's single *Xia*-class (Type 092) SSBN has reportedly never deployed on a nuclear deterrent patrol and would have had to maneuver to within about 1,700 km of its target to strike.[18] Therefore, the inclusion of the CSS-N-3 (JL-1) makes the analysis that follows inherently conservative.

In contrast to China's small and relatively unsophisticated nuclear inventory, the United States had more than 7,600 warheads deployed across a relatively balanced triad of silo-based missiles, ballistic nuclear submarines, and manned bombers. Washington's strategic arsenal included the Trident C-4 and Trident D-5 SLBMs, Minuteman II and Minuteman III ICBMs, nuclear ALCMs carried by both the B-1B Lancer and B-52H Stratofortress, and nuclear gravity bombs on B-2A Spirit aircraft. Unlike China, which had no hope of launching a successful first strike, the United States

---

target the *Jin* class (Type 094) in 2017 at Jianggezhuang Submarine Base and Yalong Submarine Base. Xiaopingdao Submarine Base is at latitude 38.817483° and longitude 121.493539°. See also *Jane's Fighting Ships*, "Type 094 (Jin Class)," February 13, 2015.

[17] We do not know the hardness of either U.S. or Chinese ballistic missile silos, so we assumed a 2,000 psi for both. Furthermore, we do not know the pressures a Chinese underground port facility is built to withstand (if any), so we assumed a 1,000 psi for existing submarine portals in the relevant time periods. We used a hardness of 20 psi, or the amount that above-ground reinforced concrete structures can withstand, for the TEL garages at Chinese mobile missile garrisons.

[18] Hans M. Kristensen, Robert S. Norris, and Matthew G. McKinzie, *Chinese Nuclear Forces and U.S. Nuclear War Planning*, Washington, D.C.: Federation of American Scientists and Natural Resources Defense Council, November 2006, p. 79; and Federation of American Scientists, "JL-1 [CSS-N-3]," web page, last updated June 10, 1998.

Table 12.2
U.S. and Chinese Strategic Nuclear Missiles and Warheads

System Name	Type	IOC	Warhead	Number	1996	2003	2010	2017	Launcher
United States									
Minuteman II Mk-11C	ICBM	1965	W56	1	Yes	No	No	No	Silo
Minuteman III Mk-12	ICBM	1970	W62	3	Yes	Yes	Yes	No	Silo
Minuteman III Mk-12A	ICBM	1979	W78	3	Yes	Yes	Yes	No	Silo
Minuteman III Mk-21/SERV	ICBM	2006	W87	1	No	No	Yes	Yes	Silo
Trident C-4	SLBM	1979	W76	6	Yes	Yes	No	No	*Ohio* class
Trident D-5 Mk-4	SLBM	1990	W76	8	Yes	Yes	Yes	No	*Ohio* class
Trident D-5 Mk-4A	SLBM	2008	W76	6	No	No	No	Yes	*Ohio* class
Trident D-5 Mk-5	SLBM	1992	W88	8	Yes	Yes	Yes	Yes	*Ohio* class
AGM-86B	ALCM	1982	W80	1	Yes	Yes	Yes	Yes	B-1B, B-52[a]
AGM-129A	ALCM	1991	W80	1	Yes	Yes	No	No	B-52[a]
B61	Bomb	1967	—	—	Yes	Yes	Yes	Yes	B-2A
B83	Bomb	1983	—	—	Yes	Yes	Yes	Yes	B-2A
China									
CSS-4 Mod 2 (DF-5A)	ICBM	1980	Unknown	1	Yes	Yes	Yes	Yes	Silo
CSS-4 Mod 3 (DF-5B)	ICBM	~2014	Unknown	3	No	No	No	Yes	Silo
CSS-10 Mod 1 (DF-31)	ICBM	2006	Unknown	1	No	Yes	Yes	Yes	TEL
CSS-10 Mod 2 (DF-31A)	ICBM	2007	Unknown	1	No	No	Yes	Yes	TEL
CSS-XX (DF-41)	ICBM	Unknown	Unknown	6	No	No	No	No	TEL

Table 12.2—Continued

System Name	Type	IOC	Warhead	Number	1996	2003	2010	2017	Launcher
China (cont.)									
CSS-N-3 (JL-1)	SLBM	1987	Unknown	1	Yes	No	No	No	*Xia* class
CSS-N-3 (JL-1A)[b]	SLBM	2000	Unknown	1	No	Yes	Yes	No	*Xia* class
CSS-N-14 (JL-2)	SLBM	Unknown	Unknown	1	No	No	No	Yes	*Jin* class

SOURCES: From *Jane's Strategic Weapons Systems*: "B61 Nuclear Bomb," June 18, 2014; "B83 Nuclear Bomb," February 5, 2015; "AGM-86 ALCM/CALCM," November 21, 2014; "AGM-129 Advanced Cruise Missile," October 13, 2011; "JL-1/-21," January 6, 2015; "JL-2 (CSS-NX-5)," June 1, 2010; "DF-5," January 6, 2015; "DF-31," March 11, 2015; "DF-41," January 6, 2015; "LGM-30F Minuteman II," October 13, 2011; "LGM-30G Minuteman III," November 28, 2014; "UGM-96 Trident C-4," October 13, 2011; and "UGM-133 Trident D-5," March 25, 2015; Robert S. Norris and William M. Arkin, 2003, 2010.

NOTES: SERV = safety-enhanced reentry vehicle. In the 2003 period the number of warheads on some Minuteman III Mk-12s was reduced from three to one. In the 2010 and 2017 periods, the number of warheads on Trident D-5 Mk-5s was reduced from eight to six.

[a] While the B-52 is capable of carrying B61 and B83 nuclear bombs, its lack of stealth and speed makes it an unlikely platform to successfully penetrate defended airspace. As a result, if employed against China, the B-52 would be effective only if equipped with nuclear ALCMs in a standoff role.

[b] The CSS-N-3 (JL-1A)'s IOC date is unknown but thought to have been sometime around 2000.

Table 12.3
U.S.-China Balance of Nuclear Forces, 1996

United States				China			
System Type	**Number**	**Warheads per System**	**Total Warheads**	**System Type**	**Number**	**Warheads per System**	**Total Warheads**
ICBMs				ICBM			
Minuteman II	14	1	14	CSS-4 Mod 2 (DF-5A)	7	1	7
Minuteman III	500	3	1,500				
SLBMs				SLBM			
Trident C-4	192	6	1,152	CSS-N-3 (JL-1)	12	1	12
Trident D-5	192	8	1,536				
Bombers				Bombers			
B-1B	93	16	1,488	None			
B-52H	93	20	1,860				
B-2A	6	16	96				
Total	1,090		7,646	**Total**	19		19

SOURCE: IISS, *The Military Balance*, 1996; Robert S. Norris and William M. Arkin, "U.S. Strategic Nuclear Forces, End of 1995," *Bulletin of the Atomic Scientists*, Vol. 52, No. 1, January–February 1996.

could conceivably have achieved a disarming first strike if China's forces were at a low enough alert level.

Chinese Vulnerability to a Disarming First Strike, 1996

In 1996, China's strategic nuclear arsenal was small and potentially susceptible to a disarming first strike, depending on its alert level.[19] If China's nuclear forces were at a low-alert level, the United States would have been able to destroy all of the nuclear forces that were susceptible to attack.[20] To achieve an 80-percent level of confidence that it could destroy all these targets, the U.S. military would have needed to use

[19] In 1996, the Chinese target set was small and consisted of the Jianggezhuang naval facility in Qingdao, where China's lone *Xia*-class (Type 092) SSBN was—and still is—located, and the seven ICBM silos housing the DF-5A (CSS-4 Mod 2), which are thought to be located in Henan province. *Jane's Strategic Weapons Systems*, "JL-1/-21," January 6, 2015; *Jane's Fighting Ships*, "Type 092 (Xia Class)," February 13, 2015; *Jane's Sentinel Security Assessment*, "Strategic Weapon System, China," April 6, 2015.

[20] The model assumes that the United States employs two W76 warheads atop Trident D-5 Mk 4 SLBMs fused for airburst at Jianggezhuang submarine base and that they are intended to strike area targets, such as piers. Nearly simultaneously, three W88 warheads atop Trident D-5 Mk 5 SLBMs attack each of the seven CSS-4 Mod 2 missile silos, for a salvo size of 21. The attack as postulated here would be comprehensive enough to destroy the *Xia*-class submarine, wherever moored, so long as it was in port. As a result, the United States would not rely on any specific intelligence to determine the location of the submarine.

23 of its 7,646 warheads. Even at the low-alert level, we assume that China's submarine would be deployed some proportion of the time, so our calculations yield an expected value of four surviving Chinese weapons.[21] In our model, these weapons are deployed away from port and therefore are not susceptible to attack. (The results of this analysis are presented in Figure 12.1.)

Of course, in reality, the single Chinese SSBN and its 12 missiles would not have been divisible, and four could not have been at sea without the other eight. The submarine could be in port (and vulnerable), at sea and trailed by U.S. attack submarines (and vulnerable), or at an unknown location at sea (and therefore effectively invulnerable). Hence, the possibility of the United States achieving a disarming first strike in 1996 would have depended heavily on the status and location of China's noisy and technically unreliable *Xia*-class SSBN. In light of strong contemporary U.S. anti-submarine warfare capabilities, the prospects would presumably offer little comfort to Beijing. Given the challenges facing Chinese leaders in contemplating second-strike survivability, it is perhaps not surprising that they subsequently embarked on a variety of programs to improve this capacity.

Figure 12.1
U.S. Nuclear Counterforce Attack, 1996

RAND *RR392-12.1*

[21] This is a conservative assumption from the U.S. perspective. The *Xia* has apparently never made an extended deterrent patrol beyond local waters. See *Jane's Missiles and Rockets*, "China's Future Jin-Class SSBN Deployment Draws on Xia-Class Experience," March 27, 2007.

U.S. Vulnerability to a Disarming First Strike, 1996

From 1996 to 2017—but especially early in that period—Chinese nuclear forces would have been inadequate to even contemplate a disarming first strike against U.S. counterforce targets. Given that reality, this section simply asks how much damage a Chinese nuclear attack could have inflicted against U.S. counterforce targets.[22] The analysis provides a baseline against which to evaluate the impact of subsequent changes in Chinese and U.S. nuclear capabilities.

At the lowest U.S. alert level, a Chinese attack could destroy more than half (56 percent) of the U.S. deployed warhead inventory (or the means to deliver them), including 40 percent of the SSBN capability and 92 percent of the U.S. nuclear bomber fleet (see Figure 12.2).[23] The remaining Chinese missiles would lack sufficient range to successfully attack remaining U.S. B-2A Spirit bombers at Whiteman AFB in Missouri or sufficient yield, accuracy, and, in some cases, range to destroy any of the 514 Minuteman II and Minuteman III ICBMs in their silos.[24] Overall, under attack at the lowest-alert level, the United States would have 3,390 warheads remaining with which to retaliate—hardly a reassuring prospect for Beijing.

Under higher-alert levels, the expected number of surviving warheads for both China and the United States would be higher. In the Chinese case, however, SSBN preparations for sea might be observed by the United States, and, subsequently, SSBNs may be trailed after they leave port. The PLA is less well positioned to track and threaten U.S. assets as its nuclear forces raise their alert levels.

China and the United States in 2003

Between 1996 and 2003, the number of Chinese nuclear missiles and warheads capable of targeting the United States more than doubled. The number of CSS-4 Mod 2 (DF-5A) missiles increased from seven to 20. By 2003, the Second Artillery had also

---

[22] One imaginable, if highly unlikely, scenario might be a situation in which Chinese leaders sought to target the most lucrative U.S. nuclear assets in an effort to increase the survivability of China's remaining nuclear systems in the face of an expected U.S. attack.

[23] To inflict maximum damage against U.S. nuclear capabilities, we assume that the CSS-4 Mod 2 (DF-5A) missiles target Kings Bay Naval Submarine Base in Georgia, Dyess AFB in Texas, and Barksdale AFB in Louisiana. If China's single *Xia*-class (Type 092) SSBN were to position itself approximately 400 km from the coast of Washington State, the CSS-N-3 (JL-1)—with a range of 2,150 km—could strike Minot AFB in North Dakota, Ellsworth AFB in South Dakota, and the former Bangor Naval Submarine Base (now Kitsap Naval Base) in Washington State. China would also still possess a small but significant nuclear arsenal deliverable by medium- and more modestly ranged ICBMs, such as the CSS-2 (DF-3), CSS-3 (DF-4), and CSS-5 (DF-21)—in addition to nuclear gravity bombs dropped from B-6 (H-6) or A-5 (Q-5) aircraft. While these forces could reach U.S. bases in Japan, South Korea, and Guam, they could not reach any U.S. counterforce nuclear targets.

[24] The CSS-N-3 (JL-1) is unable to reach any of the silos at Grand Forks AFB in North Dakota or Whiteman AFB in Missouri, though it is able to reach some silos at F. E. Warren AFB.

Figure 12.2
Chinese Nuclear Counterforce Attack, 1996

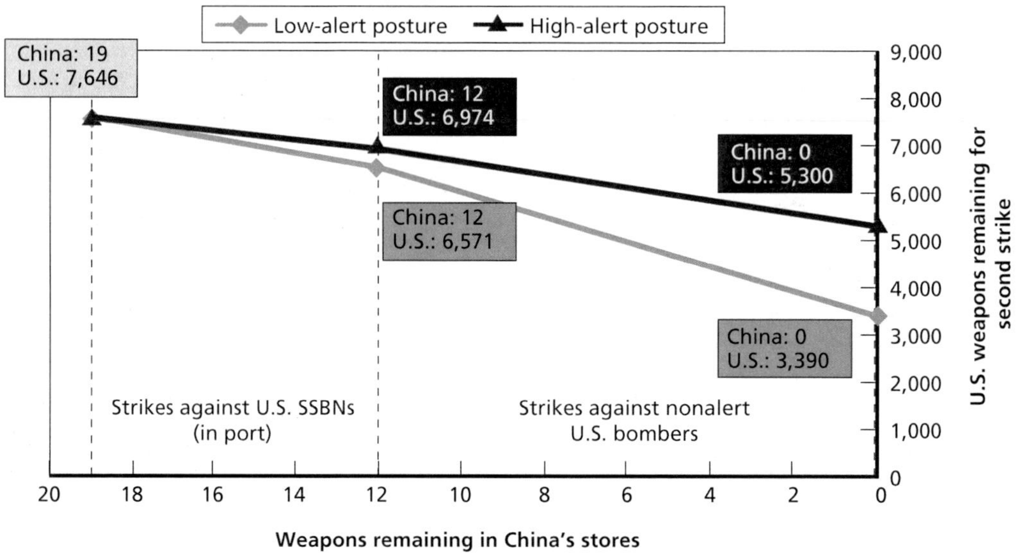

RAND *RR392-12.2*

fielded a brigade of mobile, solid-fueled CSS-10 Mod 1 (DF-31) ICBMs.[25] While the DF-31, with an estimated range of 8,000 km, could not have struck relevant targets from garrison, it could have traveled to areas in northeastern China from which it could have theoretically reached some U.S. counterforce targets. Also notable is that China improved its JL-1 (CSS-N-3) SLBM, enhancing its yield and accuracy, as well as moderately improving its range. The improved version was designated the JL-1A.

Concurrently, in compliance with 1991's START I, the United States continued to make reductions to its nuclear arsenal. Reductions by 2003 included the transition of the B-1B bomber to exclusively conventional use, the retirement of aging Minuteman II missiles, and the conversion of 150 Minuteman III missiles to single-warhead systems (from three warheads each). Some 15 new B-2As were added to the inventory, as were 42 D-5 SLBMs. In all, more than 1,000 warheads were taken out of service, though the number of platforms or delivery systems declined by a more modest 43.

Despite divergent trend lines, the U.S. nuclear inventory continued to dwarf China's. While the Second Artillery had 40 strategic missiles and the same number of warheads capable of being launched at targets in the continental United States, the U.S.

---

[25] There is some confusion in the literature regarding the NATO designation of the DF-31/DF-31A—specifically, whether it is the CSS-9 or CSS-10. We refer to the DF-31 and DF-31A as the CSS-10 Mod 1 and CSS-10 Mod 2, respectively, the designations given by the U.S. Air Force's National Air and Space Intelligence Center. See *Jane's Strategic Weapons Systems*, "DF-31," March 11, 2015, and National Air and Space Intelligence Center, *Ballistic and Cruise Missile Threat*, Wright-Patterson AFB, Ohio, NASIC-1031-0985-13, 2013, p. 3.

force numbered 1,047 delivery systems (including ICBMs, SLBMs, and bombers) and some 6,488 warheads (see Table 12.4).

Chinese Vulnerability to a Disarming First Strike, 2003

Despite the major continuing discrepancy in the relative size of U.S. and Chinese weapon inventories, China made some gains in achieving a secure second-strike capability by 2003. To strike all counterforce targets susceptible to attack with 80-percent confidence of destroying every target, the United States would have needed to employ 91 warheads in 2003 (see Figure 12.3).[26] Although this figure is significantly larger

Table 12.4
U.S.-China Balance of Nuclear Forces, 2003

United States				China			
System Type	Number	Warheads per System	Total Warheads	System Type	Number	Warheads per System	Total Warheads
ICBMs				ICBMs			
Minuteman III	150	1	150	CSS-4 (DF-5A)	20	1	20
Minuteman III	350	3	1,050	(CSS-10 Mod 1) DF-31[a]	8	1	8
SLBMs				SLBM			
Trident C-4	192	6	1,152	CSS-N-3 (JL-1A)	12	1	12
Trident D-5	240	8	1,920				
Bombers				Bombers			
B-52H	94	20	1,880	None			
B-2A	21	16	336				
Total	1,047		6,488	**Total**	40		40

SOURCES: IISS, *The Military Balance*, 2002; Robert S. Norris, William M. Arkin, Hans M. Kristensen, and Joshua Handler, "U.S. Nuclear Forces, 2002," *Bulletin of the Atomic Scientists*, Vol. 58, No. 3, May 2002; Office of the Secretary of Defense, *Annual Report to Congress: Military Power of the People's Republic of China*, 2002.

[a] The status of the DF-31 in 2003 is uncertain. IISS, *The Military Balance*, 2003, lists 8 DF-31s in service, and Jane's reports that the system entered "initial service" in 1999. By 2003, there had been nine reported test flights of the missile, but subsequent public DoD sources suggest that it did not reach full operational status until 2006. For more on these issues, see *Jane's Strategic Weapons Systems*, "DF-31," March 11, 2015.

[26] Similar to the 1996 case, the United States would strike the area targets of Jianggezhuang Submarine Base and the missile brigade garrison location with three and eight W76 warheads, respectively, from the Trident D-5 Mk-4 SLBMs. For Second Artillery brigade locations, see Sean O'Connor, "The PLA's Second Artillery Corps," *IMINT and Analysis*, Vol. 1, No. 11, December 2011. For more detail on TEL and reinforced concrete structure

Figure 12.3
U.S. Nuclear Counterforce Attack, 2003

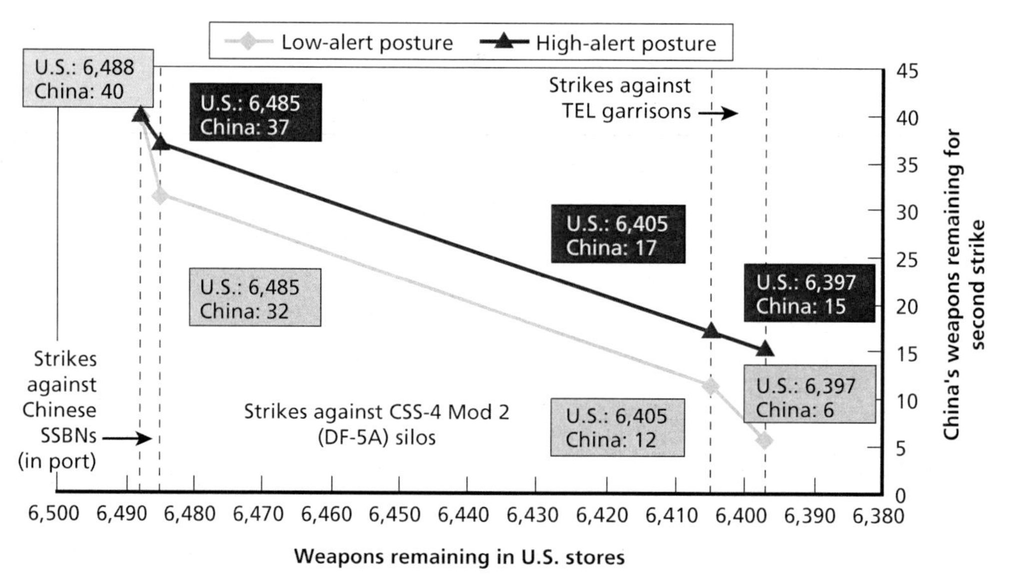

RAND *RR392-12.3*

than the 23 warheads required in 1996, the requirement would nevertheless represent only a trivial fraction (1.4 percent) of available U.S. warheads. Much more important in the context of a potential disarming first strike, the Chinese introduction of the mobile CSS-10 Mod 1 (DF-31) system raised the expected number of missiles that might be deployed to unknown locations and therefore not susceptible to attack.

Assuming a low-alert level, the expected number of missiles not susceptible to attack (and that therefore survive an attempted first strike) rose from four to six. From the Chinese perspective, the number remained both low and highly dependent on U.S. ISR assets not locating a handful of systems. The single *Xia*-class (Type 092) SSBN remained unreliable at best, and area fire by a large salvo of U.S. warheads might destroy a handful of even unlocated CSS-10 Mod 1s (DF-31s). On the other hand, with two different types of targets (an SSBN and land-based mobile missiles) potentially at large, the challenge to U.S. ISR was more substantial than it was in 1996. In other words, Beijing's outlook had improved by 2003, if not yet to the level that the leaders of most nuclear states would find satisfying.

---

hardness, see Office of Technology Assessment, *MX Missile Basing*, Washington, D.C.: U.S. Government Printing Office, PB82-108077, 1981, p. 264; and Office of Technology Assessment, *The Effects of Nuclear War*, Washington, D.C.: U.S. Government Printing Office, PB-296946, 1979, p. 18, respectively.

U.S. Vulnerability to a Disarming First Strike, 2003

Even with a doubling in the size of China's strategic nuclear inventory and a decrease in the size of the U.S. nuclear arsenal, Beijing nevertheless still lacked the ability to conduct a disarming nuclear first strike in 2003. As in the 1996 case, we examined how much damage China could achieve against counterforce targets. In our model, most of China's inaccurate but high-yield DF-5A (CSS-4 Mod 2) ICBMs are used to strike all U.S. area targets, including SSBN ports and bomber bases.[27] Remaining weapons are used to attack Minuteman III silos at Malmstrom AFB in Montana. We expect the Malmstrom attack to destroy roughly 24 U.S. warheads on eight Minuteman III missiles (see Figure 12.4).[28]

In all, a Chinese attack against low-alert-level U.S. forces would be expected to destroy up to 3,342 deployed U.S. warheads, or roughly 52 percent of the total U.S. inventory. Despite a smaller U.S. starting inventory, a larger number of Chinese attacking warheads, and improvements to Chinese CEPs and Pks, the Chinese first strike destroys fewer warheads and a lower percentage of the U.S. total than it does in 1996. By 2003, a higher percentage of U.S. warheads are deployed on submarines and single-warhead missiles, which are less lucrative targets. Some 3,146 U.S. warheads would be

Figure 12.4
Chinese Nuclear Counterforce Attack, 2003

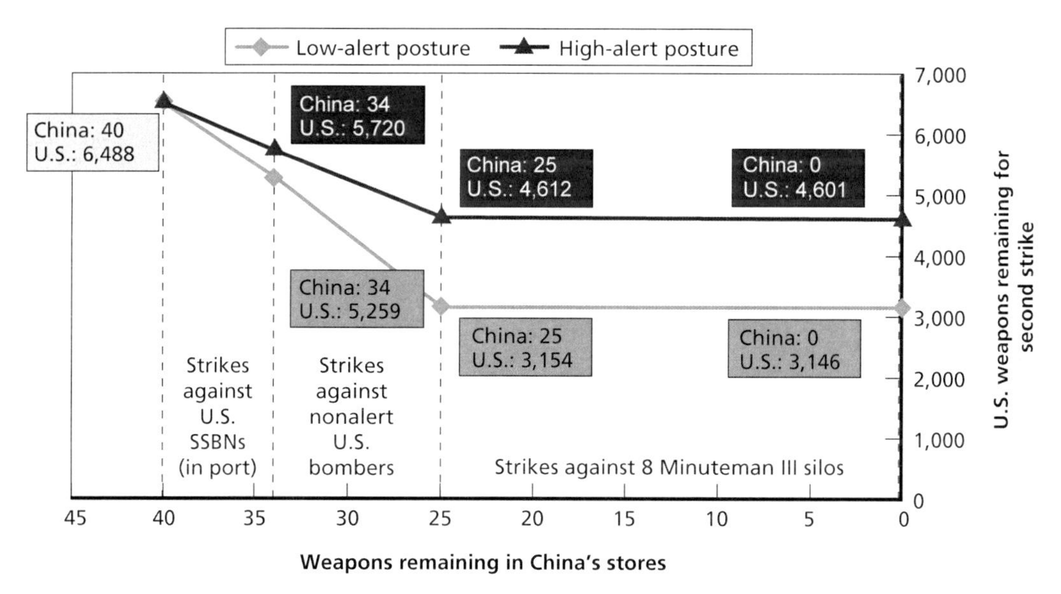

RAND *RR392-12.4*

---

[27] This includes the former Bangor Naval Submarine Base (now Kitsap Naval Base), Kings Bay Naval Submarine Base, Barksdale AFB, Minot AFB, and Whiteman AFB.

[28] This assumes that the *Xia*-class (Type 092) would launch from a location approximately 600 km off the coast of Washington State.

expected to survive the attack, only 244 fewer than in the 1996 case and more than enough to discourage any Chinese planner who might contemplate a counterforce first-strike attack.

China and the United States in 2010

In the seven years from 2003 to 2010, China's inventory of strategic nuclear weapons with the range to strike the U.S. mainland increased by almost 60 percent. Most of the additional warheads were mounted on an improved version of the CSS-10 Mod 1 (DF-31), designated the CSS-10 Mod 2 (DF-31A), with a substantially increased range.[29] Over the same period, the United States further cut its inventory of ICBMs and SLBMs, and it reduced the number of warheads deployed on many of its remaining missiles. The number of U.S. nuclear-capable bombers also fell from 115 to 96. Nevertheless, the U.S. arsenal, with 882 delivery systems and 4,806 warheads, remained far larger than China's, with 68 missiles and 68 warheads (see Table 12.5).

By 2010, the United States had also deployed a BMD system, including 24 ground-based interceptor (GBI) missiles designed to destroy a limited number of incoming ICBMs.[30] Despite reassurances from a variety of U.S. policymakers that the system is designed to defend against only limited attacks by North Korea, Chinese and Russian leaders have expressed concerns about the implications for their second-strike capability. We do not attempt to provide an assessment of the system's actual capabilities here, but we used simple modeling to examine the possible impact (given a range of probabilities of kill) on a Chinese second, or retaliatory, strike capability.[31]

PRC Vulnerability to a Disarming First Strike, 2010

Analysis of the potential for a U.S. disarming first strike in 2010 suggests that the United States could have had little confidence of success, had it considered a first strike. At the same time, Chinese planners, employing conservative (pessimistic) assumptions, might still have harbored reasonable concerns.

---

[29] The range of the CSS-10 Mod 2 (DF-31A) ICBM allows it to strike most U.S. nuclear counterforce targets from the Chinese mainland. However, various estimates of the DF-31A's range mean that the missile, along with its TEL, may have to travel some distance from garrison to strike some U.S. counterforce targets. See *Jane's Strategic Weapons Systems*, "DF-31," March 11, 2015; and National Air and Space Intelligence Center, *Ballistic and Cruise Missile Threat*, Wright-Patterson AFB, Ohio, NASIC-1031-0985-13, 2013.

[30] The U.S. government has repeatedly stated that the purpose of BMD is to counter the threat from rogue states and that the initiative is not intended to counter large numbers of Chinese or Russian missiles. See, for example, U.S. Department of Defense, *Ballistic Missile Defense Review Report*, Washington, D.C., February 2010. However, China and Russia both fear that BMD could threaten the more limited retaliatory capabilities that would remain after a hypothetical U.S. first strike.

[31] Critical factors necessary to model effectiveness would include, but are not limited to, the number of U.S. interceptors operational at any given time, and the ability of U.S. GMD to successfully detect, track, and kill incoming Chinese warheads.

Table 12.5
U.S.-China Balance of Nuclear Forces, 2010

Offensive Capabilities							
United States				China			
System Type	Number	Warheads per System	Total Warheads	System Type	Number	Warheads per System	Total Warheads
ICBMs				ICBMs			
Minuteman III	250	3	750	CSS-4 (DF-5A)	20	1	20
Minuteman III	200	1	200	CSS-10 Mod 1 (DF-31)	12	1	12
				CSS-10 Mod 2 (DF-31A)	24	1	24
SLBMs				SLBM			
Trident D-5	336	6	2,016	JL-1A	12	1	12
Bombers				Bombers			
B-52H	76	20	1,520	None			
B-2A	20	16	320				
Total	882		4,806	Total	68		68
Defensive Capabilities							
GBI	24	1	24	None			

According to the modeling, the United States could have destroyed all Chinese systems susceptible to attack using 132 U.S. warheads (see Figure 12.5).[32] As in the other cases, this calculation assumes a low-alert level for Chinese forces and 80-percent confidence of destroying the entire target set. Such an attack would have used only a fraction of the U.S. inventory (less than 3 percent), and 4,674 U.S. warheads would have remained after the attack to deter or, potentially, punish China and other states. More than twice as many missiles are not susceptible to attack in 2010 (13) as in 2003 (six).

Again, however, China's survivability depends on a small number of two types of systems, the single *Xia*-class (Type 092) SSBN and the road-mobile CSS-10 Mod 1 (DF-31) and CSS-10 Mod 2 (DF-31A) systems. Given the marginal operational capabilities of the *Xia* and the possibility that the United States could destroy the small number of mobile missiles that might be out of garrison, Chinese leaders likely

[32] Between 2003 and 2010, China's target set continued to grow, with two additional garrison locations for two DF-31A (CSS-10 Mod 2) missile brigades.

Figure 12.5
U.S. Nuclear Counterforce Attack, 2010

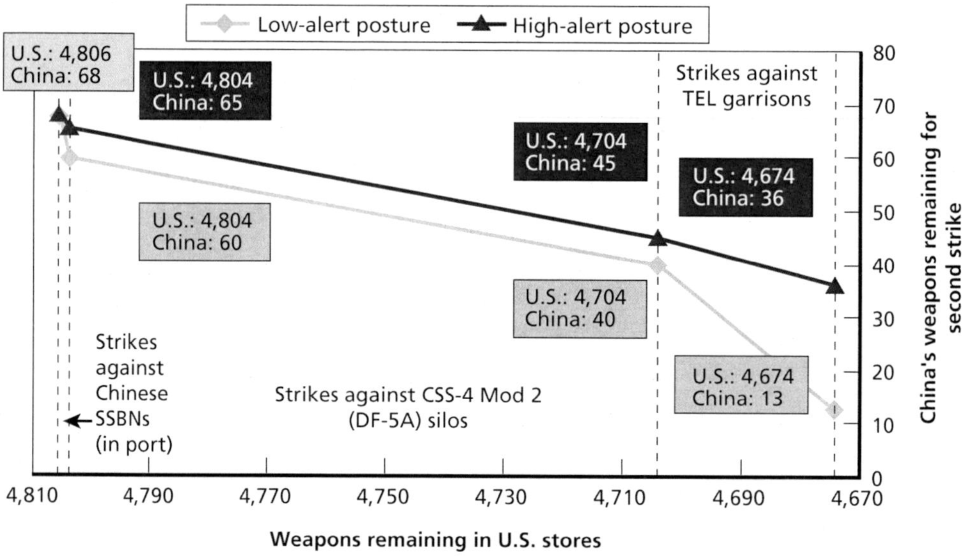

RAND *RR392-12.5*

remained concerned about the susceptibility of the arsenal to attack. Probably of even greater concern to Chinese leaders in 2010 was the uncertain impact of U.S. missile defense and whether the 24 GBI missiles then deployed could destroy China's second strike in midcourse, depriving Beijing of a retaliatory capability.

We considered a range of parameters that might shed light on the degree of risk faced by Beijing. If, for example, one assumed a 0.8 Pk for each GBI against a single missile, the expected number of missiles that survive to penetrate would be 0.84 (out of the 13 expected to survive a first strike). At a 0.9 Pk, 0.31 missiles survive, and at 0.5 Pk, 3.75 missiles survive. This range of roughly zero to four missiles surviving to retaliate would certainly be cause for concern in Beijing, especially given likely Chinese uncertainty about the actual capabilities of the system.

Penetration aids might have resulted in substantially better results for the Chinese.[33] And a shoot-look-shoot capability for the U.S. missile defense force might have made the results substantially worse for China.[34]

---

[33] China is developing and testing a variety of penetration aids. See *Jane's Sentinel Security Assessment*, "Strategic Weapons Systems—China," April 6, 2015; and *Jane's Strategic Weapons Systems*, "Ground-Based Mid-Course Defense (GMD) Segment," September 12, 2014.

[34] A shoot-look-shoot capability would enable U.S. commanders to assess the results of the first interceptor attack before allocating second and, perhaps, third interceptors. U.S. officials have discussed the development of such a capability, and the Missile Defense Agency provided funding to Northrop Grumman Corporation in 2009 for a proof of concept. See "SM-3 BMD, in from the Sea: EPAA and Aegis Going Ashore," *Defense Industry Daily*, last updated July 29, 2013.

U.S. Vulnerability to a Disarming First Strike, 2010

In 2010, China's nuclear forces remain unable to target all U.S. nuclear assets in known locations and therefore susceptible to attack. We again examine how much of the U.S. nuclear inventory China might be able to destroy. China's inaccurate but high-yield CSS-4 Mod 2 (DF-5A) systems are employed to strike U.S. area targets, including SSBN installations and air bases housing nuclear-capable bombers.[35] China uses its remaining missiles, all of which are more accurate, to strike ICBM silos. The modeling suggests that roughly 2,566 U.S. warheads (or their delivery systems) would be destroyed (see Figure 12.6). This would account for 53 percent of the U.S. total, up from 52 percent in 2003. Of the U.S. warheads neutralized, 12 would have been destroyed in Minuteman III silo attacks and the remainder lost at SSBN and bomber bases. Some 2,240 U.S. warheads would remain to deter further attacks or to retaliate.

The impact of GBI would depend largely on the tactics of both sides, which would themselves be driven by the technical capabilities of GBI missiles. If China conducted its attack in waves, then GBI missiles might be expended against a relatively small subset of the Chinese arsenal, with subsequent Chinese attacks directed against high-value area targets, now denuded of defense. If, on the other hand, China launched its weapons in a single salvo, ground-based interception might, depending on interceptor Pks and the ability to distinguish the incoming missile's aim points, be able to pro-

Figure 12.6
Chinese Nuclear Counterforce Attack, 2010

RAND *RR392-12.6*

---

[35] This list includes Kings Bay Naval Submarine Base, Kitsap Naval Base, Whiteman AFB, Barksdale AFB, and Minot AFB.

tect one or possibly two area targets. GBI missiles, then, might prevent the destruction of up to several hundred U.S. warheads (or their delivery means).

With or without ground-based interception, U.S. warhead losses would probably exceed 1,500, but this would leave the United States with more than enough to retaliate and still maintain a substantial stockpile in reserve. As in the other cases, these numbers assume a low U.S. readiness rate and a dedicated effort on the part of China to maximize damage to U.S. counterforce targets.

China and the United States in 2017

Questions about IOC dates of certain systems in the Chinese case and accounting rules under New START complicate the task of assessing the nuclear balance in 2017. We anticipate that the number of deployed U.S. warheads will decline from 4,806 (our 2010 estimate) to 2,144 in 2017 (see Table 12.6).[36] We examine two cases for China, a high estimate and a low estimate of its warhead inventory. In both cases, the number of Chinese warheads capable of targeting the United States will increase significantly, from 68 in 2010 to 106 in the lower estimate or 160 in the higher one. Modeling suggests that China's growing nuclear inventory would make only a limited difference in its ability to destroy U.S. counterforce targets in a first strike. But the number and types of systems currently entering China's inventory substantially improve the survivability of its second-strike capabilities. At the same time, the changes to China's nuclear inventory—in both qualitative and quantitative terms—raise questions about whether Beijing is placing relatively greater emphasis on nuclear forces and whether it might, in the future, adjust its nuclear policy and doctrine.

For several years, U.S. military publications have suggested that China is working on MIRVing ICBMs.[37] The 2015 DoD report on Chinese military power stipulates, for the first time, that MIRVed missiles have been deployed operationally, specifically on the CSS-4 Mod 3 (a modified version of the DF-5A).[38] In both our high and low estimates, therefore, we posit that number of Chinese CSS-4s (DF-5As) remains unchanged at 20, but that half of the force has been MIRVed, with each of the MIRVed missiles carrying three warheads.[39] It is unclear why China would begin MIRVing

---

[36] New START guidelines limit the United States to 1,550 deployed strategic warheads, but these weapons are defined in such a way that many more can actually be deployed. For example, each bomber counts as one nuclear weapon, even though U.S. and Russian bombers are capable of carrying some six to 20 warheads each. See Hans M. Kristensen, "New START Treaty Has New Counting," *FAS Strategic Security Blog*, March 29, 2010.

[37] For example, the 2014 DoD report on Chinese military power states, "China is working on a range of technologies to attempt to counter U.S. and other countries' ballistic missile defense systems, including MIRVs, decoys, chaff, jamming, and thermal shielding" (Office of the Secretary of Defense, *Military and Security Developments Involving the People's Republic of China 2014*, June 2014, p. 30).

[38] Office of the Secretary of Defense, *Military and Security Developments Involving the People's Republic of China 2015*, April 2015, p. 8.

[39] The three-warhead figure is from *Jane's Strategic Weapon Systems*, "DF-5," January 6, 2015.

Table 12.6
U.S.-China Balance of Nuclear Forces, 2017 (low and high estimates for China)

Offensive Capabilities						
United States				China		
System Type	Number	Warheads per System	Total Warheads	System Type	Lower Estimate Missiles/ Warheads	Higher Estimate Missiles/ Warheads
ICBMs				ICBMs		
Minuteman III	400	1	400	CSS-4 Mod 2 (DF-5A)	10/10	10/10
				CSS-4 Mod 3 (DF-5B)	10/30	10/30
				CSS-10 Mod 1 (DF-31)	12/12	12/12
				CSS-10 Mod 2 (DF-31A)	24/24	48/48
SLBMs				SLBM		
D-5	240	4	960	CSS-N-14 (JL-2)	30/30	60/60
Bombers				Bombers		
B-52H	44	20	528[a]	None		
B-2A	16	16	256			
Total	700		2,144	**Total**	86/106	140/160
Defensive Capabilities						
GBI	44	1	44	None		

[a] B-52H will be used to deliver ALCMs, of which the Air Force has 528. Hence, the number of warheads credited for the B-52Hs is fewer than the theoretical capacity of the aircraft. B-2s are assumed to carry B61-7 and B83-1 gravity bombs. Hans M. Kristensen and Robert S. Norris, "Nuclear Notebook: US Nuclear Forces," *Bulletin of the Atomic Scientists*, 2015.

with its force of DF-5s, which, due to their vulnerability, will not contribute much to second-strike survivability even with more warheads. Beyond 2017, it seems likely that China may MIRV its road-mobile DF-41, which is currently under development and which would yield more in terms of retaliatory credibility.[40]

[40] Given that existing mobile missiles (including the DF-31, DF-31A, and JL-2) are relatively small and would require miniaturized warheads, they are unlikely candidates. The DF-41 is said to be some 25-percent heavier than the CSS-10 Mod 2 (DF-31A). *Jane's Strategic Weapons Systems*, "DF-41 (CSS-X-10)," January 6, 2015.

Both our high and low estimates for China posit that the number of CSS-10 Mod 1s (DF-31s) will remain unchanged at 12, but in the case of CSS-10 Mod 2s (DF-31As), there is some uncertainty with regard to the size of the force, and we estimate the potential inventory at between 24 and 48.[41] There is also some uncertainty as to how many CSS-NX-14s (JL-2s) might be in service by 2017. *Jane's Fighting Ships* reports four *Jin*-class submarines in service by the end of 2015, with a fifth to be commissioned in 2017.[42] The JL-2 SLBM, carried by the *Jin*, encountered problems during testing, but DoD suggested that tests in 2012 were successful, and DoD's 2014 Annual Report to Congress on Chinese military power reported that the *Jin* would undertake its first operational patrols shortly.[43] We posit that China will have five *Jin*-class submarines in service by 2017, with between 50 percent and 100 percent of their full complement of 12 JL-2 SLBMs each (totaling between 30 and 60 missiles).[44] Both the *Jin*-class platform and the missile it carries will be a dramatic improvement over the minimal capability offered by the single *Xia*-class (Type 092) submarine in past years. We assume that the single *Xia*-class SSBN will be taken out of service. If it remains in the inventory, its operational status and usefulness will be questionable.

The U.S. nuclear arsenal size in 2017 will be constrained by the New START agreement with Russia and, as a result, will be smaller than it was in 2010. For the sake of simplicity, we use the end-point 2018 delivery system numbers for 2017, since it is uncertain when the United States will reach the New START 2018 limits (see Figure 12.7). We do not follow the New START definition of "deployed warheads." Under New START, the United States and Russia are each limited to 700 deployed strategic delivery systems (ballistic missiles and nuclear-capable bombers), with up to 100 in reserve. The treaty allows each side up to 1,550 deployed warheads. However, New START counts nuclear-capable bombers as a single deployed warhead, despite the ability to place more weapons on the platforms during wartime.[45]

Using the New START numbers for 2017 delivery vehicles but a somewhat more liberal definition of warheads, the number of U.S. delivery systems will fall from 882

---

[41] Some sources count missiles and indicate the number of DF-31As as 24, unchanged since 2010. Other sources suggest that the number of DF-31A brigades has doubled to four. If we assume 12 missiles per brigade, then the total DF-31A number would rise to 48, our upper bound. IISS, *The Military Balance*, 2015, p. 237; *Jane's Sentinel Security Assessment*, "Strategic Weapons Systems, China," April 6, 2015; and James Mulvenon and Andrew N. D. Yang, *The People's Liberation Army as Organization: Reference Volume v1.0*, Santa Monica, Calif.: RAND Corporation, CF-182-NSRD, 2002.

[42] *Jane's Fighting Ships*, "Jin class (Type 094)," February 13, 2015.

[43] Office of the Secretary of Defense, *Military and Security Developments Involving the People's Republic of China 2014*, June 2014, p. 8.

[44] The range of between 50 and 100 percent of Chinese SLBMs available in 2017 can be taken as reflecting the uncertainties associated with the operational status of the platforms (*Jin*-class submarines) and the production rate and operational status of the missile (JL-2).

[45] Amy F. Woolf, "The New START Treaty: Central Limits and Key Provisions," Washington, D.C.: Congressional Research Service, February 4, 2015.

Figure 12.7
U.S. Nuclear Counterforce Attack Against "Low Estimate" Chinese Inventory, 2017

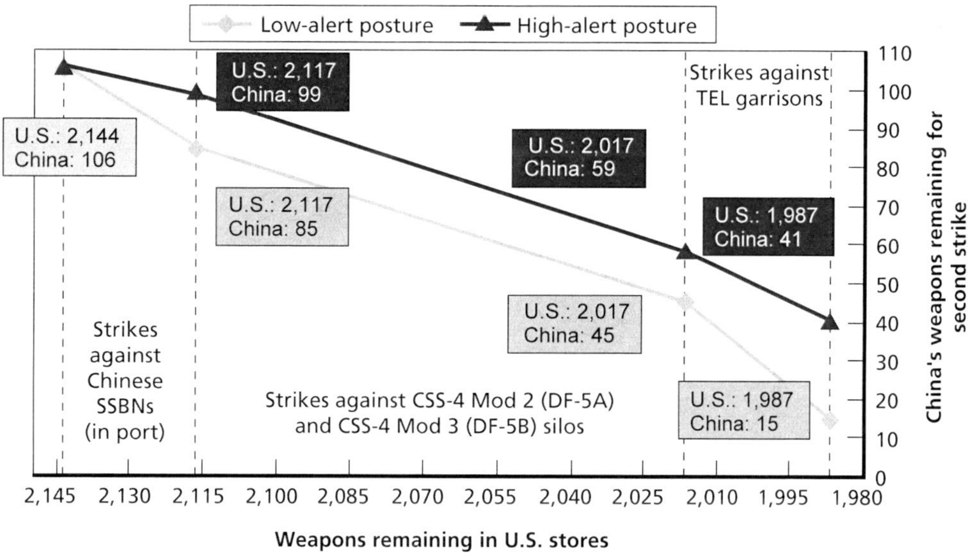

RAND *RR392-12.7*

in 2010 to 700 and the number of deployed warheads will fall from 4,806 to 2,144.[46] Based on current plans, it appears that the number of Minuteman III ICBMs will be reduced from 450 in 2010 to 400 in 2017, and all will be fitted with a single warhead. The United States will maintain its fleet of 14 *Ohio*-class SSBNs, but at least two will be in overhaul at any one time (and not count against limits), and each boat will sail with only 20 of its 24 tubes loaded with nuclear-capable Trident D-5 Mk-4A and Mk-5 SLBMs. The United States will maintain 60 nuclear-capable bombers, which we posit will include 44 B-52Hs and 16 B-2As. While we assume that there will be enough gravity bombs for each B-2A to carry a full complement of 20 weapons (for a total of 256 warheads), the load-out for B-52Hs will be limited by the inventory of air-launched cruise missiles (a total of 528 missiles and warheads). Finally, the United States expanded its number of GBIs from 24 to 30 in 2010, and, in March 2013, DoD announced a further increase to 44 by 2017.[47]

---

[46] With the exception of the GBI numbers, all data in this paragraph are derived from Hans M. Kristensen and Robert S. Norris, "Nuclear Notebook: US Nuclear Forces, 2015," *Bulletin of the Atomic Scientists*, Vol. 71, No. 2, 2015; U.S. Department of State, "The New START Treaty Aggregate Numbers of Strategic Offensive Arms," fact sheet, Washington, D.C., April 1, 2014; and U.S. Department of State, "Transparency in the U.S. Nuclear Weapons Stockpile," April 29, 2014. Note that in addition to deployed warheads, the United States currently holds a reserve of 2,680 warheads (in addition to warheads awaiting dismantlement), and it will likely continue to retain a large force in reserve.

[47] Office of the Under Secretary of Defense for Acquisition, Technology, and Logistics, *Report to Congress on Assessment of the Ground-Based Midcourse Defense Element of the Ballistic Missile Defense System*, Washington,

Chinese Vulnerability to First Strike, 2017

Even the low-end projected Chinese force structure outlined above suggests a more robust and survivable Chinese second-strike capability by 2017, though U.S. missile defense and intelligence efforts against PLAN SSBNs will likely continue to cause concerns in Beijing about the viability of its second-strike forces.

In the low Chinese case, U.S. forces must target the CSS-10 Mod 1 (DF-31), CSS-10 Mod 2 (DF-31A), CSS-4 Mod 2 (DF-5A), and CSS-4 Mod 3 (DF-5B) ICBMs and SSBN bases harboring the *Jin*-class (Type 094) that employ SLBMs. At the lowest-alert level, the United States can target and destroy all of the weapons that are located and susceptible to attack with an 80 percent chance of success using 157 weapons. While this represents approximately 7 percent of the total U.S. arsenal, it still leaves 1,987 warheads for other purposes.

The number of surviving Chinese weapons (15) in the 2017 case is marginally larger than the surviving number in the 2010 case (13). In reality, the gains to survivability should be considered more significant than this modeling might suggest. The surviving systems in the 2017 modeling include three types of relatively modern systems: the CSS-10 Mod 1 (DF-31), the CSS-10 Mod 2 (DF-31A), and the CSS-NX-14 (JL-2), associated with the *Jin*-class (Type 094) submarine. With five operational submarines and 36 road-mobile ICBM launchers, it becomes more reasonable to imagine modest alert levels that would ensure that a portion of both the land- and sea-based legs of China's nuclear dyad are kept deployed on an operational basis. While the *Jin*-class SSBNs will require a "shakedown" period, their introduction into operational service represents a major advance for Chinese nuclear forces and their survivability.

The results of attacks against the 2017 upper bound ("high estimate") case for Chinese missiles, which includes a larger number of CSS-10 Mod 2s (DF-31As) and CSS-N-14s (JL-2 missiles), are somewhat different. In this case, China possesses 160 warheads able to reach the United States prior to the attack, instead of 106. After the hypothetical U.S. first strike, 27 Chinese warheads remain, 12 more than in the "low estimate" case (see Figure 12.8). The additional Chinese missiles and the warheads atop them provide an additional buffer against the possible compromise of some portion of China's alert forces, and they provide a more convincing second-strike capability in the face of U.S. missile defenses, even when those defenses are assumed to be relatively effective in engaging targets. Note that in neither the low nor high estimate cases does the MIRVing of DF-5 missiles contribute to survivability against a first strike, since the missiles themselves are highly vulnerable to attack.

Beijing will continue to harbor concerns about U.S. missile defenses, as well about intelligence-gathering efforts directed against Chinese SSBNs. In the 2017 low Chi-

---

D.C., May 2010, p. 27; Tom Shanker, David E. Sanger, and Martin Fackler, "U.S. Is Bolstering Missile Defense to Deter North Korea," *New York Times*, March 15, 2013; Amaani Lyle, "Hagel: U.S. Bolstering Missile Defense," American Forces Press Service, March 15, 2013.

Figure 12.8
U.S. Nuclear Counterforce Attack Against "High Estimate" Chinese Inventory, 2017

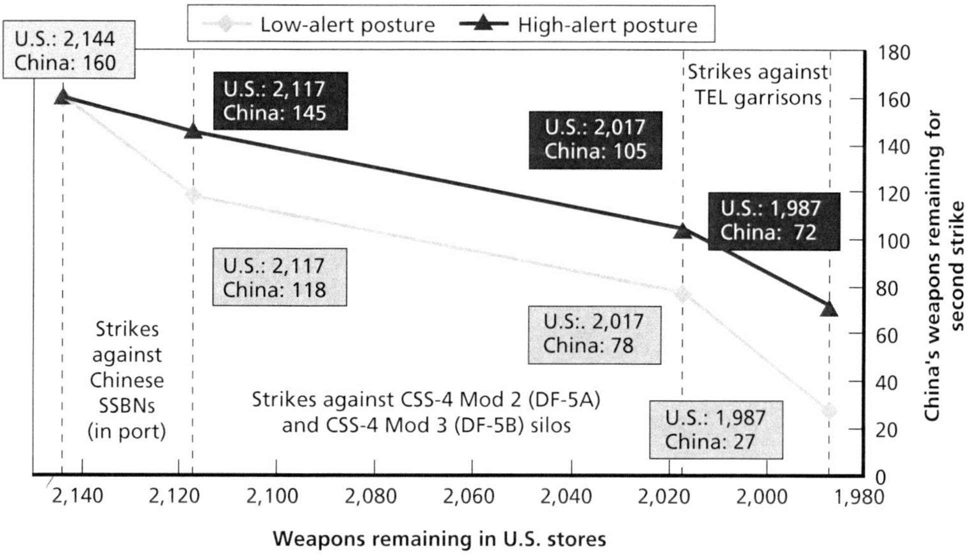

RAND *RR392-12.8*

nese case, the 44 interceptors deployed by the United States will outnumber surviving Chinese strategic warheads by nearly three to one. In the "high" case, the ratio is less than two to one. In the more distant future, if U.S. GBI missile deployment halts at 44 and each interceptor continues to carry only one warhead, the Chinese second-strike capability will ultimately be able to overwhelm U.S. defenses, especially if China puts MIRVs in the more survivable portion of its missiles. However, Chinese confidence in the effectiveness of its second-strike capability will likely depend on the extent of U.S. interceptor deployment, as well as the state of U.S. missile defense technology and advances in Chinese penetration aids.

It will also depend on the survivability of China's alert forces, especially its SSBNs. In addition to hosting large numbers of survivable missiles, SSBNs are the only Chinese weapons that can launch missiles at U.S. targets from areas where the trajectories do not take them near U.S. GBI sites in Fort Greely, Alaska. Our counterforce first strike modeling, and the sensitivity of results to the size of mobile forces, helps explain why Beijing has reacted to the activities of special oceanographic surveillance ships equipped with SURTASS, such as with the USNS *Impeccable* (T-AGOS-23), near Yalong Submarine Base in March 2009. The sonar on these ships is designed "to perform acoustic collection surveillance to help locate and identify submarines."[48] As

[48] Eric A. McVadon, "The Reckless and the Resolute: Confrontation in the South China Sea," *China Security*, Vol. 5, No. 2, Spring 2009, p. 1; "Ships, Sensors, and Weapons: Undersea Warfare Programs Target an Expeditionary Future," *Undersea Warfare*, Vol. 3, No. 3, 2001.

mentioned earlier, Yalong, with its underground submarine portal is a likely location for the new *Jin*-class (Type 094) SSBN. If the U.S. Navy can detect and trail Chinese SSBNs in the open sea, the Chinese second-strike capability once again becomes questionable, especially when U.S. anti-submarine warfare activities are combined with an active missile defense program.

This prognosis, however, only pertains to the immediate future. Looking somewhat beyond 2017, China will likely be able to achieve a highly secure second-strike capability (even with very conservative estimates from China's perspective), and further increases on its part beyond that point may raise more serious questions about its ultimate intentions.

U.S. Vulnerability to a Disarming First Strike, 2017

In the "low" case (with 106 Chinese warheads), a Chinese attack designed to maximize damage to U.S. counterforce targets could destroy 1,146 warheads (or their delivery systems), or roughly 53 percent of the 2,144 available to the United States. To achieve these results, China's CSS-4 Mod 2 (DF-5A) ICBMs would target U.S. counterforce area targets.[49] China's remaining missiles are focused on remaining silo-based targets. Because New START pushes the United States toward fewer weapons per delivery system (on average), this is the same loss percentage as that suffered in the 2010 scenario, despite a larger Chinese attacking force. In all, 998 warheads survive the Chinese first strike (see Figure 12.9). Although this total is less than half that of 2010 (2,240 warheads), it is more than sufficient to conduct a second strike that would inflict massive damage and to maintain a powerful reserve to deter other powers.

The upper-bound Chinese (or "high") case, which posits 160 Chinese warheads, yields results that differ little from the "low" case in terms of offensive potential against U.S. nuclear targets (see Figure 12.10). The only U.S. targets left to strike after attacking major nuclear-capable bomber bases and SSBN ports are Minuteman III silos. Each silo contains a single warhead, so only ten additional U.S. warheads are destroyed (for a total of 1,156 destroyed). It should be noted that in all cases back to 2003, China can do somewhat better by targeting each Minuteman III silo with only one warhead (rather than seeking the 80-percent salvo confidence level that the model requires), but even with a Pk of 1.0, fewer than 100 additional missiles and warheads would be destroyed in 2017 (or 17 in 2003).[50] The impact of missile defense is far less certain and

[49] These targets include Kitsap Naval Base, Kings Bay Naval Submarine Base, Whiteman AFB, Barksdale AFB, and Minot AFB. Because New START guidelines will likely reduce the U.S. nuclear-capable B-52H fleet from 76 to a projected number of 44, it is possible that remaining nuclear-capable B-52Hs will be consolidated at a single air force base rather than the current two (Minot AFB and Barksdale AFB). This would free additional Chinese warheads to strike other targets or be held in reserve. Because the only targets left to hit would be Minuteman III silos, the overall effect of these "extra" warheads would be minimal.

[50] Using more realistic reliability and Pk figures, we might expect 54 additional Minuteman missiles and warheads to be destroyed in 2017 (high case); 19 in the 2017 low case; 16 in 2010; and 12 in 2003. We caution, how-

Figure 12.9
Chinese Nuclear Counterforce Attack Using "Low" Chinese Estimate, 2017

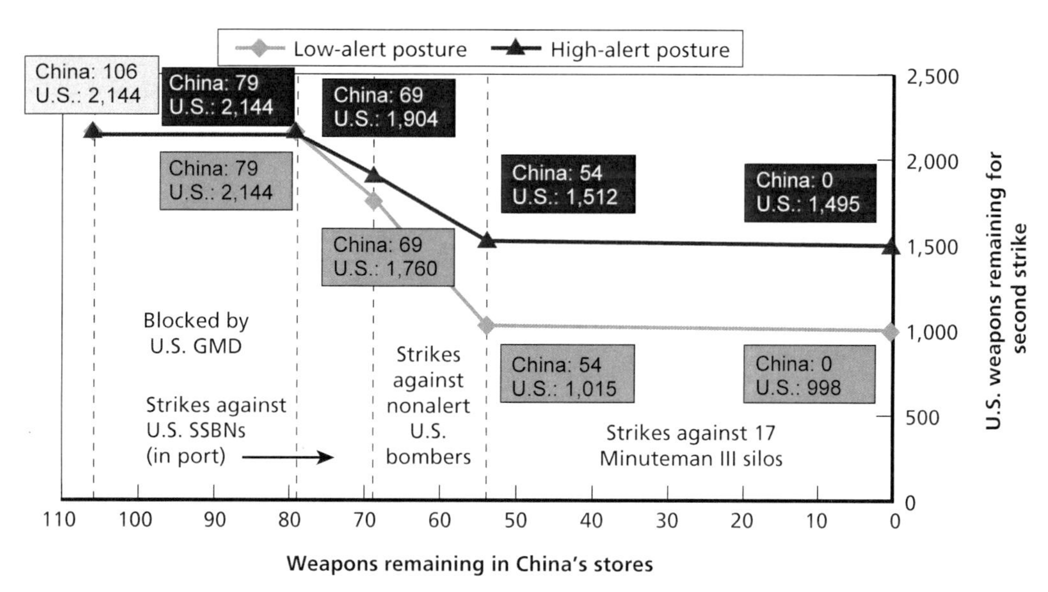

RAND *RR392-12.9*

Figure 12.10
Chinese Nuclear Counterforce Attack Using "High" Chinese Estimate, 2017

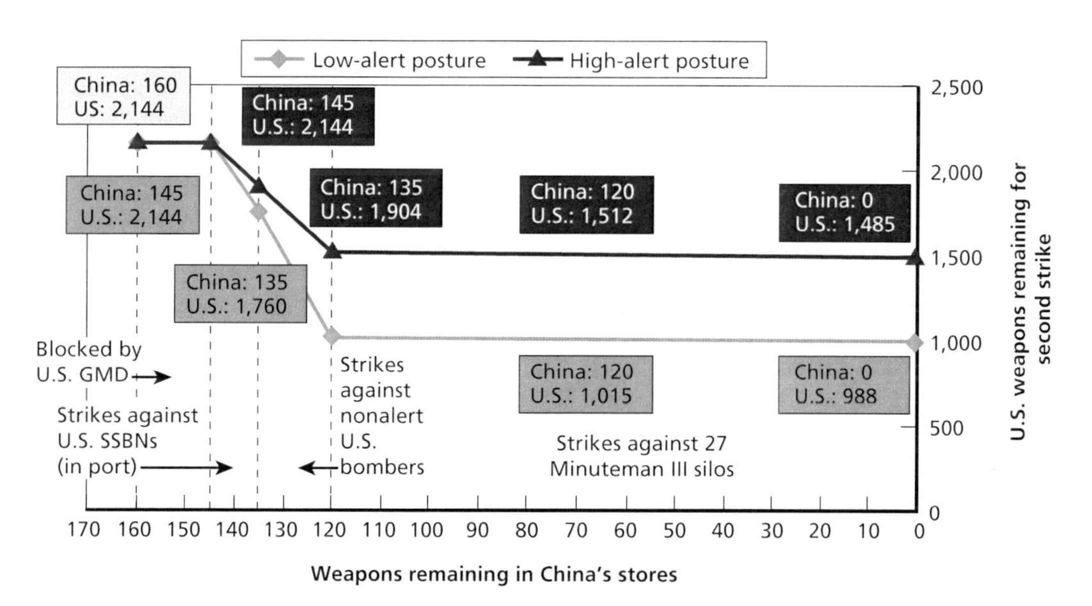

RAND *RR392-12.10*

---

ever, that adjusting assumptions and approach in this way might affect other parts of the calculations presented in the text (though much more modestly), and these figures are, therefore, provided only for heuristic purposes.

would depend on technical assumptions about firing and engagement sequencing, the quality of early warning, maneuverability of warheads, and Pks. The likely impact is small; our modeling suggests as few as four additional warheads protected. But under technical assumptions that are highly favorable to the United States, missile defenses could save several hundred warheads from destruction.[51]

Conclusions

Given the asymmetry in the respective number of nuclear weapons, it is hardly surprising that our modeling shows severe structural imbalances in outcomes (see Table 12.7). In every period, a U.S. first strike could destroy all targets with known locations with a relatively high degree of confidence, leaving a relatively small number of missiles on alert and therefore ostensibly not susceptible to attack. Given the small numbers of systems that might be deployed at any given point in time, Beijing would almost certainly have concerns about the effectiveness of its second-strike capability.

Assuming a similar level of alert for each of the periods considered, our modeling suggests that the number of Chinese warheads that might survive a first strike has increased over time, from four in 1996 to six in 2003, 13 in 2010, and between 15 and 27 in 2017. More importantly, the systems deployed have become more dependable

Table 12.7
Summary of U.S.-China Nuclear Counterforce First-Strike Results

Warheads	1996	2003	2010	2017 (low Chinese)	2017 (high Chinese)
Chinese inventory (warheads)	19	40	68	106	160
U.S. inventory (warheads)	7,646	6,488	4,806	2,144	2,144
U.S. first strike against China					
U.S. warheads employed	23	91	132	157	157
Chinese warheads surviving U.S. strike	4	6	13	15	27
U.S. GBIs	—	—	24	44	44
Chinese first strike against the United States					
Chinese warheads employed	19	40	68	106	160
U.S. warheads surviving	3,390	3,146	2,240	998	988

[51] To achieve this larger impact, U.S. missile defenses would need to completely protect one or more area targets from destruction. This might be possible if the Chinese were compelled to launch all weapons in a single salvo, if the aim points could be discerned by U.S. early-warning systems, and if the maneuverability of Chinese warheads after separation were marginal.

and survivable over time. In 1996, survivability depended entirely on the status of the *Xia*, with silo-based CSS-4 Mod 2s (DF-5As) being in fixed locations and therefore highly vulnerable to attack. Since 2003, however, new classes of road-mobile missiles, the CSS-10 Mod 1 (DF-31) and, later, the CSS-10 Mod 2 (DF-31A), have been deployed, representing a second type of mobile asset in the Chinese inventory. Unlike the CSS-4 Mod 2 (DF-5A), these missiles are solid-fueled and are therefore capable of launch on short notice. The *Jin*-class SSBN (Type 094), five of which are either complete or near completion, will replace the *Xia*-class (Type 092) by 2017 and provide a far more robust sea-based deterrent.

Despite these developments, however, Beijing continues to harbor concerns about the survivability of its nuclear forces. Chinese leaders express particular unease about the future of U.S. missile defenses and their potential impact on the nuclear balance.[52] While we did not attempt to model missile defense with any fidelity, our nuclear exchange modeling suggests that the planned number of deployed U.S. GBIs alone (without considering the SM-3 or other systems) could continue to outnumber China's expected survivable second-strike warheads as late as 2017. Any expansion of the missile defense architecture or enhancements to its capabilities, such as a move toward multiple exo-atmospheric kill vehicles on each interceptor, would further heighten concerns in Beijing about its retaliatory capability.

Moreover, Chinese planners may question whether the United States, with its sophisticated space-based and air-breathing ISR assets, might be able to locate and attack a larger portion of China's nuclear forces than our modeling suggests. Our model probably assumes a higher-alert posture for China than it presently practices, though it is true that China now has much greater capability to keep a portion of its force on alert, should it choose that course. New generations of Chinese land-based missiles (e.g., the DF-31A) are movable but not highly mobile. Chinese analysts question whether, in fact, SURTASS-equipped ships near China are seeking to collect acoustic data on PLAN SSBNs to facilitate their trail and destruction during a crisis or

[52] China's 2013 *Science of Military Strategy*, published by the Chinese Academy of Military Sciences, suggests that the nuclear security environment around China is growing more "complex" and challenging. U.S. actions play prominently in that description:

> The United States regards China as the main strategic opponent; it is speeding up construction of a missile defense system in the East Asia area; and its reliability and effectiveness for executing nuclear counterattack against China comprise an increasingly serious influence. 战略学 [*The Science of Military Strategy*], 3rd ed., 2013.

For more on Chinese thinking, see also, Wu Riqiang, "Why China Should Be Concerned with U.S. Missile Defense? How to Address It?" Georgia Institute of Technology Program on Strategic Stability Evaluation, undated; Li Bin, "The Impact of the U.S. NMD on the Chinese Nuclear Modernization," *Institute of Science and Public Affairs*, China Youth College for Political Science, undated.

conflict.[53] And Chinese concerns about advanced conventional munitions affecting the strategic balance are also increasingly pronounced.

The impact of the current balance of forces on crisis stability is uncertain. While China might fear losing its nuclear forces to the United States in a U.S. first strike, Beijing would also have little to gain—and much to lose—from striking first. Given the much improved but still relatively thin margin of China's survivable second-strike capability, however, any attrition of the PLA's nuclear capabilities during a conflict (especially attacks on one or more SSBNs, each one of which carries a dozen missiles) could provoke a nuclear "demonstration" firing to underline Beijing's resolve not be disarmed.

Issues of arms race stability also loom. Even if there is no change in China's commitment to a "lean and effective" nuclear force (which is broadly interpreted as a minimum deterrent posture), Beijing will seek to ensure that it retains a secure second-strike capability in the face of U.S. advances in the areas of ISR and missile defense. Chinese efforts will likely take two forms. First, Beijing will continue to improve the quality of its forces, with emphasis on mobility and penetration aids. Second, it will likely continue to add numbers to the force. Putting MIRVs on additional platforms, especially mobile ones, would be a relatively inexpensive way to increase the number of warheads that might survive a disarming first strike.[54] To be sure, there may also be other drivers of Chinese nuclear force structure expansion and modernization, such as bureaucratic imperatives and concerns about regional nuclear rivals (India, in particular). And outside analysts will watch with interest how China proceeds once it has achieved full confidence in its second-strike capability.

From a U.S. perspective, the model employed in this chapter suggests that increases in China's nuclear arsenal within the range of our 2017 estimates (including the "high estimate" case) produce no significant reduction in U.S. second-strike survivability, largely because of the still-sizable number of U.S. silo-based ballistic missiles. In continuing to maintain considerable numbers of silo-based ICBMs, the United States preserves a significant hedge against a surprise attack by forcing a potential attacker to spend multiple warheads to destroy a single U.S. warhead. Given the 13-to-1 warhead advantage maintained by the United States even in the 2017 "high" case for Chinese missiles, such an attack is a nonstarter from Beijing's perspective.

---

[53] See Wu Riqiang, "Survivability of China's Sea-Based Nuclear Forces," *Science and Global Security*, No. 19, Vol. 2, 2011, p. 112.

[54] Yet the mutual use of MIRVs has long been thought to undermine crisis stability since a single attacking warhead may destroy multiple adversary warheads. See, for example, Charles L. Glaser, "When Are Arms Races Dangerous? Rational Versus Suboptimal Arming," *International Security*, Vol. 28, No. 4, Spring 2004, pp. 77–79; Barry Nalebuff, "Minimal Nuclear Deterrence," *Journal of Conflict Resolution*, Vol. 32, No. 3, September 1988, p. 422; Sergey M. Rogov, "The Evolution of Strategic Stability and the Future of Nuclear Weapons," *Contemporary Security Policy*, Vol. 14, No. 2, 1993, p. 19; and Laurence S. Seidman, "Crisis Stability," *Journal of Conflict Resolution*, Vol. 34, No. 1, March 1990, p. 145.

Nevertheless, further moves by China to increase the size of its nuclear arsenal could have serious consequences. Such moves could undermine U.S. and Russian efforts to pursue further reductions in their own forces. Thus far, second-tier nuclear powers, such as China, India, and Pakistan, have had little impact on U.S. and Russian force planning—even as those second-tier powers increasingly interact with one another. However, continued growth in China's nuclear inventory will further undermine political support for cuts beyond those required by New START; this support is already highly uncertain in the wake of continuing Russian aggression in Ukraine. Washington, Moscow, or both may wish to ensure that China does not make a dash for parity while the former Cold War superpowers reduce their nuclear forces.

Even without pursuing parity, a larger Chinese nuclear inventory may complicate future U.S. and Russian planning if either contemplates simultaneous nuclear threats by China and another power. And finally, as China's second-strike capability becomes more secure—and it has many options to improve that survivability—Beijing may become bolder in its international behavior. As the other scorecards in this report indicate, the potential costs of a conventional military conflict with China are growing for the United States. And as this scorecard suggests, Chinese leaders might believe that the United States no longer has the option of escalating to the nuclear level without suffering powerful retaliation. Hence, they may believe that the United States, without absolute escalation dominance, will be less likely to intervene in the event of a regional conflict.

Despite this possibility, attempting to hold Chinese nuclear forces at risk of decapitation into the indefinite future would almost certainly be a poor idea for the United States, not least because it is likely a losing proposition. China will probably have more—and cheaper—options to improve the survivability of its second-strike capability than the United States would to threaten such survivability. Moreover, efforts to hold Chinese nuclear forces at risk would have profoundly negative consequences for arms race stability, as well as the larger political relationship with China. Engaging China on particular points of strategic concern to Washington will be critical to the long-term stability of the U.S.-China strategic relationship, as well as for regional stability more broadly.

Scorecard Coding

Figure 12.11 provides our summary coding of the results of scorecard 10. Our analysis and metrics focus primarily on whether each party might have grounds to feel confident in the survivability of its second-strike capability.

The modeling suggests that at least some Chinese strategic nuclear systems might have survived a U.S. first strike throughout the period considered. Nevertheless, in the 1996, 2003, and (to a lesser extent) 2010 cases, the expected number of surviv-

Figure 12.11
Scorecard 10 Summary Coding

	Taiwan Conflict				Spratly Islands Conflict			
Scorecard	**1996**	**2003**	**2010**	**2017**	**1996**	**2003**	**2010**	**2017**
1. Chinese attacks on air bases								
2. U.S. vs. Chinese air superiority								
3. U.S. airspace penetration								
4. U.S. attacks on air bases								
5. Chinese anti-surface warfare								
6. U.S. anti-surface warfare								
7. U.S. counterspace								
8. Chinese counterspace								
9. U.S. vs. China cyberwar								

	Country	1996, 2003, and 2010	2017
10. Nuclear stability (confidence in secure second-strike capability)	China	Low confidence	Medium confidence
	U.S.	High confidence	

NOTES: In the case of the nuclear stability scorecard, our analysis and metrics focus primarily on second-strike survivability, an important (though not the only) determinant of crisis stability. Because we evaluate survivability from both sides, and survivability, even on the U.S. side, does not necessarily correlate positively with U.S. advantage, no color-coding is employed.

Key for Scorecards 1–9

U.S. Capabilities		Chinese Capabilities
Major advantage		Major disadvantage
Advantage		Disadvantage
Approximate parity		Approximate parity
Disadvantage		Advantage
Major disadvantage		Major advantage

ing systems would have been small. More importantly, China possessed only a few types of strategic systems, and their survival would have depended on some level of Chinese alert—and probably a higher level than its normal peacetime practice—and limited U.S. ISR capability. Hence, we determined that Chinese leaders would have had low confidence in their forces' second-strike survivability through 2010. By 2017, the expected number of surviving Chinese systems grows to more than two-dozen warheads (depending on assumptions about inventory and alert levels). Combined with the fielding of larger numbers of road-mobile systems and SSBNs, Chinese leaders will have a somewhat higher level of confidence. We therefore coded 2017 as *medium confidence*, with U.S. missile defenses and ISR preventing Beijing from achieving a high level of confidence.[55] Even under the most challenging assumptions for 2017, Chinese offensive systems never threaten the survival of the U.S. second-strike capability, so we coded the entire period as one of high confidence for U.S. leaders.

---

[55] The deployment of U.S. GBI and other missile defenses prevents Chinese confidence levels from being higher in the 2017 time frame.

CHAPTER THIRTEEN

The Receding Frontier of U.S. Dominance

Not since the Vietnam War has the United States fought a sustained air superiority campaign in which U.S. aircraft were challenged by both enemy fighters and ground-based air defenses. Not since World War II has it fought an enemy capable of putting its major surface ships or submarines at risk through anything other than surprise, one-off raids. Nor since that time has it fought a high-intensity war in which its support facilities, including regional air and naval bases, were expected to operate while under systematic conventional attack. And it has never fought an opponent armed with precision standoff weapons, operationalized counterspace capabilities, or well-developed and practiced cyberwarfare capabilities, much less one armed with nuclear weapons. Yet a conflict with China would likely see all of these things—an air and naval war fought at high intensity using new generations of weapons and supported by a wide range of high-technology systems. Losses in aircraft and ships (and, quite possibly, personnel) would be higher than any seen in recent wars.

Comparisons with past wars, and especially World War II, should not be overdrawn. China is not Nazi Germany or Imperial Japan, and barring truly unforeseen circumstances, a conflict would more likely take the form of a campaign, albeit an intense one, for relatively limited objectives, rather than a multitheater conflict fought over a number of years.[1] Moreover, the U.S. military is certainly not the small, garrison force that it was in 1941. On the contrary, it is the most advanced, most experienced, and, by some standards, largest military in the world. Having noted all these qualifiers, however, conflict with China would look even less like recent wars, in which the United States established air and naval supremacy in a matter of hours or days and then proceeded to "apply force" from secure bases. Rather, this would be a war in which the United States would be challenged in the air, on (and under) the water, in space, and across the electromagnetic spectrum. U.S. forces would be hard-pressed from the start,

[1] Further study on the extent to which Chinese strategists are considering different types of conflicts is in order. Few if any states consciously embark on wars lasting years, but some nevertheless consider the possibility and take measures to prepare for it. At the other end of the spectrum, operations (such as the seizure of one of the offshore islands held by Taiwan) would present a very different and in some ways more challenging problem for the United States.

and they would probably not enjoy sanctuary in regional bases. Also unlike recent wars, the U.S. military could well sustain significant air and naval losses.

As the scorecards in this volume illustrate, the specific location and parameters of conflict would shape the extent to which PLA forces could challenge or press the United States. U.S. forces continue to maintain a number of technological, organizational, and human advantages over the PLA, and a protracted conflict would work to further favor the United States. Nevertheless, rapid budget increases, combined with organizational and doctrinal reforms, have enabled China to narrow the military gap in almost every area and move ahead in some. Moreover, in a conflict close to the Chinese mainland, the PLA would enjoy enormous geographic and positional advantages, while the United States would be handicapped in its ability to deploy and operate forces.

This chapter proceeds as follows. First, we summarize broad patterns in U.S.-China military competition, drawing on all ten scorecards. Second, we then summarize changes to the balance of capabilities in the two individual scenarios examined: a PLA invasion of Taiwan and a conflict over the Spratly Islands. And, finally, we ask whether and when Chinese and U.S. military capabilities are reaching a series of tipping points in which the United States might prevail in a protracted conflict but PLA forces might nevertheless challenge U.S. dominance during the initial battles.

Scorecard Summary Findings

Four key findings emerged from the ten scorecards presented. First, since 1996, our initial point of comparison, the PLA has made tremendous strides, and the overall capability trend lines are moving against the United States. In some areas, such as ballistic missiles, fighter aircraft, and attack submarines, improvements have come with breathtaking speed by most historical standards. Second, the trends vary by mission area and Chinese gains have not been uniform. In some areas, U.S. improvements have given the United States new options or mitigated the speed at which Chinese gains have shifted the relative balance.

Third, the geography of conflict is critical, and distances, even short distances, have a major impact on relative capabilities. Chinese power projection capabilities are improving, but the PLA's ability to control military events diminishes rapidly beyond the unrefueled range of jet fighters and diesel submarines. And fourth, China has not caught up to the U.S. military in terms of aggregate capabilities—and is not close to doing so—but it does not need to catch up to the United States to dominate its immediate periphery. The advantages conferred by proximity severely complicate U.S. military tasks. China is increasingly capable of challenging the ability of U.S. forces to accomplish mission-critical tasks in scenarios close to the Chinese mainland.

Trend Lines Are Moving Against the United States

The military equation in East Asia has changed dramatically since 1996. By that year, China had already reformed many aspects of its antiquated Maoist military system. It was beginning to take delivery of advanced military systems imported from Russia. And it stood on the cusp of rapid military budget increases that would allow it to import and, ultimately, produce a much wider array of modern military equipment. But the immediate reality for Chinese military leaders in 1996 was that the overwhelming preponderance of its equipment was obsolete. The large majority of its air and naval platforms were only partially modernized versions of systems that had entered Soviet service in the 1950s and early 1960s. Since 1996, however, China has increased its real (inflation-adjusted) military spending by an annual average of 11 percent. It has concurrently reduced the size of its forces and streamlined procurement, enabling the PLA to modernize virtually every aspect of its capabilities.

As this report has emphasized, with increased resources, the PLA has rapidly replaced much of its obsolete equipment with modern systems. The proportion of modern, fourth-generation fighters in the inventory rose from less than 1 percent in 1996 to 29 percent in 2010 and 51 percent in 2015.[2] There is less agreement on what constitutes "modern" naval craft, but according to one (arguably liberal) definition, the proportion rose from 7 percent of the surface fleet in 1996 to 41 percent in 2010 and 68 percent in 2015.[3] The newest Chinese warships are armed with a variety of modern close- and long-ranged SAMs, sophisticated supersonic ASCMs, and improved (though still not state-of-the-art) anti-submarine warfare systems.

In terms of attack submarines, the proportion of modern boats rose from less than 3 percent in 1996 to 48 percent by 2010 and to 66 percent in 2015.[4] China has also

---

[2] Aircraft generations are commonly defined against the most advanced aircraft of particular eras, beginning with the advent of the first jet fighters. See, for example, Joe Yoon, "Fighter Generations," *Aerospaceweb.org*, June 27, 2007.

[3] Here, we employ a relatively liberal definition of *modern*, categorizing the *Sovremenny* (DDG), *Luhai* (DDG), *Luyang* (DDG), *Luzhou* (DDG), *Jiangwei* (FFG), and *Jiangkai* (FFG) classes as modern ships, and the *Luda* (DDG), *Luhu* (DDG), *Chengdu* (FFG), and *Jianghu* (FFG) classes as legacy ships. Modern naval platforms must have sensors and weapons to survive and fight in a combat environment. Such systems generally include advanced radar systems, such as planar arrays; long-range SAMs; and point defenses capable of protecting the ship from aircraft and missile attack. Most modern ships also have at least some defenses against submarines, including active and passive sonar and helicopters that can attack subs. Flexible and rapid-reloading vertical-launch missile systems for both defensive and offensive missiles are also becoming standard. There is, however, no standard definition of *modern* in the naval realm, so we focused primarily on modern, long-range air defenses and anti-ship missiles.

[4] We categorize *Romeo*-, *Ming*-, and *Han*-class boats as legacy craft and *Kilo*- (877 and 636), *Song*-, *Yuan*-, and *Shang*-class boats as modern. Apart from age and noise levels, China's legacy diesel submarines had angular hulls (designed for cruising on the surface), rather than the teardrop shape associated with most boats designed since the mid-1950s, and they lacked the cruise missiles that would allow long-range engagements. The *Han*-class SSN lacks adequate shielding around its reactor and poses a health threat to its crew. China's newer classes of SSKs borrow a variety of design features from the Russian *Kilo* class and are armed with modern ASCMs. For more

added capabilities in entirely new areas. In 1996, the PLA was just beginning to develop conventionally armed ballistic missiles and had only a few dozen relatively inaccurate conventionally armed short-range missiles that could reach targets in Taiwan or Korea, but not U.S. bases in Japan. By 2015, however, the PLA's Second Artillery had deployed more than 1,200 conventionally armed ballistic missiles, including the DF-21C with the range to attack targets throughout Japan. Perhaps more importantly, accuracy had increased dramatically, with CEPs falling from several hundred meters in the 1990s to as few as 15 feet, transforming China's theater ballistic missiles from largely indiscriminate weapons to systems capable of severely and reliably damaging U.S. facilities.[5] By 2015, China had also deployed hundreds of ground- and air-launched cruise missiles, further complicating the defense of U.S. and allied rear areas.

The United States has also improved its military capabilities over the period considered in our study. U.S. military spending surged after the September 11, 2001, attacks on the United States, rising from $316 billion in 2001 to $691 billion (including supplemental spending) in 2010, before falling to $560 billion in 2015. Much of the additional budget over this period went to fund operations in the Middle East, and U.S. military procurement priorities over this period have focused largely on systems designed for low-intensity conflict, not the kinds of high-intensity operations likely in East Asia. Increased funding since 2001 did, nevertheless, bring improvements to U.S. conventional warfighting capabilities, including significant advances in net-centric warfare and joint operations. In several areas, the U.S. military has deployed next-generation military equipment, while Chinese modernization has focused more on systems comparable to U.S. equipment that came of age during the 1980s and 1990s.

Although China has not closed the gap with the United States, it has narrowed it—and it has done so quite rapidly. Even for many of the contributors to this report, who track developments in the Asian military situation on an ongoing basis, the speed of change revealed by the analysis of retrospective data was striking. Typically, U.S. military equipment has advanced by one generation over the entire period (e.g., from fourth- to fifth-generation fighter aircraft), while Chinese capabilities advanced farther from a more primitive base (e.g., from second- to fourth-generation aircraft).

Trends Vary by Mission Area

Although overall trends are running against the United States, these trends are not uniform. Moreover, some scorecard themes are more important than others and have significant spillover effects into other areas. Two areas of particular concern are threats

---

details on Chinese submarines generally and noise levels in particular, see Office of Naval Intelligence, *People's Liberation Army Navy*, 2009.

[5] Vitaliy O. Pradun, "From Bottle Rockets to Lightning Bolts: China's Missile Revolution and PLA Strategy Against U.S. Military Intervention," *Naval War College Review*, Vol. 64, No. 2, Spring 2011, p. 14.

to U.S. forward air bases and the U.S. surface fleet, especially aircraft carriers. Accurate ballistic missiles and surface- and air-launched cruise missiles pose a serious threat to U.S. bases in Korea and Japan, as well as some threat to Andersen Air Base on Guam (see Chapter Three, scorecard 1). U.S. aircraft at these locations are largely unprotected, and Chinese missiles could cause widespread destruction against deployed assets. Similarly, the modernization of the Chinese submarine fleet, combined with maritime strike aircraft and improved maritime ISR, poses increasing challenges for U.S. surface ships operating within 1,500 km of the Chinese coast (see Chapter Seven, scorecard 5).

While the U.S. military will develop countermeasures against threats to both air bases and aircraft carriers, the PLA will also continue to improve its offensive power. Geography and the PLA's development of redundant capabilities make it unlikely that, barring a revolutionary technological breakthrough, these threats will diminish. The PLA's ability to hold land bases and carriers at risk compounds the problems faced by U.S. forces in the air superiority battle, which would be challenging even in the absence of threats to bases (see Chapter Four, scorecard 2). U.S. aircraft will either face the risk of destruction at forward bases or fly from more distant bases to get into the fight, reducing the total number of sorties and loiter times once on station. As the air superiority battle becomes more competitive and airspace is contested, a host of other missions will also become more difficult. To cite but one example, U.S. anti-submarine warfare efforts, which rely heavily on large, lightly defended airborne platforms, would be substantially hampered in contested airspace.

Although the trend lines are negative in most areas, they do vary substantially. In some cases, U.S. relative capabilities remain robust, either because the Chinese have made relatively less effort in those areas or because the United States has taken measures to mitigate or reverse Chinese gains. For instance, as discussed in Chapter Five, scorecard 3, the Chinese introduction of double-digit SAMs and fourth-generation fighters has compromised the ability of U.S. legacy aircraft to penetrate Chinese airspace. But the combination of stealth and a larger, more capable inventory of precision standoff weapons has resulted in a net improvement in the U.S. capability to attack certain types of targets in mainland China, though these gains might not hold in a protracted conflict that exhausts the available stock of standoff weapons (see Chapter Six, scorecard 4). Similarly, despite improvements to Chinese anti-submarine warfare capabilities, the U.S. submarine fleet remains capable of doing extensive damage to China's surface fleet (see Chapter Eight, scorecard 6). Although expected U.S. submarine losses have grown somewhat over the period considered, they remain low, at least in the context of a war between major powers.[6]

[6] Expected U.S. losses in a seven-day campaign against a Chinese amphibious fleet in the Taiwan Strait rose from roughly 0.5 submarines in a hypothetical 2003 conflict to 1.8 submarines in 2017. These modeling results should be treated as merely illustrative of general trends, rather than precise predictions.

Finally, for some scorecards on which overall U.S. capabilities are slipping relative to those of China, developments at the system level are better than others. For example, although the air superiority battle is becoming more challenging, the introduction of the F-22 and F-35 will likely ensure that, in combat between forces approaching even numbers, U.S. air forces will continue to achieve high kill ratios. Similarly, while Chinese counterspace capabilities are improving overall, the threat to some satellite constellations may be significantly mitigated by ongoing improvements to particular U.S. satellite systems, especially those associated with PNT and missile warning.

Distances (Even Relatively Short Distances) Matter

Not only do the results vary by mission area, but they also vary across the two scenarios. The assessment of the Taiwan and Spratly Islands scenarios suggests that Chinese power diminishes rapidly across even relatively modest distances. The U.S. basing structure is not optimized for operations in the South China Sea, and both sides would be able to bring less to the Spratly Islands fight than they would in the case of Taiwan. But Chinese capabilities suffer relatively more, as China currently lacks the support structure necessary to sustain significant combat forces at a distance from its coast, and its current land reclamation efforts and the construction of new infrastructure on the islands would be of only modest benefit in a high-intensity war.

Although China is extending the range of its conventionally armed cruise and ballistic missile forces, the numbers it can deliver at longer ranges will be far smaller than those capable of striking closer targets in the Taiwan scenario. Modeling of the air superiority battle suggests that the U.S. fighter inventory required to prevail in a South China Sea scenario would be roughly half that required in a Taiwan scenario. And as the analysis of U.S. attacks on Chinese air bases suggests (see Chapter Six, scorecard 4), China's basing options would be severely constrained in the case of a Spratly Islands scenario, enabling the United States to focus its high-value aircraft and munitions on a smaller target set, increasing the impact of its attacks.

As in the case of the Taiwan scenario, the trend lines are moving in a negative direction for the United States in the Spratly Islands case. Further improvements in relative Chinese capabilities can be expected, but with less asymmetry in the geographic dimension, these improvements may come at a higher cost to China. They will require China to invest relatively more in support capabilities (e.g., tankers, SATCOM, basing infrastructure), which will compete for defense resources with fighters, warships, and other combat assets.

China Can Pose Problems for the United States Without Catching Up

To say that the United States would have difficulty achieving a variety of critical missions in particular scenarios is not to imply that the Chinese military has "caught up" to the U.S. military in overall quality, sophistication, or numbers of high-end systems.[7] By many standards, the PLA continues to lag far behind the U.S. military. It is only now readying its first aircraft carrier, while the United States operates ten full-sized carriers plus nine additional amphibious assault ships capable of supporting fixed-wing aircraft. The U.S. Air Force's first purpose-built stealth combat aircraft, the F-117, entered IOC in 1983. As of 2015, the United States had deployed ten squadrons of F-22s—by far the most advanced fifth-generation aircraft in the world. The PLAAF, for its part, conducted its first test flight of a stealth aircraft, designated the J-20, in January 2011 and tested a second model, the J-31, for the first time in October 2012. Despite these tests, China is likely years from fielding an actual capability, much less one that matches the F-22.

The United States also maintains far more (and far more capable) support aircraft, such as tankers and AWACS, and enjoys a similar lead in deployed satellite capabilities, attack submarines, and anti-submarine platforms. To take one example, China operates a total of ten tanker aircraft (all converted H-6 airframes) and is currently taking delivery of an additional three Il-78M tankers from Ukraine. The U.S. Air Force has 475 larger and more capable tanker aircraft in its active and reserve components.

However, the overall balance of forces is only an abstract concept. When armies clash in the physical world, operational factors, such as the objectives of the two sides, the available time for mobilization, the distance between various operationally relevant points, and the movement speeds of the assets involved, can have a decisive impact. The scorecards show not that China has caught up to the United States but that it does not have to do so to mount a serious challenge to U.S. forces near the Chinese mainland.

Taipei is roughly 11,000 km from San Diego (and more than 8,000 km from Honolulu) but only 160 km from the closest point on mainland China. Fighting in China's front yard endows Chinese forces with enormous advantages. From secure bases on the mainland, China can target U.S. forward bases in Asia with large numbers of accurate ballistic and air-launched cruise missiles. While the United States could retaliate with air and missile attacks of its own, there are 39 PLA air bases within 800 km of Taipei (roughly the range of unrefueled fighter aircraft), whereas there is only a single U.S. Air Force base (Kadena AB) within that distance—and only three within 1,500 km. The effective loss of a single U.S. air base, therefore, would be far more consequential to the United States than the loss of a single Chinese air base (or even several) would be to the PLA. Proximity would give China advantages in other realms as well, such as communication and logistical support. While U.S. forces would be largely dependent on satellite links for their communication, Chinese forces in a

[7] The language of this section title is borrowed from Thomas Christensen, though it is used here to highlight a different range of issues than those raised in Christensen's article. See Christensen, "Posing Problems Without Catching Up," 2001.

Taiwan conflict could rely on land-based communication, which is far less vulnerable to disruption.

As the scorecards illustrate, the Chinese military does not necessarily have to overtake the U.S. military in terms of quality, or even the number of high-end naval or air systems, to challenge it and potentially emerge victorious. Indeed, in the Taiwan scenario, the U.S. military would find itself hard-pressed even today.

Coding the Scorecard Results

We coded most of the scorecards (nine out of ten) using a five-color stoplight scheme to denote major or minor U.S. advantage, a competitive situation, or major or minor Chinese advantage. *Advantage*, in this case, means that one side is able to achieve its primary objectives in an operationally relevant time frame while the other side would have trouble in doing so.[8] A number of assumptions and caveats are associated with the coding, and we emphasize that the stoplight coding be considered only in the context of the larger and more detailed scorecard analysis presented in Chapters Three through Twelve. We did not evaluate scorecard 10 (on nuclear stability) using the stoplight rating system because the issue addressed in that scorecard is not amenable to assessing advantage to one side or the other; rather, we considered both sides' confidence in the survivability of their second-strike forces.

Several of the scorecards depict one side or the other in a more operationally offensive role. In such cases, the other side is measured in terms of its ability to achieve relevant operational objectives in the face of those attacks. For example, in scorecard 5, which looks at Chinese anti-surface warfare, the capability of the U.S. side is measured by its ability to avoid or foil attack, or to destroy Chinese offensive systems, putting it in a position to bring its own surface capabilities to bear offensively.

As noted earlier, *advantage* is also assessed in the context of operationally relevant time frames. In the two scenarios examined, we evaluated relative capabilities for the first 21 days of combat. The duration considered is critical to coding. For example, in none of the cases (scenarios or time periods) would the United States "lose" an indefinitely protracted battle for air superiority against China. But we coded the air superiority battle (scorecard 2) based on the U.S. ability or inability to gain air superiority and begin attacking Chinese strike aircraft within the first 21 days.

In our judgment, Chinese forces would either prevail within several weeks of the start of hostilities or run into increasingly severe logistical problems (with the progressive destruction of their lift capability) that would ultimately doom the assault. So, for

[8] For example, even if the U.S. military could clear the skies of Chinese escort fighters with minimal friendly losses, the air superiority scorecard could be coded as "Chinese advantage" if the United States cannot prevail while the invasion hangs in the balance. If U.S. forces cannot move on to focus on destroying attacking strike and bomber aircraft, they cannot contribute to the larger mission of protecting Taiwan.

example, if U.S. air forces could not achieve air superiority in that given time frame, U.S. airpower would be unable to influence the outcome of the larger contest before it is effectively decided one way or the other. The logic in the Spratly Islands case is somewhat different but also points to a relatively short duration. In a conflict over limited objectives, as a Spratly conflict would be, both sides would face intense pressure to terminate hostilities quickly.

Nevertheless, while there is a logic to the temporal conditions in our analysis, the duration of an actual conflict would be affected by a number of political-military variables in addition to the course of military operations. A longer time frame could produce different results. In general, a longer conflict would favor the United States, as the U.S. military would be able to bring additional assets into the theater. However, this generalization is becoming somewhat less true over time as U.S. air and naval forces become more dependent on a finite number of standoff missiles (see Chapters Six and Eight, scorecards 4 and 6). Given the potential magnitude of a conflict with China and the large number of potential targets, limited U.S. munition inventories would figure heavily in a long war.

The final and perhaps most important caveat is that although the coding is informed by the qualitative and quantitative analysis in the relevant chapters, there is inevitably a subjective component to deciding what constitutes "advantage." In evaluating PLA attacks on U.S. air base runways, for example, we can assess how long China might be able to close the Kadena AB (or other air bases) to U.S. fighter operations. Stipulating how many days of base closure would constitute Chinese "advantage," however, is more difficult. Conceptually, the question is whether Chinese attacks can substantially reduce U.S. air operations, but is five days of closure or ten days of closure the right threshold? In each chapter, we present a brief summary of how we assessed advantage, but there is also a subjective component to the evaluation.

It is important to keep these stipulations in mind when interpreting the results of the ten scorecards, summarized in Figure 13.1. The figure shows relative U.S. and Chinese capabilities in each of the two scenarios in each of the four snapshot years (1996, 2003, 2010, and 2017) addressed in our study.

In the Taiwan scenario, we coded seven out of nine relevant scorecards as having either decisive or moderate U.S. advantage in 1996.[9] The degree of U.S. dominance declined somewhat by 2003, but the United States continued to enjoy advantages in most areas.[10] In 2010, however, U.S. forces retained clear advantages in only four of nine areas, with relative parity in four and U.S. disadvantage in one. By 2017, we project that China will hold the advantage in two areas, with rough parity in four and

---

[9] Because China had no military space assets at that time and the United States had no operationalized counterspace capability, we did not code U.S. counterspace capabilities for that year.

[10] And while U.S. counterspace capabilities were weak, limited contemporary Chinese satellite capabilities made this scorecard relatively unimportant.

Figure 13.1
Summary Coding of Scorecard Results

Scorecard	Taiwan Conflict				Spratly Islands Conflict			
	1996	2003	2010	2017	1996	2003	2010	2017
1. Chinese attacks on air bases								
2. U.S. vs. Chinese air superiority								
3. U.S. airspace penetration								
4. U.S. attacks on air bases								
5. Chinese anti-surface warfare								
6. U.S. anti-surface warfare								
7. U.S. counterspace								
8. Chinese counterspace								
9. U.S. vs. China cyberwar								

	Country	1996, 2003, and 2010	2017
10. Nuclear stability (confidence in secure second-strike capability)	China	Low confidence	Medium confidence
	U.S.	High confidence	

NOTES: To prevail in either Taiwan or the Spratly Islands, China's offensive goals would require it to hold advantages in nearly all operational categories simultaneously. U.S. defensive goals could be achieved by holding the advantage in only a few areas. Nevertheless, China's improved performance could raise costs, lengthen the conflict, and increase risks to the United States.

Key for Scorecards 1–9

U.S. Capabilities	Chinese Capabilities
Major advantage	Major disadvantage
Advantage	Disadvantage
Approximate parity	Approximate parity
Disadvantage	Advantage
Major disadvantage	Major advantage

U.S. advantage in three. In other words, the initial stages of a Taiwan scenario will be extremely competitive by 2017, with China able to challenge U.S. capabilities in a wide range of areas.

U.S. forces fare significantly better in our analysis of a Spratly Islands conflict, though trend lines are moving against the United States in that case as well. In 1996, the United States had major advantages in eight out of nine relevant areas (the ninth being U.S. counterspace, in which U.S. capabilities were weak but China had little to attack). By 2003, the U.S. margin of advantage declined only slightly, and, even as late as 2010, the United States continued to enjoy advantages in seven of nine areas, with rough parity in two. Finally, the analysis suggests that, by 2017, the United States will hold the advantage in five areas, with rough parity in four.

Overall advantage—and the ability to prevail in the conflict—cannot be judged by the scorecard results. Maintaining advantage in a majority of areas does not suggest that one side will necessarily "win." In general, the offensive strategic goals outlined for China in the two scenarios are more demanding than U.S. defensive ones. Therefore, China might have to hold advantage in a large majority of operational areas to achieve its objectives. Moreover, not all the scorecards are of equal importance to the outcome. To take but the most obvious example, it is conceivable that the United States could prevail in a conflict if it retained a decisive edge in anti-surface warfare (and could therefore sink a Chinese landing force), even if China held the advantage in all other areas.

We also remind the reader that, as discussed in Chapter One, the analysis is primarily centered on systems interactions and takes only limited account of differences in training and proficiency. Overall, U.S. training remains significantly more realistic and advanced than that of China, suggesting that the United States will do somewhat better in most areas than our coding suggests. However, it should also be noted that the PLA has undertaken a variety of organizational, manpower, and training reforms that have progressively narrowed the gap. Given this improvement, the difference between results yielded by the systems approach taken in this report and more qualitative measures has almost certainly narrowed significantly over the period assessed.

Taken as a whole, the scorecard coding indicates that the U.S. military would face increasing challenges in a Taiwan invasion scenario. In 1996, the United States was dominant in all areas. War being unpredictable, the United States might have suffered losses from individual events that might have come as both a surprise and shock to the military or, especially, to the U.S. public. Nevertheless, the outcome would not have been in doubt. By 2017, a Taiwan invasion could look dramatically different. At a minimum, the U.S. military would have to mount a substantial effort—certainly much more so than in 1996—if it hoped to prevail, and losses to U.S. forces would likely be heavy.

Moreover, although not all scorecards are equally important, they are all interrelated. Our coding considers each area largely independently of the others. In reality, advantage would often work to compound advantage and disadvantage in other areas. The next section of this chapter addresses interrelationships between the scorecards in the context of the two scenarios considered in the study.

Evaluating the Scenarios

Here, we shift our focus from the individual scorecards to how they would interact in the context of our two scenarios: a Taiwan invasion and a conflict over the Spratly Islands. For each scenario, we assess relative U.S. and Chinese capabilities in 1996, 2010, and 2017. The discussion draws selectively on modeling presented in the scorecard chapters. As in all the modeling undertaken for this project, these results should be taken as indicators of general performance and of change over time rather than as fully developed or precise predictions.

Taiwan Scenario

The United States and China have many good strategic reasons to avoid military conflict. In the case of a Taiwan scenario, both also have good operational reasons. From the U.S. perspective, the defense of Taiwan would place U.S. forces at an acute operational disadvantage, given the immediate proximity to the mainland and the optimization of Chinese forces for the tasks at hand. U.S. forward air bases would be in range of Chinese land-based ballistic missiles, and even defensive U.S. combat air patrols over or near Taiwan could, at times, be forced to operate within the engagement envelopes of SAM systems located on the mainland. U.S. naval forces, too, would face the dilemma of operating within relatively easy striking range of Chinese submarines or, alternatively, positioning themselves too far from the fight to contribute effectively to the campaign.

The summary coding of the Taiwan scenario analysis suggests that the United States moved from a situation in which it could dominate a Taiwan conflict in virtually all respects in 1996 to one in which it could be severely tested in a number of areas. Yet these results do not necessarily translate into great promise for a Chinese effort to take the island by force, even by 2017. While proximity makes the scenario challenging to U.S. forces, the task of occupying a large, heavily populated island in the face of opposition by some of the world's best air and naval forces (even if small in number at the outset) would be extraordinarily daunting. While the United States might experience losses and setbacks in a number of areas, Chinese leaders face a situation in which failure in even one area could spell catastrophe.

While we cannot predict a "winner," some things can be said with great confidence: The problem of defending Taiwan has become significantly more difficult for the United States since 1996. Relative capabilities are likely to continue shifting against the United States, at least as long as economic trends favor China. And a war for Taiwan would be a short, sharp, and probably desperate affair with significant losses on both sides.

Taiwan Scenario in 1996

The scorecard analysis suggests that, in 1996, U.S. military forces enjoyed overwhelming advantages in almost all operational areas. U.S. land-based airpower could operate freely from forward bases in Japan, Korea, and Guam. With the possible exception of bases in Korea, U.S. forward bases were effectively secure from attack.[11] In the air, all but a very small handful of Chinese fighters were second- and third-generation aircraft. Many of them were modernized versions of Soviet designs that first entered service in the 1950s.[12] Tactical and operational-level modeling of the air war suggests that, in the first seven days of a conflict, the United States could have gained air superiority using less than a single air wing's worth of contemporary fighters.[13]

The United States could exploit air superiority, penetrate Chinese airspace, and strike critical targets. In 1996, Chinese airspace was defended by antiquated SA-2 SAMs with a maximum range of roughly 32 km. Only 32 of more than 500 systems deployed by China were modern, so-called double-digit SAMs (SA-10s or better), with maximum ranges of 100 km or more. Our modeling suggests that, in a hypothetic conflict, nonstealthy U.S. aircraft flying at high altitude could have struck about 38 percent of a notional Chinese target set on the mainland opposite Taiwan (within 800 km of Taipei) with moderate risk to the attacking aircraft.[14] The use of stealth, SEAD, and low-altitude attack would have further expanded the accessible target set.[15]

The maritime prospects were equally propitious. The PLA had no dedicated space-based ISR or long-range ground-based radar, giving the PLA virtually no ability to find and target U.S. surface ships beyond visual range. Even if targets could be located, the Chinese military had few assets capable of actually attacking U.S. surface forces. PLA fighters and strike aircraft had very limited combat radius (as well as little ability to navigate over water). Its bomber fleet consisted of 15 H-6 (Tu-16) medium bombers and 130 H-5 light bombers. The most capable Chinese air-launched ASCMs were YJ-81s, which had a maximum range of 70 km—well within range of U.S. naval air defenses, not to mention carrier-based aircraft.

In the submarine realm, the PLAN had taken delivery of its first two *Kilo*-class boats from Russia the previous year. The PLAN's other 78 submarines were obsolete,

---

[11] Bases in Korea could have been hit by Chinese SRBMs in 1996, though these missiles were limited in number and too inaccurate to pose a serious threat even to those bases. Bases in Japan were beyond the range of Chinese missiles and were under threat only from Chinese bombers. Given the U.S. ability to achieve air superiority quickly, Chinese bombers had scant prospects for successful attack.

[12] The preponderance of Chinese fighters at that time were J-6s (a modernized Chinese variant of the Mig-19), J-7s (the modernized Chinese variant of the Mig-21), and J-8s (a Chinese design that combined elements of the Mig-21 and Su-15). In addition, IISS, *The Military Balance*, 1996, indicates that the PLAAF had taken delivery of its first four Su-27s from Russia.

[13] This wing could be provided by either the Air Force or Navy and is standardized to 72 fighters or two U.S. aircraft carrier wings (which each embark roughly 44 F-18s, some of which would be held back for the defense of the carrier).

[14] See Table 5.4 in Chapter Five. The set of 823 targets identified in that chapter included both military and civilian infrastructure locations, but it was not intended to represent an actual target list. Rather, the intention was to provide an analytical tool for examining the density and quality of air defenses over time.

[15] The United States had deployed stealth aircraft by 1996, including 40 F-117s.

noisy vessels with design features that were decades old. Our modeling of a Chinese submarine campaign against U.S. carriers suggests that these submarines would have only very small odds of engaging U.S. carriers.[16] At the same time, the Chinese surface fleet's lack of long-range air defenses and anti-submarine capabilities made it highly vulnerable to both air and submarine attack in 1996. Modeling of U.S. submarine attacks against the amphibious fleet suggests a 100-percent loss rate for the latter over a seven-day period.[17] Although less is known about the status of Chinese counterspace research efforts, there is no evidence in the public record that China had deployed systems that could significantly compromise the functioning of U.S. satellites by 1996.

Taiwan Scenario in 2010

Improvements in Chinese capabilities allowed the PLA to narrow the gap in almost all areas by 2010, despite substantial investments and a number of important technological developments on the U.S. side. The Chinese deployment of conventionally armed DF-21C ballistic missiles brought U.S. bases in Japan into range of Chinese missile attack, and Chinese ALCMs presented at least some threat to bases as far away as Guam. Modeling of attacks on Kadena AB using DF-21Cs suggests that China might have been able to close the base to fighter operations for somewhere between four and ten days, depending on assumptions about the size of China's available missile inventory and the percentage of missiles employed.

The U.S. Air Force deployed the world's first operational fifth-generation fighter squadron in 2005, when the 27th Fighter Squadron, equipped with F-22s, was activated at Langley AFB. Despite the very impressive advantages enjoyed by the F-22, China's much more extensive replacement of second- and third-generation aircraft with fourth-generation fighters more than offset the impact of U.S. advances between 1996 and 2010.[18] Although our tactical air assessment gives U.S. fifth-generation fighters significant effectiveness and survivability advantages over China's most advanced fighters (J-10s and J-11s), modeling of the larger campaign indicates that the challenges to the U.S. side grew substantially between 1996 and 2010. Our air campaign model suggests that the force structure required in theater to gain air superiority within seven days rose from less than one wing in 1996 to between roughly three and 4.5 air wings by

---

[16] Without cueing, they would achieve an expected 0.03 attacks against a single carrier over the course of a seven-day campaign. With cueing, Chinese submarines would do better, but given poor Chinese ISR capabilities at that time, cueing would have been unlikely. For the purposes of this analysis, we assumed that only one carrier would be operating within the operational area (within 1,000 nm of Taiwan).

[17] This does not mean that no forces are landed. Many ships are lost after making at least one trip to deliver PLA soldiers to Taiwan. Nevertheless, losses are severe.

[18] In 2010, 192 of the 542 Chinese fighter aircraft involved in air operations over Taiwan are fourth-generation aircraft (J-10s and J-11s).

2010, depending on U.S. basing assumptions. Fewer air wings (roughly two) would be required in theater to prevail in 21 days, up from about a third of an air wing in 1996.[19]

Not all aspects of the air balance have deteriorated from the U.S. perspective. Although the PLA dramatically improved its air defenses between 1996 and 2010, the United States improved its capability to strike and destroy certain target sets. By 2010, the PLA had replaced roughly 200 of its aging and outmoded SA-2 SAM launchers with modern long-range double-digit SAMs (or domestic equivalents), making penetration by legacy U.S. bombers and their escorting fighters extremely difficult and risky. On the other hand, the introduction of additional U.S. stealth aircraft and, especially, a larger number and variety of conventionally armed air- and sea-launched cruise missiles gave the United States new options and made attacks far more effective.

Illustrative modeling of a campaign to attack Chinese air bases suggests that although penetrating Chinese air defenses became more difficult by 2010, the damage that could be done increased substantially.[20] The improvements came between 1996 and 2003, however, not between 2003 and 2010, when the net balance remained largely static (even as both sides improved their absolute capabilities). And although U.S. net capabilities did improve after 1996, U.S. forces would have increasingly relied on a growing but nevertheless finite supply of standoff missiles to achieve those results. Hence, the same positive results might not have held in a war lasting more than a few weeks, though in a longer war, direct attacks might have become more feasible if Chinese air defenses had been sufficiently degraded.

A similar pattern held on the maritime front. Of most concern was the growing threat to the U.S. surface fleet, particularly aircraft carriers. By 2010, a small but growing constellation of surveillance satellites (including three imaging satellites and four SAR satellites), combined with OTH radar, provided the Chinese with the ability to find and, potentially, target the U.S. surface fleet. Although the prospects for neutralizing parts or all of this ISR system were good (given its limited scale), these Chinese systems nevertheless represent a problem for U.S. planners that did not exist in 1996.[21]

Chinese strike capabilities improved rapidly over this period. By 2010, the PLA had activated its first squadron of fourth-generation strike fighters in the form of 24 Su-

---

[19] In each case, the range of values reflects different basing possibilities. The smaller figure assumes that the preponderance of aircraft can base at Kadena AB (on Okinawa) or an equivalent distance from Taiwan. The larger figure would be required if aircraft were based primarily at Andersen AFB (Guam) or an equivalent distance.

[20] The magnitude of change depends on assumptions about specific target sets (e.g., whether the runways are struck in an effort to close the bases to operations or whether parking ramps are struck in an effort to destroy aircraft) and about the technical parameters of U.S. aircraft (especially the RCS levels of U.S. stealth aircraft, which were parameterized to reflect a range of different possibilities). In the case of runway attacks, our modeling suggests that, in 1996, U.S. attacks could close the 40 Chinese air bases within 1,000 km of Taipei for roughly 5 percent of the total "base days" over the first seven days of a conflict. By 2010, this percentage increased to between 29 and 44 percent, depending on assumptions about the RCS levels of U.S. stealth bombers.

[21] One of the most effective measures that the U.S. military could have taken would have been to neutralize Chinese ground-based OTH radar, either by kinetic attack or by sustained jamming.

30Mk2 Flankers and had increased its medium bomber force from 15 to 30 H-6s. To go along with these platforms, China developed improved ALCMs, including models with speeds of Mach 2.5 and others with ranges of 1,500 km. On the maritime front, the modernization of China's submarine fleet posed an increasingly serious challenge to the U.S. surface fleet. Our modeling suggests that, without cueing, the number of times one or more Chinese submarines might be expected to come within engagement range of a single U.S. carrier operating within 1,000 nm of Taiwan rose from 0.04 over a seven-day period in 1996 to 0.42 in 2010. Potential engagement numbers rose by almost another order of magnitude (to 3.25 engagements in 2010) with even modest cueing from external sources about the location of U.S. carriers.[22]

Despite relative Chinese gains in some areas of the maritime competition, U.S. submarine capabilities against Chinese amphibious forces and surface action groups remained robust. This can be explained, in part, by China's seeming inattention to the development of its anti-submarine capabilities. Although there was some increase in the number and quality of Chinese anti-submarine warfare assets, they remained relatively modest. For example, the PLAN had deployed no acoustic towed arrays and very few variable depth sonars on its ships.[23] Modeling of U.S. submarine attacks on Chinese amphibious forces shows a substantial drop in the percentage of the transport ships destroyed between 1996 and 2010. However, this is primarily driven by an increase in the total number of amphibious ships fielded rather than improvements to Chinese anti-submarine warfare capabilities. And despite a drop in the proportion of the amphibious fleet destroyed by 2010, the level of destruction from submarines alone (73 percent of the fleet in days, with a 38 percent reduction in the number of forces delivered across the Strait) would likely have had crippling effects on a Chinese landing. U.S. aircraft and surface ships could have inflicted additional damage.

In 2010, U.S. forces would almost certainly have prevailed in a Taiwan scenario, but the battle would have been hotly contested in a number of areas.

Changes to the Taiwan Scenario in 2017

Changes through 2017 are somewhat uncertain, given that not all development and procurement choices are clear. However, our analysis suggests that the U.S. ability to achieve its objectives in a Taiwan scenario have continued to erode over this period. The speed and magnitude of change is likely to be as great or greater between 2010 and 2017 as it was between 2003 and 2010. Although Chinese defense budget growth may be slowing somewhat, the PLA either has deployed or is preparing to deploy a number of new systems during this period. These include the world's first ASBM, new classes of fighter and strike aircraft, improved nuclear and conventional submarines, and larger

---

[22] The analysis *does* consider U.S. anti-submarine warfare capabilities and the sinking of Chinese attack submarines.

[23] Bussert and Elleman, *People's Liberation Army Navy*, 2011, p. 127.

and more capable surface warships. The United States has also deployed a number of new systems, including several extended-range cruise missiles, new classes of warships (such as the *Zumwalt*-class destroyer), and the F-35.

Despite the disproportionate media attention to new weapons or types of weapons, shifting relative capabilities is determined at least as much, if not more, by the rate at which existing modern designs are produced and incorporated into the force structure. Defense industrial reforms put into practice by the Chinese in the 1980s and 1990s had, by the 2000s, produced designs across a wide range of areas that the leadership deemed good enough for series production. By mid-2015, series production on an even broader range of systems (including, for example, large amphibious assault ships, *Luyang III* destroyers, new classes of cruise missiles, and several types of aircraft) had either begun or looked imminent. Although the United States significantly expanded its production of long-range cruise missiles, a variety of planned air and naval platforms encountered problems that resulted in delayed or curtailed production.

Based on our 2017 estimates of China's deployed ballistic and cruise missiles, the PLA's ability to disrupt the operation of U.S. air bases in the Western Pacific, especially at Kadena and other U.S. air bases in Japan, will likely increase substantially by 2017. If the Chinese deploy a conventionally armed IRBM, as recent reports suggest they will, the threat to Andersen AB on Guam will also increase dramatically.[24] U.S. countermeasures (including improved BMD, improved runway repair capabilities, the hardening of base facilities, and the dispersion of U.S. aircraft to a broader range of existing or new air bases) could ultimately mitigate the threat, but these measures are likely to be only selectively adopted—and to only marginal effect—by 2017. Our modeling of air base attacks indicates that, without these improvements, the Chinese ability to close Kadena AB to fighter operations could increase several-fold between 2010 and 2017.

Similarly, China will come close to replacing the last of its second- and third-generation fighters with fourth-generation fighters during this period, making U.S. efforts to secure air control more demanding. Assuming that U.S. fighters could continue to operate primarily from Okinawa, our modeling shows that the number of U.S. fighter wings necessary in theater to gain air superiority in a 21-day campaign would increase from roughly two in 2010 to between three and four in 2017 (depending on basing). Given the Chinese missile threat to forward bases and the large number of support aircraft (such as tankers) that would also be required, this would begin to strain existing U.S. basing capacity in Japan and Guam.

Modeling of the U.S. capability to strike key targets on the mainland suggests that the gains that accrued to the U.S. side between 1996 and 2010 with the addition of new standoff weapons will erode somewhat by 2017 but remain significantly greater than it was in 1996. The erosion of the U.S. ability in this dimension is driven primarily

[24] Office of the Secretary of Defense, *Military and Security Developments Involving the People's Republic of China 2013*, May 2013, p. 42.

by China's ability to cover additional ingress routes with newly deployed SAM systems, improvements to the quality of Chinese defensive combat air patrol aircraft, improvements to the range and accuracy of deployed SAM systems, and the increased range at which U.S. tanker aircraft must operate in the face of higher threat levels. Nevertheless, the deployment of additional U.S. fifth-generation aircraft and, especially, more standoff weapons will provide the U.S. military with robust strike capabilities.

In the maritime dimension, the risks to the U.S. Navy's surface fleet near Taiwan will continue to increase through 2017. Chinese satellite ISR capabilities have improved significantly since 2010, with both an acceleration in the rate of launches as well as new types of and more capable satellites. By 2017, the PLA will have further refined the DF-21D ASBM, and it may have conducted overwater tests against a maneuvering target. Our analysis suggests that technical uncertainties make the degree of threat posed by ASBMs difficult to predict.

Much clearer is that the threat posed by Chinese air and submarine attacks will increase. The PLA Air Force has assumed a larger maritime role since 2010, and both the PLA Navy aviation and PLAAF have added significant numbers of new strike aircraft. With the deployment of additional *Yuan* SSKs and improved *Shang* SSNs, more than two-thirds of the Chinese submarine fleet will be composed of modern boats by 2017, up from half in 2010. Even with expected improvements to U.S. anti-submarine warfare capabilities—including the deployment of P-8 maritime patrol aircraft—modeling shows significant gains for Chinese submarines operating against U.S. aircraft carriers.

By 2017, the Taiwan scenario will become extremely demanding, with a significant risk of major losses. The task could potentially be eased if the United States initially operates at greater distances from the mainland until critical Chinese capabilities are neutralized. However, stretching out the campaign in this way would come with political and military costs. If it became clear, for example, that the United States intended to cede ground close to the Chinese mainland during the initial stages of a conflict, the confidence and commitment of allies and partners might be shaken. Political risks might grow in a longer war as both countries face growing international (and possibly domestic) pressure to terminate hostilities.

Spratly Islands Scenario

The United States and China are arguably less likely to go to war willfully over differences in the South China Sea than they are to fight over Taiwan. Nevertheless, the two disagree on important maritime issues, most importantly on the interpretation of the UN Convention on the Law of the Sea.[25] China has become increasingly assertive in

---

[25] Specifically, the two differ on the permissibility of different types of military activities within exclusive economic zones, defined by the convention. From the U.S. perspective, these issues form part of the larger question of freedom of navigation.

asserting its territorial claims there, most recently through its land reclamation activities, and the United States has grown increasingly wary of Chinese intentions. U.S.-China maritime issues intersect with territorial ones involving China and its neighbors, including several U.S. allies and partners.[26]

Given this complex web of issues, it is not unthinkable that a local crisis could lead unintentionally to an armed conflict between Beijing and Washington. Indeed, some believe that because the two sides do not anticipate a war in the South China Sea, unexpected developments are more likely, making conflict in this area a more dangerous problem than Taiwan. Decisionmaking under normal and crisis conditions will be heavily influenced by perceptions of the military balance, making it important for policymakers to understand the potential dynamics of operations in the South China Sea. At the same time, parallel and comparative analysis of the relative capabilities in both the Taiwan and Spratly Islands scenarios facilitates the assessment of the impact of distance on U.S. and Chinese military capabilities more broadly.

The same Chinese military improvements that are making the outcome of a Taiwan scenario increasingly questionable over time also undermine U.S. advantage in the South China Sea. Nevertheless, the U.S. military remains in a much stronger position relative to the PLA in the Spratly Islands scenario. With all of its regional main operating bases located in Northeast Asia, U.S. forces are not well positioned for combat in Southeast Asia. But the difference between the two scenarios is even greater for China, particularly in terms of the position of Chinese bases relative to the locations of the two scenarios. Power projection assets, such as aerial tankers, satellite-based support, and long-range heavy bombers, could be important to the Chinese in a Taiwan conflict, but are far more so in the Spratly case. While the PLA is developing improved power projection capabilities, they are still relatively weak. The United States, on the other hand, has a mature set of power projection capabilities that it could use in any East Asian scenario.

Spratly Islands Scenario in 1996

In 1996, the PLA had virtually no capability to enter into a contest with U.S. military forces near the Spratly Islands. The United States held decisive advantages in virtually every area examined. PLA conventionally armed ballistic missiles could not reach any of the relevant U.S. bases, either in southwest Japan or in Southeast Asia. The PLAAF had a total of 24 fighters (plus 80 bombers) that could reach the Spratly Islands in

---

[26] Mark J. Valencia, *Foreign Military Activities in Asian EEZs: Conflict Ahead?* Seattle, Wash.: National Bureau of Asia Research, Special Report, No. 27, May 2011; Michael D. Swaine and M. Taylor Fravel, "China's Assertive Behavior, Part Two: The Maritime Periphery," *China Leadership Monitor,* No. 35, September 2011; Alexander Nicoll and Sarah Johnstone, eds., "Behind Recent Gunboat Diplomacy in the South China Sea," *IISS Strategic Comments,* Vol. 17, No. 28, August 2011; Murray Hiebert, Phuong Nguyen, and Gregory B. Poling, eds., *Perspectives on the South China Sea: Diplomatic, Legal, and Security Dimensions of the Dispute*, Washington, D.C.: Center for Strategic and International Studies, September 2014.

1996, compared with the more than 600 fighters (plus more than 400 bombers) that might have been available for a Taiwan scenario.[27]

Modeling suggests that roughly one-third of a U.S. air wing equivalent (supplied by either the Air Force or the Navy) would have been sufficient to gain air superiority in a seven-day campaign over the Spratly Islands, or roughly one-third of that required to prevail in the contemporary Taiwan scenario. U.S. maritime dominance was also near complete. Chinese surface ships, highly vulnerable in any scenario in the 1996 time frame, would have had to venture farther from ground-based air and SAM protection and would have become even more vulnerable than in the Taiwan case to air or submarine attack. U.S. aircraft carriers and other surface ships could be positioned farther from the Chinese coast in this scenario and would therefore be less vulnerable to Chinese submarine, air, and missile attack.

Spratly Islands Scenario in 2010

The U.S. military advantage diminished somewhat by 2010, but U.S. forces nevertheless retained important advantages. By 2010, Chinese DF-21C MRBMs could reach U.S. air bases in southwestern Japan. They could also target bases on the Philippine's Luzon island, should the United States have opted to place forces there. (The scenario assumes access to Philippine locations for the course of the conflict.) But bases on Guam and Mindanao remained beyond the range of attacks from all except air-launched DH-10 cruise missiles.[28]

Between 1996 and 2010, the number of bombers that could reach the Spratly Islands increased from 80 to 148 (including 100 H-6s and 48 JH-7s), while the number of fighters increased from 24 to 233 (including 97 Su-27s or J-11s and 136 Su-30MKs). In 1996, the Chinese had no operational tanker aircraft, but by 2010, it had converted roughly ten H-6 bomber aircraft to tankers. Modeling of the air superiority battle suggests that, even without attacks on Chinese bases, the United States would need roughly two air wing equivalents to gain air superiority within the first seven days of the conflict, even if the PLA committed most of its modern combat aircraft to the fight. While this number is up from roughly one-third of an air wing in 1996, it is still between a half and two-thirds of what would have been required in the 2010 Taiwan scenario.

If the U.S. military were permitted to strike the few air bases from which China's longest-range fighters could reach the Spratly Islands in unrefueled flight, U.S. air attacks aimed at cutting runways could shut all of those bases for the entire first week of a conflict. That 100-percent closure rate in this case compares with a 29- to

---

[27] The Taiwan numbers include roughly half of the total PLAAF fighter and bomber inventory, with the rest held for defensive missions or a reserve capability.

[28] In 2010, the DH-10 threat is potentially serious, but the Chinese H-6 bombers were (and are) themselves vulnerable in launching these missiles. In addition, U.S. forces could receive substantially more warning time than in the case of ballistic missile attacks.

44-percent rate against relevant bases in a 2010 Taiwan contest. A smaller number of Chinese bases within striking distance of the theater and weaker air defenses around those bases explain the greater U.S. effectiveness in the Spratly case. While U.S. political leaders would be extremely reluctant to strike mainland targets in a limited conflict, the option might become more palatable if the Chinese struck U.S. bases in Japan or elsewhere. In a South China Sea scenario, the capability to shut Chinese bases (or destroy aircraft on the ground) would provide the United States with escalation dominance in the air and missile realms.

Chinese naval capabilities improved between 1996 and 2010 but remained disadvantaged by distance in this scenario. Although Chinese destroyers and frigates enjoyed better air defenses and, to a much lesser extent, anti-submarine capability than they did in 1996, PLAN surface ships remained highly vulnerable beyond the effective cover of Chinese land-based aircraft and air defenses. Submarines, however, posed a real threat, increasing their effectiveness (measured in number of expected engagements) by an order of magnitude between 1996 and 2010.[29]

Nevertheless, although the Spratly scenario had become more difficult for U.S. forces by 2010, continuing U.S. advantages in most areas would have assured military success, probably at modest cost.

Changes to the Spratly Islands Scenario by 2017

Through 2017, the U.S. military will almost certainly continue to enjoy the upper hand in most areas, though the degree of advantage will continue to erode. The United States will probably still retain the ability to attack and close all of the Chinese air bases relevant to a Spratly Islands scenario in 2017. But assuming that the PLA deploys additional double-digit SAMs in southern China by 2017, strikes by legacy aircraft may become risky, forcing the United States to rely, at least initially, on its much smaller force of stealthy aircraft and its limited supply of cruise missiles. In the maritime realm, both sides may be able to target the other's surface warfare forces in the confined spaces of the South China Sea, creating substantial areas that are high-risk for both sides.

Although the PLA continues to place a high priority on Taiwan-related capabilities, it has, since 2004, also been preparing to execute "new historic missions."[30] This formulation, which calls on the PLA to protect China's national interests and play a role in supporting world peace and development, is less narrowly focused than previous guidelines (which addressed the defense of territorial sovereignty and domestic tasks). As such, it supports the acquisition of additional power projection capability.[31] Greater

---

[29] Our modeling of Chinese submarine attacks against U.S. surface elements yielded an expected 0.3 Chinese engagement opportunities (or 2.5 with daily cueing) against a single U.S. carrier during a seven-day campaign, up substantially from 0.03 in 1996.

[30] Mulvenon, "Chairman Hu and the PLA's 'New Historic Missions,'" 2009.

[31] Previous mission guidance to the PLA varied over time but was more narrowly focused. For a discussion of the possible implications of China's "new historic missions" for the PLAN, see Cortez A. Cooper, *The PLA*

emphasis on the acquisition of support capabilities, such as tankers and airlift, could significantly improve the Chinese capability to conduct operations in the South China Sea in the years beyond 2017. Nevertheless, it will always be more difficult for China to fight farther from its shores than closer to home.

Conclusions

The analysis presented in this chapter indicates that while the United States maintains unparalleled military forces overall, it faces a progressively receding frontier of military dominance in Asia. Chinese military modernization, combined with the advantages conferred by geography, have endowed China with a strong military position vis-à-vis the United States in areas close to its own territory. As a result, the balance of power between the United States and China may be approaching a series of tipping points, first in contingencies close to the Chinese coast (e.g., Taiwan) and possibly later in more distant locations (e.g., the Spratly Islands).

These tipping points may not give China ultimate victory in a war with the United States. Indeed, the United States is likely to maintain important advantages in a longer conflict. They do, however, represent points at which PLA forces could gain local or temporary air and naval superiority during the initial battles, and at which ultimate U.S. success might entail sustained combat and significant losses. It is difficult to state with precision when these points might be reached, but a tipping point in a Taiwan conflict might come as early as 2020, while tipping points in more distant scenarios might lie a decade or more beyond that.

Factors beyond the control of U.S. military strategists, such as the future health of the Chinese and U.S. economies and the development of dual-use technologies, will influence when such points are reached (and, in the case of more distant scenarios, whether they are reached). Other factors, including U.S. national military strategy, procurement, force posture, and operational practices, are more amenable to manipulation and could also affect outcomes by limiting losses in early hostilities and better positioning U.S. forces to regain the initiative more quickly than might otherwise be the case. Proactive adjustments in these areas, to which we turn in the next chapter, will buttress deterrence and improve U.S. prospects in the event of a conflict.

Navy's "New Historic Missions": Expanding Capabilities for a Re-Emergent Maritime Power, testimony before the U.S.-China Economic and Security Review Commission, Santa Monica, Calif.: RAND Corporation, CT-332, June 11, 2009.

CHAPTER FOURTEEN

Implications and Recommendations

As long as the Chinese economy continues to grow faster than that of the United States and Beijing continues to make military modernization a priority, the challenges facing U.S. military planners in Asia will grow more severe over time. Given the size and technical sophistication of the U.S. arsenal, together with the accumulated experience and resiliency of its military personnel and commanders, the United States will remain capable of fighting and winning a protracted air and naval battle against China—assuming that U.S. political positions remain firm. However, our scorecard analyses suggest that the two sides might be reaching a series of tipping points, whereby PLA forces could severely challenge the United States in the initial battles of a war and impose heavy costs on the U.S. military, first in scenarios near the Chinese coast and, later, in more distant locations.

In the long run, economic, political, and technological events could change the equation and provide one side or the other with a more decisive advantage. In the meantime, the United States should strengthen its relevant military capabilities and enhance deterrence without increasing (and hopefully while decreasing) the probability of unintended military incidents or, in the event of war, vertical escalation. Moreover, this tall order must be filled in a fiscally constrained environment that will likely mean less money, and certainly no more, for defense. By shifting resources and adjusting strategy, however, it should be possible to increase the uncertainties facing any Chinese leader who might contemplate adventure, reduce the vulnerability of U.S. forces in the opening stages of a conflict, and impose costs on China should it elect to fight.

In this chapter, we offer several suggestions, including measures to (1) increase efforts to shape perceptions and strengthen deterrence, (2) refocus procurement to better meet the challenge, (3) adjust concepts of operations, (4) develop regional military relationships that increase strategic depth, and (5) establish conditions that minimize the risks of vertical (especially nuclear) escalation in the event of a crisis or war.

Shaping Perceptions

We know very little about how the Chinese government or the PLA evaluates military capabilities, much less how Chinese leaders view their country's current capability to undertake particular missions in the face of U.S. opposition. Nevertheless, in both their public and private comments, senior PLA officers have expressed far greater confidence in Chinese power than they once did.[1] Indeed, there is some danger that Chinese officers may overestimate their relative capabilities. Chinese civilian leaders take a broad view of national interests and are extremely cautious in the use of force. Nevertheless, crises may place them at the center of cross-cutting pressures, and their perceptions of military realities may, under those circumstances, become a critical variable in determining war or peace.

The most important of these realities involves measures to strengthen U.S. military and strategic capabilities, topics to which we turn later in this chapter. Here, we focus on strategic signaling and communication. Deterrence requires not only capabilities but also the communication of those capabilities, as well as the resolve to use them. Public discussions of relative military capabilities, whether detailed analyses (such as this present report) or more limited commentaries, should not only present challenges facing the United States but also, where appropriate, highlight the uncertainties and dangers that would confront Chinese planners. While a degree of conservatism may be in order in some types of analyses, care must be taken not to reinforce false or exaggerated impressions of Chinese capability.

An effective strategy would highlight the uncertainties facing China. In the event of a conflict, the United States has the capability to expand the scope of war laterally to other locations or to extend the duration of a conflict beyond what China could sustain without heavy damage to its economy. For Chinese military leaders, achieving their objectives and concluding hostilities quickly would be essential in a maritime or air conflict with the United States. Any doubt that success in an initial plan of operation would bring an end to hostilities could have a powerful impact on Chinese thinking and crisis behavior. To be sure, a protracted conflict would not be good for either side. With production chains passing through China, the global economy would be badly disrupted. And as noted in several places in this report, U.S. stocks of key weapons, especially standoff weapons, could be exhausted in a protracted fight.

Nevertheless, China has far more to fear from a long conflict than does the United States. From an operational perspective, Chinese conventional ballistic missiles, though numerous, are not in infinite supply, and China lacks the number of systems maintained by the United States in most other categories. From an economic

---

[1] Although there are a variety of views within the PLA about the precise nature of the current military balance, we agree with U.S. observers who note that PLA officials have been expressing relatively greater confidence than they once did. There is also anecdotal evidence of this shift in the printed record. For more on this topic, see Andrew Scobell, "Is There a Civil-Military Gap in China's Peaceful Rise?" *Parameters*, Summer 2009.

perspective, most plausible scenarios would be centered somewhere in its backyard, bringing significantly more disruption to its own commercial activity and economy than to the world as a whole (or to the United States). As a percentage of GDP, China's total trade is almost twice that of the United States, which generates relatively more from domestic activity. China is more dependent on imported oil, and it is far more dependent on oil imports through the waterways most likely to be affected. Perhaps most importantly, protracted military conflict and the economic disruption that would result would not fundamentally challenge the U.S. social order or political system, while the Communist Party of China would have no such assurance. The point is not that the United States should or would welcome a protracted conflict but that it should design strategies to buttress deterrence by plausibly extending a conflict in space or time.

At the same time, U.S. analysts should also emphasize the operational options open to the United States. To an extent, it already does this. U.S. Pacific Command periodically surges forces, in part to demonstrate U.S. power. In 2006, Valiant Shield brought three CSGs, 30 warships, and 280 aircraft together for the largest naval exercise in the Pacific since the Vietnam War. Not incidentally, Valiant Shield was also the first time that PLA officers were invited to observe U.S. exercises. In addition to such displays, the U.S. military could also highlight technologies (including the use of air- or submarine-launched missiles) that might wreak havoc on a landing fleet or beachhead even in the absence of the commitment of a massive fleet.

Analysts and policymakers must weigh the objectives of demonstrating U.S. capabilities and options against other strategic and diplomatic goals. China is not an enemy, and demonstrations of capabilities or resolve must walk a fine line lest they be taken as offensive in nature. Nevertheless, "teachable moments" should not be neglected, especially when they can be harmonized with other goals. At the end of the day, messaging may be relatively less important than capabilities, but it is also relatively inexpensive, and it complements other aspects of the larger deterrence effort.

Refocusing Procurement and Force Structure

Barring major economic or political turmoil in China, Chinese defense budgets are likely to increase much more quickly than those of the United States for the foreseeable future. And although the U.S. defense budget will remain higher than China's for many years, the changing balance of material resources, combined with the tyranny of distance, poses increasingly significant challenges to the United States in its planning for Asian contingencies. How might the United States refocus its military efforts to compensate? The scorecard analysis points to a number of adjustments to procurement priorities.[2]

Given both the high costs and the dramatic capabilities associated with new aircraft (e.g., the F-22 and F-35), these systems tend to garner enormous and disproportionate attention in the media and in public debate. While aircraft performance and capabilities are undeniably important in modern warfare, their ultimate impact is largely shaped or affected by a much larger and often less expensive but more mundane set of capabilities. As the air and missile scorecards highlight, the number, availability, and hardening of U.S. Air Force bases, as well as repair capabilities associated with them, are critical in determining outcomes. Some of the measures that might be taken to strengthen bases are relatively inexpensive, at least compared with the flyaway costs of combat aircraft.[3]

The U.S. Navy confronts parallel challenges. Ships designed for the less threatening post–Cold War environment, such as the LCS, may not be defensible in a hypothetical China scenario. At the other end of the spectrum, large carriers, which are primary targets for China's growing array of anti-access capabilities, may also be suboptimal. The U.S. Navy might productively explore smaller fast carriers, which could be escorted into harm's way by more capable (and stealthier) destroyers, which might borrow promising technologies imbedded in the *Zumwalt* class. The scorecards demonstrate that U.S. submarines would be at a premium in a war with China. In a largely denied environment, especially at the outset of a conflict, submarines provide the ability to approach China's coast to interdict Chinese warships and submarines or attack ground targets with cruise missiles.

Munitions tend to receive less attention than platforms but also have a critical impact on outcomes. Stealth aircraft are an important addition to the force mix but only partially mitigate the threat posed by increasingly dense arrays of modern (double-digit) SAM systems. As noted in Chapter Six (scorecard 4) with regard to U.S. air attacks on ground targets, increasing the number, range, speed, and capability of cruise missiles, such as the JASSM-ER, is at least equally important in maintaining the U.S. ability to strike surface targets. In practice, standoff weapons and stealth platforms provide complementary capabilities, with standoff weapons able to strike targets during the first days of a conflict, making it easier for stealth (and in some cases legacy) aircraft to penetrate opposing air defenses.

To maximize its capabilities in a war with China, the U.S. military would want to continue adjusting its current mix of standoff weapons so that, among other things,

---

[2] In practice, scientific rigor is often missing from force structure decisions, even where scientific language is employed to justify decisions. Politics often intervene, states and localities fight to garner procurement dollars, and costs are often dramatically higher than bids might suggest, all of which compromises the ability to assess cost-effectiveness. Moreover, assessing different military scenarios will produce different outcomes, making scenario selection subject to intensive debate and interested lobbying.

[3] "F-22 Raptor: Program and Events," *Defense Industry Daily*, April 9, 2013.

they are "network enabled, but not network dependent."[4] Faced with more capable Chinese combat aircraft, warships, and SAMs, the United States will want to redouble its efforts to develop and deploy new generations of ASCMs, such as the Long Range Anti-Ship Missile (LRASM), that can match those fielded by Russia and China. It will need to deploy HARM missiles that can be carried inside the F-22 and F-35 and, perhaps, a larger variety of HARM options, including some with longer ranges or faster speeds than the current generation.

Budgetary pressures often produce sharper interservice debates, and in the coming years, there are likely to be arguments between proponents of land-based airpower and carrier-based airpower. The scorecard analysis suggests that such debates miss the mark. Land-based and carrier-based airpower provide different advantages, with U.S. Navy carriers frequently able to arrive on station and begin operations relatively early, while land-based airpower can bring scale and a fuller range of support functions. More importantly, in light of Chinese anti-access capabilities, maintaining both land-based and carrier-based air will spread (and decrease) risk in the face of uncertainty. Such a move would represent a hedge against unforeseen failure or vulnerability in one area.

Although the scorecards do not address "red" versus "green" combat (i.e., combat between Chinese and U.S. allies or partners), the results of our U.S.-China modeling nevertheless speak to the importance of partner and allied capabilities. For example, the Taiwanese ability to extend the duration of a contest has a substantial impact on some scorecard results. To the extent that Taiwan can prolong the duration of the conflict, U.S. force requirements could be eased to a large degree.[5] It is easier for the United States to employ its air and naval power to influence events on the ground in a longer war than in a shorter one. The United States should strongly encourage Taiwan to undertake (or complete) defense reforms that will maximize its odds of avoiding quick defeat.[6]

In an era of budgetary austerity, any discussion of procurement needs would be remiss in not addressing where savings might be achieved. The optimal capabilities for Asian scenarios will have to be weighed against other global demands and contingencies. However, given the extent to which Asia is becoming more important in U.S. defense thinking, further cuts to ground forces (the U.S. Army and, to a lesser extent,

---

[4] U.S. Naval Air Systems Command spokesman, quoted in *Jane's Defence Weekly*, "Striking Out at Sea," October 10, 2012.

[5] Modeling for the air-to-air scorecard shows that the U.S. force structure required to gain air superiority in 21 days is roughly half that required to gain superiority within seven days (using the 2017 case).

[6] Prominent reform measures include the transition to volunteer military services, a procurement and organizational structure that emphasizes highly mobile systems, and the deployment of assets in ways that exploit Taiwan's naturally defense-dominant terrain. Unfortunately, changes to date have increased costs and resulted in a sharply reduced force structure, which may more than offset the gains in effectiveness. Completing and consolidating reforms, further shifts in procurement priorities, and, to the extent economic conditions allow, increasing defense budgets will be necessary to improve capabilities.

the U.S. Marine Corps) offer the greatest potential for savings. The U.S. Air Force might draw down its legacy fighters faster or purchase fewer F-35s than planned. The U.S. Navy, for its part, could spread risk and increase flexibility by moving to smaller carriers, as well as save resources. While all of these reductions would weaken the capability to conduct certain types of missions, the reallocation of resources to more urgent priorities would increase overall capability.

Concepts of Operations

Adjusting procurement priorities to optimize for East Asia contingencies may help mitigate current trends or even reverse them temporarily in some areas, but so long as the Chinese economy grows faster than that of the United States, the U.S. military will be unable to address the entire range of challenges through these means alone. The rate of growth in Chinese military budgets may decrease over the next decade, but U.S. military spending will likely decline relative to China's—and quite possibly in absolute terms. Under these circumstances, examining and adjusting operational concepts on a sustained and ongoing basis will be critical in maximizing the synergies and, possibly, shifting the terms of the contest.

The introduction of the AirSea Battle concept is one effort to achieve better results from the same level of resource allocation. (In January 2015, the concept name was officially changed to Joint Concept for Access and Maneuver in the Global Commons, or JAM-GC, but here we use the term *AirSea Battle concept*, by which it is better known.[7]) Inspired partly by the AirLand Battle concept, which was adopted in 1982 to stop Soviet tank armies in Central Europe, AirSea Battle is intended to gain synergies between the U.S. Navy and U.S. Air Force to defeat the anti-access strategies of regional powers. Few authoritative documents outline the specific components of the AirSea Battle concept, but a number of possibilities have been raised: Air Force counterspace operations could blind PLA space-based ocean surveillance systems to prevent the PLA from targeting U.S. surface craft. U.S. Aegis-equipped warships might, in turn, provide additional missile defense capabilities to protect U.S. forward air bases.[8] As the concept is developed and refined, it may feed into future procurement priorities and decisions.

U.S. strategic planners regularly review Asia-specific plans and concepts. In the coming years, as the challenges in Asia grow, this process will become even more

[7] The name change is, in part, intended to make the concept more inclusive and enable the U.S. Army and U.S. Marine Corps to play more significant roles in the wider effort.

[8] For more detail on this and other examples, see Jan Van Tol, Mark Gunzinger, Andrew F. Krepinevich, and Jim Thomas, *AirSea Battle: A Point-of-Departure Operational Concept*, Washington, D.C.: Center for Strategic and Budgetary Assessment, 2010, and Jose Carreno, Thomas Culora, George Galdorisi, and Thomas Hone, "What's New About the AirSea Battle Concept," *Proceedings*, Vol. 136, No. 8, August 2010.

important. Every effort should be made to promote an active dialogue between different types of strategic reviews (e.g., those done at military educational institutions and at the unified combat command level) and to free the process, wherever possible, from bureaucratic constraints. The Navy's continuous development and review of War Plan Orange (the plan for war with Imperial Japan) prior to World War II provides a worthy historical example of innovative and effective planning. Not only did the architects of War Plan Orange operate in a relatively open intellectual environment, but they also faced many of the same problems that confront today's planners, including the question of how to support friendly states located much closer to the potential threat than to the United States.[9]

In the current context, an active denial strategy should be explored.[10] The scorecard results indicate that Chinese military capabilities atrophy across even relatively modest distances. An active denial strategy, designed to capitalize on that reality, would have three primary features. The first is a resilient force posture, better able to withstand initial attacks and continue operating. This would include dispersion and the initial deployment of forces in depth, such as positions Asian locations well away from China, as well as the maintenance of forward positions with reduced force levels and improved missile defenses. The second feature is a robust capability to counter Chinese power projection, with relatively greater emphasis on defeating PLA assets beyond China's shores than on striking mainland targets. And the third is leveraging allies' strengths, including geographic position and niche military capabilities. Mobility and the preservation of combat power would enable U.S. forces to absorb initial blows and fight their way back toward areas requiring protection. Depending on the specific design, this strategy could prove more affordable than those associated with today's forward-leaning posture. And it would have the added benefit of strengthening crisis stability by reducing both sides' incentives for preemptive attack.

More radical departures from current concepts could include lateral escalation to other areas where China's ability to defend its interests might be less developed, or a distant blockade strategy implemented independently or in conjunction with forward defense strategies.[11] However, these operational concepts could have important strategic and diplomatic consequences for relations with China, as well as for U.S. relations with allies and partners. U.S. strategic planners should weigh the potential political

---

[9] On War Plan Orange, see Edward S. Miller, *War Plan Orange: The U.S. Strategy to Defeat Japan, 1897–1945*, Annapolis, Md.: Naval Institute Press, 1991.

[10] For a discussion of active denial possibilities, see Eric Heginbotham and Jacob L. Heim, "Deterring Without Dominance: Discouraging Chinese Adventurism Under Austerity," *Washington Quarterly*, Vol. 38, No. 1, Spring 2015.

[11] For arguments in favor of a distant blockade strategy, see T. X. Hammes, "Offshore Control: A Proposed Strategy for an Unlikely Conflict," International Institute for Strategic Studies, Strategic Forum, June 2012; and Douglas C. Peifer, "China, the German Analogy, and the New Air-Sea Operational Concept," *Orbis*, Vol. 55, No. 1, Winter 2011.

impact carefully before introducing deterrence options that rely on implicit escalatory threats.

U.S. Diplomacy and the Search for Strategic Depth

U.S. diplomacy and foreign influence may contribute much to mitigating emergent challenges. Whether or not the U.S. military adopts new concepts of operation, such as the active denial approach described above, it will look to expand basing access at greater distances from China. In November 2011, Washington and Canberra announced that U.S. Marines would rotate through a base in Darwin, Australia. By April 2015, a contingent of 1,150 Marines was conducting the fourth six-month rotation. In April 2014, the United States and the Philippines signed the Enhanced Defense Cooperation Agreement that allows the United States to station military forces on Philippine territory on a rotational basis.[12] In April 2015, Washington asked Manila for access to eight bases in the Philippines, including the former American Clark Air Base and Subic Bay Naval Base.[13] The U.S. military is also considering additional basing in the Mariana Islands.

Despite these recent headlines, however, developing robust basing access will require intensified diplomatic efforts with regional states—particularly those in Southeast Asia—that have heretofore taken second place in America's Asian strategy. During the Cold War, U.S. military forces in East Asia were concentrated in a handful of discrete "main operating bases" in Japan, South Korea, the Philippines, and, for 15 years, Thailand. After most U.S. forces departed Thailand in 1975 and the Philippines in 1991, remaining U.S. forces in Asia were overwhelmingly located in Northeast Asia, mostly in Japan and South Korea, close to Chinese territory. The United States also maintained bases on Guam, an unincorporated U.S. territory in the Mariana Island chain. The scorecard analysis suggests that Chinese efforts to build anti-access capabilities have increasingly put bases and forces close to the mainland at risk, making it critical for the United States to seek greater "strategic depth" in its efforts to hedge against Chinese power.

This does not imply that bases in the Republic of Korea and Japan should be abandoned but, rather, that the U.S. military will want to operate from multiple locations at varying distances from China. Strategic depth will provide a number of strategic and operational advantages, especially from the perspective of air warfare. Strategically, dispersing and operating from bases in a variety of countries will present Chinese leaders with difficult choices. On the one hand, allowing U.S. aircraft to

---

[12] Juliet Eilperin, "US, Philippines to Sign 10-Year Defense Agreement Amid Rising Tensions," *Washington Post*, April 28, 2014.

[13] "US Seeks Access to Philippine Bases as Part of Asian Pivot," Reuters, April 25, 2015.

operate freely from various nearby countries during a conflict would cede a degree of military advantage to the United States. On the other hand, striking these bases could bring additional states into a war against China, increasing the diplomatic, economic, and (possibly) military costs.

From an operational perspective, expanding U.S. strategic depth in Asia will allow U.S. planners to concentrate their basing in different locations in different phases of a campaign. As noted earlier, this would allow them to better weather the first blows in a conflict while mustering U.S. forces for a counterstrike. Greater depth would allow the Air Force (and possibly the Marine Corps and Navy) to fly different types of aircraft from different locations. Fighter aircraft might be dispersed to a variety of forward bases, while larger and more vulnerable support aircraft might be held at rear locations until the missile and air threat is contained. Finally, greater strategic depth might enable planners to play a military version of the shell game, forcing Chinese planners to guess where U.S. aircraft are located. To be sure, such a strategy would require greater investment in mobility and logistical capability and diminished economies of scale, though such economies may prove to be false anyway in the context of Asian military conflict.

The U.S. military should also continue efforts to expand its basing infrastructure on islands in the West and Southwest Pacific, Australia, and Southeast Asia. China's economic presence in East Asia is growing even faster than its military presence, and some might suggest that economic power will translate into diplomatic influence. But popular and elite opinion in East Asia continues to look first and foremost to the United States as the ultimate guarantor of regional stability and security.[14] To be sure, in the eyes of many Asian leaders, prudence demands cautious policy and avoiding statements or activities that look too much like "taking sides" against China. Nevertheless, China's rise, and its increasingly assertive behavior in the South China Sea and East China Sea, has heightened insecurity in much of Asia, and that has markedly increased the willingness of most regional states to engage the United States on security issues.

Within Southeast Asia, the United States may want to shift the balance of its engagement activities. It will be important for Washington to continue strengthening the diplomatic, economic, and security relationships it has built with states in the northern part of Southeast Asia (especially the Philippines and Vietnam), but it should also intensify engagement with the states of Southeast Asia's southern tier (such as Indonesia and Malaysia). These may provide some of the most secure initial operating locations in the future, especially as China develops conventionally armed ballistic

[14] Even prior to the latest demonstrations of Chinese assertiveness, U.S. regional allies and partners were wary of Chinese power and strongly inclined to support a strong U.S. role in regional security. See Evan S. Medeiros, Keith Crane, Eric Heginbotham, Norman D. Levin, Julia F. Lowell, Angel Rabasa, and Somi Seong, *Pacific Currents: The Responses of U.S. Allies and Security Partners in East Asia to China's Rise*, Santa Monica, Calif.: RAND Corporation, MG-736-AF, 2008.

missiles and attack aircraft with longer ranges. In most cases, and especially in Southeast Asia's southern tier, expanding security cooperation will necessarily be incremental and justified through a variety of stability and security missions. These states remain, to differing degrees, reluctant to take sides between China and the United States. Other, non–China-related rationales for cooperation should not and need not be purely rhetorical. Washington has a wide range of political and security interests, many of which would be well served by a more robust network of basing options throughout Asia.

Although the greatest political-military challenge may be gaining access in new locations from which to conduct a defense in depth, a receding frontier of U.S. dominance in Asia will also challenge U.S. diplomatic relations with existing allies and partners closer to China. As the U.S. military signals a relative shift in emphasis within Asia, reassuring long-standing allies, such as Japan, South Korea, and the Philippines, may become more difficult. Those states will likely ask for increasingly specific statements of U.S. commitment to particular areas or scenarios, raising the possibility that the United States will be drawn into a conflict not of its choosing. Indeed, as maritime tensions around China's periphery become increasingly heated, U.S. reassurance could embolden these allies and partners to take provocative measures to strengthen their own positions vis-à-vis China.

Even without such provocation, statements by U.S. officials intended to reassure Asian partners could stoke Chinese fears of U.S. interference or containment. Hence, U.S. reassurance will require a delicate balancing act that weighs multiple competing priorities in U.S. foreign policy.

Minimizing the Risks of Vertical Escalation

Finally, some mention should be made of measures to reduce the possibility of nuclear escalation. The scorecard results suggest that although the chances of crossing the nuclear threshold would remain low, even in the event of war, those risks may be growing. China maintains a nuclear no-first-use policy, and the United States has also pledged to reduce the role of nuclear weapons in its national security planning.[15] Nevertheless, history suggests that certain structural conditions can increase the probability of states taking military actions during crises that they did not originally anticipate. In the contemporary East Asian context, the threat of vertical escalation has been increased by two developments: first and foremost, the blurring of conventional and

[15] The 2010 *U.S. Nuclear Posture Review Report* states, "The United States will continue to reduce the role of nuclear weapons in deterring non-nuclear attack." It pledges that the United States will not use nuclear weapons against non-nuclear states that are in compliance with Nuclear Nonproliferation Treaty obligations. With regard to other states, the report explains, "[T]he United States wishes to stress that it would only consider the use of nuclear weapons in extreme circumstances." U.S. Department of Defense, *Nuclear Posture Review Report*, 2010, pp. 15–17.

nuclear boundaries and, second, growing incentives for U.S. forces to strike targets on the mainland in the event of war. Although there are no quick or easy solutions, it is in the interests of both parties to address and minimize those risks.

The blurring of the conventional and nuclear realms has been driven in part by the development of two types of "crossover" capabilities on both sides: variants of systems originally designed for nuclear missions but also suited to conventional attack roles and conventional systems designed to attack nuclear weapons. In the U.S. case, the "conventional prompt global strike" (CPGS) capabilities are being pursued to buttress U.S. efforts to deter or defeat adversaries by enabling the United States to attack high-value or fleeting targets at the beginning of a conflict. Congress blocked funding for deploying conventional warheads on SLBMs, but the United States is pursuing development of a hypersonic glide delivery vehicle that would be deployed on a modified Peacekeeper ICBM (a system called the conventional strike missile, or CSM).[16] In the Chinese case, the PLA is already heavily invested in conventionally armed ballistic missiles. It is currently developing a new generation of ballistic missiles capable of attacking ships at sea.

These systems pose two potential problems. The first and admittedly lesser possibility is that the launch of one or more of these missiles could be taken as a possible nuclear attack, compelling the receiving side to launch a nuclear counterattack. This danger is, in the current context, probably more theoretical than real, since a small salvo of ballistic missiles would be unlikely to prompt an adversary to launch a warning. (Also, China has only limited early-warning capabilities and would be unlikely to know it was under attack until ballistic missiles landed.)

A more significant danger is that the use (or possibility of use) of conventionally armed ballistic missiles would make those systems high-value targets in the event of war. Attacks on such systems could inadvertently jeopardize the survival of the targeted side's nuclear forces. Given the very substantial threat posed by Chinese ballistic missiles, particularly DF-21Cs and DF-21Ds, U.S. military planners would have very high incentives to find and destroy these missiles. However, conventionally armed DF-21s may be difficult or impossible to distinguish from nuclear-armed DF-21s, and the hunt for conventionally armed missiles could result in the attrition of China's nuclear-capable missile force. This could ultimately create a "use-them-or-lose-them" dilemma for Chinese strategic planners, particularly if other parts of China's strategic system (such as SSBNs) were under attack.

The scorecard results suggest that the incentives for U.S. forces to strike a range of targets on the mainland in the event of a conflict, particularly a Taiwan scenario, are growing. In a difficult fight with potentially high losses to the U.S. side, military leaders will look to strike high-value targets. They might, for example, look to strike

[16] Amy F. Woolf, *Conventional Prompt Global Strike and Long-Range Ballistic Missiles: Background and Issues*, Washington, D.C.: Congressional Research Service, February 6, 2015.

command-and-control facilities, satellite control or downlink facilities, and OTH radar arrays. Many of these targets did not exist in 1996, and the United States might have prevailed in some scenarios even without striking those that did exist. Today, the incentives are much higher. Modeling of an air superiority campaign over Taiwan illustrates the change. Our model showed that striking air bases on the mainland would not have lowered the overall force requirements needed to prevail in 1996, but by 2010, such attacks would have contributed significantly.

Striking air bases or other targets on the mainland might require first neutralizing Chinese air defense early-warning radars and long-range SAMs, another substantial set of targets. Ostensibly, China's nuclear no-first-use policy foreswears the use of nuclear weapons except in cases in which the other side has used nuclear weapons against China first. However, Chinese strategic planners acknowledge internal discussions about whether conventional attacks on Chinese strategic systems (including nuclear weapons, command and control, and early-warning systems) might force China to modify its policy and include such activities as potential triggers.[17] If China were engaged in a military conflict in which its strategic capabilities were being rapidly attrited or degraded, PLA leaders could feel intense pressure to escalate. Given the massive U.S. advantage in the nuclear realm, any nuclear use by China would almost certainly be calibrated to signal Chinese resolve while minimizing the risk of further escalation by, for example, preparing nuclear weapons for launch or detonating a weapon in an unoccupied area or against an isolated military target.

Although the chances of escalation to the use of nuclear weapons would be small even in the event of a conflict, the stakes are such that it is in the interests of both parties to minimize such possibilities. The United States should make every effort to understand the key elements of the Chinese strategic system. It should engage the Chinese in an active dialogue on the dangers of deploying and using systems that are difficult to distinguish from nuclear weapons in conventional operations. Even if such discussions are unlikely to affect Chinese procurement decisions (which will be decided much more by considerations of military efficacy), they may spark conversations within the Chinese government about the danger and unpredictability inherent in war.

Conclusions and Suggestions for Future Research

Chinese military capabilities have improved relative to those of the United States over the last two decades. Nevertheless, the PLA entered the 1990s so far behind the U.S. military in so many areas that Chinese modernization had few immediate consequences. It was not until the mid-2000s that the cumulative weight of PLA modern-

---

[17] These perspectives are derived from our discussions with Chinese government and think-tank representatives, Beijing, November 9, 2010.

ization began to coalesce into a significant challenge to U.S. regional primacy. With Chinese capabilities continuing to mature, meeting this challenge will require new focus by the United States. Unless and until the United States adjusts its regional commitments, maintaining the capability to meet those commitments will require the ongoing review of procurement priorities, concepts of operations, basing posture, and political-military relationships with regional partners—as well as engagement with Beijing to reduce the likelihood that China and the United States will ever turn weapons on one another.

We hope this report stimulates further research on the relative U.S.-China military balance and its implications. Other operational areas that were either not treated in this study or treated as subordinate elements of one of the ten scorecards could and should be explored in greater detail. Areas of further research would include (but are not limited to) ground combat, information warfare, anti-submarine warfare, landing operations, and ship-to-ship combat. The scorecards analyzed for this study could be explored from other angles or in greater detail. For example, future research assessing the impact of more complex types of SAM or SEAD operations, as well as a wider range of air superiority operations, would help fill out the picture provided here.

The contribution or impact of "green" (i.e., third-country) forces should also be systematically evaluated. In some operational areas, the difference in outcomes between our seven- and 21-day campaign modeling highlights the importance of green forces. For example, in a Taiwan invasion, if local forces could resist effectively and buy time for the United States to marshal its forces, U.S. intervention would be more effective and less costly than if U.S. forces had to act more hastily to stave off disaster. Military elements from other third countries, such as Japan, might be able to assist with important tasks (e.g., submarine or, especially, anti-submarine warfare), even if a conflict did not initially directly threaten its territory.

Future research could also model warfare at other levels. Perhaps the most direct follow-on to this study would be the creation of a unified model to assess the interrelationships between the different scorecards. Such an effort could explore alternative operational concepts, such as blockade operations by one side or the other. And the impact of different force structures could also be productively examined. In short, we hope the current study makes a significant contribution to the examination of U.S.-China military issues, but we do not present it as the final word. On the contrary, we expect that it will prompt others to delve deeper into the subject.

Bibliography

Addabbo, Rep. Joseph P., "B-210596 L/M," legal decision, Washington, D.C.: U.S. Government Accountability Office, February 1, 1983.

Aid, Matthew, *The Secret Sentry: The Untold History of the National Security Agency*, New York: Bloomsbury Press, 2009.

Air Force Doctrine Document 2-2, *Space Operations*, Washington, D.C., November 27, 2006.

Air Force Doctrine Document 2-2.1, *Counterspace Operations*, Washington, D.C., August 2, 2004.

Air Force Global Strike Command, "B-52 Stratofortress," fact sheet, May 2014. As of June 17, 2015: http://www.af.mil/AboutUs/FactSheets/Display/tabid/224/Article/104465/b-52-stratofortress.aspx

Air Force Instruction 11-202, Vol. 3, *General Flight Rules*, October 22, 2010.

Air Force Space Command, "The 4th Space Control Squadron," fact sheet, Peterson AFB, Colo., undated. As of June 3, 2015:
http://www.peterson.af.mil/library/factsheets/factsheet_print.asp?fsID=4707

———, "The 76th Space Control Squadron," fact sheet, Peterson AFB, Colo., undated. As of June 3, 2015:
http://www.peterson.af.mil/library/factsheets/factsheet_print.asp?fsID=4808

———, "Fact Sheet: Advanced Extremely High Frequency System," March 25, 2015. As of June 23, 2015:
http://www.afspc.af.mil/library/factsheets/factsheet.asp?id=7758

"Alleged Laser Test Sparks Debate on U.S./China Space Cooperation," *Aerospace Daily and Defense Report*, Vol. 219, No. 3, October 2, 2006.

Allen, Kelly, "Navy Answers: Anti-Ship Cruise Missiles," *Navy Live*, December 13, 2012. As of June 3, 2015:
http://navylive.dodlive.mil/2012/12/13/navyanswers-anti-ship-cruise-missiles

Allen, Kenneth W., *The Ten Pillars of the People's Liberation Army Air Force: An Assessment*, Washington, D.C.: Jamestown Foundation, April 2011.

Allen, Kenneth W., Glenn Krumel, and Jonathan D. Pollack, *China's Air Force Enters the 21st Century*, Santa Monica, Calif.: RAND Corporation, MR-580-AF, 1995. As of June 3, 2015: http://www.rand.org/pubs/monograph_reports/MR580.html

Andress, Jason, and Steve Winterfield, *Cyber Warfare: Techniques, Tactics and Tools for Security Practitioners*, Waltham, Mass.: Syngress, 2011.

Andrew, Martin, "China's Anti-Ballistic Missile Test: Much Ado About Nothing," *Air Power Australia NOTAM*, No. 55, January 14, 2010. As of June 3, 2015:
http://www.ausairpower.net/APA-NOTAM-140110-1.html

Anthony, Sebastian, "Windows 10 Will Be a Free Upgrade for All Users Worldwide," *Ars Technica*, March 18, 2015. As of June 23, 2015:
http://arstechnica.com/information-technology/2015/03/
windows-10-will-be-a-free-upgrade-for-genuine-and-non-genuine-users/

Aspin, Les, *Report on the Bottom-Up Review*, Washington, D.C.: Office of the Secretary of Defense, October 1993.

Axe, David, "China's Fighters Won't Match US," *The Diplomat*, March 10, 2011. As of June 3, 2015:
http://thediplomat.com/flashpoints-blog/2011/03/10/chinas-fighters-wont-match-us

Balle, Joakim Kasper Oestergaard, "About the THAAD System," *Aerospace and Defense Intelligence Report*, October 22, 2014. As of June 19, 2015:
https://www.bga-aeroweb.com/Defense/THAAD.html

———, "SBIRS: Space-Based Infrared," *Aerospace and Defense Intelligence Report*, December 29, 2014. As of June 23, 2015:
https://www.bga-aeroweb.com/Defense/SBIRS.html

Bamford, James, *The Shadow Factory: The NSA from 9/11 to the Eavesdropping on America*, New York: Random House, 2008.

Barbosa, Rui C., "China Launch YaoGan Weixing-9, Announce Increase in Vehicle Production," *NASASpaceFlight.com*, March 5, 2010. As of June 3, 2015:
http://www.nasaspaceflight.com/2010/03/china-yaogan-weixing-9-increase-in-vehicle-production

Belasco, Amy, *The Cost of Iraq, Afghanistan, and Other Global War on Terror Operations Since 9/11*, Washington, D.C.: Congressional Research Service, March 29, 2011.

Bell, Thaddeus G., *Probing the Oceans for Submarines: A History of the AN/SQS-26 Long-Range Echo-Ranging Sonar*, Newport, R.I.: Naval Undersea Warfare Center, 2010.

Bender, Bryan, "Army Successfully Fires MIRACL Laser at Satellite," *Defense Daily*, October 21, 1997.

Bianco, Lucien, *Origins of the Chinese Revolution, 1915–1949*, Stanford, Calif.: Stanford University Press, 1971.

Blasko, Dennis J., *The Chinese Army Today: Tradition and Transformation for the 21st Century*, New York: Routledge, 2012.

Bodmer, Sean, Max Kilger, Gregory Carpenter, and Jade Jones, *Reverse Deception Organized Cyber Threat Counter-Exploitation*, New York: McGraw-Hill, 2012.

Boeing, "Boeing Awarded Contract for Next-Generation Harpoon Block III Missile," press release, January 31, 2008. As of June, 2015:
http://boeing.mediaroom.com/index.php?s=20295&item=143

Bolkcom, Christopher, *Military Suppression of Enemy Air Defenses (SEAD): Assessing Future Needs*, Washington, D.C.: Congressional Research Service, January 24, 2005.

Broad, William J., "Administration Researches Laser Weapon," *New York Times*, May 3, 2006.

Broadcasting Board of Governors, "BBG Condemns Cuba's Jamming of Satellite TV Broadcasts to Iran," press release, July 15, 2003. As of June 3, 2015:
http://www.bbg.gov/blog/2003/07/15/bbg-condemns-cubas-jamming-of-satellite-tv-broadcasts-to-iran

Brodkin, Jon, "Ballmer to Hu: 90% of Microsoft Customers in China Using Pirated Software," *Network World*, January 21, 2011. As of June 3, 2015: http://www.networkworld.com/news/2011/012111-ballmer-hu-china-software-piracy.html

Bussert, James C., and Bruce Elleman, *People's Liberation Army Navy: Combat Systems Technology, 1949–2010*, Annapolis, M.D.: Naval Institute Press, 2011.

Butler, Amy, "Bomber in a Pinch," *Aviation Week and Space Technology*, Vol. 172, No. 21, May 31, 2010.

———, "PTSS Kill Leaves Hole in Missile Defense Sensor Plan," *Aviation Week and Space Technology*, Vol. 175, No. 14, April 29, 2013.

Butler, Amy, Michael Bruno, David A. Fulghum, and John M. Doyle, "Ambiguous Intercept: Impact of the Satellite Shootdown Both Lauded and Damned," *Aviation Week and Space Technology*, Vol. 168, No. 8, February 25, 2008.

Butt, Yousaf, *Satellite Laser Ranging in China*, technical working paper, Cambridge, Mass.: Union of Concerned Scientists, January 8, 2007.

Cadirci, Serdar, *RF Stealth (Or Low Observable) and Counter-RF Stealth Technologies: Implications of Counter-RF Stealth Solutions for Turkish Air Force*, thesis, Monterey, Calif.: Naval Postgraduate School, March 2009.

Campbell, Matthew, "Chinese 'Death Ray' Threatens U.S. Satellites," *Sunday Times* (London), December 6, 1998.

Capaccio, Tony, "China's Ballistic Missile, Stealth-Fighter Advances Draw Attention of U.S.," *Bloomberg*, January 6, 2011.

Carr, Jeffrey, *Inside Cyber Warfare: Mapping the Cyber Underworld*, 2nd ed., Sebastopol, Calif.: O'Reilly Media, 2011.

Carreno, Jose, Thomas Culora, George Galdorisi, and Thomas Hone, "What's New About the AirSea Battle Concept," *Proceedings*, Vol. 136, No. 8, August 2010.

Carter, Ashton B., "Defense Management Challenges for the Next American President," *Orbis*, Vol. 53, No. 1, 2009, pp. 41–53.

Cavas, Christopher P., "USAF, U.S. Navy to Expand Cooperation: Air-Sea Battle Will Close Gaps, Boost Strength," *Defense News*, November 9, 2009.

Central Intelligence Agency, "Soviet Forces in the Far East," National Intelligence Estimate 11-14/40-81, October 1985.

Cereijo, Manuel, *China and Cuba and Information Warfare (IW), Signals Intelligence (SIGINT), Electronic Warfare (EW), and Cyber-Warfare*, Coral Gables, Fla.: Cuban-American Military Council, 2003. As of June 3, 2015: http://www.camcocuba.org/html/ADDITIONAL%20PAGES/CEREIJO%20E/cereijo-1/CEREIJO-48-E.html

Chalmers, Malcolm, and Lutz Unterseher, "Is There a Tank Gap? Comparing NATO and Warsaw Pact Tank Fleets," *International Security*, Vol. 13, No. 1, Summer 1988, pp. 5–49.

Chandrashekar, S., and Soma Perumal, *China's Constellation of Yaogan Satellites and the Anti-Ship Ballistic Missile—An Update*, Bangalore, India: International Strategic and Security Studies Programme, National Institute of Advanced Studies, January 2015.

Chang, Andrei, "Analysis: China Attains Nuclear Strategic Strike Capability," United Press International Asia, September 8, 2007.

Chase, Michael S., Jeffrey G. Engstrom, Tai Ming Cheung, Kristen A. Gunness, Scott Warren Harold, Susan Puska, and Samuel K. Berkowitz, *China's Incomplete Military Transformation: Assessing the Weaknesses of the People's Liberation Army (PLA)*, Santa Monica, Calif.: RAND Corporation, RR-893-USCC, 2015. As of June 5, 2015:
http://www.rand.org/pubs/research_reports/RR893.html

Chase, Michael S., Andrew S. Erickson, and Christopher Yeaw, "Chinese Theater and Strategic Missile Force Modernization and Its Implications for the United States," *Journal of Strategic Studies*, Vol. 32, No. 1, February 2009, pp. 67–114.

Cheng, Dean, "Chinese Views on Deterrence," *Joint Force Quarterly*, No. 60, 1st Quarter 2011, pp. 92–94.

Cheung, Tai Ming, ed., *China's Emergence as a Defense Technological Power*, New York: Routledge, 2014.

"China Develops New Light Weapons—Hong Kong Press," BBC Summary of World Broadcasts, January 13, 2000.

"China's Land Reclamation in Disputed Water Stokes Fears of Military Ambitions," *The Guardian*, May 8, 2015. As of June 26, 2015:
http://www.theguardian.com/world/2015/may/08/
china-land-reclamation-south-china-sea-stokes-fears-military-ambitions

"China Signs Arms-Sales Deal with Russia," United Press International, March 25, 2013. As of June 3, 2015:
http://www.upi.com/Top_News/World-News/2013/03/25/China-signs-arms-sales-deal-with-Russia/UPI-29041364269020/

"China Signs Contract to Purchase Russian S-400 Missile Systems," *Russia Beyond the Headlines*, April 14, 2015. As of June 12, 2015:
http://rbth.com/news/2015/04/14/china_signs_contract_to_purchase_russian_s-400_missile_systems_45211.html

"China Targets U.S. Satellite," *Courier Mail* (Australia), October 7 2006, p. 45.

Christensen, Thomas J., "Posing Problems Without Catching Up: China's Rise and Challenges for U.S. Security Policy," *International Security*, Vol. 25, No. 4, Spring 2001, pp. 5–40.

Cilluffo, Frank J., Director, Homeland Security Policy Institute, and Co-Director, Cyber Center for National and Economic Security, George Washington University, "Cyber Threats from China, Russia and Iran," testimony before the Committee on Homeland Security, U.S. House of Representatives, Washington, D.C., March 20, 2013.

Claremont Institute, "S-400 (SA-20 Triumf)," *Missile Defense Systems*, 2009.

Clark, Colin, "Norway's Joint Strike Missile Tempts Aussies; Raytheon Likes It Too," *Breaking Defense*, July 16, 2014. As of June 17, 2015:
http://breakingdefense.com/2014/07/norway-joint-strike-missile/

Clarke, Richard, and Robert Knake, *Cyber War: The Next Threat to National Security and What to Do About It*, New York: HarperCollins, 2010.

Cliff, Roger, "The Development of China's Air Force Capabilities," testimony before the U.S.-China Economic and Security Review Commission, Santa Monica, Calif.: RAND Corporation, CT-346, May 20, 2010. As of June 3, 2015:
http://www.rand.org/pubs/testimonies/CT346.html

Cliff, Roger, Mark Burles, Michael S. Chase, Derek Eaton, and Kevin L. Pollpeter, *Entering the Dragon's Lair: Chinese Antiaccess Strategies and Their Implications for the United States*, Santa Monica, Calif.: RAND Corporation, MG-524-AF, 2007. As of June 3, 2015:
http://www.rand.org/pubs/monographs/MG524.html

Cliff, Roger, John Fei, Jeff Hagen, Elizabeth Hague, Eric Heginbotham, and John Stillion, *Shaking the Heavens and Splitting the Earth: Chinese Air Force Employment Concepts in the 21st Century*, Santa Monica, Calif.: RAND Corporation, MG-915-AF, 2011. As of June 3, 2015:
http://www.rand.org/pubs/monographs/MG915.html

Cliff, Roger, Phillip C. Saunders, and Scott Harold, *New Opportunities and Challenges for Taiwan's Security*, Santa Monica, Calif.: RAND Corporation, CF-279-OSD, 2011. As of June 3, 2015:
http://www.rand.org/pubs/conf_proceedings/CF279.html

Cohen, Eliot A., "Toward Better Net Assessment: Rethinking the European Conventional Balance," *International Security*, Vol. 13, No. 1, Summer 1988, pp. 50–89.

Cole, Bernard D., *The Great Wall at Sea: China's Navy in the Twenty-First Century*, 2nd ed., Annapolis, Md.: Naval Institute Press, 2010.

Commission on the Roles and Missions of the Armed Forces, *Directions for Defense*, Washington, D.C., May 1995.

Cooper, Cortez A., *The PLA Navy's "New Historic Missions": Expanding Capabilities for a Re-Emergent Maritime Power*, testimony before the U.S.-China Economic and Security Review Commission, Santa Monica, Calif.: RAND Corporation, CT-332, June 11, 2009. As of June 3, 2015:
http://www.rand.org/pubs/testimonies/CT332.html

Cordesman, Anthony H., and Martin Kleiber, *Chinese Military Modernization: Force Development and Strategic Capabilities*, Washington, D.C.: Center for Strategic and International Studies, 2007.

Costlow, Terry, "Meeting NATO's Satcom Needs Is No Simple Task," *Defense Systems*, February 25, 2011. As of June 3, 2015:
http://defensesystems.com/articles/2011/02/28/cover-story-sidebar-nato-satcom.aspx

Cote, Owen R., Jr., *Assessing the Undersea Balance Between the U.S. and China*, MIT Security Studies Working Paper, Cambridge, Mass.: Massachusetts Institute of Technology, February 2011.

Covault, Craig, "Navigation Warfare," *The Year in Defense: Aerospace Edition*, Summer 2010. As of June 3, 2015:
http://www.defensemedianetwork.com/stories/navigation-warfare

Crane, Keith, Roger Cliff, Evan S. Medeiros, James C. Mulvenon, and William H. Overholt, *Modernizing China's Military: Opportunities and Constraints*, Santa Monica, Calif.: RAND Corporation, MG-260-AF, 2005. As of June 3, 2015:
http://www.rand.org/pubs/monographs/MG260-1.html

Crawford, Neta C., *U.S. Costs of Wars Through 2014: $4.4 Trillion and Counting*, 2014. As of June 8, 2015:
http://costsofwar.org/sites/default/files/articles/20/attachments/Costs%20of%20War%20Summary%20Crawford%20June%202014.pdf

Crothers, Brian, Jeff Lanphear, Brian Garino, Paul P. Konyha III, and Edward P. Byrne, "U.S. Space-Based Intelligence, Surveillance, and Reconnaissance," *Space Primer*, Maxwell AFB, Ala.: Air University, 2009, pp. 167–181.

Cunningham, Randy, "Suppression of Enemy Air Defenses: Improvements Needed," Washington, D.C.: Electronic Warfare Working Group, U.S. House of Representatives, Issue Brief No. 7, June 11, 2001.

Custer, C., "China's 'Home-made' Operating System Isn't Home Made at All, But Maybe That's OK," *TechinAsia*, December 23, 2014. As of June 25, 2015:
https://www.techinasia.com/chinas-homemade-operating-system-home

Dai Qingmin, "On Integrating Network Warfare and Electronic Warfare," *China Military Science*, Winter 2002.

Danchev, Dancho, "China's 'Secure' OS Kylin—A Threat to U.S. Offensive Cyber Capabilities?" *ZDNet*, May 13, 2009. As of June 3, 2015:
http://www.zdnet.com/article/chinas-secure-os-kylin-a-threat-to-u-s-offensive-cyber-capabilities

Davis, Lynn Etheridge, and Warner R. Schilling, "All You Ever Wanted to Know About MIRV and ICBM Calculations but Were Not Cleared to Ask," *Journal of Conflict Resolution*, Vol. 17, No. 2, June 1973, pp. 207–242.

Deagle.com, "YJ-18," web page, last updated May 5, 2015. As of June 18, 2015:
http://www.deagel.com/Anti-Ship-Missiles/YJ-18_a002884001.aspx

DigitalGlobe, homepage, undated. As of June 3, 2015:
http://digitalglobe.com

Dill, Catherine, "Korla Missile Test Complex Revisited," *Arms Control Wonk* (Blog), March 26, 2015.

Divis, Dee Ann, First GPS III Launch Slips to FY17," *Inside GNSS*, 2014. As of June 23, 2015:
http://www.insidegnss.com/node/4270

Dollar, David, "Why So Little Investment Between the United States and China?" Brookings Institution, February 26, 2015. As of June 24, 2015:
http://www.brookings.edu/blogs/up-front/posts/2015/02/26-investment-between-us-and-china-dollar

Donley, Michael B., Secretary of the Air Force, speech presented at the Air Force Association Global Warfare Symposium, Los Angeles, Calif., November 16, 2012.

Dou, Eva, "China's Great Firewall Gets Taller," *Wall Street Journal*, January 30, 2015. As of June 25, 2015:
http://www.wsj.com/articles/chinas-great-firewall-gets-taller-1422607143

Dragon in Space, "Ziyuan 2," web page, last updated April 3, 2012. Archived, as of August 9, 2015:
http://archive.is/iAKZ3

———, "Yaogan," web page, last updated May 29, 2012. Archived, as of August 9, 2015:
http://archive.is/06HwD

Drew, Christopher, "Drones Are Weapons of Choice in Fighting Al-Qaeda," *New York Times*, March 16, 2009.

Dunn, John E., "Internet Hijack Claims Denied by Techcom," *Techworld*, November 18, 2010. As of June 3, 2015:
http://news.techworld.com/security/3249346/internet-hijack-claims-denied-by-china-telecom

Dunnigan, James F., *How to Make War: A Comprehensive Guide to Modern Warfare in the 21st Century*, 4th ed., New York: HarperCollins, 2003.

Easton, Ian, *The Assassin Under the Radar: China's DH-10 Cruise Missile Program*, Arlington, Va.: Project 2049 Institute, 2009.

———, *China's Evolving Reconnaissance Strike Capabilities: Implications for the U.S.-Japan Alliance*, Arlington, Va.: Project 2049 Institute and Japan Institute of International Affairs, February 2014.

Eaton, Kit, "China Behind Yesterday's YouTube, Facebook, Twitter Outage," *Fast Company*, March 26, 2010. As of June 3, 2015:
http://www.fastcompany.com/1598176/china-green-dam-censorship-dns-error-sweden-world-google-filtering

Eberhardt, Brian, Kenneth Kemmerly, and Paul Konyha III, "Satellite Communications," *Space Primer*, Maxwell AFB, Ala.: Air University, 2009, pp. 183–199.

中国空军百科全书编审委员会 [Editorial Committee of the People's Liberation Army Air Force Encyclopedia], 中国空军百科全书 [*China Air Force Encyclopedia*], Beijing: Aviation Industry Press, 2005.

Efroymson, Rebecca A., Winifred Hodge Rose, Sarah Nemeth, and Glenn W. Suter II, *Ecological Risk Assessment Framework for Low-Altitude Overflight by Fixed-Wing and Rotary-Wing Military Aircraft*, Oak Ridge, Tenn.: Oak Ridge National Laboratory, 2000. As of June 3, 2015:
http://www.esd.ornl.gov/programs/ecorisk/documents/overflight-e1.pdf

Eighth Air Force Historical Society, "WWII 8thAAF Combat Chronology, January 1944 Through June 1944," web page, undated. As of June 3, 2015:
http://www.8thafhs.org/combat1944a.htm

Eilperin, Juliet, "US, Philippines to Sign 10-Year Defense Agreement Amid Rising Tensions," *Washington Post*, April 28, 2014.

Elder, Miriam, and Tania Branigan, "Russia Arrests Chinese 'Spy' in Row Over Defence Weapons," *The Guardian*, October 5, 2011. As of June 4, 2015:
http://www.theguardian.com/world/2011/oct/05/russia-arrests-chinese-translator-spying

Epstein, Joshua M., *Measuring Military Power: The Soviet Air Threat to Europe*, Princeton, N.J.: Princeton University Press, 1984.

———, *Strategy and Force Planning: The Case of the Persian Gulf*, Washington, D.C.: Brookings Institution Press, 1987.

Erickson, Andrew S., "China Channels Billy Mitchell: Anti-Ship Ballistic Missile Alters Region's Military Geography," *China Brief*, Vol. 13, No. 5, March 2013.

———, Chinese Anti-Ship Ballistic Missile Development: Drivers, Trajectories, and Strategic Implications, Washington, D.C.: Jamestown Foundation, May 2013.

Erickson, Andrew S., Abraham M. Denmark, and Gabriel Collins, "Beijing's 'Starter Carrier' and Future Steps: Alternative and Implications," *Naval War College Review*, Vol. 65, No. 1, Winter 2012, pp. 15–55.

"F-22 Raptor: Program and Events," *Defense Industry Daily*, May 14, 2015. As of June 4, 2015:
http://www.defenseindustrydaily.com/f22-raptor-procurement-events-updated-02908

Federation of American Scientists, "B-1 Lancer," web page, undated. As of June 4, 2015:
http://fas.org/nuke/guide/usa/bomber/b-1b.htm

———, "B-2 Spirit," web page, undated. As of June 4, 2015:
http://fas.org/nuke/guide/usa/bomber/b-2.htm

———, "B-52 Stratofortress," web page, undated. As of June 4, 2015:
http://fas.org/nuke/guide/usa/bomber/b-52.htm

———, "F-16 Fighting Falcon," web page, undated. As of June 4, 2015:
http://fas.org/man/dod-101/sys/ac/f-16.htm

———, "SSN-637 Sturgeon Class," web page, undated. As of June 4, 2015:
https://fas.org/man/dod-101/sys/ship/ssn-637.htm

———, "U.S. Smart Munitions," web page, undated. As of October 22, 2013: http://www.fas.org/programs/ssp/man/uswpns/ussmartmunitions.html

———, "Low Frequency Active (LFA)," web page, June 21, 1997. As of June 4, 2015: http://www.fas.org/irp/program/collect/lfa.htm

———, "JL-1 [CSS-N-3]," web page, last updated June 10, 1998. As of June 4, 2015: http://www.fas.org/nuke/guide/china/slbm/jl-1.htm

Feng, "2012 in Review," *Information Dissemination*, December 28, 2012. As of June 4, 2015: http://www.informationdissemination.net/2012/12/2012-in-review.html

Ferguson, Robyn E., *Information Warfare with Chinese Characteristics: China's Future View of Information Warfare and Strategic Culture*, thesis, Ft. Leavenworth, Kan.: U.S. Army Command and General Staff College, 2002.

Finkelstein, David M., "Thinking About the PLA's Revolution in Doctrinal Affairs," in James Mulvenon and David M. Finkelstein, eds., *China's Revolution in Doctrinal Affairs: Emerging Trends in the Operational Art of the Chinese People's Liberation Army*, Arlington, Va.: Center for Naval Analyses, 2005, pp. 1–27.

———, "China's National Military Strategy Revisited: An Overview of the 'Military Strategic Guidelines,'" in Andrew Scobell and Roy Kamphausen, eds., *Right-Sizing the People's Liberation Army: Exploring the Contours of China's Military*, Carlisle, Pa.: U.S. Army War College, 2007, pp. 69–140.

Flaherty, Mary Pat, Jason Samenow, and Lisa Rein, "Chinese Hack U.S. Weather Systems, Satellite Network," *Washington Post*, November 12, 2014: http://www.washingtonpost.com/local/chinese-hack-us-weather-systems-satellite-network/2014/11/12/bef1206a-68e9-11e4-b053-65cea7903f2e_story.html

Fletcher, Owen, "China Game Boss Sniped Rivals, Took Down Internet," *IDG News Service*, August 28, 2009. As of June 4, 2015: http://www.networkworld.com/news/2009/082809-china-game-boss-sniped-rivals.html

Fravel, M. Taylor, and Evan S. Medeiros, "China's Search for Assured Retaliation: The Evolution of Chinese Nuclear Strategy and Force Structure," *International Security*, Vol. 35, No. 2, Fall 2010, pp. 48–87.

Freedberg, Sydney J., "Hagel Lists Key Technologies for US Military; Launches Offset Strategy,'" *Breaking Defense*, November 16, 2014.

Friedman, Norman, *The Naval Institute Guide to World Naval Weapons Systems, 1991/92*, Annapolis, Md.: Naval Institute Press, 1991.

———, *The Naval Institute Guide to World Naval Weapon Systems*, 5th ed., Annapolis, Md.: Naval Institute Press, 2006.

Fuell, Lee, Technical Director for Force Modernization and Employment, National Air and Space Intelligence Center, testimony before the U.S.-China Economic and Security Review Commission, January 30, 2014. As of June 11, 2015: http://www.uscc.gov/sites/default/files/Lee%20Fuell_Testimony1.30.14.pdf

Gady, Franz-Stefan, "China Just Doubled the Size of Its Amphibious Mechanized Infantry Divisions," *The Diplomat*, January 9, 2015. As of June 26, 2015: http://thediplomat.com/2015/01/china-just-doubled-the-size-of-its-amphibious-mechanized-infantry-divisions/

Gereffi, Gary, Vivek Wadhwa, Ben A. Rissing, and Ryan Ong, "Getting the Numbers Right: International Engineering Education in the United States, China, and India," *Journal of Engineering Education*, Vol. 97, No. 1, January 2008, pp. 13–25.

Gertler, Jeremiah, *U.S. Unmanned Aerial Systems*, Washington, D.C.: Congressional Research Service, 2012. As of June 11, 2015:
https://www.fas.org/sgp/crs/natsec/R42136.pdf

Giarra, Paul S., President, Global Strategies and Transformation, *A Chinese Anti-Ship Ballistic Missile: Implications for the USN*, statement before the U.S.-China Economic and Security Review Commission, Washington, D.C., June 11, 2009. As of June 4, 2015:
http://www.uscc.gov/sites/default/files/6.11.09%20Giarra.pdf

Gilboy, George J., and Eric Heginbotham, *Chinese and Indian Strategic Behavior: Growing Power and Alarm*, Cambridge, UK: Cambridge University Press, 2012.

Glaser, Charles L., "When Are Arms Races Dangerous? Rational Versus Suboptimal Arming," *International Security*, Vol. 28, No. 4, Spring 2004, pp. 44–84.

Glasstone, Samuel, and Philip J. Dolan, eds., *The Effects of Nuclear Weapons*, 3rd ed., Washington, D.C.: U.S. Department of Defense and U.S. Energy Research and Development Administration, 1977.

Globalsecurity.org, "AGM-158 JASSM Program Developments," web page, last updated July 7, 2011. As of June 4, 2015:
http://www.globalsecurity.org/military/systems/munitions/jassm-program.htm

———, "AGM-86C/D Conventional Air-Launched Cruise Missile," web page, last updated July 7, 2011. As of June 4, 2015:
http://www.globalsecurity.org/military/systems/munitions/agm-86c.htm

———, "BGM-109 Tomahawk: Tomahawk Operational Use," web page, last updated July 7, 2011. As of June 4, 2015:
http://www.globalsecurity.org/military/systems/munitions/bgm-109-operation.htm

———, "H-6D Bomber," web page, last updated July 11, 2011. As of June 5, 2015:
http://www.globalsecurity.org/military/world/china/h-6d.htm

———, "B-2 Operations," web page, last updated July 24, 2011. As of June 5, 2015:
http://www.globalsecurity.org/wmd/systems/b-2-ops.htm

———, "RT-21M/SS-20 SABER Specifications, web page, last updated July 24, 2011. As of June 11, 2015:
http://www.globalsecurity.org/wmd/world/russia/rt-21m-specs.htm

———, "Chinese Warships," web page, last updated April 8, 2015. As of June 4, 2015:
http://www.globalsecurity.org/military/world/china/navy.htm

———, "HQ-19," Anti-Ballistic Missile Interceptor," web page, last updated April 20, 2015. As of June 12, 2015:
http://www.globalsecurity.org/space/world/china/hq-19.htm

Glosny, Michael A., "Strangulation from the Sea? A PRC Blockade of Taiwan," *International Security*, Vol. 28, No. 4, Spring 2004, pp. 125–160.

Gons, Eric Stephen, *Access Challenges and Implications for Airpower in the Western Pacific*, dissertation, Santa Monica, Calif.: Pardee RAND Graduate School, RGSD-267, 2011. As of June 4, 2015:
http://www.rand.org/pubs/rgs_dissertations/RGSD267.html

Goodin, Dan, "Massive Denial-of-Service Attack on GitHub Tied to Chinese Government," *Ars Technica*, March 31, 2015. As June 25, 2015: http://arstechnica.com/security/2015/03/massive-denial-of-service-attack-on-github-tied-to-chinese-government

Gorman, Siobhan, "Electricity Grid in U.S. Penetrated by Spies," *Wall Street Journal*, April 8, 2009.

Gormley, Dennis M., Andrew S. Erickson, and Jingdon Yuan, *A Low-Visibility Force Multiplier: Assessing China's Cruise Missile Ambitions*, Washington, D.C.: National Defense University Press, 2014.

GPS.gov, "Space Segment," web page, June 17, 2015. As of June 23, 2015: http://www.gps.gov/systems/gps/space

Graham, William, "ULA Atlas V Launch with SBIRS GEO-2 Successful," *NASASpaceflight*, March 19, 2013. As of June 4, 2015: http://www.nasaspaceflight.com/2013/03/ula-atlas-v-launch-sbirsgeo

Grant, Rebecca, *The Radar Game: Understanding Stealth and Aircraft Survivability*, Arlington, Va.: Mitchell Institute Press, 2010.

Greenert, Jonathan, "Sea Change: The Navy Pivots to Asia," *Foreign Policy*, November 14, 2012.

Greenert, Jonathan, and Mark Welsh, "Breaking the Kill Chain: How to Keep America in the Game When Our Enemies Are Trying to Shut Us Out," *Foreign Policy*, May 16, 2013.

Grow, Brian, Keith Epstein, and Chi-Chu Tschang, "The New E-spionage Threat," *BusinessWeek*, April 21, 2008, pp. 32–41.

Gruss, Mike, "Report: SBIRS Tech Update Would Be Costly," *Space News*, August 14, 2014.

Hagt, Eric, and Matthew Durnin, "China's Antiship Ballistic Missile: Developments and Missing Links," *Naval War College Review*, Vol. 62, No. 4, Fall 2009, pp. 87–117.

———, "Space, China's Tactical Frontier," *Journal of Strategic Studies*, Vol. 34, No. 5, October 2011, pp. 733–761.

Hallion, Richard P., Roger Cliff, and Phillip C. Saunders, eds., *The Chinese Air Force: Evolving Concepts, Roles, and Capabilities*, Washington, D.C.: National Defense University Press, 2012.

Hammes, T. X., *Offshore Control: A Proposed Strategy for an Unlikely Conflict*, Tel Aviv, Israel: Institute for National Security Studies, 2012.

Hanemann, Thilo, and Cassie Gao, "Chinese FDI in the United States: Q4 and Full Year 2014 Update," *Rhodium Group*, January 15, 2015. As of June 24, 2015: http://rhg.com/notes/chinese-fdi-in-the-united-states-q4-and-full-year-2014-update

Hardy, James, and Sean O'Connor, "China Building Airstrip-Capable Island on Fiery Cross Reef," *Jane's Defence Weekly*, November 20, 2014.

Harley, Jeff, *Space Control and Information Operations*, Huntsville, Ala.: U.S. Army Space and Missile Defense Command/Army Forces Strategic Command, 2002.

Harris, Francis, "Beijing Secretly Fires Lasers to Disable U.S. Satellites," *Daily Telegraph* (London), September 26, 2006.

Hartling, William D., "It's Time to Sink the Littoral Combat Ship," *Defense One*, August 25, 2014.

Headquarters Air Mobility Command, *Airfield Suitability and Restrictions Report*, Scott AFB, Ill., December 4, 2007.

Heginbotham, Eric, and Jacob L. Heim, "Deterring Without Dominance: Discouraging Chinese Adventurism Under Austerity," *Washington Quarterly*, Vol. 38, No. 1, 2015, pp. 185–199.

"HEL MD Laser Continues Testing, Moves Towards 60 kW System," *Jane's Defence Weekly*, September 10, 2014.

Hernandez, Adolfo J., *Military Role in Space Control: A Primer*, Washington, D.C.: Congressional Research Service, September 23, 2004.

Hewitt, William A., *Planting the Seeds of SEAD: The Wild Weasel in Vietnam*, thesis, Maxwell AFB, Ala.: Air University, 1993.

Higgins, Kelly Jackson, "Flaws in the 'Aurora' Attacks," *DarkReading*, January 25, 2010. No longer available online.

Hildreth, Steven, *Cyberwarfare*, Washington, D.C.: Congressional Research Service, June 19, 2001.

Hillestad, Richard J., Bart E. Bennett, and Louis R. Moore, *Modeling for Campaign Analysis: Lessons for the Next Generation of Models, Executive Summary*, Santa Monica, Calif.: RAND Corporation, MR-710-AF, 1996. As of June 4, 2015:
http://www.rand.org/pubs/monograph_reports/MR710.html

Hoyler, Marshall, "China's 'Antiaccess' Ballistic Missiles and U.S. Active Defenses," *Naval War College Review*, Vol. 63, No. 4, Fall 2010, pp. 84–105.

IISS—*See* International Institute for Strategic Studies.

Inkster, Nigel, "The Chinese Intelligence Agencies: Evolution and Empowerment in Cyberspace," in John R. Lindsay, Tai Ming Cheung, and Derek S. Reveron, eds., *China and Cybersecurity: Espionage, Strategy, and Politics in the Digital Domain*, London: Oxford University Press, 2015, pp. 29–50.

International Institute for Strategic Studies, *The Military Balance*, London: Routledge, 1978.

———, *The Military Balance*, London: Routledge, 1989.

———, *The Military Balance*, London: Routledge, 1991.

———, *The Military Balance*, London: Routledge, 1996.

———, *The Military Balance*, London: Routledge, 2002.

———, *The Military Balance*, London: Routledge, 2003.

———, *The Military Balance*, London: Routledge, 2010.

———, *The Military Balance*, London: Routledge, 2012.

———, *The Military Balance*, London: Routledge, 2013.

———, *The Military Balance*, London: Routledge, 2014.

———, *The Military Balance*, London: Routledge, 2015.

International Laser Ranging Service, "Stations Site Listing," NASA, June 3, 2015. As of June 4, 2015:
http://ilrs.gsfc.nasa.gov/network/stations/active/index.html

International Monetary Fund, World Economic Outlook Database, 2015. As of May 19, 2015:
https://www.imf.org/external/pubs/ft/weo/2015/01/weodata/index.aspx

Jane's 360, "Navy League 2015: Boeing Developing Kit to Upgrade Harpoon Missiles for Extended Range," April 15, 2015. As of June 19, 2015:
http://www.janes.com/article/50754/navy-league-2015-boeing-developing-kit-to-upgrade-harpoon-missiles-for-extended-range

Jane's Aircraft Upgrades, "Northrop Grumman (Northrop) B-2A Spirit," September 16, 2014.

———, "Lockheed Martin (Lockheed) P-3 Orion," March 9, 2015.

Jane's Air-Launched Weapons, "Penetrating and Area Denial Bombs," March 18, 2005.

———, "Chinese Laser-Guided Bombs (LGBs)," January 22, 2010.

———, "YJ-8K (C-801K), YJ-82 (C-802AK/KD) and YJ-83 (C-803)," December 31, 2012.

———, "YJ-91, KR-1 (Kh-31P)," December 31, 2012.

———, "KD-63 (YJ-63), K/AKD-63," January 28, 2014.

———, "Joint Dual Role Air Dominance Missile (JDRADM), T3 and Next Generation Missile (NGM)," August 19, 2014.

———, "AGM-158A JASSM (Joint Air-to-Surface Standoff Missile), AGM-158B JASSM-ER and LRASM," August 28, 2014.

———, "AGM-84E SLAM, AGM-84H/K SLAM-ER," October 22, 2014.

———, "AGM-86 Air-Launched Cruise Missile (ALCM) and CALCM," October 22, 2014.

———, "AGM-88 (High-Speed Anti-Radiation Missile)," October 22, 2014.

———, "GBU-39/B Small Diameter Bomb (SDB I), GBU-39B/B Laser SDB," January 2, 2015.

———, "Fei Teng Guided Bombs (FT-1, FT-2, FT-3, FT-5)," January 16, 2015.

———, "LT-2 Laser-Guided Bomb (LS-500J)," February 26, 2015.

———, "AGM-84 Harpoon," March 25, 2015.

———, "AGM-158A JASSM, AGM-158B JASSM-ER and LRASM," May 26, 2015.

Jane's All the World's Aircraft, "Northrup Grumman B-2 Spirit," June 20, 2006.

———, "Airborne ASW Platforms (China)," June 15, 2011.

———, "Lockheed Martin (645) F-22 Raptor," April 8, 2013.

———, "SAC (Sukhoi Su-27) J-11," June 30, 2014.

———, "SAC (Sukhoi Su-27SK) J-11A," June 30, 2014.

———, "Sukhoi Su-30M," August 26, 2014.

———, "XAC H-6," October 20, 2014.

———, "SAC Shen Fei," January 7, 2015.

———, "XAC JH-7," January 7, 2015.

———, "XAC Y-20 Kunpeng," January 7, 2015.

———, "CAC J-20," February 3, 2015.

———, "SAC Y-8 and Y-9 (Special Mission Versions)," February 3, 2015.

Jane's Amphibious and Special Forces, "Sea Lift (China)," April 22, 2014.

Jane's Armour and Artillery, "Analysis: Shenyang's 'Shen Fei' Breaks Cover," September 19, 2012.

———, "NORINCO Type 98/Type 99 (ZTZ-98/ZTZ-99) MBT," April 1, 2014.

Jane's C4ISR and Mission Systems, "Boeing EA-18G Growler," July 28, 2014.

———, "Space-Based Infrared System," November 20, 2014.

Jane's Electronic Mission Aircraft, "AGM-88 High-Speed Anti-Radiation Missile," January 11, 2011.

———, "Raytheon AGM-88 High-Speed Anti-Radiation Missile," October 22, 2014.

Jane's Electro-Optic Systems, "Northrop Grumman Skyguard," September 3, 2010.

———, "Northrop Grumman Mid-InfraRed Advanced Chemical Laser (MIRACL)," September 21, 2010.

Jane's Fighting Ships, "Spruance Class: Destroyers," March 7, 2006.

———, "Spreadsheet: World Naval Ship Fleets," February 12, 2015.

———, "Han Class (Type 091/091G)," February 13, 2015.

———, "Jin Class (Type 094)," February 13, 2015.

———, "Luzhou Class (Type 051C)," February 13, 2015.

———, "Ming Class (Type 035), February 13, 2015.

———, "Shang Class (Type 093/093A)," February 13, 2015.

———, "Song Class (Type 039/039G)," February 13, 2015.

———, "Type 094 (Jin Class)," February 13, 2015.

———, "Xia Class (Type 092)," February 13, 2015.

———, "Yuan Class (Type 041)," February 13, 2015.

———, "Jiangkai II (Type 054A) Class," February 16, 2015.

———, "Luyang II (Type 052C) Class," February 16, 2015.

———, "Luyang III (Type 052D) Class," February 16, 2015.

———, Jiangdao (Type 056/056A) Class," March 11, 2015.

———, "Arleigh Burke (Flights I and II) Class," March 24, 2015.

———, "Ticonderoga Class," March 24, 2015.

———, "Virginia Class," March 24, 2015.

———, "Zumwalt (DDG 1000) Class," March 24, 2015.

———, "Los Angeles Class," March 24, 2015.

———, "Arleigh Burke (Flight IIA) Class," April 2, 2015.

Jane's Land Warfare Platforms, "China Precision Machinery Import and Export Corporation (CPMIEC) 300mm (10-Round) A100 Multiple Rocket System," March 5, 2012.

———, "China Continues to Test J-20 Engine and Airframe," May 24, 2012.

———, "China Precision Machinery Import and Export Corporation (CPMIEC) 302mm WS-1B (4-Round) Artillery Rocket System," July 22, 2013.

———, "NORINCO Type 63A Light Amphibious Tank," October 15, 2014.

———, "Skyguard (Laser Air Defense)," March 10, 2015.

———, "Chinese Amphibious Assault Vehicle ZBD-2000 (ZBD-05)," March 13, 2015.

———, "NORINCO Airborne Assault Vehicle (AAV) ZLC2000 (ZBD-03)," March 13, 2015.

———, "NORINCO Type 77 Armoured Personnel Carrier," March 13, 2015.

———, "Artillery and Air Defense S-400," April 17, 2015.

Jane's Defence Weekly, "China Accepts Su-30MK2 Fighters," March 31, 2004.

———, "Striking Out at Sea," October 10, 2012.

Jane's Intelligence Review, "Breaching Protocol: The Threat of Cyberespionage," February 11, 2010.

———, "Space Invaders—China's Space Warfare Capabilities," July 3, 2014.

———, "Back Into the Blue: LRASM Honed for Extended Reach, Precision Punch," September 10, 2014.

Jane's International Defence Review, "Storm Force Warning: China's Anti-Ship Missile Range Spreads Its Wings," April 17, 2013.

Jane's Land-Based Air Defence, "S-75 Family (SA-2 'Guideline')," August 5, 2011.

———, "S-300P," February 19, 2014.

Jane's Land Warfare Platforms, "China Precision Machinery Import and Export Corporation (CPMIEC) 300mm (10-Round) A100 Multiple Rocket System," March 5, 2012.

———, "Tactical High-Energy Laser (THEL)," August 14, 2012.

———, "China Precision Machinery Import and Export Corporation (CPMIEC) 302mm WS-1B (4-Round) Artillery Rocket System," July 22, 2013.

———, "China Precision Machinery Import and Export Corporation (CPMIEC) 320mm (4-Round) WS-1 Artillery Rocket System," July 22, 2013.

———, "NORINCO AR2 300mm (12-Round) Multiple Launch Rocket System," July 22, 2013.

———, "NORINCO Type 63A Light Amphibious Tank," October 15, 2014.

———, "HQ-9/FT-2000," February 20, 2015.

———, "THAAD," March 3, 2015.

———, "Chinese Amphibious Assault Vehicle," ZBD-2000 (ZBD-05)," March 13, 2015.

———, "NORINCO Airborne Assault Vehicle (AAV) ZLC2000 (ZBD-03)," March 13, 2015.

———, "NORINCO Type 77 Armoured Personnel Carrier," March 13, 2015.

Jane's Missiles and Rockets, "B-2 Spirit Releases 80 JDAMs in Test," October 17, 2003.

———, "China's Future Jin-Class SSBN Deployment Draws on Xia-Class Experience," March 27, 2007.

Jane's Naval Weapons Systems, "Mk 48 (YU-6)," July 29, 2011.

Jane's Navy International, "Striker Beneath the Sea," March 18, 2003.

———, "Rising Star: Upgrades Fuel Poseidon's Ascent," September 29, 2011.

———, "China Unveils ASW Version of Z-18 Helicopter," August 20, 2014.

Jane's Sentinel Security Assessment, "China and Northeast Asia Procurement," February 16, 2012.

———, "China and Northeast Asia: Navy," March 2015.

———, "Air Force, China," March 5, 2015.

———, "China: Procurement," March 5, 2015.

———, "Procurement, United States," April 2, 2015.

———, "Strategic Weapon Systems, China," April 6, 2015.

Jane's Space Systems and Industry, "Wideband Global SATCOM," August 27, 2009.

———, "ZiYuan-2/JianBing-3 Series," November 24, 2014.

———, "China-Brazil Earth Resources Satellite (CBERS)/Ziyaun Series," December 17, 2014.

———, "Yaogan Series," January 20, 2015.

———, "NAVSTAR Global Positioning System (GPS)," March 6, 2015.

———, "Beidou/Compass Series," April 13, 2015.

Jane's Strategic Weapons Systems, "Weapon Inventories—Offensive/Defensive Weapons Tables, China," December 2, 2010.

———, "AGM-129 Advanced Cruise Missile," October 13, 2011.

———, "LGM-30F Minuteman II," October 13, 2011.

———, "LGM-118 Peacekeeper," October 13, 2011.

———, "UGM-96 Trident C-4," October 13, 2011.

———, "C-801 (CSS-N-4 'Sardine'/YJ-1/-8/-81), C-802 (CSSC-8 'Saccade'/YJ-2/-21/-22/-82/-85), and C-803 (YJ-3/-83/-88)," February 7, 2012.

———, "S-400 Triumf (SA-21 'Growler')," July 17, 2013.

———, "Smerch/Shtil-1/-2 (SA-N-7B/C or SA-N-12 'Grizzly')," July 23, 2013.

———, "Urugan (SA-N-7 'Gadfly')," July 23, 2013.

———, "RIM-66/-67/-156 Standard SM-1/-2, RIM-161 Standard SM-3, and RIM-174 Standard SM-6," July 30, 2013.

———, "HQ-9/-15 and HHQ-9 (RF-9/-15, FD-2000 and FT-2000)," August 9, 2013.

———, "HQ-6/HHQ-6 (RF-6, SD-1 and CSA-N-2)," August 19, 2013.

———, "HQ-16/-17 (HHQ-16/-17 and MD-2000)," August 19, 2013.

———, "Multirole ASM (KD-88)," March 6, 2014.

———, "C-602 (HN-1/-2/-3/YJ-62/X-600/DH-10/CJ-10/HN-2000)," March 24, 2014.

———, "RGM/UGM-109 Tomahawk," May 6, 2014.

———, "DF-3 (CSS-2)," May 7, 2014.

———, "B61 Nuclear Bomb," June 18, 2014.

———, "DF-21 (CSS-5)," June 24, 2014.

———, "AGM/RGM/UGM-84 Harpoon/SLAM/SLAM-ER," July 14, 2014.

———, "Ship-Based Laser," July 25, 2014.

———, "Ground-Based Mid-Course Defense (GMD) Segment," September 12, 2014.

———, "AGM-86 ALCM/CALCM," November 21, 2014.

———, "JL-2," November 28, 2014.

———, "LGM-30G Minuteman III," November 28, 2014.

———, "DF-11 (CSS-7/M-11)," December 11, 2014.

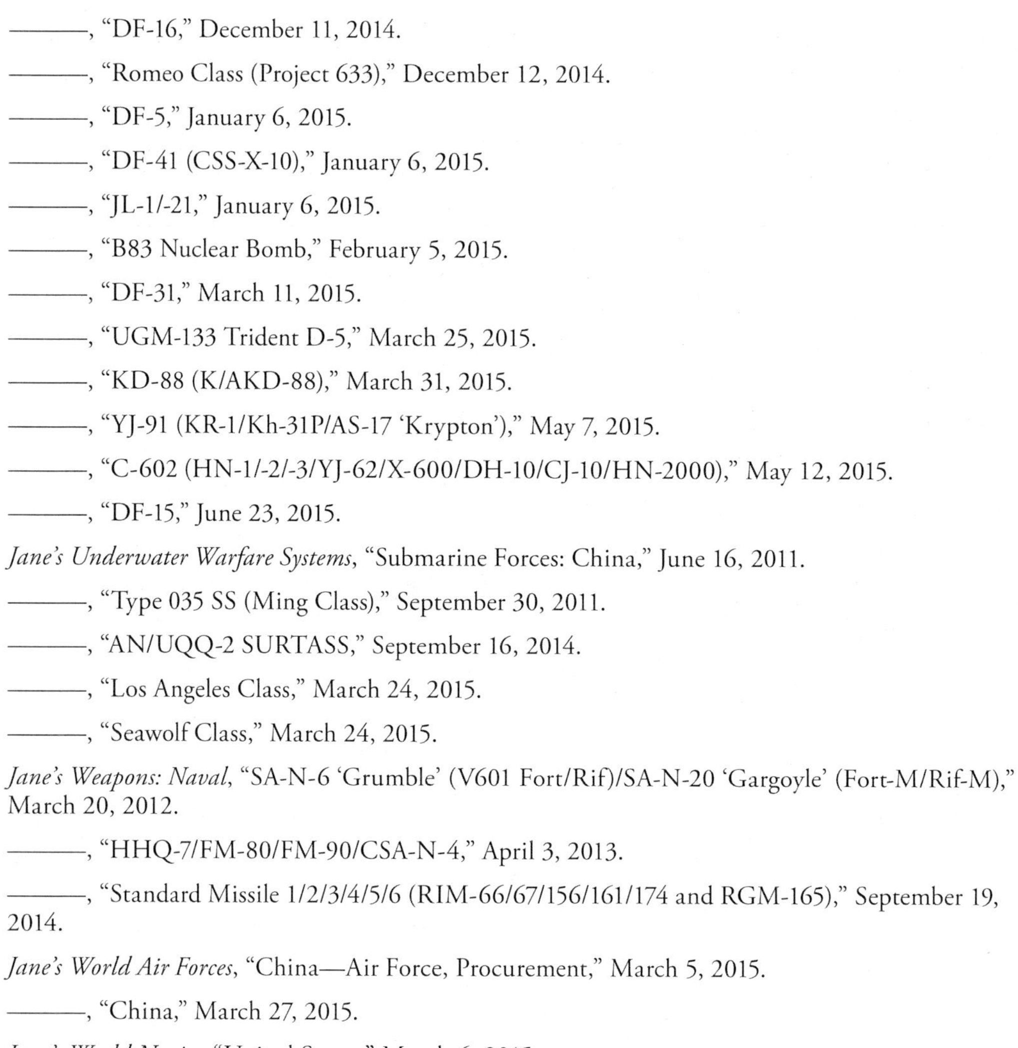

———, "DF-16," December 11, 2014.

———, "Romeo Class (Project 633)," December 12, 2014.

———, "DF-5," January 6, 2015.

———, "DF-41 (CSS-X-10)," January 6, 2015.

———, "JL-1/-21," January 6, 2015.

———, "B83 Nuclear Bomb," February 5, 2015.

———, "DF-31," March 11, 2015.

———, "UGM-133 Trident D-5," March 25, 2015.

———, "KD-88 (K/AKD-88)," March 31, 2015.

———, "YJ-91 (KR-1/Kh-31P/AS-17 'Krypton')," May 7, 2015.

———, "C-602 (HN-1/-2/-3/YJ-62/X-600/DH-10/CJ-10/HN-2000)," May 12, 2015.

———, "DF-15," June 23, 2015.

Jane's Underwater Warfare Systems, "Submarine Forces: China," June 16, 2011.

———, "Type 035 SS (Ming Class)," September 30, 2011.

———, "AN/UQQ-2 SURTASS," September 16, 2014.

———, "Los Angeles Class," March 24, 2015.

———, "Seawolf Class," March 24, 2015.

Jane's Weapons: Naval, "SA-N-6 'Grumble' (V601 Fort/Rif)/SA-N-20 'Gargoyle' (Fort-M/Rif-M)," March 20, 2012.

———, "HHQ-7/FM-80/FM-90/CSA-N-4," April 3, 2013.

———, "Standard Missile 1/2/3/4/5/6 (RIM-66/67/156/161/174 and RGM-165)," September 19, 2014.

Jane's World Air Forces, "China—Air Force, Procurement," March 5, 2015.

———, "China," March 27, 2015.

Jane's World Navies, "United States," March 6, 2015.

Johnson, David E., Jennifer D. P. Moroney, Roger Cliff, M. Wade Markel, Laurence Smallman, and Michael Spirtas, *Preparing and Training for the Full Spectrum of Military Challenges: Insights from the Experiences of China, France, the United Kingdom, India, and Israel*, Santa Monica, Calif.: RAND Corporation, MG-836-OSD, 2009. As of July 25, 2013:
http://www.rand.org/pubs/monographs/MG836.html

Kamphausen, Roy, David Lai, and Travis Tanner, eds., *Learning by Doing: The PLA Trains at Home and Abroad*, Carlisle, Pa.: Strategic Studies Institute, U.S. Army War College, 2012.

Kamphausen, Roy, and Andrew Scobell, eds., *Right-Sizing the People's Liberation Army: Exploring the Contours of China's Military*, Carlisle, Pa.: Strategic Studies Institute, U.S. Army War College, September 2007.

Kan, Shirley, *China's Anti-Satellite Weapon Test*, Washington, D.C.: Congressional Research Service, April 23, 2007.

Karotin, Jesse L., Senior Intelligence Office for China at the Office of Naval Intelligence, "Trends in China's Naval Modernization," testimony before the U.S.-China Economic and Security Review Commission, January 30, 2014.

Keenan, T. D., and S. J. Anderson, "Some Examples of Surface Wind Field Analysis Based on Jindalee Sky Wave Radar Data," *Australian Meteorological Magazine*, Vol. 35, December 1987, pp. 153–161.

Kent, Glenn A., and David E. Thaler, *First-Strike Stability: A Methodology for Evaluating Strategic Forces*, Santa Monica, Calif.: RAND Corporation, R-3765, 1989. As of May 28, 2015: http://www.rand.org/pubs/reports/R3765.html

Kerchner, R. M., et al., *The TAC Brawler Air Combat Simulation Analyst Manual*, Rev. 3.0, Decision Science Applications Report No. 668, 1985.

Kessler, Glenn, "Bachmann's Claim that China 'Blinded' U.S. Satellites," *Washington Post*, October 4, 2011. As of June 22, 2015: http://www.washingtonpost.com/blogs/fact-checker/post/bachmanns-claim-that-china-blinded-us-satellites/2011/10/03/gIQAHvm7IL_blog.html

Kimmage, Daniel, "Up in Arms Over Iraqi Arms," *Russia Weekly*, No. 251, April 3, 2003.

Kirk, Jeremy, "Microsoft Finds New Computers in China Preinstalled with Malware," *IDG News Service*, September 13, 2012. As of May 28, 2015: http://www.pcworld.com/article/262308/microsoft_finds_new_computers_in_china_preinstalled_with_malware.html

Koh, Harold Hongju, Legal Advisor U.S. Department of State, "U.S. Position on Conventional Weapons Negotiations on Cluster Munitions Protocol," Special Briefing, Washington, D.C., November 16, 2011.

Kopp, Carlo, "Surviving the Modern Integrated Air Defence System," *Air Power Australia, Analysis 2009-02*, February 3, 2009. As of May 28, 2015: http://www.ausairpower.net/APA-2009-02.html

———, "Advances in PLA C4ISR Capabilities," *China Brief*, Vol. 10, No. 4, February 18, 2010, pp. 5–8. As of May 28, 2015: http://www.jamestown.org/uploads/media/cb_010_20.pdf

———, *High Energy Laser Directed Energy Weapons*, Air Power Australia, Technical Report APA-TR-2008-0501, updated April 2012. As of May 28, 2015: http://www.ausairpower.net/APA-DEW-HEL-Analysis.html

———, *Russian/PLA Low Band Surveillance Radars*, Air Power Australia, Technical Report APA-TR-2007-0901, updated April 2012. As of May 28, 2015: http://www.ausairpower.net/APA-Rus-Low-Band-Radars.html

———, *People's Liberation Army Air Force and Naval Air Arm Air Base Infrastructure*, Air Power Australia, Technical Report APA-TR-2007-0103, updated April 3, 2012. As of May 28, 2015: http://www.ausairpower.net/APA-PLA-AFBs.html

———, *Almaz S-300P/PT/PS/PMU/PMU1/PMU2; Almaz-Antey S-400 Triumf; SA-10/20/21 Grumble/Gargoyle*, Air Power Australia, Technical Report APA-TR-2006-1201, updated January 2014. As of May 28, 2015: http://www.ausairpower.net/APA-Grumble-Gargoyle.html

Kopp, Carlo, and Martin Andrew, *PLA Cruise Missiles, PLA Air-Surface Missiles*, Air Power Australia, Technical Report APA-TR-2009-0803, updated January 2014. As of May 28, 2015: http://www.ausairpower.net/APA-PLA-Cruise-Missiles.html

Krekel, Bryan, *Capability of the People's Republic of China to Conduct Cyber Warfare and Computer Network Exploitation*, McLean, Va.: Northrop Grumman Corporation, October 9, 2009.

Krebs, Gunter Kirk, "Spacecraft by Country," *Gunter's Space Page*, last updated July 17, 2013. As of May 28, 2015:
http://space.skyrocket.de/directories/sat_c.htm

Krepinevich, Andrew F., *Operation Iraqi Freedom: A First-Blush Assessment*, Washington, D.C.: Center for Strategic and Budgetary Analysis, 2003.

———, *Why Air-Sea Battle?* Washington, D.C.: Center for Strategic and Budgetary Assessment, 2010.

Krimmage, Daniel, "Up in Arms Over Iraqi Arms," *Russian Weekly*, No. 251, April 3, 2003.

Kristensen, Hans M., "New START Treaty Has New Counting," *FAS Strategic Security Blog*, March 29, 2010. As of May 29, 2015:
http://blogs.fas.org/security/2010/03/newstart

———, "Chinese Nuclear Developments Described (and Omitted) by DoD Report," *FAS Security Blog*, May 14, 2013. As of May 29, 2015:
http://blogs.fas.org/security/2013/05/china2013

Kristensen, Hans M., and Robert S. Norris, "Chinese Nuclear Forces, 2013," *Bulletin of the Atomic Scientists*, Vol. 69, No. 6, 2013.

———, "Nuclear Notebook: US Nuclear Forces, 2015," *Bulletin of the Atomic Scientists*, Vol. 71, No. 2, 2015.

Kristensen, Hans M., Robert S. Norris, and Matthew G. McKinzie, *Chinese Nuclear Forces and U.S. Nuclear War Planning*, Washington, D.C.: Federation of American Scientists and Natural Resources Defense Council, November 2006.

Krawetz, Neal, "Kylin Time," *Hacker Factor Blog*, May 23, 2009. As of May 29, 2015:
http://www.hackerfactor.com/blog/index.php?/archives/284-Kylin-Time.html

Kumar, A. Vinod, *The Dragon's Shield: Intricacies of China's BMD Capability*, New Delhi, India: Institute for Defence Studies and Analysis, February 25, 2010.

Lachow, Irving, *The Global Positioning System and Cruise Missile Proliferation: Assessing the Threat*, Discussion Paper 94-04, Cambridge, Mass.: Kennedy School of Government, Harvard University, June 1994.

LaGrone, Sam, "U.S. Navy Allowed to Use Persian Gulf Laser for Defense," *USNI News*, December 10, 2014. As of June 22, 2015:
http://news.usni.org/2014/12/10/u-s-navy-allowed-use-persian-gulf-laser-defense

Lambeth, Benjamin S., "Kosovo and the Continuing SEAD Challenge," *Air and Space Power Journal*, Vol. 16, No. 2, 2002, pp. 8–21.

Lanzit, Kevin, "Education and Training in the PLAAF," in Richard P. Hallion, Roger Cliff, and Phillip C. Saunders, eds., *The Chinese Air Force: Evolving Concepts, Roles, and Capabilities*, Washington, D.C.: National Defense University Press, 2012, pp. 235–254.

Lewis, George N., and Theodore B. Postol, "Long-Range Nuclear Cruise Missiles and Stability," *Science and Global Security*, Vol. 3, 1992, pp. 49–99.

Lewis, John Wilson, and Xue Litai, *China Builds the Bomb*, Stanford, Calif.: Stanford University Press, 1988.

———, *Imagined Enemies: China Prepares for Uncertain War*, Stanford, Calif.: Stanford University Press, 2006.

———, "Making China's Nuclear War Plan," *Bulletin of the Atomic Scientist*, Vol. 68, No. 5, September 2012, pp. 45–65.

Li Bin, "The Impact of the U.S. NMD on the Chinese Nuclear Modernization," Institute of Science and Public Affairs, China Youth College for Political Science, undated. As of June 25, 2015: http://www.emergingfromconflict.org/readings/bin.pdf

"Libya Jamming 'Exposed Vulnerability,'" BBC News, January 13, 2006. As of May 29, 2015: http://news.bbc.co.uk/2/hi/science/nature/4602674.stm

Lieggi, Stephanie, and Erik Quam, "China's ASAT Test and the Strategic Implications of Beijing's Military Space Policy," *Korean Journal of Defense Analysis*, Vol. 19, No. 1, Spring 2007, pp. 5–27.

Lim, Louisa, "Plagiarism Plague Hinders China's Scientific Ambition," National Public Radio, August 3, 2011. As of May 29, 2015:
http://www.npr.org/2011/08/03/138937778/plagiarism-plague-hinders-chinas-scientific-ambition

Lin, Jeffrey, and P. W. Singer, "China Flies Its Largest Ever Drone: The Divine Eagle," *Popular Science*, February 6, 2015. As of June 17, 2015:
http://www.popsci.com/china-flies-its-largest-ever-drone-divine-eagle

———, "Divine Eagle, China's Enormous Stealth Hunting Drone, Takes Shape," *Popular Science*, May 28, 2015. As of June 12, 2015:
http://www.popsci.com/divine-eagle-chinas-enormous-stealth-hunting-drone-takes-shape

Lindsay, Jon R., "China and Cybersecurity: Controversy and Context," in Jon R. Lindsay, Tai Ming Cheung, and Derek S. Reveron, eds., *China and Cybersecurity: Espionage, Strategy, and Politics in the Digital Domain*, London: Oxford University Press, 2015, pp. 1–28.

"LRASM Missiles: Reaching for a Long-Range Punch," *Defense Industry Daily*, last updated February 11, 2015. As of May 29, 2015:
http://www.defenseindustrydaily.com/lrasm-missiles-reaching-for-a-long-reach-punch-06752

Lu Linzhi, "Preemptive Strikes Crucial in Limited High-Tech Wars," *Jiefangjun Bao* [*PLA Daily*], February 14, 1996.

Lyle, Amaani, "Hagel: U.S. Bolstering Missile Defense," American Forces Press Service, March 15, 2013.

Lyons, Donna, "Construction Contract Awarded for USS Michael Monsoor (DDG 1001)," *Defense Media Network*, July 28, 2011. As of May 29, 2015:
http://www.defensemedianetwork.com/stories/
construction-contract-awarded-for-uss-michael-monsoor-ddg-1001/

Mahnken, Thomas G., "China's Anti-Access Strategy in Historical and Theoretical Perspective," *Journal of Strategic Studies*, Vol. 34, No. 3, June 2011, pp. 299–343.

Majumdar, Dave, "Report Reveals Undisclosed F-35 Problems," *Defense News*, January 18, 2011.

Mandiant Corporation, *APT1: Exposing One of China's Cyber Espionage Units*, Alexandria, Va., February 19, 2013.

Mann, Charles C., "The Mole in the Machine," *New York Times Sunday Magazine*, July 25, 1999, pp. 32–35.

Manson, George Patterson III, "Cyberwar: The United States and China Prepare for the Next Generation of Conflict," *Comparative Strategy*, Vol. 30, No. 2, 2011, pp. 121–133.

Markoff, John, and David Barboza, "Academic Paper in China Sets Off Alarms in U.S.," *New York Times*, March 20, 2010.

May, Michael M., George F. Bing, and John D. Steinbruner, "Strategic Arsenals After START: The Implications of Deep Cuts," *International Security*, Vol. 13, No. 1, Summer 1988, pp. 90–133.

McCauley, Kevin, "The PLA's Three-Pronged Approach to Achieving Jointness in Command and Control," *China Brief*, Vol. 12, No. 6, March 2012, pp. 12–16.

McDonough, David, "Unveiled: China's New Naval Base in the South China Sea," *The National Interest*, March 20, 2015.

McVadon, Eric A., "The Reckless and the Resolute: Confrontation in the South China Sea," *China Security*, Vol. 5, No. 2, Spring 2009, pp. 1–15.

Mearsheimer, John J., "Numbers, Strategy, and the European Balance," *International Security*, Vol. 12, No. 4, Spring 1988, pp. 174–185.

Medeiros, Evan S., Keith Crane, Eric Heginbotham, Norman D. Levin, Julia F. Lowell, Angel Rabasa, and Somi Seong, *Pacific Currents: The Responses of U.S. Allies and Security Partners in East Asia to China's Rise*, Santa Monica, Calif.: RAND Corporation, MG-736-AF, 2008. As of May 29, 2015:
http://www.rand.org/pubs/monographs/MG736.html

Mehuron, Tamar A., "2009 Space Almanac: The US Military Space Operation in Facts and Figures," *Air Force Magazine*, August 2009. As of June 23, 2015:
http://www.space-library.com/0908AFM_SpaceAlmanac.pdf

Military.com, "SSN774 Virginia-Class Fast Attack Submarine," web page, undated. As of May 29, 2015:
http://tech.military.com/equipment/view/138675/ssn774-virginia-class-fast-attack-submarine.html

Mills, Elinor, "Facebook Detour Through China: Accident or Not?" *CNet*, March 24, 2011. As of May 29, 2015:
http://news.cnet.com/8301-27080_3-20046338-245.html

———, "Web Traffic Redirected to China in Mystery Mix-Up," *CNet*, March 25, 2010. As of May 29, 2015:
http://news.cnet.com/8301-27080_3-20001227-245.html

Miller, Edward S., *War Plan Orange: The U.S. Strategy to Defeat Japan, 1897–1945*, Annapolis, Md.: Naval Institute Press, 1991.

Minnick, Wendell, "Russia: No Deal on Sale of Fighters, Subs to China," *Defense News*, March 25, 2013. As of May 29, 2015:
http://www.defensenews.com/article/20130325/DEFREG03/303250014/

Ministry of Foreign Affairs of the People's Republic of China, "China and the World Meteorological Organization (WMO)," April 3, 2012.

Missile Defense Agency, "Terminal High Altitude Area Defense," fact sheet, May 2014. As of June 19, 2015:
http://www.mda.mil/global/documents/pdf/thaad.pdf

"More Problems Cited in F-35 JSF Program," United Press International, April 1, 2013.

Morgan, Forrest E., *Deterrence and First-Strike Stability in Space: A Preliminary Assessment*, Santa Monica, Calif.: RAND Corporation, MG916-AF, 2010. As of May 29, 2015:
http://www.rand.org/pubs/monographs/MG916.html

Mozur, Oaul, and Shanshan Wang, "China, a Fish Barrel for Cybercriminals," *New York Times*, December 2, 2014. As of June 25, 2015:
http://bits.blogs.nytimes.com/2014/12/02/china-a-fish-barrel-for-cybercriminals/?_r=0

Mulvenon, James, "The PLA and Information Warfare," in James Mulvenon and Richard Yang, eds., *The People's Liberation Army in the Information Age*, Santa Monica, Calif.: RAND Corporation, CF-145-CAPP/AF, 1999. As of May 29, 2015:
http://www.rand.org/pubs/conf_proceedings/CF145.html

———, "Chinese Information Operations Strategies in a Taiwan Contingency," testimony before the U.S.-China Economic and Security Review Commission Hearing, Washington, D.C., September 15, 2005.

———, "PLA Computer Network Operations: Scenarios, Doctrines, Organizations, and Capability," in Roy Kamphausen, David Lai, and Andrew Scobell, eds., *Beyond the Strait: PLA Missions Other Than Taiwan*, Carlisle, Pa.: Strategic Studies Institute, U.S. Army War College, 2009, pp. 253–285.

———, "Chairman Hu and the PLA's 'New Historic Missions,'" *China Leadership Monitor*, No. 27, January 2009. As of May 29, 2015:
http://www.hoover.org/research/chairman-hu-and-plas-new-historic-missions

Mulvenon, James, and David M. Finkelstein, eds., *China's Revolution in Doctrinal Affairs: Emerging Trends in the Operational Art of the Chinese People's Liberation Army*, Alexandria, Va.: Center for Naval Analyses, 2005.

Mulvenon, James, and Andrew N. D. Yang, *The People's Liberation Army as Organization: Reference Volume v1.0*, Santa Monica, Calif.: RAND Corporation, CF-182-NSRD, 2002. As of June 5, 2015:
http://www.rand.org/pubs/conf_proceedings/CF182.html

Muncaster, Phil, "China Is the World's Most Malware-Ridden Nation," *The Register*, February 7, 2013. As of May 29, 2015:
http://www.theregister.co.uk/2013/02/07/panda_china_most_infected_pcs

Murray, William S., "Revisiting Taiwan's Defense Strategy," *Naval War College Review*, Vol. 61, No. 3, Summer 2008, pp. 13–38.

Myers, Steven Lee, "U.S. Seeks to Curb Israeli Arms Sales to China," *New York Times*, November 11, 1999.

n2yo.com, *Satellite Database*, data collected December 1, 2011. As of May 29, 2015:
http://www.n2yo.com/database

Nalebuff, Barry, "Minimal Nuclear Deterrence," *Journal of Conflict Resolution*, Vol. 32, No. 3, September 1988, pp. 411–425.

National Aeronautics and Space Administration, "Old and New Satellite Breakups Identified," *Orbital Debris Quarterly News*, Vol. 14, No. 2, 2010, pp. 3–4. As of June 26, 2015:
http://orbitaldebris.jsc.nasa.gov/newsletter/pdfs/ODQNv14i2.pdf

National Air and Space Intelligence Center, *Ballistic and Cruise Missile Threat*, Wright-Patterson AFB, Ohio, NASIC-1031-0985-09, 2009.

———, *People's Liberation Army Air Force 2010*, Wright-Patterson AFB, Ohio, August 1, 2010.

National Bureau of Statistics of China, *China Statistics Yearbook*, Beijing, 2006.

National Defense Panel, *Transforming Defense: National Security in the 21st Century*, Washington, D.C., December 1997.

National Intelligence Council, *Foreign Missile Developments and the Ballistic Missile Threat Through 2015: An Unclassified Summary of a National Intelligence Estimate*, Washington, D.C., December 2001.

National Priorities Project, "Cost of National Security Counters," 2015. As of June 11, 2015:
https://www.nationalpriorities.org/cost-of/resources/notes-and-sources/

National Science Board, *Science and Engineering Indicators: 2010*, Arlington, Va.: National Science Foundation, 2010. As of May 29, 2015:
http://www.nsf.gov/statistics/seind10/pdf/front.pdf

National Security Agency, Central Security Service, "U.S. Military Academy Wins NSAs 14th Annual Cyber Defense Exercise," press release, April 11, 2014.

National Space Policy of the United States of America, Washington, D.C.: White House, June 28, 2010. As of May 29, 2015:
http://www.whitehouse.gov/sites/default/files/national_space_policy_6-28-10.pdf

National Space Studies Center, *AU-18 Space Primer*, Maxwell, Ala.: Air University Press, September 2009.

Naughton, Barry, "The Third Front: Defence Industrialization in the Chinese Interior," *China Quarterly*, Vol. 115, September 1988, pp. 351–386.

Naval Air Systems Command, "MQ-4C Triton," undated. As of June 18, 2015:
http://www.navair.navy.mil/index.cfm?fuseaction=home.displayPlatform&key=F685F52A-DAB8-43F4-B604-47425A4166F1

Naval Network Warfare Command Public Affairs, "Navy Transitions to Wideband Global System," *CHIPS Magazine*, Vol. 26, No. 3, September 2008, p. 43. As of May 29, 2015:
http://www.doncio.navy.mil/uploads/0131XHF11967.pdf

Nicoll, Alexander, and Sarah Johnstone, eds., "Behind Recent Gunboat Diplomacy in the South China Sea," *IISS Strategic Comments*, Vol. 17, No. 28, August 2011.

Nielson, J. T., "CALCM: The Untold Story of the Weapon Used to Start the Gulf War," *Aerospace and Electronic Systems Magazine*, Vol. 9, No. 7, July 1994, pp. 18–22.

Norris, Robert S., and William M. Arkin, "U.S. Strategic Nuclear Forces, End of 1995," *Bulletin of the Atomic Scientists*, Vol. 52, No. 1, January–February 1996, pp. 62–63.

Norris, Robert S., William M. Arkin, Hans M. Kristensen, and Joshua Handler, "U.S. Nuclear Forces, 2002," *Bulletin of the Atomic Scientists*, Vol. 58, No. 3, May 2002, pp. 70–75.

O'Connor, Sean, "Dragon's Fire: The PLA's 2nd Artillery Corps," *IMINT and Analysis Blog*, June 26, 2010. May 29, 2015:
http://geimint.blogspot.com/2009/04/dragons-fire-plas-2nd-artillery-corps.html

———, "The PLA's Second Artillery Corps," *IMINT and Analysis*, Vol. 1, No. 11, December 2011, pp. 1–51.

———, *PLA Second Artillery Corps*, Air Power Australia, Technical Report APA-TR-2009-1204, April 2012. As of May 29, 2015:
http://www.ausairpower.net/APA-PLA-Second-Artillery-Corps.html

———, "Worldwide SAM Site Overview," *IMINT and Analysis*, June 2, 2013. As of May 29, 2015:
http://geimint.blogspot.com/2008/06/worldwide-sam-site-overview.html

Office of Management and Budget, *Terminations, Reductions, and Savings, Budget of the U.S. Government, FY 2010*, Washington, D.C., 2010.

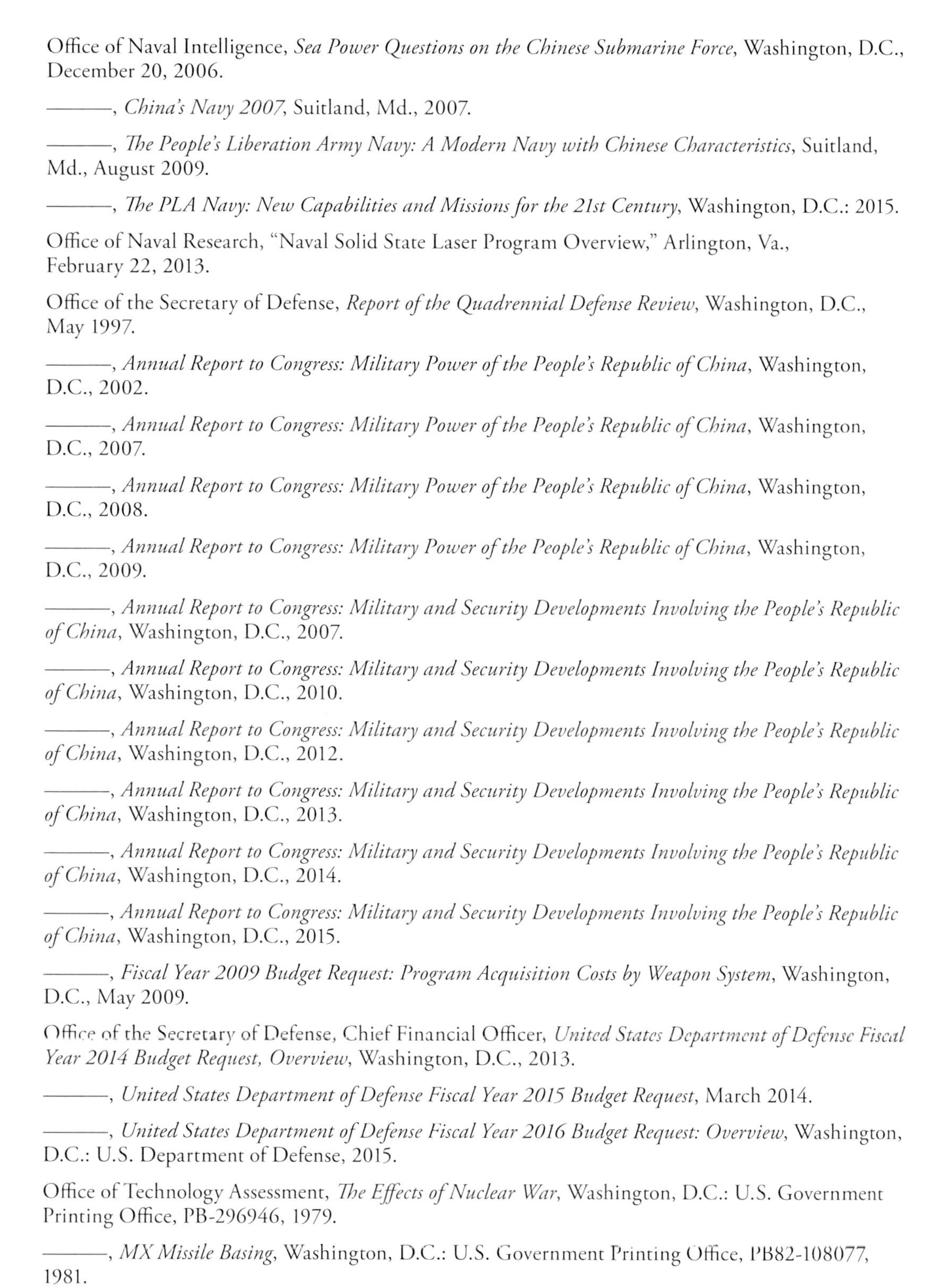

Office of Naval Intelligence, *Sea Power Questions on the Chinese Submarine Force*, Washington, D.C., December 20, 2006.

———, *China's Navy 2007*, Suitland, Md., 2007.

———, *The People's Liberation Army Navy: A Modern Navy with Chinese Characteristics*, Suitland, Md., August 2009.

———, *The PLA Navy: New Capabilities and Missions for the 21st Century*, Washington, D.C.: 2015.

Office of Naval Research, "Naval Solid State Laser Program Overview," Arlington, Va., February 22, 2013.

Office of the Secretary of Defense, *Report of the Quadrennial Defense Review*, Washington, D.C., May 1997.

———, *Annual Report to Congress: Military Power of the People's Republic of China*, Washington, D.C., 2002.

———, *Annual Report to Congress: Military Power of the People's Republic of China*, Washington, D.C., 2007.

———, *Annual Report to Congress: Military Power of the People's Republic of China*, Washington, D.C., 2008.

———, *Annual Report to Congress: Military Power of the People's Republic of China*, Washington, D.C., 2009.

———, *Annual Report to Congress: Military and Security Developments Involving the People's Republic of China*, Washington, D.C., 2007.

———, *Annual Report to Congress: Military and Security Developments Involving the People's Republic of China*, Washington, D.C., 2010.

———, *Annual Report to Congress: Military and Security Developments Involving the People's Republic of China*, Washington, D.C., 2012.

———, *Annual Report to Congress: Military and Security Developments Involving the People's Republic of China*, Washington, D.C., 2013.

———, *Annual Report to Congress: Military and Security Developments Involving the People's Republic of China*, Washington, D.C., 2014.

———, *Annual Report to Congress: Military and Security Developments Involving the People's Republic of China*, Washington, D.C., 2015.

———, *Fiscal Year 2009 Budget Request: Program Acquisition Costs by Weapon System*, Washington, D.C., May 2009.

Office of the Secretary of Defense, Chief Financial Officer, *United States Department of Defense Fiscal Year 2014 Budget Request, Overview*, Washington, D.C., 2013.

———, *United States Department of Defense Fiscal Year 2015 Budget Request*, March 2014.

———, *United States Department of Defense Fiscal Year 2016 Budget Request: Overview*, Washington, D.C.: U.S. Department of Defense, 2015.

Office of Technology Assessment, *The Effects of Nuclear War*, Washington, D.C.: U.S. Government Printing Office, PB-296946, 1979.

———, *MX Missile Basing*, Washington, D.C.: U.S. Government Printing Office, PB82-108077, 1981.

Office of the Under Secretary of Defense for Acquisition, Technology, and Logistics, *Report to Congress on Assessment of the Ground-Based Midcourse Defense Element of the Ballistic Missile Defense System*, Washington, D.C., May 2010. As of May 29, 2015:
http://missiledefenseadvocacy.org/wp-content/uploads/2015/04/2010-DOD-GMD-Report.pdf

O'Rourke, Ronald, *Cruise Missile Inventories and NATO Attacks on Yugoslavia: Background Information*, Washington, D.C.: Congressional Research Service, 1999.

———, *Navy Force Structure and Shipbuilding Plans: Background and Issues for Congress*, Washington, D.C.: Congressional Research Service, December 10, 2012.

———, *Naval Shipboard Lasers for Surface, Air, and Missile Defense: Background and Issues for Congress*, Washington, D.C.: Congressional Research Service, June 27, 2013.

———, *China Naval Modernization: Implications for U.S. Navy Capabilities*, Washington, D.C.: Congressional Research Service, August 8, 2013.

———, *Navy Ford (CVN-78) Class Aircraft Carrier Program: Background and Issues for Congress*, Washington, D.C.: Congressional Research Service, March 24, 2015.

———, *Navy Virginia (SSN-774) Class Attack Submarine Procurement: Background and Issues for Congress*, Washington, D.C.: Congressional Research Service, June 1, 2015.

———, *Navy AEGIS Ballistic Missile Defense (BMD) Program: Background and Issues for Congress*, Washington, D.C.: Congressional Research Service, June 12, 2015.

———, *Navy DDG-51 and DDG-1000 Destroyer Programs: Background and Issues for Congress*, Washington, D.C.: Congressional Research Service, June 12, 2015.

Panda Security, *PandaLabs Annual Report 2014*, Maitland, Fla.: Panda Security, 2015.

Parsch, Andreas, "BLU-42/B," *Designations of U.S. Aeronautical and Support Equipment*, last updated October 23, 2012. As of June 2, 2015:
http://www.designation-systems.net/usmilav/asetds/u-b.html#_BLU42

Pearlman, Michael, Carey Noll, Jan McGarry, Werner Gurtner, and Erricos Pavlis, "The International Laser Ranging Service," undated. As of June 2, 2015:
http://ilrs.gsfc.nasa.gov/docs/AOGS_ILRS_0708.pdf

Peebles, Curtis, *High Frontier: The United States Air Force and the Military Space Program*, Washington, D.C.: Air Force History and Museums Program, 1997.

Peifer, Douglas C., "China, the German Analogy, and the New Air-Sea Operational Concept," *Orbis*, Vol. 55, No. 1, 2011, pp. 114–131.

中国人民解放军空军 [People's Liberation Army Air Force], 中国空军百科全书 [*China Air Force Encyclopedia*], Beijing: 航空工业出版社 [Aviation Industry Press], 2005.

Pickler, Nedra, "US May Split Command of Spy and Cyber Agencies," Associated Press, November 7, 2013.

Pietrucha, Michael W., "The Comanche and the Albatross: About Our Neck Was Hung," *Air and Space Power Journal*, May–June 2014. As of June 12,2015:
http://www.airpower.maxwell.af.mil/digital/pdf/articles/2014-May-Jun/F-Pietrucha.pdf

中国人民解放军第二炮兵部队 [PLA Second Artillery], 第二炮兵战役学 [*Science of Second Artillery Campaigns*], Beijing: People's Liberation Army Press, 2004.

Planeman, "Bluffer's Guide: Fortress China: Main Area Defense Systems," discussion forum posts, *SinoDefence.com*, February 3, 2009.

Pollack, Jonathon D., and Dennis J. Blasko, "Is China Preparing for a 'Short, Sharp War' Against Japan?" Brookings Institution, February 25, 2014.

Pollpeter, Kevin, "Towards an Integrative C4ISR System: Informationization and Joint Operations in the People's Liberation Army," in Roy Kamphausen, David Lai, and Andrew Scobell, eds., *The PLA at Home and Abroad: Assessing the Operational Capabilities of China's Military*, Carlisle, Pa.: Strategic Studies Institute, U.S. Army War College, 2010, pp. 193–235.

Polmer, Norman, *Naval Institute Guide to the Ships and Aircraft of the U.S. Fleet*, Annapolis, Md.: Naval Institute Press, 2005.

———, "Chinese Writings on Cyberwarfare and Coercion," in Jon R. Lindsay, Tai Ming Cheung, and Derek S. Reveron, eds., *China and Cybersecurity: Espionage, Strategy, and Politics in the Digital Domain*, London: Oxford University Press, 2015, pp. 138–162.

Posen, Barry R., "Measuring the European Conventional Balance," *International Security*, Vol. 9, No. 3, Winter 1984–1985, pp. 47–88.

———, "Is NATO Decisively Outnumbered?" *International Security*, Vol. 12, No. 4, Spring 1988, pp. 186–202.

———, *Inadvertent Escalation: Conventional War and Nuclear Risks*, Ithaca, N.Y.: Cornell University Press, 1992.

Powell, Colin, "Building the Base Force: National Security for the 1990s and Beyond," annotated briefing, September 1990.

Pradun, Vitaliy O., "From Bottle Rockets to Lightning Bolts: China's Missile Revolution and PLA Strategy Against U.S. Military Intervention," *Naval War College Review*, Vol. 64, No. 2, Spring 2011, pp. 7–39.

Puttre, Michael, "Facing the Shoulder-Fired Threat," *Journal of Electronic Defense*, Vol. 24, No. 4, April 2001, pp. 38–46.

Putz, Catherine, "Sold: Russian S-400 Missile Defense System to China," *The Diplomat*, April 14, 2015. As of June 12, 2015:
http://thediplomat.com/2015/04/sold-russian-s-400-missile-defense-systems-to-china/

PwC Technology Institute, *PwC Global 100 Software Leaders: The Growing Importance of Apps and Services*, March 2014. As of June 24, 2015:
https://www.pwc.com/gx/en/technology/assets/pwc-global-100-software-leaders-2014.pdf

Qiao Liang, and Wang Xiangsui, *Unrestricted Warfare: China's Master Plan to Destroy America*, Panama City: Pan American Publishing Company, 2002.

Rajagopalan, Megha, "Eyeing Exports, China Steps Up Research into Military Drones," Reuters, April 29, 2015. As of June 18, 2015:
http://www.reuters.com/article/2015/04/29/us-china-drones-idUSKBN0NK2SK20150429

Rauhala, Emily, "Activists Are Finding New Ways Around China's Great Firewall," *Time*, November 21, 2013. As of June 25, 2014:
http://world.time.com/2013/11/21/activists-are-finding-new-ways-around-chinas-great-firewall

Richardson, Doug, "China Plans 4,000 km–Range Conventional Ballistic Missile," *Jane's Missiles and Rockets*, March 1, 2011.

———, "China Deploys DF-16 Ballistic Missile, Claims Taiwan," *Jane's Missiles and Rockets*, March 21, 2011.

Rip, Michael Russell, and James M. Hasik, *The Precision Revolution: GPS and the Future of Aerial Warfare*, Annapolis, Md.: U.S. Naval Institute Press, 2002.

Rogov, Sergey M., "The Evolution of Strategic Stability and the Future of Nuclear Weapons," *Contemporary Security Policy*, Vol. 14, No. 2, 1993, pp. 5–22.

"Russia Arms Deal to Supply China with S-400 Air Defense Systems," *Sputnik International*, April 13, 2015. As of June 12, 2015:
http://sputniknews.com/world/20150413/1020809219.html

Samaan, Jean-Loup, "Beyond the Rift in Cyber Strategy," *Strategic Insights*, Vol. 10, No. 1, Spring 2011, pp. 4–14.

Sanger, David E., *Confront and Conceal: Obama's Secret Wars and Surprising Use of American Power*, New York: Random House, 2012.

———, "U.S. Blames China's Military Directly for Cyberattacks," *New York Times*, May 6, 2013.

Saunders, Phillip C., and Joshua K. Wiseman, "China's Quest for Advanced Aviation Technologies," in Richard P. Hallion, Roger Cliff, and Phillip C. Saunders, eds., *The Chinese Air Force: Evolving Concepts, Roles, and Capabilities*, Washington, D.C.: National Defense University Press, 2012, pp. 271–324.

Saunders, Phillip C., Christopher Yung, Michael Swaine, and Andrew Nien-Dzu Yang, eds., *The Chinese Navy: Expanding Capabilities, Evolving Roles*, Washington, D.C.: National Defense University Press, 2011.

Schactman, Noah, *Pirates of the ISPs: Tactics for Turning Online Crooks into International Pariahs*, Washington, D.C.: Brookings Institution Press, 2011.

Schneier, Bruce, *Crypto-Gram Newsletter*, May 15, 2002. As of June 2, 2015:
https://www.schneier.com/crypto-gram-0205.html

SCImago Journal and Country Rank, home page, 2015. As of June 24, 2015:
http://www.scimagojr.com/index.php

Scobell, Andrew, "Is There a Civil-Military Gap in China's Peaceful Rise?" *Parameters*, Vol. 39, No. 2, Summer 2009, pp. 4–22.

Scott, Richard, "Surface Navy 2015: SM-3 Block IIA Program Set of CTV Test," *Jane's 360*, January 13, 2015.

Seidman, Laurence S., "Crisis Stability," *Journal of Conflict Resolution*, Vol. 34, No. 1, March 1990, pp. 130–150.

Shalal-Esa, Andrea, "U.S. Sees China Launch as Test of Anti-Satellite Muscle: Source," Reuters, May 16, 2013.

Shambaugh, David, Modernizing China's Military: Progress, Problems, and Prospects, Berkeley, Calif.: University of California Press, 2002.

Shanker, Thom, David E. Sanger, and Martin Fackler, "U.S. Is Bolstering Missile Defense to Deter North Korea," *New York Times*, March 15, 2013. As of June 25, 2015:
http://www.nytimes.com/2013/03/16/world/asia/us-to-bolster-missile-defense-against-north-korea.html

Shear, David, Assistant Secretary of Defense for Asian and Pacific Security Affairs, testimony before the Senate Committee on Foreign Relations at the hearing "Safeguarding American Interests in the South and East China Seas," Washington, D.C., May 13, 2015.

Sheldon, Robert, and Joe McReynolds, "Civil-Military Integration and Cybersecurity: A Study of Chinese Information Warfare Militias," in Jon R. Lindsay, Tai Ming Cheung, and Derek S. Reveron, eds., *China and Cybersecurity Espionage, Strategy, and Politics in the Digital Domain*, London: Oxford University Press, 2015, pp. 188–224.

Shi, Ting, "China Says Third Missile-Defense Test in Four Years Successful," *Bloomberg Business*, July 24, 2014. As of June 22, 2014:
http://www.bloomberg.com/news/articles/2014-07-24/china-says-third-missile-defense-test-in-four-years-successful

"Ships, Sensors, and Weapons: Undersea Warfare Programs Target an Expeditionary Future," *Undersea Warfare*, Vol. 3, No. 3, 2001.

Shlapak, David A., "Equipping the PLAAF: The Long March to Modernity," in Richard P. Hallion, Roger Cliff, and Phillip C. Saunders, eds., *The Chinese Air Force: Evolving Concepts, Roles, and Capabilities*, Washington, D.C.: National Defense University Press, 2012, pp. 191–212.

Shlapak, David A., David T. Orletsky, Toy I. Reid, Murray Scot Tanner, and Barry Wilson, *A Question of Balance: Political Context and Military Aspects of the China-Taiwan Dispute*, Santa Monica, Calif.: RAND Corporation, MG-888-SRF, 2009. As of June 2, 2015:
http://www.rand.org/pubs/monographs/MG888.html

Shlapak, David A., David T. Orletsky, and Barry A. Wilson, *Dire Strait? Military Aspects of the China-Taiwan Confrontation and Options for U.S. Policy*, Santa Monica, Calif.: RAND Corporation, MR-1217-SRF, 2000. As of June 2, 2015:
http://www.rand.org/pubs/monograph_reports/MR1217.html

寿晓松 [Shou Xiaosong], ed., 《战略学》 [The Science of Military Strategy], 3rd ed., Beijing: Military Science Press, 2013.

Sim, Chow Yen Desmond, *The Propagation of VHF and UHF Radio Waves Over Sea Paths*, thesis, Leicester, UK: Leicester University, November 2002.

SinoDefence.com, "KongJing-2000 Airborne Warning and Control System," web page, last updated December 15, 2013. As of June 2, 2015:
http://www.sinodefence.com/airforce/specialaircraft/kj2000.asp

———, "Project 956/EM Sovremenny Class Missile Destroyer," web page, last updated February 28, 2009. No longer available.

Sklar, Marc, "Taking Aim at the Future," *Boeing Frontiers*, August 2008.

"SM-3 BMD, in from the Sea: EPAA and Aegis Ashore," *Defense Industry Daily*, last updated October 13, 2014. As of June 2, 2015:
http://www.defenseindustrydaily.com/Land-Based-SM-3s-for-Israel-04986

Smith, Charles, "China Take's Aim at U.S. GPS," *Newsmax*, November 20, 2007.

Snakenberg, Mark K., "Junior Leader PME in the PLA: Implications for the Future," *Joint Force Quarterly*, Vol. 62, No. 3, 3rd Quarter 2011, pp. 104–109.

Solomon, Jonathan F., *Defending the Fleet from China's Anti-Ship Ballistic Missile: Naval Deception's Role in Sea-Based Missile Defense*, thesis, Washington, D.C.: Georgetown University, April 15, 2011.

Space-Track.org, homepage, undated. As of June 2, 2015 (requires login):
https://www.space-track.org/auth/login

Stares, Paul B., *The Militarization of Space: U.S. Policy, 1945–1984*, Ithaca, N.Y.: Cornell University Press, 1985.

State Council Information Office of the People's Republic of China, *China's Military Strategy*, Beijing, May 2015.

Sternstein, Aliya, "Threat of Destructive Coding on Foreign-Manufactured Technology Is Real," *Nextgov*, July 7, 2011. As of June 2, 2015:
http://www.nextgov.com/cybersecurity/2011/07/
threat-of-destructive-coding-on-foreign-manufactured-technology-is-real/49363

———, "The Military's Cybersecurity Budget in 4 Charts," *Defense One*, March 16, 2015. As of June 23, 2015:
http://www.defenseone.com/management/2015/03/militarys-cybersecurity-budget-4-charts/107679

Stillion, John, and David T. Orletsky, *Airbase Vulnerability to Conventional Cruise-Missile and Ballistic-Missile Attacks: Technology, Scenarios, and U.S. Air Force Responses*, Santa Monica, Calif.: RAND Corporation, MR-1028-AF, 1999. As of June 2, 2015:
http://www.rand.org/pubs/monograph_reports/MR1028.html

Stockholm International Peace Research Institute (SIPRI), "SIPRI Arms Transfers Database," 2015. As of June 11, 2015:
http://www.sipri.org/databases/armstransfers

Stokes, Mark A., *China's Strategic Modernization: Implications for the United States*, Carlisle, Pa.: Strategic Studies Institute, U.S. Army War College, September 1999.

———, *China's Evolving Conventional Strategic Strike Capability: The Anti-Ship Ballistic Missile Challenge to U.S. Maritime Operations in the Western Pacific and Beyond*, Arlington, Va.: Project 2049 Institute, September 2009.

———, *China's Nuclear Warhead Storage and Handling System*, Arlington, Va.: Project 2049 Institute, March 12, 2010.

———, "The Chinese People's Liberation Army Computer Network Operations Infrastructure," in Jon R. Lindsay, Tai Ming Cheung, and Derek S. Reveron, eds., *China and Cybersecurity: Espionage, Strategy, and Politics in the Digital Domain*, London: Oxford University Press, 2015, pp. 163–187.

Stokes, Mark A., and Ian Easton, *Evolving Aerospace Trends in the Asia-Pacific Region: Implications for Stability in the Taiwan Strait and Beyond*, Arlington, Va.: Project 2049 Institute, May 2010.

Strategic Defense Intelligence, "U.S. Awards Communications Systems Upgrade Contract to Harris," November 12, 2012.

"Striking Out at Sea," *Jane's Defence Weekly*, October 10, 2012.

Strohm, Chris, and Michael A. Riley, "China-Linked Hacking Foiled by Private-Sector Sleuthing," *Bloomberg Business*, October 28, 2014. As of June 24, 2015:
http://www.bloomberg.com/news/articles/2014-10-28/china-linked-hacking-foiled-by-private-sector-sleuthing

Sullivan, Brian, ed., *Project 2015: Power and Progress*, Washington, D.C.: National Defense University Press, 1996.

Sullivan, Michael J., *Joint Strike Fighter: Restructuring Added Resources and Reduced Risk, but Concurrency Is Still a Major Concern*, statement before the Subcommittee on Tactical Air and Land Forces, Committee on Armed Services, U.S. House of Representatives, Washington, D.C., GAO-12-525T, March 20, 2012.

Swaine, Michael D., and M. Taylor Fravel, "China's Assertive Behavior, Part Two: The Maritime Periphery," *China Leadership Monitor*, No. 35, September 2011. As of June 5, 2015:
http://carnegieendowment.org/files/CLM35MS.pdf

"Telecom Experts in Guangzhou Doubling as Militia Information Warfare Elements," *Guofang*, PLA Academy of Military Science, September 15, 2003.

Tellis, Ashley J., "China's Military Space Strategy," *Survival*, Vol. 49, No. 3, September 2007, pp. 41–72.

Tellis, Ashley J., and Travis Tanner, eds., *Strategic Asia 2012–13: China's Military Challenge*, Seattle, Wash.: National Bureau of Asian Research, 2012.

Thomas, Timothy L., "China's Electronic Long-Range Reconnaissance," *Military Review*, November–December 2008, pp. 47–54.

Thornborough, Anthony M., and Frank B. Mormillo, *Iron Hand: Smashing the Enemy's Air Defenses*, London: JH Haynes and Co., 2002.

"Thuraya Satellite Telecom Says Jammed by Libya," Reuters, February 24, 2011.

Tirpak, John A., "Here Comes Adversary Stealth," *Air Force Magazine*, Vol. 95, No. 12, December 2012.

Tucholski, Edward, "Regions of the Sound Velocity Profile," course handout for SP411, Underwater Acoustics and Sonar, U.S. Naval Academy, Annapolis, Md., undated. As of June 5, 2015: http://usna.edu/Users/physics/ejtuchol/documents/SP411/Chapter5.pdf

Turnbull, Grant, "The P-8 Poseidon Adventure: Delivering a New-Era of Maritime Aircraft," *Naval-Technology.com*, January 28, 2014. As of June 18, 2015: http://www.naval-technology.com/features/feature-p-8-poseidon-adventure-new-era-maritime-aircraft

Twing, Shawn L., "U.S. Defense Intelligence Agency Report Accuses Israel of Laser Technology Transfer to China," *Washington Report on Middle East Affairs*, April–May 1999, pp. 44–45.

Twomey, Christopher P., and Taylor M. Fravel, "Projecting Strategy: The Myth of Chinese Counter-Intervention," *Washington Quarterly*, Vol. 37, No. 4, 2015, pp. 171–187. As of June 8, 2015: https://twq.elliott.gwu.edu/sites/twq.elliott.gwu.edu/files/downloads/Fravel_Twomey.pdf

Union of Concerned Scientists, Exhibit R-2, RDT&E Budget Item Justification, PE No. 0604421F, "Counterspace Systems," February 2004. As of June 5, 2015: http://www.dtic.mil/descriptivesum/Y2005/AirForce/0604421F.pdf

———, UCS Satellite Database, data collected at various points through August 31, 2013. As of June 5, 2015: http://www.ucsusa.org/nuclear_weapons_and_global_security/space_weapons/technical_issues/ucs-satellite-database.html

"United States Seeks Access to Philippine Bases as Part of Asian Pivot," Reuters, April 25, 2015.

U.S. Air Force, Exhibit R-2, RDT&E Budget Item Justification, PE No. 0604421F, "Counterspace Systems," February 2004.

———, "EC-130H Compass Call," fact sheet, May 27, 2005. As of July 15, 2015: http://www.af.mil/AboutUs/FactSheets/Display/tabid/224/Article/104550/ec-130h-compass-call.aspx

———, *Airfield Damage Repair Operations*, Air Force Pamphlet 10-219, Vol. 4, May 28, 2008.

———, "Global Positioning System," fact sheet, August 2010. As of June 5, 2015: http://www.afspc.af.mil/library/factsheets/factsheet_print.asp?fsID=4856&page=1

———,"Starfire Optical Range at Kirtland Air Force Base," fact sheet, March 9, 2012. As of June 5, 2015: http://www.kirtland.af.mil/library/factsheets/factsheet.asp?id=15868

———, *Fiscal Year 2016 Budget Overview*, Washington, D.C.: U.S. Department of Defense, 2015. As of June 8, 2015: http://www.saffm.hq.af.mil/shared/media/document/AFD-150421-011.pdf

———, "Sixth Wideband Global SATCOM Satellite Launched," press release, August 7, 2013.

U.S.-China Economic and Security Review Commission, *2014 Report to Congress*, Washington, D.C.: U.S. Government Printing Office, 2014.

U.S. Department of Defense, *Airfield Damage Repair*, Unified Facilities Criteria 3-270-07, draft, Washington, D.C., June 30, 2003.

———, *Quadrennial Defense Review Report*, Washington, D.C., February 6, 2006.

———, *Fiscal Year 2009 Budget Request: Program Acquisition Costs by Weapon Syste*m, Washington, D.C., May 2009.

———, *Ballistic Missile Defense Review Report*, Washington, D.C., February 2010.

———, *Nuclear Posture Review Report*, Washington, D.C., April 2010.

———, *Sustaining U.S. Global Leadership: Priorities for 21st Century Defense*, Washington, D.C., January 2012.

———, *Joint Operational Access Concept (JOAC)*, version 1.0, Washington, D.C., January 17, 2012.

———, *The Guidelines for U.S.-Japan Defense Cooperation*, Washington, D.C., April 27, 2015. As of June 11, 2015:
http://www.defense.gov/pubs/20150427_--_GUIDELINES_FOR_US-JAPAN_DEFENSE_COOPERATION.pdf

———, *The Department of Defense Cyber Strategy*, Washington, D.C., April 2015. As of June 25, 2015:
http://www.defense.gov/home/features/2015/0415_cyber-strategy/Final_2015_DoD_CYBER_STRATEGY_for_web.pdf

U.S. Department of Justice, "U.S. Charges Five Chinese Military Hackers for Cyber Espionage Against U.S. Corporations and a Labor Organization for Commercial Advantage," press release, Washington, D.C., May 19, 2014.

U.S. Department of State, *Transparency in the U.S. Nuclear Weapons Program*, Washington, D.C., April 29, 2014.

U.S. Department of State, Bureau of Arms Control, Verification, and Compliance, *New START Treaty Aggregate Numbers of Strategic Offensive Arms*, fact sheet, Washington, D.C., April 1, 2014.

U.S. General Accounting Office, *Combat Air Power: Funding Priority for Suppression of Enemy Air Defenses May Be Too Low*, Washington, D.C., GAO/NSIAD-96-128, April 10, 1996.

———, *Precision-Guided Munitions: Acquisition Plans for the Joint Air-to-Surface Standoff Missile*, Washington, D.C., GAO/NSIAD-96-144, June 28, 1996.

———, *Military Training: Limitations Exist Overseas but Are Not Reflected in Readiness Reporting*, Washington, D.C., GAO-02-525, April 30, 2002a.

———, *Electronic Warfare: Comprehensive Strategy Still Needed for Suppression of Enemy Air Defenses*, Washington, D.C., GAO-03-51, November 25, 2002b.

U.S. Government Accountability Office, *Electronic Warfare: Comprehensive Strategy: Actions Needed to Improve Operating Cost Estimates and Mitigate Risks in Implementing New Concepts*, Washington, D.C., GAO-10-257, February 2, 2010.

———, *Defense Acquisitions: Assessments of Selected Weapon Programs*, Washington, D.C., GAO-13-294SP, March 28, 2013.

———, *F-35 Joint Strike Fighter: Restructuring Has Improved the Program, but Affordability Challenges and Other Risks Remain*, GAO-13-690T, Washington, D.C., June 19, 2013.

———, *F-35 Sustainment: Need for Affordable Strategy, Greater Attention to Risks, and Improved Cost Estimates*, GAO-14-778, Washington, D.C., September 23, 2014.

U.S. Joint Chiefs of Staff, *Joint Vision 2020*, Washington, D.C.: U.S. Government Printing Office, 2000.

———, *Space Operations*, Joint Publication 3-14, Washington, D.C., May 29, 2013.

———, *DoD Dictionary of Military Terms*, Joint Publication 1-02, Washington, D.C., as amended through March 15, 2015. As of June 5, 2015:
http://www.dtic.mil/doctrine/dod_dictionary

U.S. Marine Corps, *Marine Aviation Plan 2015*, Washington, D.C., 2014.

"US MDA to Take Delivery of First CE II Block 1 Interceptor by End of December," *Jane's Defence Weekly*, December 16, 2014.

"U.S. Military Creating Cyberwarfare Teams," United Press International, March 13, 2013. As of June 5, 2015:
http://www.upi.com/Top_News/US/2013/03/13/US-military-creating-cyberwarfare-teams/UPI-22631363156200

U.S. Naval Institute News, "Video: Tomahawk Strike Missile Punches Hole Through Moving Maritime Target," February 9, 2015.

U.S. Navy, *Report to Congress on Annual Long-Range Plan for Construction of Naval Vessels for FY 2011*, Washington, D.C., February 2010.

———, "Attack Submarines—SSN," fact sheet, November 27, 2012. As of June 5, 2015:
http://www.navy.mil/navydata/fact_display.asp?cid=4100&ct=4&tid=100

———, "Destroyers—DDG," fact sheet, April 4, 2013. As of June 5, 2015:
http://www.navy.mil/navydata/fact_display.asp?cid=4200&tid=900&ct=4

"U.S. Test-Fires 'MIRACL' at Satellite Reigniting ASAT Weapons Debate," *Arms Control Today*, October 1997. As of June 5, 2015:
http://www.armscontrol.org/act/1997_10/miracloct

"USA Upgrades Submarine Fleet Acoustics Under A-RCI Program," *Defense Industry Daily*, April 30, 2012.

Valencia, Mark J., *Foreign Military Activities in Asian EEZs: Conflict Ahead*? Seattle, Wash.: National Bureau of Asia Research, Special Report No. 27, May 2011.

Van Tol, Jan, Mark Gunzinger, Andrew F. Krepinevich, and Jim Thomas, *AirSea Battle: A Point-of-Departure Operational Concept*, Washington, D.C.: Center for Strategic and Budgetary Assessment, 2010.

Viebeck, Elise, "Pentagon Moves to Hire 3k Cyber Workers," *The Hill*, March 8, 2015. As of June 23, 2015:
http://thehill.com/policy/cybersecurity/234951-pentagon-moves-to-hire-3k-cyber-workers

Wall, Robert, "Directed-Energy Threat Inches Forward," *Aviation Week and Space Technology*, Vol. 153, No. 18, October 30, 2000.

Wang Fengshan, Yang Jianjun, and Chen Jiesheng, 《信息时代的国家防控》 [*National Air Defense in the Information Age*], Beijing: Aviation Industry Press, 2004.

Wang Houqing, and Zhang Xingye, *The Science of Military Campaigns*, Beijing: National Defense University Press, 2000.

Wang, Jian-Wei, and Li-Li Rong, "Cascade-Based Attack Vulnerability on the U.S. Power Grid," *Safety Science*, Vol. 47, No. 10, 2009, pp. 1332–1336.

Weeden, Brian, "The Space Review Implications of Missile Defense," *Space Review*, September 28, 2009.

———, *Through a Glass, Darkly: Chinese, American, and Russian Anti-Satellite Testing in Space*, Secure World Foundation, March 17, 2014.

Weisgerber, Marcus, "Lockheed Working to Extend Range of U.S. Missile Interceptors," *Defense One*, January 7, 2015. As of June 19, 2015:
http://www.defenseone.com/threats/2015/01/
pentagon-wants-extend-range-one-its-missile-interceptors/102444/

Wilkening, Dean A., "A Simple Model for Calculating Ballistic Missile Defense Effectiveness," *Science and Global Security*, Vol. 8, No. 2, 1999, pp. 183–215.

Winnefeld, James A., Preston Niblack, and Dana J. Johnson, *A League of Airmen: U.S. Air Power in the Gulf War*, Santa Monica, Calif.: RAND Corporation, MR-343-AF, 1994. As of June 5, 2015:
http://www.rand.org/pubs/monograph_reports/MR343.html

Wohlstetter, Albert J., Fred S. Hoffman, R. J. Lutz, and Henry S. Rowen, *Selection and Use of Strategic Air Bases*, Santa Monica, Calif.: RAND Corporation, R-266, 1954. As of July 17, 2013:
http://www.rand.org/pubs/reports/R0266.html

Woolf, Amy F., *Conventional Prompt Global Strike and Long-Range Ballistic Missiles: Background and Issues*, Washington, D.C.: Congressional Research Service, February 6, 2015.

———, *The New START Treaty: Central Limits and Key Provisions*, Washington, D.C.: Congressional Research Service, February 4, 2015.

Work, Robert O., Deputy Secretary of Defense, "The Third U.S. Offset Strategy and Its Implications for Partners and Allies," speech at the Willard Hotel, Washington, D.C., January 28, 2015.

Wright, David, "Response to 'Space, China's Tactical Frontier' by Eric Hagt and Matthew Durnin," *Journal of Strategic Studies*, Vol. 34, No. 5, October 2011, pp. 763–773.

Wu Riqiang, "Why China Should Be Concerned with U.S. Missile Defense? How to Address It?" Atlanta: Georgia Institute of Technology Program on Strategic Stability Evolution, undated. As of June 5, 2015:
http://posse.gatech.edu/sites/posse.gatech.edu/files/China_Concerned_WuRiqiang_3.pdf

———, "Survivability of China's Sea-Based Nuclear Forces," *Science and Global Security*, No. 19, Vol. 2, 2011, pp. 91–120.

Wu, Sofia, "China Forms Missile Brigade for South China Sea," Focus Taiwan News Channel, July 2, 2012. As of June 5, 2015:
http://focustaiwan.tw/news/atod/201207020037.aspx

Wussler, Col Donald E., Vice Commander, Space and Missile Systems Center, "Space Superiority Systems Wing," briefing, Space and Missile Systems Center Industry Days, El Segundo, California, April 18, 2007. As of June 5, 2015:
http://www.smcindustrydays.org/2007/wussler.pdf

Xing, Yuqing, "The People's Republic of China's High-Tech Exports: Myth and Reality," ABDI Working Paper, April 2012. As of June 24, 2015:
http://www.eaber.org/sites/default/files/documents/2012.04.25.wp357.prc_.high_.tech_.exports.myth_.reality.pdf

薛兴林 [Xue Xinglin], ed., 战役理论学习指南 [*Campaign Theory Study Guide*], Beijing: National Defense University Press, 2001.

Yang Fumin, "Current Status and Future Plans for the Chinese Satellite Laser Ranging Network," *Surveys in Geophysics*, Vol. 22, Nos. 5–6, 2001, pp. 465–471.

Yoon, Joe, "Fighter Generations," Aerospaceweb.org, June 27, 2007. As of June 5, 2015:
http://www.aerospaceweb.org/question/history/q0182.shtml

Zaloga, Steven J., *Red SAM: The SA-2 Guideline Anti-Aircraft Missile*, New York: Osprey Publishing, 2007.

Zhang Han, and Huang Jingjing, "New Missile Ready by 2015," *People's Daily Online*, February 18, 2011.

张玉良 [Zhang Yuliang], ed., 战役学 [*The Science of Military Campaigns*], 2nd ed., Beijing: National Defense University Press, 2006.

Ziegler, David W., "Safe Heavens: Military Strategy and Space Sanctuary," in Bruce M. DeBlois, ed., *Beyond the Paths of Heaven: The Emergence of Space Power Thought*, Maxwell AFB, Ala.: Air University Press, September 1999, pp. 185–245.